Plant Responses to Xenobiotics

Anita Singh • Sheo Mohan Prasad
Rajeev Pratap Singh
Editors

Plant Responses to Xenobiotics

Springer

Editors
Anita Singh
Ranjan Plant Physiology
and Biochemistry Laboratory
Department of Botany
University of Allahabad
Allahabad, UP, India

Sheo Mohan Prasad
Ranjan Plant Physiology
and Biochemistry Laboratory
Department of Botany
University of Allahabad
Allahabad, UP, India

Rajeev Pratap Singh
Faculty of Science
Institute of Sustainable Development
Banaras Hindu University
Varanasi, UP, India

ISBN 978-981-10-2859-5 ISBN 978-981-10-2860-1 (eBook)
DOI 10.1007/978-981-10-2860-1

Library of Congress Control Number: 2016962721

Printed on acid-free paper

This Springer imprint is published by Springer Nature
The registered company is Springer Nature Singapore Pte Ltd.
The registered company address is: 152 Beach Road, #22-06/08 Gateway East, Singapore 189721, Singapore

Dedicated to Our Parents

Foreword

I am pleased to write the foreword for this book, edited by Dr. Anita Singh, Dr. Sheo Mohan Prasad, and Dr. Rajeev Pratap Singh. They have focused this book on an extremely important area of active research – the effects of environmental xenobiotics on plant function. Plants play an important role in the functioning of all terrestrial and aquatic ecosystems, as well as providing a critical food source for human populations. Unfortunately, human activity can result in soil, air, and water contamination by compounds including metals, radionuclides, pesticides, and other trace organic compounds, as well as newly identified contaminants such as engineered nanoparticles. These contaminants are introduced to the environment through industrial and municipal waste discharges, agricultural activities, and mining practices. The influence of these environmental contaminants on plants can include uptake of the contaminants into the plant, including the edible portion of food crops, as well as effects on plant physiological and biochemical processes. This book highlights important emerging research focusing on the responses of plants to these xenobiotic inputs, including chapters focusing on mechanisms for plant uptake and accumulation of xenobiotics, regulation and degradation of xenobiotics in planta, and plant toxicity to xenobiotics. Additionally, this book addresses the influence of critical emerging practices such as the use of municipal or agricultural wastewater as an irrigation water source. The use of this practice will only increase as we continue to face water shortages; however, the effect of using wastewater as an irrigation source water requires that we fully understand the environmental and human health impacts of this practice. Finally, the book presents information on alternative biodegradable thermoplastics, which could reduce the use of traditional plastics such as polystyrene and polyethylene terephthalate. I believe this book makes an important contribution to our understanding of the impacts of environmental contaminants on plants, and its focus on mechanistic studies and risk assessment will be of interest to researchers as well as policy makers.

Department of Civil Engineering,
University of Nebraska-Lincoln,
Lincoln, NE, USA

Shannon L. Bartelt-Hunt

Preface

Environmental pollution is one of the major problems due to fast pace of industrialization and uncontrolled exploitation of natural resources. It leads to the enhanced exposure of ecosystems to the substances called xenobiotics. The term xenobiotic is derived from the Greek words ξένος (xenos) = foreigner, stranger and βίος (bios, vios) = life. So, xenobiotics are any chemical or other substance that is not normally found in the ecosystems or that is present at the concentrations harmful to all biological organisms. They include organic contaminants such as pesticides, solvents, and petroleum products and inorganic contaminants such as heavy metals, nonmetals, metalloids, and simple soluble salts. They affect each and every component of ecosystem, and the plants which are the keystone of this system also get affected by the presence of xenobiotics in the environment. In this volume different kinds of xenobiotics are discussed in detail, and their effect and how the plants cope with such situation are explained by different contributions through the following chapters.

Chapter 1. *Environmental Xenobiotics and Its Effects on Natural Ecosystem*: Environmental xenobiotic is a global issue due to several activities, and one of them is the release of pharmaceutical residues in surface water. Recently, the demand for pharmaceuticals versus population growth has placed the public at risk. In addition, the making of unlawful drugs has led to the discharge of harmful carcinogens into the water system. The release of these harmful pollutants results in numerous short- and long-term effects to the natural ecosystem. This chapter takes a critical look at the various forms of environmental xenobiotics present in our ecosystem. In subsequent subheadings, their classification, sources, and routes of exposure to man, animals, and plants have been discussed. Lastly environment-friendly approaches to prevention had been discussed in a broader view and recommendations proffered.

Chapter 2. *Heavy Metal and Their Regulation in Plant System: An Overview*: Environmental pollution due to heavy metals is a threatening issue in the present scenario. In the past few decades, rapid and unplanned industrialization has caused the contamination of land and water. Plants exposed to such disturbed environment experience several physiological and biochemical alterations. However, some plants have acclimatized to the changed situation and developed defense mechanism like immobilization, compartmentalization, etc., to withstand the stressed conditions. This chapter gives an insight to the various physiological and biochemical alterations and mechanisms evolved by these plants growing in contaminated environment.

Chapter 3. *Regulation of Xenobiotics in Higher Plants: Signalling and Detoxification*: Xenobiotics are the chemical compounds that are not internally produced in plants, and their exposure is continuously increasing in plants, due to enhanced industrial pollutants. It can affect the growth, physiology, and other metabolic changes in every organism alone and/or in combination, which varies from species to species. Plants already have a versatile detoxification system to combat these changes (phytotoxicity) arising from a wide variety of natural and synthetic chemicals-xenobiotics present in the environment. One of the important detoxification mechanisms is chemical modification of xenobiotics by the covalent linkage to endogenous glutathione. The reactions involved in chemical modification have two phases: phase I (activation) reactions, which usually involve hydrolysis or oxidation, and phase II (conjugation) reactions, which are involved in synthesis. The resulting glutathione conjugates are exported from cytosol to vacuole via ATP-dependent tonoplast transporter. In addition to this, agrochemicals like safeners are also known to protect the crop plants from herbicide damage without reducing the activity in target weed species by elevating the expression of xenobiotic detoxifying enzymes, such as glutathione-S-transferases (GSTs). So, the present chapter gives valuable information on the different fates of xenobiotics as well as provides better understanding in the field of xenobiotic action in plants.

Chapter 4. *Metabolic Responses of Pesticides in Plants and Their Ameliorative Processes*: This chapter is of great relevance to students, researchers, scientists, and even the general public (lacking any scientific background) at large. Large anthropogenic inputs of pesticides for enhancing agricultural productivity and also for amelioration of vector-borne diseases have led to serious health implications to man and his environment. This is a comprehensive chapter which provides collective information of different types of pesticides, their sources, disadvantages, and toxicity symptoms in target as well as non-target organisms. It also deals with uptake, transport, and metabolism of pesticides in plants. Simultaneously it highlights the detoxification mechanisms adopted by plants to protect themselves from the adverse effects of pesticides. Another important aspect of this chapter is its holistic approach in hinting towards practices which can aid us in designing efficient and cheap vegetative treatment systems for remediation of contaminated soil and water.

Chapter 5. *Assessment of Antioxidant Potential of Plants in Response to Heavy Metals*: This chapter encompasses the complete comprehensive coverage on heavy metal occurrence, translocation, and their toxicities in plants and also the different mechanisms of antioxidants acting on the plant during oxidative stress. Here the chapter incorporates the detailed discussion on the scavenging mechanism of various antioxidants. The chapter further includes the information on traditional antioxidant like enzymatic, non-enzymatic, and secondary metabolites, and it also focuses on the method of assessment of antioxidant potential in *in vivo* and *in vitro* condition. Further, a brief attempt has also been made to provide contemporary and relevant collections of different studies ongoing on the antioxidant potential of plant.

Chapter 6. *Impact of Heavy Metals on Physiological Processes of Plants: With Special Reference to Photosynthetic System*: The present chapter highlights the effect of heavy metals on photosynthetic pigments, photosynthetic apparatus, and light and dark reactions. It helps us to understand in detail how exposure of plants to heavy metals led to generation of ROS and ROS-induced lipid peroxidation which destroy their cell membrane and their associated organelles. This chapter also highlights detail mechanism of the (i) impact of heavy metals on chlorophyll biosynthesis and also on various enzymes involved in chlorophyll biosynthesis and (ii) their effect on electron transport in light reactions and how they affect various enzymes in dark reactions.

Chapter 7. *Impact, Metabolism, and Toxicity of Heavy Metals in Plants*: Heavy metal contamination is a serious problem worldwide. These metals are the major inorganic contaminants of soil, and a considerable large area of land is contaminated with them due to anthropogenic activities. Contamination of agricultural soil by heavy metals has become a critical environmental concern due to their long-term persistent nature and potential harmful ecological effects. Therefore, it is important to study about the entry of these toxic metals in food chain. In this chapter we describe how far heavy metals enter and transport in plants, what are the different strategies of plants at different levels (from binding at cell wall to synthesis of some proteins to bind) to overcome toxic condition, and what are the heavy metal specific toxic effects on plants when exposed to metal-enriched environment.

Chapter 8. *Heavy Metal Accumulation Potential and Tolerance in Tree and Grass Species*: Identification of the role of higher plants in monitoring and in remediation of xenobiotics is important for polluted environments. Heavy metals are most widespread and one of the most toxic constituents of our environment. For selection of trees and grasses for monitoring and remediation purposes, it is necessary to identify plants from diverse environment and classify them based on their tolerance. Thus plants with higher accumulation potential have special tolerance mechanisms that enable them to survive and accumulate metals in higher concentrations compared to other plants. Considering these points, we have tried to identify tree and grass species based on their metal accumulation potential and tolerance so that academicians, researchers, plant breeders, urban planners, and environmental managers can utilize these findings to improve the understanding of the topic and may use the information for sustainable environmental management practices.

Chapter 9. *Microbial-Mediated Management of Organic Xenobiotic Pollutants in Agricultural Lands*: Exposure of plants to organic xenobiotic pollutants leads to several biochemical and molecular alterations producing serious variations in plant physiology resulting in deformed growth and development, ultimately hampering the productivity. As the concentration of these xenobiotics increases, their detrimental effects on plants are also pronounced. Soil microorganisms play crucial role in the management of the detrimental effects of such organic xenobiotic pollutants. They act as a barrier between the plant and the xenobiotic, and with their effective abilities to degrade the xenobiotic, they restrict the entry of these compounds in the plants and hence maintain plant growth and productivity. In this context the present chapter proves significant as it discusses the possible mechanisms for mitigation of the organic xenobiotic pollution load in the agricultural fields and opens up newer areas of research.

Chapter 10. *Metals from Mining and Metallurgical Industries and Their Toxicological Impacts on Plants*: Mining and metallurgical industries have been of great importance in the economic development of a country. However, solid wastes generated from such industries are characterized by elevated levels of essential and non-essential metals which may pose toxic effects on plants growing in and around the dumping sites. Minimum to maximum concentrations of twenty predominantly occurring metals in solid wastes from mining and metallurgical industries have been illustrated in this chapter. Adverse effects of metals exceeding their phytotoxic thresholds on growth performance and physiological and biochemical parameters of crop and medicinal plants have been discussed. The study also emphasizes the impacts of metals on plant community structure in the vicinity of industrial areas. The chapter suggests some future prospects in plausible and better management options of industrial wastes including site restoration by rehabilitation and phytoremediation using native and medicinal plant species.

Chapter 11. *The Risk Associated with the Xenobiotics Released Through Wastewater Reuse*: This chapter addresses the problems arising due to repeated use of wastewater on living system. The application of wastewater releases xenobiotic compounds that include heavy metal, pharmaceutical, pesticides, personal care products, etc. Here in this chapter, the authors have provided an overview of risk assessment arising from entrance of xenobiotic in the environment and also given a brief description of regulatory bodies involved in managing the risk.

Chapter 12. *Silver Nanoparticle in Agroecosystem: Applicability on Plant and Risk-Benefit Assessment*: In the current era, food security, life sustainability, and climate change are the most emerging challenges for researchers. Applicability of nanotechnology represents a novel step in the development and improvements in agricultural sectors to cope with scarcity of increasing food demand. The agricultural sectors can be seen to be clearly benefiting from nanotechnology. In particular, silver nanoparticles are reported to show a prime role in crop protection, antimicrobial applications, supplementation of required nutrients, and pesticide delivery in an optimized and controlled way. However, the negative effects due to the excessive use of these nanoparticles on biological life may also not be overlooked. This chapter focuses on applicability of silver nanoparticles in agricultural sector and points out their risk value to clearly instruct the use of these nanoparticles in a regulated and managed way.

Chapter 13. *The Significance of Plant-Associated Microbial Rhizosphere for the Degradation of Xenobiotic Compounds*: Presently human populations are increasing day by day causing pollution of various xenobiotic compounds in the environment to degrade the soil fertility and health. These xenobiotic compounds (heavy metals and hydrocarbons, pesticides, persistent organic pollutants, POPs), present in soils and waters, create many human and animal diseases (like immunosuppression, hormone disruption, reproductive abnormalities, and cancer). Degradation of xenobiotic pollutants by conventional approaches based on physicochemical methods is economically and technically challenging. Rhizo-remediation and microbial remediation techniques based on plant roots and their associated microbes are the most promising, efficient, cost-effective, environment-friendly, and sustainable technology. A variety of chemicals like organic acids, amino acids, and phenolic compounds are secreted by such plants as root exudates. These compounds play a significant role in communication between plant root and microbes and also are helpful to stimulate the remediation and the efficiency of microbes against xenobiotic pollutants. This book chapter heightens the degradation of xenobiotic compounds with the help of rhizosphere microbes that can be associated with plants to enhance the plant growth and yields as well as degrade the xenobiotic compound into elemental form that can be taken by the plant and microbes as nutrient or carbon source. The application of rhizospheric microorganisms (like bacteria, fungi, and actinomycetes) that interact with plant roots helps in the degradation of xenobiotic compounds without causing any environmental problems and also provides efficient, economic, and sustainable green remediation technology.

Chapter 14. *Biodegradable Polyhydroxyalkanoate Thermoplastics Substituting Xenobiotic Plastics: A Way Forward for Sustainable Environment*: Plastic, a miracle material of modern world, is indispensible and ubiquitous. The stability, durability, and low cost of plastics have attributed for their wide adaptability. Durability and resistance to degradation are desirable features when plastics are in use. Nevertheless, they cause serious problems to the environment when disposed due to their xenobiotic nature. All these issues of xenobiotic plastics translated into the need and concern for the production of biodegradable plastics/bioplastics. Amongst bioplastics, the completely biodegradable polyhydroxyalkanoates (PHAs) have received increasing research and commercial interest owing to their eco-friendly, optically active, elastomeric, and piezoelectric properties, renewable compounds, high degree of polymerization, non-toxicity, biocompatibility, hydrophobicity, and material properties comparable to conventional plastics. The foremost obstacle facing triumphant commercialization of PHA bioplastics is the high price of bacterial fermentation. Photoautotrophic hosts such as plants and cyanobacteria are being explored across the globe for low-cost PHA production. However, large-scale production is still a constraint. Presently, major attempt has been devoted to making PHA production process cost-effectively more feasible by changing the substrate from expensive to cost-effective, engineering efficient microorganisms, improving fermentation and separation processes, or applying mutational approaches/genetic engineering techniques.

So, overall this book contains all the valuable information related to the different kinds of xenobiotics and their impact upon plant physiology and metabolism. It will definitely be useful for the scientists, academicians, researchers, as well as students of different streams.

Allahabad, UP, India — Anita Singh
Allahabad, UP, India — Sheo Mohan Prasad
Varanasi, UP, India — Rajeev Pratap Singh

Acknowledgment

The editors acknowledge the Head of the Department of Botany, University of Allahabad, for providing necessary facilities. The editor Dr. Anita Singh is thankful to the Science and Engineering Research Board (SERB) for providing fellowship under Start-Up Research Grant as a DST Young Scientist (Grant No. SB/YS/LS/-228/2013). The editor Dr. Rajeev Pratap Singh is thankful to DST (P-45/18) and ICMR for providing the fund and the Director and Head of Institute of Environment and Sustainable Development, Banaras Hindu University, for providing necessary facilities.

Contents

About the Editors

Currently **Dr. Anita Singh** is working as DST-Young Scientist in the Department of Botany, University of Allahabad, Allahabad, India. Her expertise falls under the domain of sustainable agriculture as well as metal remediation. She has done Ph.D. on the topic entitled "Effect of irrigation water on heavy metal contamination of vegetable Crops" under the supervision of Prof. Madhoolika Agrawal from Banaras Hindu University, Varanasi, India. She has worked on Metal remediation, impact of abiotic stress on plant physiology and biochemistry, assessment of Florescence Transients and other plants metabolic activities. Her research capability can be judged from some very good publications in reputed journals. Till date, She has 40 publications including research article, review and chapters to her credit in the reputed international journal/books. During her research period she got International Green Talent Award organized by BASF, Germany. She also got an International Alice J. Murphy Outstanding Achievement Award for excellence in Research and Education leading to better Understanding of the Ecology of the Tropics.

Dr. Sheo Mohan Prasad is Professor in the Department of Botany, University of Allahabad. He obtained his M.Sc. and Ph.D. degrees from the Banaras Hindu University, Varanasi, India. Prof. Prasad has authored more than 138 scientific publications. He has guided 20 Ph.D. students and is supervising 7 research students and 3 P.D.F. on varied aspects of physiology and biochemistry of plants and cyanobacteria exposed to abiotic stresses: heavy metal, pesticide, salinity, high light with special reference to the role of ROS as signalling as well as oxidants. Prof. Prasad is also working as editor and reviewers of several reputed international journals and he is also the life member and Fellow of the Indian Botanical Society and member of The National Academy of Sciences. India.

Dr. Rajeev Pratap Singh is working as an Assistant Professor at Institute of Environment and Sustainable Development (IESD), Banaras Hindu University (INDIA). Dr. Singh completed his M.Sc. in Environmental Science at Lucknow University and his Ph.D. at Banaras Hindu University. His major research interests are solid waste management, bio-composting, and green technology. He has received several National and International awards, such as the "Green Talent" award from the Federal Ministry of Education and Research (BMBF), Germany; the Prosper. Net Scopus Young Scientist award; Water Advanced Research and Innovation (WARI) Fellowship from Department of Science and Technology (DST), Govt. of India and Indo-US Science and Technology Forum (IUSSTF); etc. Dr. Singh also serves as a reviewer and editorial team member for several important scientific journals and has co-authored more than 40 highly cited research and review articles on solid waste management.

List of Abbreviations

AAS	Absorption spectroscopy
ABC	ATP-binding cassette
AChE	Acetylcholinesterase
AgNPs	Nanosilver particles
AMPA	Aminomethyl-phosphonic acid
ANCA	Anti-neutrophil cytoplasmic antibody
APX	Ascorbate peroxidase
BOD	Biological oxygen demand
BRs	Brassinosteroids
CAT	Catalase
CDF	Cation diffusion facilitator
CGWB	Central Ground Water Board
CM-H2DCFDA	Chloromethyl-2',7'-dichlorohydrofluorescein diacetate
COD	Chemical oxygen demand
COX	Cyclooxygenase enzyme
CP	Carboxypeptidase
CPF	Chlorpyrifos
CYPs	Cytochrome P450 mixed function oxidases
Cysteine	(Cys)
DAB	Diaminobenzidine
DDE	Dichlorodiphenyldichloroethylene
DHAP	Dihydroxyacetone phosphate
DHAR	Dehydroascorbate reductase
DHE	Dihydroethidium
DP	Dipeptidase
EDCs	Endocrine disrupting compounds
EDTA	Ethylenediaminetetraacetate
EDX	Energy-dispersive X-ray analysis
EPPP	Environmental Pharmaceutical Persistent Pollutants
EPS	Extracellular polymeric substances
ET	Electron transfer
EU	European Union
FAO	Food and Agriculture Organization
GABA	Gated chloride channel blockers

GAPDH	Glyceraldehyde 3-phosphate dehydrogenase
GDP	Gross domestic product
GR	Glutathione reductase
GSR	Glutathione-disulfide reductase
GSSG	Oxidized glutathione
GSTs	Glutathione-*S*-transferases
GT	Glycosyltransferases
HAT	Hydrogen atom transfer
HDL	High-density lipoproteins
HM	Heavy metals
HPPD	4-Hydroxyphenylpyruvate dioxygenase
IA	Image analyzer
IARC	International Agency for Research on Cancer
IMDH	Isopropylmalate dehydrogenase
IRA	Infrared analysis
IRT	Iron-regulated transporter
LIMS	Laser ionization mass spectrometer
LPIC	Lipid peroxidation inhibition capacity
MDA	Malondialdehyde
MATE	Multidrug and toxin efflux
MDHA	Monodehydroascorbate
MTs	Metallothioneins
MPL	Maximum permissible limit
MRP	Multidrug resistance-associated protein
NACHR-	Nicotinic acetylcholine receptor
NBT	Nitro blue tetrazolium
NHPs	Natural health products
NA	Nicotianamine
NPP	Net primary productivity
NRAMP	Natural resistance-associated macrophage protein
NSAID	Non-steroidal anti-inflammatory drugs
OP	Organophosphate
ORAC	Oxygen radical absorbance capacity
PAHs	Polycyclic aromatic hydrocarbons
PCBs	Polychlorinated biphenyls
PCE	Perchloroethylene
PCS	Phytochelatin synthetase
PDR	Pleiotropic drug resistance
PGA	Phosphoglyceric acid
PhACs	Pharmaceutical active compounds
PHE	Phenanthrene
PIXE	Proton-induced X-ray emission
PHA	Polyhydroxyalkanoate
POPs	Persistent organic pollutants
POX	Peroxidases

PPCPs	Pharmaceuticals and personal care products
RA	Rheumatoid arthritis
ROS	Reactive oxygen species
RuBisCO	Ribulose bisphosphate carboxylase
SA	Salicylic acid
SEM	Scanning electron microscopy
SIMS	Secondary ion mass spectrometer
SLE	Systematic lupus erythematosus
SOD	Superoxide dismutase
SSc	Systemic sclerosis
TBT	Tributyltin
TCE	Trichloroethene
THMs	triethylene, trihalomethanes
TLR	Toll-like receptors
TRAP	Total radical trapping antioxidant parameter
UTUC	Urinary tract urothelial cell carcinoma
VLDL	Very low-density lipoprotein
WHO	World Health Organization
XRD	X-ray diffraction
YSL	Yellow-stripe-like
ZRT	Zinc-regulated transporter

Environmental Xenobiotics and Its Effects on Natural Ecosystem

1

Asha Embrandiri, S. Katheem Kiyasudeen, Parveen Fatemeh Rupani, and Mahammad Hakimi Ibrahim

Abstract

Environmental contamination by xenobiotics is a worldwide phenomenon as a result of human activities resulting from rise in urbanization and population growth. There are numerous sources of xenobiotics ranging from pharmaceuticals to agriculture. Recently, the demand for pharmaceuticals versus population growth has placed the public at risk. In addition, the making of unlawful drugs has led to the discharge of harmful carcinogens into the water system. The release of these harmful pollutants results in numerous short- and long-term effects to the natural ecosystem. This review takes a look at the sources of xenobiotics, their fate in the ecosystem and means of action with possible prevention methods.

Keywords

Xenobiotics • Carcinogens • Degradation

1.1 Introduction

Since the time of the Industrial Revolution, scientific and technological developments permitted humans in the over utilization of resources creating disturbance to the natural ecosystem (Sikandar et al. 2013). The generation of huge amount of toxic substances released from industrial processes caused widespread contamination of the ecosystem. The major contaminants are halogenated and nitrated hydrocarbons (Jain et al. 2005). Several herbicides, insecticides and fertilizers used in

A. Embrandiri (✉) • S.K. Kiyasudeen • P.F. Rupani • M.H. Ibrahim
School of Industrial Technology, Universiti Sains Malaysia,
Gelugor, Penang 11800, Malaysia
e-mail: ashanty66@gmail.com

A. Singh et al. (eds.), *Plant Responses to Xenobiotics*,
DOI 10.1007/978-981-10-2860-1_1

agricultural activities as well as synthetic compounds are produced from industrial activities, namely, pharmaceuticals, agrochemicals, dyes, solvents, halogenated compounds, hydraulics, fire retardants, pigments, etc. (Reineke and Knackmuss 1988). Pharmaceutical wastes have become well known sources of prolonged environmental contamination due to the continous use in anthropoid and veterinary medications. These chemicals are believed to have specific mode of action in the body. The chemical composition of pharmaceuticals lead to possible effects on aquatic flora and fauna as compared to other chemical compounds. However, pharmaceuticals are proposed to pose only a low risk for acute toxicity in the environment. For chronic effects, the situation may be different; nevertheless, there is a considerable lack of information for the chronic effects and its risk of toxicity. In addition, there is little or no informationregarding multi generational life cycle effects, knowing that exposure to toxicity in many aquatic organisms happen during their entire life (Fent et al. 2006). Accordingly, various environmental analyses reported that the drug residues in surface water and treated wastewater are widespread. These chemical compounds that arise from industries are xenobiotics. Xenobiotic compound is persistent in the environment with toxicity effect making them potential health hazards leading to significant impacts on the ecosystem. Therefore, researchers need to focus more on effects of pollution and its prevention techniques.

1.2 Xenobiotics: Sources and Types

Originally, the term xenobiotic comes from the Greek word xenos, which means foreign or strange, and 'bios', which means life. Xenobiotics are chemical compounds exhibiting abnormal structural characteristics (Fetzner 2002). The unusual presence of any substance in high concentrations can also be regarded as xenobiotics, for instance, the presence of antibiotic drugs in the human body which may not be produced by the body itself nor is a normal part of diet. At times, a natural substance can be defined as a xenobiotic if it found its way into humans or other animals. Bonjoko (2014) proposed the word 'xenobiotic' based on the physiological and biological effects of exogenous substances whether natural or synthetic (drugs, chemicals) on the cells, tissues or organs of the organisms.

Many xenobiotics are potientially hazardous to the organisms which are exposed to them in the environment. However, bioavailabilities of such substances are dependent upon the characteristics of the organism, the chemical, and the environment. Maenpaa (2007) reported that the toxicity of any xenobiotic is related to the bioaccumulated chemical residue in the organism. Xenobiotics may persevere for long term (months to years) in the environment. For example, the polymer structure of lignin, or the constituents of the cell wall of the spores of a few fungi (melanin polymers), may not degrade rapidly in the natural environment (Fetzner 2002). Similarly, in aquatic environments, hydrophobic pollutants which are eventually stored in sediments become hazardous on exposure to benthic organisms. Any exposure to the sediments contaminated by xenobiotics possibly affects the lower trophic levels. It may also result in biomagnification or more serious toxic effects at higher trophic levels (Landrum and Robbins 1990; Lee II 1992; Streit 1992; Newman 1998).

New technologies to determine trace polar compounds have helped to give new insights on the removal of xenobiotics. In the beginning, pharmaceutical products were reported in treated wastewater in the USA, with the range of about 0.8–2 μg/L (Garrison et al. 1976). Thereafter, the UK reported 1 μg/L of clofibric acid in the rivers (Richardson and Bowron 1985). In 1986, Rogers (from Canada) identified the concentration of naproxen and ibuprofen in wastewaters. Accumulation of diclofenac, a pain killer which was used by veterinarian to treat cattle, significantly reduced the population of Asian white-backed and Indian vultures (9 from 150 in 1997 to 25 in 2010) nesting in Keoladeo Natural Park in North Western India. The Geological Survey Department (United States) reported traces of many different drugs and toiletries as well as steroids, insect repellants and phthalates in the water supply. Even though the concentrations were in traces, the effect of chronic exposure can be unpredictable. For example, production of bulk drugs has been recently identified as an important source of environmental pollution which consists of active pharmaceutical elements in certain locations (Gunnarsson et al. 2009; Fick et al. 2010). Also, there are raising concerns worldwide on the pharmaceutical residues found in surface water which can have effects on aquatic organisms. Therefore, there is a major challenge in developing lucid strategy for prioritizing drugs on which to focus the most extensive environmental research efforts for (Fick et al. 2010).

1.3 Sources of Pharmaceutical-Based Xenobiotics

There are different synthesized chemicals present in the environment which may have different interactions with the exposure to humans and the ecosystem. However, the details of these impacts are not adequately studied or understood. Among the different pharmaceutical substances, pharmaceutical active compounds (PhACs) are xenobiotic-based elements that entered the environment as the parent compound or as pharmacologically active metabolites (Bonjoko 2014). PhACs are considered as potentially toxic compounds that are largely used in agriculture and industry. However, for many years the researches were based on the pharmaceutical regulations which were of interest by drug organizations, and less attention was paid on the toxicity and its environmental issues (Jones et al. 2001).

Environmental pharmaceutical persistent pollutants (EPPPs) are the components which are available in waterbodies all over the world. Not much literature is available about the possible negative effects and impacts of EPPP in humans and the environment. Bonjoko (2014) reported that EPP's exposure may cause extinction of species and imbalance of sensible ecosystems (EPPS affect the reproductive systems of, e.g. frogs, fish and mussels). For example, in sewage plants of pharmaceutical industries, large amount of antibiotics and other pharmaceutical compounds have been found. According to the European Union (EU), about 3000 different substances were found in downstream that were used in human medicine such as antibiotics, beta-blockers, analgesics and anti-inflammatory drugs and many others. Likewise, a large number of pharmaceuticals are used in veterinary medicine such as antibiotics and anti-inflammatory (Fent et al. 2006). Bonjoko (2014) explained the potential routes of entry of pharmaceutical and household care products in the environment. It includes:

1. Through patient excretion
2. Direct release into the wastewater system from manufacturing, hospitals or disposed through toilets and sinks
3. Terrestrial depositions, i.e. irrigation with treated and untreated wastewater, sludge application to land, leaching from solid waste landfills
4. Non-pharmaceutical industrial sources, i.e. plastic products
5. Agricultural wastes such as herbicides, pesticides and fertilizers
6. Through ageing infrastructures, i.e. synthetic compounds such as analgesics and antihistamines which were exposed in streams and rivers
7. Drugs associated with plant health
8. Herbal preparations and their interaction with the environment

1.4 Fate/Biodegradation of Xenobiotic Compounds

Xenobiotics with the presence of microbes can undergo biodegradation process depending on the microbe's species and the xenobiotic compounds.

Xenobiotic metabolism undergoes a biochemical modification of pharmaceutical substances (xenobiotics) by living organisms, which usually occurs through specialized enzymatic systems. Enzyme like cytochrome P_{450} secreted in the liver helps in the degradation process and thus excreted by urination, exhalation, sweating and defecation. Biodegradation and oxidation of a parent compound happen to form carbon dioxide and water. Each stages in the degradation pathway is catalysed by a specific enzyme produced by the degrading cell. However, degradation of some xenobiotics depends on its specific compound structure, which includes the required enzymes, for example, oxygenases. These enzymes are metabolized to provide energy as well as reducing equivalent for the degradation process (Bonjoko 2014). Figure 1.1 illustrates the possible environmental fate of a xenobiotic compound.

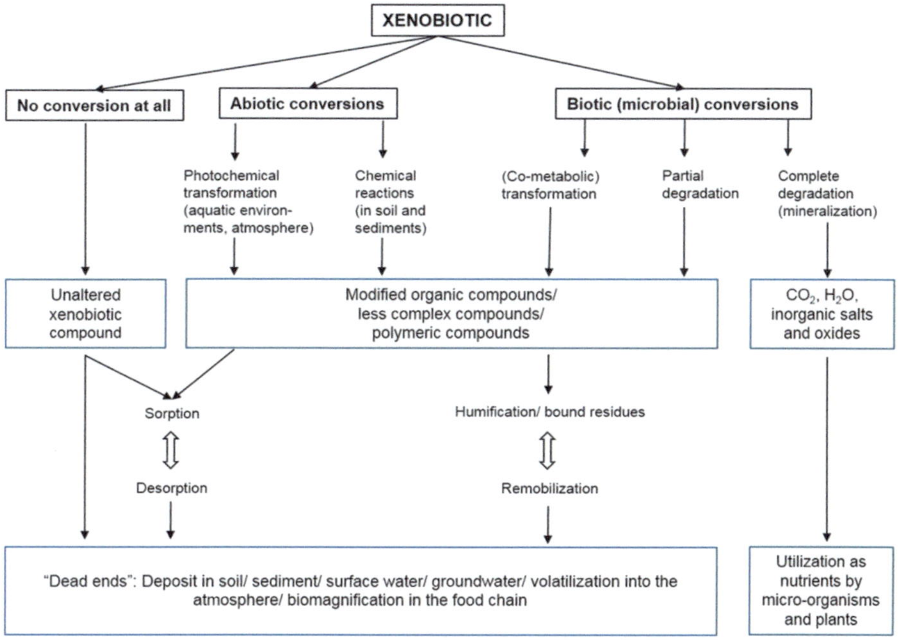

Fig. 1.1 Possible environmental fate of a xenobiotic compound

1.5 Common Xenobiotic Compounds Based on Its Course of Action

Xenobiotics are designed to target specific metabolic and molecular pathways in humans and animals in the ecosystem. However, when xenobiotics are introduced into the environment, they may affect the same pathways in animals having identical or similar target biomolecules, organs, tissues or cells. The current ecotoxicological effects of pharmaceuticals deal mainly with the acute toxicity in standardized tests and it is generally focused on aquatic organisms. The influence of environmental parameters such as pH on toxicity has only rarely or not yet been investigated. More studies have focused on acidic pharmaceuticals that may induce different toxicities depending on speciation at different ambient pH. Moreover, till date less research has been done on the effects of drug metabolites. The following are the common pharmaceutically based xenobiotic compounds that pose such environmental concerns. Figure 1.2 illustrates the different types of pharmaceutical xenobiotics.

1.5.1 Analgesics and Non-steroidal Anti-inflammatory Drugs (NSAIDs)

The widely used non-steroidal anti-inflammatory drugs (NSAIDs) ibuprofen, naproxen and diclofenac and some of their metabolites such as hydroxyl-ibuprofen and carboxy-ibuprofen are widely used and usually can be detected in surface and

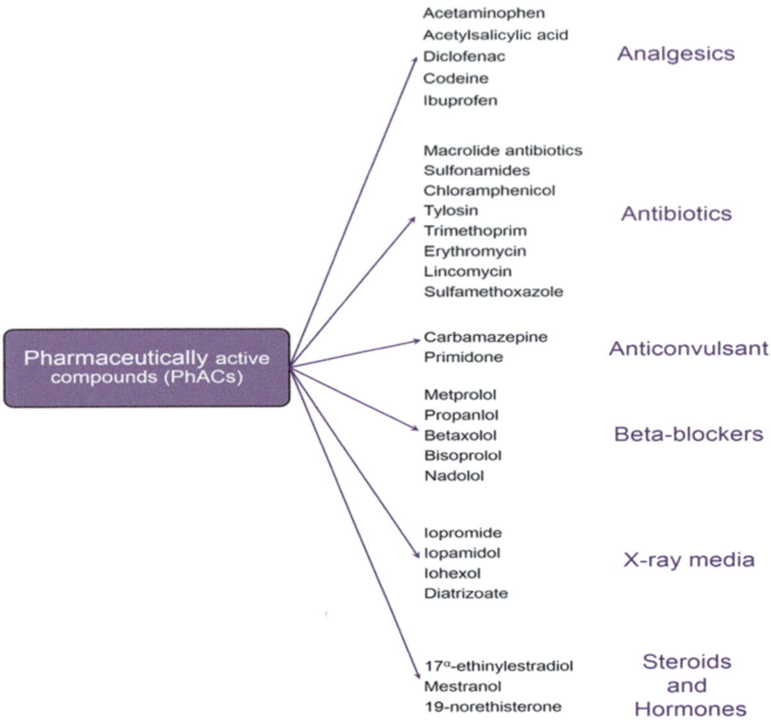

Fig. 1.2 Different types of pharmaceutical xenobiotics

water sewage. Gross et al. (2004) reported that NSAID levels exceed in sewage system to 1 μg/L, and it can exceed the concentration to 0.1 μg/L in the effluent of conventional sewage plants (mechanical clarification and biological treatment) in the USA. The deacylated which is a more active form of acetylsalicylic acid has detected in many municipal wastewaters at the levels up to 4.1 μg/L, 13 μg/L or even 59.6 μg/L. Similar to acetylsalicylic acid, acetaminophen (paracetamol) is well removed from STP. However, Kolpin et al. (2002) reported that up to 10 μg/L (median 0.11 μg/L) acetaminophen is spotted in 24 % of samples from US streams (Kolpin et al. 2002). Also, the analgesic codeine was detected in 7 % of samples (median 0.01 μg/L). Moreover, in many countries, diclofenac was frequently detected in the wastewater and also in lower levels of surface water.

Wiegel et al. (2004) have reported that in Norway, ibuprofen and its metabolites have been found in all sewage samples and in seawater at the concentrations of 0.1–20 μg/L. Kolpin et al. (2002) detected ibuprofen in 10 % of stream water samples in high concentrations up to 1 μg/L (median 0.2 μg/L). Moreover, several other NSAID compounds have been detected in sewage and surface water as well as in drinking water samples and groundwater.

1.5.1.1 Mode of Action

NSAIDs (non-steroidal anti-inflammatory drugs) are frequently used to treat inflammation and pain and to relieve fever, and sometimes they are also used for long-term treatment of rheumatic diseases. NSAIDs act by inhibiting factor either reversibly or irreversibly one isoform of the cyclooxygenase enzyme (COX-1 and COX-2), which catalyse the synthesis of different prostaglandins from arachidonic acid. COX-1 and COX-2 inhibit by classical NSAID at different degrees, whereas new NSAID act more selectively on COX-2, the inducible form which is responsible for the inflammatory reactions. Differences in binding site size are in charge for the selectivity of these drugs. NSAID inhibit nonspecifically.

In the kidney, prostaglandins are elaborate in maintenance of the equilibrium between vasoconstriction and vasodilatation of the blood vessel that supply glomerular filtration.

At times, after chronic NSAID treatments, renal damages or renal failure seems to be triggered by the lack of prostaglandins in vasodilatation-induction. Inhibition of both COX isoforms can cause gastric damages. In contrast, liver damages are apparently due to building of reactive metabolites (e.g. acyl glucuronides) rather than inhibition of prostaglandin synthesis (Bjorkman 1998). The mode of action of paracetamol is not yet fully elucidated. However, it has been found that this drug acts mainly by inhibiting the cyclooxygenase of the central nervous system, and it does not have anti-inflammatory effects, because of the lack of inhibition of peripheral cyclooxygenase involved in inflammatory processes. Adverse effects of paracetamol mainly occur when the availability of glutathione is diminished in liver cells which could be due to formation of hepatotoxic metabolites, primarily *N*-acetyl-*p*-benzoquinone.

1.5.2 Blood Lipid Regulators

The most frequently reported pharmaceutical in monitoring studies is clofibric acid which is an active metabolite from a widely used blood lipid regulators such as clofibrate, etofylline clofibrate and etofibrate. These compounds have been found in numerous wastewaters, surface waters and seawater and at rather high concentrations in drinking water (0.07–0.27 μg/L) and groundwater (4 μg/L). Bezafibrate and gemfibrozil which are lipid-lowering agent have been found in maximal concentrations of up to 4.6 and 0.79 μg/L, respectively, in wastewater and surface water, respectively (Kolpin et al. 2002). In addition, other drugs which act as metabolite of fenofibrate such as gemfibrozil, clofibric acid and fenofibric acid have also been detected in sewage up to the μg/L level and in surface water (Heberer 2002).

1.5.2.1 Mode of Action

There are basically two types of antilipidemic drugs, namely, statins and fibrates, which are used to decrease the concentration of cholesterol (statins and fibrates) and triglycerides (fibrates) in the blood plasma. These drugs have been targeted analytically more often in the aquatic environment. Statins as inhibitors of cholesterol synthesis act by inhibiting the 3-hydroxymethylglutaril coenzyme A reductase (HMG-CoA), responsible for the limiting step in the cholesterol synthesis, namely, the conversion of HMG-CoA to mevalonate. Due to interactions of statins with mevalonate metabolism, multiple additional effects occur (anti-inflammatory, anti-oxidative). Studies also show that statins affect juvenile hormone synthesis in insects as fluvastatin completely suppressed its biosynthesis in vitro and in the mandibular organ of lobsters. The effects of fibrates lead to alterations in transcription of genes encoding for proteins controlling lipoprotein metabolism and they also activate the lipoprotein lipase enzyme, which is mainly responsible for the conversion of very-low-density lipoprotein (VLDL) to high-density lipoproteins (HDL), decreasing therefore plasma triglyceride concentration.

Moreover, fibrates stimulate cellular fatty acid uptake by conversion to acetyl-CoA derivatives and catabolism by the beta-oxidation pathways. Hence, these processes are combined with a reduction in fatty acid and triglyceride synthesis that decreases in VLDL production. Studies on animal (rat) show that hepatic damages can occur after chronic exposure to fibrates and which could be due to the inhibition of mitochondrial oxidative phosphorylation. Fibrates caused in rodents a massive proliferation of peroxisomes. Strong correlation between fibrate exposure and hepatocarcinogenicity in rodents was found, while this was not observed in humans (Cajaraville et al. 2003). These findings increase the interest for ecotoxicological impact of this therapeutic class of drugs. Activators of PPARα genes (found in fishes) include different endogenously present fatty acids, leukotrienes and hydroxyeicosatetraenoic acids and drugs, such as fibrates.

1.5.3 Beta-Blockers

Several beta-blockers such as bisoprolol, propranolol and metoprolol were identified in wastewater showing 0.59, 2.9 and 2.2 μg/L, respectively. Also in lower concentration, other beta-blockers, namely, nadolol (in surface water) and betaxolol (0.028 μg/L in surface water), were detected (Ternes 1998). Moreover, Sacher et al. (2001) reported the presence of propranolol, metoprolol and bisoprolol in surface water and sotalol in groundwater.

1.5.3.1 Mode of Action

Beta-blockers act by competitively inhibiting beta-adrenergic receptors. They are employed for the treatment of high blood pressure (hypertension) and the prevention of heart attacks in high-risk patients. Regular body processes like heartbeat regulation and oxygen supply, vasodilatation of blood vessels and bronchodilation are the functions of the adrenergic system. Furthermore, it is important for the metabolism of carbohydrates and lipids in cases of starvation. Beta-blockers could selectively impede one or more β-receptor types based on their needs. For instance, β2-blockers are employed for the treatment of hypertension preventing impending cardiac arrests, as this receptor subtype is not present in the heart. Beta-blocker propranolol, a beta1-adrenoceptor antagonist, has the ability to stabilize cell membranes, unlike metoprolol which does not have that property (Doggrell 1990). Side effects of these beta-blockers are mostly bronchoconstriction and disturbed peripheral circulations. They are supposed to pass the blood-brain barrier and to act in the central nervous system because of their lipophilicity (Heberer 2002). Clenbuterol or ractopamine that functions in mammals as β-agonist had a different reaction in rainbow trouts. The different structures and function of the receptors may be responsible for varied affinity with β-blockers and mechanisms triggered by these drugs.

1.5.4 Neuroactive Compounds (Antiepileptics and Antidepressants)

Antiepileptic carbamazepine was detected most frequently and in highest concentration in wastewater (up to 6.3 μg/L) (Ternes 1998) and at lesser concentrations in other media (Heberer 2002). Carbamazepine was found in all effluent samples of the Canadian STP at concentration up to 2.3 μg/L. This substance was reported to be present in all samples of German river Elbe and streams (Wiegel et al. 2004), exceeding 1 μg/L in other surface waters (Ternes 1998; Heberer 2002) and also occurred in groundwater (Sacher et al. 2001). Carbamazepine was also reported at average levels of 20.9 ng/mg solids of STP. Diazepam was noted in 8 out of 20 treatment plants in Germany at relatively low concentrations of up to 0.04 μg/L (Ternes 1998), whereas in Belgium it was recorded at concentration up to 0.66 μg/L (van derVen et al. 2004). The antidepressant fluoxetine was also recorded in Canadian effluent samples, and in US streams, median concentrations of 0.012 μg/L were estimated (Kolpin et al. 2002). In addition to these, an antiepileptic drug, primidone (0.6 μg/L), was also identified in sewage (Heberer 2002).

1.5.4.1 Mode of Action

Antiepileptic drugs decrease the overall neuronal activity. This can be achieved either by blocking voltage-dependent sodium channels of excitatory neurons (e.g. carbamazepine) or by enhancing inhibitory effects of the GABA neurotransmitter by binding on an exact site in the gamma subunit of the corresponding receptor (e.g. diazepam, member of benzodiazepine family). The uptake of serotonin is inhibited by a very common antidepressant, fluoxetine. It is a neurotransmitter that has to do with hormonal and neuronal mechanisms, and it is vital for sexual behaviour and food intake. Fluoxetine, sertraline, norfluoxetine and desmethylsertraline have been discovered in fish sampled from the wild in the USA and therefore reflect a bioaccumulation potential (Brooks et al. 2005).

1.5.5 Various Other Compounds

Effluents of the sewage treatment plants and surface waters which have been contaminated by drugs are comprised of caffeine and cotinine (a nicotine metabolite).

In the USA, caffeine was found in streams at high levels of 6.0 μg/L (median 0.1 μg/L) (Kolpin et al. 2002) which can serve as an anthropogenic marker in aquatic systems as a result of its ubiquity in surface water, in seawater (Wiegel et al. 2004) and also in groundwater. Cimetidine and ranitidine (antacids) were estimated to occur at concentrations of 0.58 and 0.01 μg/L, respectively, in streams in the USA (Kolpin et al. 2002). Iopamidol has been detected in municipal wastewater at very high concentrations (15 μg/L), in surface water (0.49 μg/L) and in groundwater.

The antidiabetic compound metformin was observed in 5 % of stream water samples with estimated levels of 0.11 μg/L (Kolpin et al. 2002). Bronchodilators (β2-sympathomimetics terbutaline and salbutamol) were also detected in sewage not exceeding 0.2 μg/L (Ternes 1998).

1.5.5.1 Mode of Action

Cimetidine and ranitidine are compounds, which act by hindering the histamine receptor type 2 in the gastric system, thus inhibiting the acid secretion (antacid). These drugs are for the treatment of gastric ulcer. Metformin is an antidiabetic agent, whose mechanisms of actions have not been fully studied. It has been reported that this drug acts by increasing the cellular use of glucose and inhibiting the gluconeogenesis. Metformin acts on insulin receptor by direct stimulation of the insulin receptor or indirectly through inhibition of tyrosine phosphatase (Holland et al. 2004).

1.5.6 Steroidal Hormones

Steroidal hormones have been reported in wastewater and surface waters in a number of countries in Europe, Canada, the USA, Japan, Brazil, etc. A study in the USA showed that the average oestrogen concentration was 73 ng/L and levels of mestranol were 74 ng/L (Kolpin et al. 2002). They were detectable in 16 and 10 % of the streams sampled. Typical wastewater effluent concentrations are 0.5 ng/L and they are even lower in surface water.

1.6 Effects of Xenobiotics on Ecosystem

Every year, more than 13 million deaths and 24 % of world diseases are said to be as a result of environmental pollutants/exposures which can be avoided. Today, detectable levels of pharmaceutical preparations either as parent drug or metabolite are present in foodstuffs and water, i.e. both rivers and seas (Bonjoko 2014). Medications for humans and animals have severe consequences extending far beyond the traditional objectives of conventional medical care. The healthcare industry is the major source of active pharmaceutical ingredients (API) from medications, residues of which could lead to environmental pollution.

1.6.1 Effects on Aquatic Ecosystem

Aquatic organisms are significant biological indicators of pollution. Fent et al. (2006) conducted a comprehensive study on the occurrence; end result of pharmaceuticals in the aquatic environment, discussed potential mechanisms of action based on knowledge from mammalian studies and described the acute and chronic ecotoxicological effects on organisms. Pharmaceuticals are most often released back into the environment either in their original form or as metabolites. In humans, the main pathway is ingestion following excretion and disposal via wastewater. Municipal wastewater is the largest source of human pharmaceuticals. Hospital wastewater, wastewater from manufacturers and landfill leachates may contain significant concentrations of pharmaceuticals. Pharmaceuticals that are nondegradable in the sewage treatment plant (STP) are being released into treated effluents resulting in the contamination of rivers, lakes, estuaries and, rarely, groundwater and eventually drinking water. There is also likelihood of contamination when sewage is applied in agriculture. In addition, drugs meant for animals enter the waterways during surface application for agriculture purposes and runoff and also via direct application in fish farming. Pharmaceuticals of environmental significance oftentimes have high production volume in addition to environmental persistence and biological activity, especially after long-term exposure.

In recent studies, it has been observed that the increasing amounts of pharmaceuticals found in surface waters worldwide have raised concerns especially with respect to their effects on the aquatic flora and fauna. It would therefore be a huge task to initiate a strategy for prioritizing drugs on which to focus the most expensive environmental research efforts on. Among aquatic organisms, fish most often share drug targets with humans. Not much is known about the long-term effect of drugs in aquatic organisms. Diclofenac influences the expression of genes in fish and organ histology when exposed to a concentration of 1 μg/L of this drug (Cuklev et al. 2012). A study in India on surface water from 27 locations of the Kaveri velar and Tami rapani rivers in southern India revealed the presence of a number of nonsteroidal anti-inflammatory drugs (NSAIDs): naproxen, ibuprofen, diclofenac, acetylsalicylic acid and ketoprofen. This situation poses risks of direct toxicity to all consumers of the water (Shanmugan et al. 2013). Another case, likewise, effluents from a treatment plant in Hyderabad, India, was observed to be the cause of deleterious effects on water organisms. An embryo toxicity test that was carried out observed that as little as 0.2 % of the effluent reduced tadpole growth by 40 %; however, zebra fish (*Danio rerio*) growth was not impeded. Although the study focused on fish, it also increased knowledge about how aquatic vertebrates are possibly affected by effluent exposures, which substances in the effluent are causing the toxic effects and their threshold dilutions (Shanmugan et al. 2013).

Streams and rivers have been identified to be exposed to combinations of different drugs. Antidiabetic and antihistamine diphenhydramines were observed to cause significant disruption to the biofilm community which is important to the ecosystem. Biofilms are aggregates of microorganisms in which cells that are frequently embedded within a self-produced matrix of extracellular polymeric substances

(EPS) adhere to each other and or to a surface. Biofilms serve as the important food source for invertebrates that in turn feed larger animals like fish. The effects of diphenylamines on biofilm could therefore have repercussion for animals in stream food web such as insects and fish (Rosi-Marshall 2013). The use of antidepressants disrupts the aquatic equilibrium by activating early spawning in some shellfish. Furthermore, propranolol and fluoxetine were observed to have deleterious effects on zooplankton and benthic organisms (Hoffman et al. 2005). Factors such as cellular recognition of specific or non-specific attachment sites, nutritional cues or the exposure of planktonic cells to subinhibitory concentration of antibiotics lead to the formation of biofilms by microbes (Hoffman et al. 2005; Karatan and Watrick 2009). Masculinization (imposex) had been witnessed in female marine snails exposed to tributyltin (TBT). The dog whelk (*Nucella lapillus*), 'a species of predatory sea snail', is particularly sensitive, and imposex has resulted in decline or extinction of local populations worldwide, including coastal areas all over Europe and the North Sea.

DDE (dichlorodiphenyldichloroethylene)-induced eggshell thinning in birds is probably the best example of reproductive impairment causing several population declines in a number of raptor species in Europe and North America. Gradual exposure to the DDT complex (dichlorodiphenyltrichloroethane) has been linked to ovotestis in male western gulls. EDCs (endocrine disruptors) have negatively affected a variety of fish species exposed to effluents causing reproductive problems. Turtles have also been affected in a similar manner (Cleuvers 2003; Le Page et al. 2011). Triclosan (TCS) is a broad-spectrum antimicrobial compound that is contained in most of the cleaning products for the prevention of bacterial, fungal and mildew growth. Triclosan enters into water streams from domestic wastewater, leaking sewerage and sewage overflows. The continuous use of these antibiotics leads to the emergence of resistant bacteria that could diminish the usefulness of important antibiotics (Drury et al. 2013).

1.6.2 Effects on Animals

More commonly observed effects of EDC are impaired reproduction and development in aquatic animals (Kid et al. 2007). A number of brain targets for EDC present in environmentally relevant concentrations in surface waters had been identified from recent surveys. In mammals, field studies on Baltic grey and ringed seals and Wadden Sea harbour seals revealed that reproduction and immune function were impaired by PCBs (polychlorinated biphenyls). Other mammals in the food chain that have the likelihood of being affected include the polar bear, rabbit and guinea pig. In Florida, due to a pesticide spill, alligators were found to have inhibited genital developments. Furthermore, the oestrogenic and androgenic effects observed have been linked to experimental studies with alligator eggs to the DDT complex.

1.6.3 Effects of Plants

With the growing influence of herbal drugs worldwide, botanical plants with pharmacological properties should be cautiously handled in order not to contaminate crops, vegetables and surface water. A common herb which may contaminate the ecosystem is marijuana with the potential ability to interfere with the biological system of aquatic organisms. In addition, herbs like St. John's wort were noted to cause modulation of cytochrome p450 and may interfere with prescribed therapeutic agents (Guengerich 1997). *Aristolochia* plants are commonly used in traditional herbal preparation as health supplements and remedies for various health problems including weight loss, menstrual symptoms and rheumatism. In the 1990s, epidemiological studies revealed AA exposure was associated with a high risk of nephrotoxicity and upper urinary tract urothelial cell carcinoma (UTUC) (Grollman et al. 2007; Debelle et al. 2008) caused by the ability of AA to bind DNA, forming DNA adduct (Schmeiser et al. 1998). These findings eventually resulted in the ban on *Aristolochia*-containing herbal preparations in Europe and North America since 2001 and in Asia since 2003 (Debelle et al. 2008). Currently, AA is classified in the International Agency for Research on Cancer (IARC) monograph as a group 1 human carcinogen (IARC 2012). There is the likelihood that AA may contaminate surface water, grain and vegetables during the processing of *Aristolochia*-containing herbs and the disposal of its waste. TP 53 mutation signature in urothelial tumours and the existence of aristolactam-DNA adducts in the renal cortex are defined in the course of research as a robust biomarker of exposure to this potent nephrotoxin and human carcinogen (Moriya et al. 2014).

1.6.3.1 Pesticides

Ninety percent of pesticides currently in use are synthetic, but in the past 20 years, there had been conscious attempts to develop safe and environmentally friendly pesticides. Organic or natural pesticides have received the most acclaim and certain have the endorsement of environmentalists. Pesticides such as fungicides, herbicides and rodenticides can be very helpful in the sense that they protect man's health by killing germs, animals or plants that can hurt us. On the other hand, a good number of pesticides can be injurious to human and animals. The proper disposing of pesticides is very vital for the protection of the environment. Biologically based pesticides are becoming more popular as they are safer than traditional pesticides. They come in the form of pheromones and microbial pesticides. To insects and rodents, pesticides are inherently toxic. Organophosphate and organochlorine insecticides (synthetic pesticides) have been linked to a wide range of ailments from cancer to neurological disorders and lung irritations in humans. A diversity of pesticides like mineral oil, malathion, sulphur dimethylamine and many others are used to control fungi and insects on wheat and cereals. Chlorinated hydrocarbons present in synthetic pesticides such as methoxychlor, endosulfan and captain accumulate in fatty tissue because it is not completely filtered from the system.

1.6.4 Effects on Humans

The possible exposure pathway of endocrine disruptors in humans includes direct exposure at the workplace and via consumer products such as food, certain plastic, paints, detergents and cosmetics as well as indirect exposure via the environment, viz. air, water and soil. Apart from the drug DES (synthetic oestrogens), environmental oestrogens were not proven to cause human health problems. Lead is regarded as being hepatotoxic, while cadmium is a well-known nephrotoxic agent. Health effects of pesticides are irritation to the eye, nose and throat, injury to the central nervous system and kidney and augmented risk of cancer. Symptoms of pesticide toxicity include nausea, muscular weakness, headache and dizziness, whereas chronic exposure to certain pesticides could result in liver, kidney, endocrine and nervous system damage. Exposure to elevated levels of cyclodiene pesticides associated with improper use caused various symptoms, including headaches, dizziness, muscle twitching, weakness, tingling sensation and nausea. It is assumed that cyclodienes might cause long-term damage to the liver and the central nervous system as well as a heightened risk of cancer.

Steroid receptors for oestrogens and androgen functions of the brain and the cardiovascular, the skeletal and the urogenital system are regulated by these hormones and can therefore be affected by EDCs. EDCs have the potential to cause reduced quality of semen and low sperm counts, low ejaculate volume and high number of abnormal spermatozoa motility. Other effects may include testicular cancer and malformed reproductive tissue, viz. undescended testes, small penis size, prostate disease and other unrecognized abnormalities of male reproductive tissues. Bisphenol A, a component used for plastic products, binds to the local anaesthetic receptor site to block the human cardiac sodium channel. There are currently putative links between EDC and some female diseases including breast and reproductive organ tissue cancers, fibrocystic disease of the breasts, polycystic ovarian syndrome, endometriosis, uterine fibroid and pelvic inflammatory diseases. Phthalates, most often used in cosmetics like nail polish, are reported to affect the endocrine system and are being investigated for a link with infertility in women. EDCs have been associated to impaired behaviour, mental, immune and thyroid functions in developing children. Others include precocious puberty, osteoporosis, foetal growth and obesity (Meeker 2012). Children are most prone to environmental contaminants from foodstuff to drug and plastic toys. They are similarly quite vulnerable to poisoning from unprescribed medications.

1.6.4.1 Autoimmune Diseases

Environmental exposures play a role in the development and/or the exacerbation of autoimmune diseases (Ritz 2010). Autoimmune diseases collectively afflict approximately 24.5 million Americans with women disproportionally affected. Autoimmune diseases such as rheumatoid arthritis (RA), systemic sclerosis (SSc), systematic lupus erythematosis (SLE) and anti-neutrophil cytoplasmic antibody (ANCA)-related vasculitis, solvent exposure and the development of SSc and smoking and the development of seropositive RA, and an inverse relationship between ultraviolet radiation exposure and the risk of development of multiple sclerosis (MS) can be caused. The mechanisms by which environmental factors alter basic biological processes to induce autoimmune diseases continue to be examined but remain largely unknown. Several studies point to many mechanisms likely involved in environmental exposure-based autoimmunity and include a role for xenobiotics in the activation of Toll-like receptors (TLRs), B-cell activation, impairment of T helper 17 (Th 17) and T-regulatory (T-reg) cell immune function, modifications of self-antigens and alteration of DNA methylation profiles. Despite growing advances in the field, knowledge of the interactive roles of the environment and genetics in the autoimmune process is still lacking, and additional progress is needed on many fronts.

1.7 Other Forms of Environmental Pollutants

Nickel, arsenic, chromium and lead are well-known environmental toxicants. Many household products like insecticides, paints, cosmetics, cleaning fluids and nanomaterial-based items are known to contain some of these toxicants. Even though they may not be directly toxic, their interaction with cellular organelles results in cancer. Many household products (oven and drain cleaners, laundry powder, floor polish, paint and pesticides) are potentially dangerous substances. Arts and craft supplies and gardening products can be hazardous. Many household products can be harmful to kids and pets and end up in the ecosystem if not properly disposed. Exposure via inhalation, swallowing or absorption through the skin is a potential killer. Recent evidence had shown that natural health products (NHPs) therapies are gradually more suggested by health providers, including conventional physicians leading to increased consumption of vitamins and many herbal agents worldwide. The WHO (World Health Organization) reports that the current annual need for medicinal plants is approximately US $14 billion, and it is estimated to likely increase to about $5 trillion by 2050. Cultivators of herbs for medicinal uses are usually ignorant of the WHO regulations, and these products may be infected with banned pesticides and microbial agents like fungi, heavy metals and chemical toxins which may cause unfavourable outcomes such as sensorineural defects, congenital paralysis and liver and kidney damage. In addition Wong et al. (1993) also reported concentration of heavy metals such as cadmium, cobalt, copper, iron, manganese, nickel lead, zinc and mercury in Chinese herbal drugs. Chloramines and chlorine dioxides are well-known disinfection methods to eradicate harmful microorganisms. Chlorine reacts with organic compounds in water to form potentially harmful chemical by-products. These by-products include triethylin, trihalomethanes (THMs) and haloacetic acids (HAAs) are carcinogenic in large quantities and are regulated in the USA by the Environmental Protection Agency. Chloroform, dichloroacetic acid (DCA) and trichloroacetic acid (TCA) which are known liver and kidney carcinogens are by-products of chlorine disinfection found in drinking water. Trihalomethanes, viz. chloroform, bromo-dichloromethane, chloro-bromomethane and bromoform, are regulated organic contaminants in drinking water. Methylation in the promoter region of the c-myc gene was reduced by the trihalomethanes, a process consistent with carcinogenic activities.

1.8 Conclusion

With population increase and urbanization, there is high likelihood of xenobiotic contamination in our food and water. Right from our daily care products to agricultural uses, the existence of harmful xenobiotics has been detected. Even though there are a number of sewage treatment and detection methods, xenobiotics is fast becoming a peril to our ecosystem as over long term there is bound to be repercussions. Long-term effects are autoimmune disorders, liver and kidney damage, cardiac problems and eventually cancer as a result of prolonged consumption of these pollutants in food or drinks. It is therefore a great challenge to the environment health researchers to address this issue. Research is being carried out on antibiotic resistance in sewage as the current trend will lead to a worldwide disaster. The earlier remedies or preventive methods are established, the better it is for the natural ecosystem.

Acknowledgement The author Katheem Kiyasudeen acknowledges University Sains Malaysia (USM) for funding and research facilities via RUI grant (Grant Number: 1001/PTEKIND/811254) and USM fellowship-2015 award for academic support'.

References

Bjorkman D (1998) Nonsteroidal anti-inflammatory drug associated toxicity of the liver, lower gastrointestinal tract, and esophagus. Am J Med 105(5, Suppl. 1):7S–21S

Bonjoko B (2014) Environmental pharmacology: an overview. In: Pharmacology and therapeutics (Monograph on the internet). Intech, pp 133–178. Available from: http://www.intechopen.com/books/pharmacology-andtherapeutics/environmental-pharmacology-an-overview

Brooks BW, Chambliss CK, Stanley JK et al (2005) Determination of select antidepressants in fish from an effluent-dominated stream. Environ Toxicol Chem 24(2):464–469

Cajaraville MP, Cancio M, Ibabe A, Orbea A (2003) Peroxisome proliferation as a biomarker in environmental pollution assessment. Microsc Res Tech 61:191–202

Cleuvers M (2003) Aquatic ecotoxicity of pharmaceuticals including the assessment of combination effects. Toxicol Lett 142(3):185–194

Cuklev F, Fick J, Cvijovic M et al (2012) Does ketoprofen or diclofenac pose the lowest risk to fish? J Hazard Mater 229–230:100–106

Debelle FD, Wanherghem JL, Nortier JL (2008) Aristolochic acid nephropathy: a worldwide problem. Kidney Int 74:154–169

Doggrell SA (1990) The membrane stabilizing and beta1- adrenoceptor blocking activity of (+) – and (−) -propranolol on the rat left atria. Gen Pharmacol Vasc Sci 21(5):677–680

Drury R, Scott J, Rosi – Marshall EJ, Kelly JJ (2013) Triclosan exposure increases triclosan resistance and influences taxonomic composition of benthic bacterial communities. Environ Sci Technol 47(15):8923–8930

Fent K, Weston AA, Caminada D (2006) Ecotoxicology of human pharmaceuticals. Aquat Toxicol 76:122–159

Fetzner S (2002) Biodegradation of xenobiotics. In Encyclopedia of Life Support Systems (EOLSS) Publishers, developed under the Auspices of the UNESCO. Biotechnology, Edited by Doelle and Da Silva. EOLSS Oxfor, U.K. p 32

Fick J, Lindberg RH, Tysklind M, Larsson DG (2010) Predicted critical environmental concentrations for 500 pharmaceuticals. Regul Toxicol Pharmacol 58(3):516–523

Garrison AW, Pope JD, Allen FR (1976) Analysis of organic compounds in domestic wastewater. In: Keith CH (ed) Identification and analysis of organic pollutants in water. Ann Arbor Science, Michigan, pp 517–566

Grollman AP, Shibutani S, Moriya M, Muller F, Wu L, Moil U, Swzulai N, Fernandes A, Rosenqiust T, Medeverec Z, Jakovinak BB, Slade N, Turesky RJ, Goodenough AK, Rieger R, Nukelic M, Jelakovic B (2007) Aristolochic acid and the etiology of endemic. (Balkan) nephropathy. Proc Natl Acad Sci U S A 104:12129–12134

Gross B, Montgomery-Brown J, Naumann A, Reinhard M (2004) Occurrence and fate of pharmaceuticals and alkylphenol ethoxylate metabolites in an effluent-dominated river and wetland. Environ Toxicol Chem 23(9):2074–2083

Guengerich FP (1997) Role of cytochrome p450 enzymes in drug – drug interaction. Adv Pharmacol 47:7–35

Gunnarsson L, Kristiannsson E, Rutgersson C, Sturve J, Fick J, Forlin L, Larsson DGJ (2009) Pharmaceutical industry effluent diluted 1:500 affects global gene expression, cytochrome 1A activity and plasma phosphate in fish. Environ Toxicol Chem 28(12):2639–2647

Heberer T (2002) Occurrence, fate, and removal of pharmaceutical residues in the aquatic environment: a review of recent research data. Toxicol Lett 131(1/2):5–17

Hoffman LR, D'Argenio DA, MacCoss MJ et al (2005) Aminoglycoside antibiotics induce bacterial biofilm formation. Nature 436(7054):1171–1175

Holland W, Morrison T, Chang Y et al (2004) Metformin (glucophage) inhibits tyrosine phosphatase activity to stimulate the insulin receptor tyrosine kinase. Biochem Pharmacol 67(11):2081–2091

IARC (2012) IARC working group on the evaluation of carcinogenic risks to humans; pharmeceuticals. Volume 100A. A review of human carcinogens. IARC Monogr Eval Carcinog Risks Num 100:101–401

Jain RK, Kapur M, Labana S, Lal B, Sarma PM et al (2005) Microbial diversity: application of micro-organisms for the biodegradation of xenobiotics. Curr Sci 89(1):101–112

Jones OAH, Voulvoulis N, Lester JN (2001) Human pharmaceuticals in the aquatic environment. Environ Technol 22:1383–1394

Karatan E, Watrick P (2009) Signals, regulatory networks and materials that build and break bacterial biofilms. Microbiol Mol Biol Rev 73(2):310–347

Kid KA, Blanchfield PJ, Mills KH et al (2007) Collapse of a fish population after exposure to a synthetic estrogen. Proc Natl Acad Sci U S A 104(21):8897–8901

Kolpin DW, Furlong ET, Meyer MT et al (2002) Pharmaceuticals, hormones, and other organic wastewater contaminants in U.S. streams 1999–2000: a national reconnaissance. Environ Sci Technol 36(6):1202–1211

Landrum PF, Robbins JA (1990) Bioavailability of sediment-associated contaminants to benthic invertebrates. In: Baudo R, Giesy JP, Muntau H (eds) Sediments: chemistry and toxicity of in-place pollutants. Lewis Publishers Inc, Chelsea

Le Page Y, Vosges M, Servil A, Brown F, Kah O (2011) Neuroendocrine effects of endocrine disruptors in teleost fish. J Toxicol Environ Health B Crit Rev 14(5–7):370–386

Lee II H (1992) Models, muddles, and mud: predicting bioaccumulation of sediment-associated pollutants. In: Burton Jr. GA (ed) Sediment toxicity assessment. Chelsea

Maenpaa KA (2007) The toxicity of xenobiotics in an aquatic environment: connecting body residues with adverse effects. PhD Dissertation, University of Joensuu, Finland

Meeker JD (2012) Expensive to environmental endocrine disruptions and child development. Arch Pediatr Adolesc Med 166(10):952–958

Moriya M, Slade N, Brder B, Medverec Z et al (2014) TP53 mutational signature for aristolochic acid: an environmental carcinogen. Int J Cancer 129:1532–1536

Newman MC (1998) Fundamentals of ecotoxicology. Sleeping Bear/Ann Arbor Press, Chelsea

Reineke W, Knackmuss HJ (1988) Microbial degradation of haloaromatics. Annu Rev Microbiol 42:263–287

Richardson ML, Bowron JM (1985) The fate of pharmaceutical chemicals in the aquatic environment. J Pharm Pharmacol 37(1):1–12

Ritz SA (2010) Air pollution as a potential contributor to the 'epidemic' of autoimmune disease. Med Hypothesis 74(1):110–117

Rosi-Marshall E (2013) Streams stressed by pharmaceutical pollution. www. Environmental change. nd. edu/events/2 Last visited 10-08-2013

Sacher F, Lange FT, Brauch HJ, Blankenhorn I (2001) Pharmaceuticals in groundwaters: analytical methods and results of a monitoring program in Baden-Wurttemberg, Germany. J Chromatogr A 938(1/2):199–210

Schmeiser HH, Shoepe KB, Wiessler M (1998) DNA adduct formation of aristolochic acid I and II in vitro and in vivo. Carcinogenesis 9:297–303

Shanmugan G, Sampath S, Selvaraj KK et al (2013) Non-steroidal anti inflammatory drugs in Indian rivers. Environ Sci Pollut Res 21(2):921–931

Sikandar A, Shehzadi K, Arshad Q, Munir K (2013) Phytoremediation: an analytical technique for the assessment of biodegradation of organic xenobiotic pollutants: a review. Int J Sci Res 4(2):2250–2253

Streit B (1992) Bioaccumulation processes in ecosystems. Experientia 48:955–970

Ternes TA (1998) Occurrence of drugs in German sewage treatment plants and rivers. Water Res 32(11):3245–3260

van derVen K, Van Dongen W, Maes BUW et al (2004) Determination of diazepam in aquatic samples by capillary liquid chromatography–electrospray tandem mass spectrometry. Chemosphere 57(8):967–973

Wiegel S, Aulinger A, Brockmeyer R et al (2004) Pharmaceuticals in the river Elbe and its tributaries. Chemosphere 57(2):107–126

Wong MK, Tan P, Wee YC (1993) Heavy metals in some Chinese herbal plants. Biol Trace Elem Res 36(2):135–142

Heavy Metal and Their Regulation in Plant System: An Overview

2

Dhananjay Kumar, D.P. Singh, S.C. Barman, and Narendra Kumar

Abstract

Unplanned industrialization and improper waste disposal have resulted in the release of enormous quantities of inorganic toxicants like metal, metalloids, and radionuclides in the biosphere. Since, metals are non-biodegradable and tend to bioaccumulate via food chain, they pose threat to human health. Indiscriminate disposal of industrial waste to the environment causes adverse impact on ecosystem. Plants growing on metal-contaminated sites display several disturbances related to physiology and biochemical process like gaseous exchange, CO_2 fixation, respiration, nutrient absorption, etc. These disturbances subsequently cause reduction in plant growth and lower biomass production. Although being an essential micronutrient, some heavy metals at lower concentrations are vital for plant growth; however, at higher concentrations they become very toxic. To cope up with the metal toxicity, plants have developed various mechanisms like immobilization, exclusion, chelation, and compartmentization. Plants have distinct cellular mechanism such as chelation and vacuolar compartmentization of metals to withstand the metal toxicity. Phytochelatins, the thiol peptides, potentially chelate metals and form complexes in cytoplasm; subsequently these metal-thiol complexes are sequestrated into vacuole via ATP-binding cassette transporters (ABC transporters). In the last couple of

D. Kumar • D.P. Singh • N. Kumar (✉)
Department of Environmental Science, Babasaheb Bhimrao Ambedkar University, Lucknow 226025, Uttar Pradesh, India
e-mail: narendrakumar_lko@yahoo.co.in

S.C. Barman
Environmental Monitoring Division, Indian Institute of Toxicology Research, Lucknow 226001, Uttar Pradesh, India

A. Singh et al. (eds.), *Plant Responses to Xenobiotics*,
DOI 10.1007/978-981-10-2860-1_2

decades, the role of phytochelatin synthetase (PCS) and phytochelatins (PCs) in metal detoxification has been proven. In present scenario, there is a great need of sound and intensified research for better understanding of metal toxicity and its metabolism in plants to maintain our ecological harmony.

Keywords
Heavy metals • Toxicity • Contamination • Metabolism

2.1 Introduction

Accelerated industrialization and modernization throughout the world has led to the emergence of various luxurious facilities and goods; however, such industrialization and modernization also releases a considerable amount of undesirable xenobiotic or toxic wastes to various components of environment, i.e., soil, air, and water (Adriano 1992; McIntyre 2003; Kumar et al. 2013). Even the most pristine environment like the Arctic Circle and Antarctic has not been spared by the globally transported xenobiotics (AMAP 2002). Although environmental pollution is a natural process, the human activities like improper waste management practices, landfill operations, mining, the use of chemical fertilizers, application of sewage sludge, etc. have accelerated the level, rate, and types of contamination to soil, air, and water. Further, application of sewage sludge and discharge of industrial effluent containing inorganic chemicals like heavy metals to agricultural lands intensifies the problem of soil pollution. The problem of contamination turns to be more complex when the effluents are discharged directly because of the heterogeneity in the quality and quantity of discharged effluents (Srivastava et al. 1994; Kara 2005; Singh et al. 2010).

Heavy metals (e.g., Cr, Cd, As, Fe, Ni, Pb, Hg, Zn, etc.) are an important class of environmental pollutants, and many of these are highly toxic in soluble forms. Since, heavy metals are non-biodegradable in nature and persist in soil for a long time, they tend to bioaccumulate in ecological food chain through uptake at primary producer level and subsequently via consumption at secondary and tertiary levels (Sakakibara et al. 2011; Bauddh and Singh 2012; Lu et al. 2014; al. 2016). Therefore, removal of the metals and other contaminants is an important concern and a major policy priority globally.

Chemically, the term heavy metal refers to any metallic element with a specific gravity greater than 5 (Venugopal and Luckey 1978). Heavy metals are naturally found in dispersed form in rock. However, industrialization and urbanization have increased the heavy metals in biosphere. Further, the major availability of heavy metals is in the soil and aquatic system, and relatively a smaller portion is available in the atmosphere in the form of particulate or vapors. Plants growing on land contaminated with heavy metals display several disturbances related to physiology and biochemical processes like gaseous exchange, CO_2 fixation, respiration, nutrient absorption, etc. These disturbances subsequently cause reduction in plant growth and lower biomass production. The toxicity due to heavy metals in plants varies with several factors, viz., plant species, concentration of the metal and its chemical form, soil composition, and pH (Nagajyoti et al. 2010).

2.2 Impact of Heavy Metals on Plant Growth and Development

Like all living creature, plants are also sensitive to both deficiency and excess of micronutrient including trace elements. Some heavy metals are essential in nature as they are required for normal growth of plant. Essential heavy metals like Cu, Zn, Fe, Mn, Mo, and Ni play important roles in biochemical and physiological functions in plants (Reeves and Baker 2000). Being essential micronutrients, Cu and Zn are very important for normal plant growth as they either serve as a cofactor and activators of enzyme reactions or exert a catalytic property such as prosthetic group in metalloproteins (Mildvan 1970). These essential heavy metals are also involved in redox reactions, electron transfer, basic functions in nucleic acid metabolisms, and, as direct participant, being integral part of several enzymes. The availability of these essential metals in growing medium at certain concentration is very important, but their excess concentration leads to several toxic effects (Blaylock and Huang 2000; Monni et al. 2000). Due to their presence in trace in environmental matrices, these heavy metals are also known as trace elements ($10mgKg^{-1}$ or mgL^{-1} of metal in soil/aquatic medium) or ultra-trace element ($1\mu gKg^{-1}$ or μgL^{-1} of metal in soil/aquatic medium). Besides these essential trace elements, another category of heavy metals, class B metals, that are considered as non-essential trace elements like Hg, Ag, Pb, Ni, etc. are very toxic in nature (Nieboer and Richardson 1980). In terrestrial system, plants are stationary, and the their roots are the main contact sites for trace metal ions, while in aquatic system, the entire plant body is exposed, and metal ions are absorbed directly due to particle deposition on leaves' surfaces.

2.2.1 Copper

Copper is considered as an essential micronutrient for plants and algae especially because of its crucial role in photosynthesis, CO_2, and ATP synthesis (Thomas et al. 1998; Chatterjee et al. 2006; Mahmood and Islam 2006). Copper is an important component of several proteins like plastocyanin of photosystem and cytochrome oxidase of respiratory electron transport chain (Demirevska-kepova et al. 2004). It is a primary electron donor in photosystem I. Copper plays a crucial role as a cofactor for enzymes involved in the elimination of superoxide radicals (superoxide dismutase and ascorbate oxidase) and also of oxidase and mono- and dioxyegenase (amine oxidases, ammonia monoxidase, ceruloplasmin, lysyl oxidase). Further, the toxicity due to the exposure of excess copper to plants has been well reported by several researchers (Moreno-Caselles et al. 2000; Singh and Tewari 2003; Keller et al. 2015). The presence of excess copper in soil displays cytotoxic role, induces stress, and causes damage to plants which leads to several deformities including retardation in plant growth and leaf chlorosis (Lewis et al. 2001; Adrees et al. 2015a, b). Excess of copper also generates ROS causing oxidative stress by damaging macromolecule disturbance in metabolic pathways (Hegedus et al. 2001; Habiba et al. 2015). In combination with Cd, Cu has been reported for its adverse effects on seed germination, length of seedling, and number of lateral roots (Neelima and Reddy 2002).

2.2.2 Cadmium

Cadmium is a non-essential heavy metal for plants. Cadmium has been ranked 7th among the top 20 toxins due to its great solubility in water and high toxicity (Yang et al. 2004). Cd has been reported as an extremely significant pollutant among the class of heavy metal pollutants (Das et al. 1997; Rizwan et al. 2016). Cadmium toxicity is easily identifiable in the form of stunt growth, chlorosis, browning of root tips, and finally plant death (Das et al. 1997; Wojciek and Tukiendorf 2004; Mohanpuria et al. 2007; Guo et al. 2008). Excess of Cd in growing soil can cause leaf chlorosis; however, it may be due to deficiency of iron and the interaction with toxic metals. Chlorosis may appear due to direct or indirect interaction with Fe present in leaves. The presence of excess cadmium in growing medium causes suppression in uptake of iron (Haghiri 1973). Cadmium induces the inhibition of root Fe(III) reductase which leads to deficiency of iron, severely affecting plant photosynthesis (Alcantara et al. 1994). Cd-induced chlorosis may be attributed to the changes in Fe/Zn ratio. It has been reported that Cd also interfere with the uptake, transport, and use of various essential elements like Ca, Mg, P, K, etc. and water (Das et al. 1997; Asgher et al. 2015). Further, Cd also inhibits the nitrate reductase activity which reduces the absorption of nitrate and its transport from roots to shoots (Hernandez et al. 1996). Cadmium has also been reported for its cytotoxic effects in the form of swelling, vacuolization, degeneration of mitochondria, inhibition in cell proliferation, and a low mitotic index (Silverberg 1976; Rosas et al. 1984; Khan et al. 2016). The chromosomal aberrations have been also reported in onions, beans, peas, and barley on exposure to excess Cd (Oehlkers 1953; Von Rosen 1954; Degreave 1981). Rosas et al. (1984), reported that the plant exposed to Cd at concentration of 1.5 to 10 mgL^{-1} for 24 h had caused physiological and genetical damages. They also reported that Cd inhibits the cell division and alters the chromosome. Further, they also mentioned that the inhibition of cell proliferation, shown by low mitotic index, was proportional to the concentration and time of exposure (Rosas et al. 1984). Moreover, exposure to Cd causes decrease in nitrogen fixation and primary ammonia assimilation in the root nodules (Balestrasse et al. 2003).

2.2.3 Zinc

Zinc is considered as an essential micronutrient for plants because at optimal concentration it is essential for normal functioning of cell metabolism as well as for plant growth (Dhankhar et al. 2012; Broadley et al. 2007). It plays crucial role as a cofactor in many physiological processes such as metabolism of several biomolecules, gene expression and regulation, enzyme activation, protein synthesis, and reproductive development (Cakmak 2000). However, accumulation of Zn in plant at higher concentration (>300μgg^{-1} in dry weight) causes physiological alteration and growth inhibition (Foy et al. 1978). High level of exposure of Zn in growing medium inhibits several plant metabolic functions, results in stunted growth, and causes senescence. Zn toxicity restricts the growth of roots and shoots (Choi et al. 1996; Fontes and cox 1998). At high concentration, it also causes chlorosis in premature leaves, which can extend to older leaves on prolonged high exposure. Excess of Zn also causes deficiency of other essential elements, viz., Mn and Cu, in shoots which hinders the transfer of these essential micronutrients from root to shoot. The possible reason for this hindrance of transfer of these micronutrients is the concentration of Fe and Mn in plant grown in Zn-rich media is greater in root than the shoot (Ebbs and Kochian 1997).

2.2.4 Arsenic

Arsenic (As) is a metalloid and considered as a nonessential and toxic element for plants (Zhao et al. 2009). Mobility and availability of As in soil depends upon its ionic form. As(III) is very toxic in nature but less mobile than As(V). Arsenate, i.e., As(V), is the most stable form found in the soil, and hence its availability for plant is greater than As(III). The availability and mobility of As in soil highly depends on soil pH. As commonly forms complexes with calcium at high pH (pH 6–8) while it frequently binds with iron at low pH (pH 4) (Fayiga and Ma 2006). Further, the presence of Fe and MnO in soil also increases the availability and mobility of As (Zavala and Duxbury 2008). In plants, it mainly accumulates in roots and to very less extent in shoots. Generally, plants uptake arsenic as As(V) and translocate it via the xylem along with water and minerals as As(III)-S compound (Wang et al. 2002). Chemically, As(V) is analog to PO_4^{3+} and hence competes with PO_4^{3+} uptake in root and interferes metabolic processes like ATP synthesis, oxidative phosphorylation, and transport across the plasma membrane through phosphate transport channels (Meharg and Macnair 1992; Tripathi et al. 2007; Stoeva and Bineva 2003). The presence of excess As in growing medium causes physiological changes, interference with metabolic processes, growth inhibition, ultimate reduction in crop productivity, and finally death (Miteva 2002; Stoeva et al. 2004; Anjum et al. 2016). Arsenic toxicity may be seen as a consequence of binding of As with sulfhydryl (SH) group of protein, leading to inhibition of protein activity or structural disruption, or replacing the essential element resulting in deficiency effects (Assche and Clijsters 1990; Delnomdedieu et al. 1994; Kumar et al. 2015). Arsenic may also stimulate the formation of free radicals and reactive oxygen species like $O_2^{\cdot-}$, $OH^{\cdot}$, and H_2O_2 which are strong oxidizing agents and cause oxidative damage to biomolecules like lipids and protein and finally cell death (Dietz et al. 1999; Molassiotis et al. 2006; Gunes et al. 2009).

2.2.5 Nickel

Nickel is considered as an essential element at lower concentration (0.01 to 5μgg^{-1}) for plants. The uptake of Ni from growing medium takes place mainly via passive diffusion and active transport. Plants passively absorb the soluble Ni compounds via cation transport system. The chelated Ni compounds are taken and transported via active-transported-mediated system using transport proteins like permeases. Moreover, the insoluble Ni compounds are absorbed in root cells through endocytosis and easily transported to shoots through the xylem via transpiration stream and can get accumulated in newly developed buds, fruits, and seeds. Ni is an essential component of several metalloenzymes such as superoxide dismutase, NiFe hydrogenases, methyl coenzyme M reductase, urease, acetyl Co-A synthase, carbon monoxide dehydrogenase, hydrogenases, and RNase-A. Further, the high exposure of Ni in growing medium affects the activities of amylases, proteases, and ribonucleases subsequently affecting the digestion and metabolization of food reserves in germinating seeds (Ahmad and Ashraf 2011). High concentration of Ni in growing medium causes alteration in physiological process and diverse toxicity symptoms such as chlorosis, necrosis, and wilting (Zornoza et al. 1999; Rao and Sresty 2000; Nakazawa et al. 2004). Plants growing in excess Ni medium show negative effects on photosynthesis, mineral nutrients, sugar transport, and water balance (Samarakoon and Rauser 1979; Tripathy et al. 1981; Parida et al. 2003; Sethy and Ghosh 2013). Decrease in uptake of water is an indicator of the increasing Ni toxicity in plants (Pandey and Sharma 2002; Gajewska et al. 2006). Ni toxicity has also been attributed for the impairment of nutrient balance, disturbance of lipid composition, and H-ATPase activity resulting in the cell membrane dysfunctions (Ros et al. 1992). Exposure of high level of Ni increases MDA concentration which might disturb membrane function and cytoplasmic ion balance, particularly K^+; the most mobile ion across the cell membrane.

2.2.6 Chromium

Chromium (Cr) is considered as a non-essential metal for plant. Chromium has been well reported for its toxicity to plant growth and development (Huffman and Alloway 1973; Vikram et al. 2011). On high exposure (1–5 mgL^{-1}), Cr causes chlorosis and alteration in several metabolic processes, *viz.*, growth inhibition and decline in the chlorophyll synthesis (Dube et al. 2003; Ahemad 2015). Some plants have been reported with potential to accumulate Cr without showing any symptoms of Cr toxicity. Chromium enters and accumulate in root cells by the symplastic pathway. Plants uptake chromium in its trivalent form, i.e., Cr(III) by passive mechanism, while uptake of Cr(VI) is inhibited by SO_4^{2-} and Ca^{2+} (Zayed and Terrey 2003). Hexavalent ions, i.e., Cr(VI), damage the root membranes due to their high oxidation power. Cr enters into plant roots by reduction and/or complexation with root exudates, which enhance the solubility and mobility via root xylem (Shanker et al. 2005; Bluskov et al. 2005). Although accumulation and mobilization of Cr inside the storage tissue depends on its ionic form, however, it accumulates mainly in roots and translocated poorly to shoots (James and Barlett 1983). Like cadmium, Cr(VI) also reduces the uptake of many essential elements like Fe, Mg, Mn, Ca, P, and K resulting in many negative effects on plant growth (Gardea-Torresdey et al. 2005; Peralta-Videa et al. 2009). Seed germination of *Phaseolus vulgaris* was reduced by 48 % on exposure of Cr(VI) at concentration of 500 ppm (Parr and Taylor. 1982). Reduction in seed germination was observed in seeds of *Medicago sativa* by 23 % at 40 ppm of Cr(VI) (Peralta et al. 2001). Adverse effect of Cr on photosynthesis has been also well documented (Assche and Clijsters 1983; Vikram et al. 2011). Chromium affects the photosynthesis in the form of reduction in photosynthetic pigments and inhibition in photophosphorylation, electron transport, and enzyme activities (Clijsters and Assche 1985; Vikram et al. 2011). Furthermore, it also causes disorganization of ultrastructure of chloroplasts, which are the primary site for photosynthesis (Vazques et al. 1987; Ahemad 2015).

2.2.7 Lead

Lead (Pb) is a non-essential and one of the most ubiquitously distributed toxic elements in the soil. Plant gets lead mainly from soil and aerosol (Sharma and Dubey 2005). In plants, roots have greater ability to accumulate Pb; however, its subsequent translocation to aerial parts is highly restricted (Lane and Martin 1977). It was also reported that lead could be translocated and accumulated in leaves in a concentration-dependent manner (Miller and Koeppe 1971). Further, the extent of Pb uptake by plant from aerial sources, through leaves, depends on the ability and specific leaf morphology (Godzik 1993). Availability of lead in soil highly depends on soil conditions like soil pH, particle size, and cation exchange capacity. Moreover, the availability and uptake of Pb is also affected by some other factors such as root surface area, root exudation, mycorrhization, and degree of transpiration (Davies 1995). Absorption of Pb from soil increases with the increase in pH from 3 to 8.5, while at pH 5.5 to 7.5, its solubility is controlled by phosphate or carbonate ions (Sharma and Dubey 2005). Plants' root absorbs the Pb through apoplastic pathway or via Ca^{2+} permeable channels (Rudakova et al. 1988; Pourrut et al. 2011). After uptake, it accumulates primarily in root cells, due to the blockage by the Casparian strips inside the endodermis. Further, lead is also trapped by the negative charges that exist on the roots' cell wall (Seregin and Ivaniov 1997, 2001). At root surface, Pb binds to carboxyl groups of mucilage uronic acids which restrict the Pb uptake into the root and form an important barrier to protect root system (Morel et al. 1986). Plant growing in Pb-contaminated medium exerts several adverse effects. Accumulation of lead in plants exerts several deleterious effects on morphological, physiological, and biochemical function of plants, either directly or indirectly. When Pb enters inside the cells, it causes toxicity by altering cell membrane permeability, by reacting with active groups of metabolic enzymes, by replacing essential ions, and by complex formation with phosphate group of ADP or ATP. Lead toxicity causes inhibition of enzyme activities, disturbed mineral nutrition, water imbalance, hormonal disturbances, inhibition of ATP production, lipid peroxidation, change in membrane permeability, and DNA damage by overproduction of reactive oxygen species (ROS) (Sharma and Dubey 2005; Pourrut et al. 2011; Sethy and Ghosh 2013). Further, high concentration of Pb in growing medium causes inhibition of seed germination, root and stem elongation, and leaf expansion (Morzck and Funicclli 1982; Gruenhage and Jager 1985). The extent of inhibition of root elongation depends on the concentration and ionic composition of lead and pH of the growing medium (Gruenhage and Jager 1985).

2.2.8 Manganese

Manganese (Mn) is an essential element for plant with a key role in various physiological processes particularly in photosynthesis and as an enzyme antioxidant cofactor. In plant cell, it exists as a cation in several complexes and can form metalloproteins in which Mn is tightly bound, probably to produce an appropriate protein conformation. Deficiency of Mn also affects the photosynthesis by affecting water-splitting system of photosystem II, which provides necessary electrons for photosynthesis (Buchanan et al. 2000). Mn deficiency occurs mostly in severely weathered sandy and organic soil having pH more than 6 (Alloway 2008). Mn has low phloem mobility, resulting in typical leaf symptoms of Mn deficiency which initially develops into premature leaves. In biological system Mn exists in many states preferably as II, III, and IV. In soil divalent state, i.e., Mn(II), is the most soluble form, while Mn(III) and Mn(IV) are very less soluble (Guest et al. 2002). The bioavailability of Mn in soil is influenced by soil pH and redox potential of Mn. Lower pH (<5.5) and increased redox potential of Mn increase the amount of soluble Mn(II) in soil (Kogelmann and Sharpe 2006; Watmough et al. 2007). Higher soil pH (up to 8) favors chemical autoxidation of Mn(II) causing the formation of MnO_2, Mn_2O_3, Mn_3O_4, and Mn_2O_7 which are normally unavailable (Ducic and Polle 2007; Humphries et al. 2007). Moreover, high pH also causes adsorption of Mn on soil particles, thereby decreasing their bioavailability to plants (Fageria et al. 2002). Mn is transported from root to aerial parts via the transpiration stream and accumulates in leaves which did not re-mobilize to other aerial parts through the phloem (Loneragan 1988). Accumulation of high concentration of Mn in leaves causes reduced rate of photosynthesis (Kitao et al. 1997a, b). Mn toxicity causes necrotic brown spots on leaves which start from the lower leaves and progresses with the time toward upper leaves (Horiguchi 1988; Wu 1994). Furthermore, with the time, the number and size of necrotic spots increase, resulting in necrotic lesions, leaf browning, and finally death (Elamin and Wilcox 1986a, b). Mn toxicity has also been attributed for the crinkled leaf, chlorosis, and browning of the youngest leaf, petiole, and stem tissues (Wu 1994; Bachman and Miller 1995). Probably, Mn-induced iron deficiency is the possible reason for chlorosis in younger leaves (Horst 1988). Mn toxicity is also associated with the brown coloring and sometimes cracks in roots (Bot et al. 1990a, b; Foy et al. 1995). Accumulation of Mn in leaves inhibits synthesis of chlorophyll by blocking iron, a concerning process resulting in the decrease in photosynthesis (Clarimont et al. 1986).

2.3 Heavy Metal Tolerances

Roots are the primary contact sites in terrestrial plants with exposure to metal. In case of aquatic plants, the whole plant body is exposed to metal present in growing medium. The growing medium contains essential and non-essential metals which on excess become toxic resulting in inhibition of growth and development and even death of the plant. In order to survive, plants have evolved some efficient and specific mechanisms to deal with the heavy metal stress. The adaptive mechanism evolved by plants to cope up with metal stress includes immobilization, plasma membrane exclusion, restriction of uptake and transport, synthesis of specific heavy metal transporters, induction of stress proteins, chelation and sequestration by specific ligands, etc. (Cobbett et al. 2000; Clemens 2006; Dalcorso et al. 2008; Hossain et al. 2009; Hossain and Fujita 2009; Sharma and Dietz 2009; Hossain et al. 2012a, b; Adrees et al. 2015a, b). Cellular mechanism for metal tolerance involves two basic approaches to keep low concentration of toxic metal ions in cytoplasm by preventing metal from being transported across the plasma membrane. It can be achieved either by increasing binding of metal ions to cell wall or by pumping out the metal from cell by active efflux pumps. Another approach is detoxification of toxic metal ions by inactivation via chelation or conversion of toxic metal ion into less toxic forms (Zhu et al. 2004).

2.3.1 Cellular Exclusion of Heavy Metals

Cellular exclusion of heavy metals is an important adaptive system for plants to tolerate the heavy metal toxicity. A large fraction of heavy metals are found in the apoplastic space in plant roots. Tice et al. (1992), defined apoplastic and symplastic aluminum fraction in root tips of Al-intoxicated wheat (i.e., Al-sensitive and Al-tolerant wheat cultivars) and reported that at equal external Al concentrations, a sensitive wheat cultivar had more symplastic Al than a tolerant cultivar suggesting exclusion mechanism. They also suggested that the distribution of Al in two cultivars did not support a symplastic detoxification hypothesis, but the role of cytoplasmic exclusion remains disturbed. The transporter proteins are potentially involved in the cellular exclusion of toxic metal ions from the symplastic to apoplastic space. Further, cytoplasmic exclusion could be accomplished through selective permeability of plasma membrane, formation of a plant-induced pH barrier in the rhizosphere, immobilization of metal on the cell wall, or exudation of chelating ligands (Taylor 1991; Tice et al. 1992).

2.3.2 Heavy Metal Complexation at Cell Wall-Plasma Membrane

When a plant cell is exposed with heavy metals, the cell wall-plasma membrane interface accumulates large portion of heavy metals. Iwasaki et al. (1990), reported that about 60 % of the total root Cu was bound to the root cell walls and plasma membranes in Italian ryegrass (*Lolium multiflorum*) and red clover (*Trifolium pratense* L.). Exchange site present on the cell wall determines the cation exchange capacity (CEC). Masion and Bertsch (1997), reported that a sensitive wheat cultivar have low cell wall CEC concentration and show less tolerance to Al, while a tolerant cultivar have a high concentration of cell wall CECs and show high tolerance to Al. Further, the sensitive wheat cultivars showed a higher affinity for aluminum than tolerant cultivars which indicate that tolerance mechanism is based on the cell wall permeability.

2.3.3 Sequestration Within Vacuoles

Vacuole is commonly considered as the main storage cell organelle for metals in plant, and there is evidence that phytochelatin-metal complexes are driven into vacuole (Salt and Rauser 1995). There are several studies showing that the vacuole is the site for the accumulation of heavy metals (Ernst et al. 1992; De 2000). Once a plant cell is exposed to any toxic metal ions, it mechanizes various strategies to cope with the metal toxicity. Intracellular sequestration or vacuole compartmentalization is also one of them, in which toxic metals are transported either out of the cell sequestrated into vacuole, thereby removing it from the cytosol or other cellular compartments where sensitive metabolic activities take place (Clemens 2006; Dalcorso et al. 2010). In some hyperaccumulator plants, vacuole compartmentalization of metal is also a part of tolerance mechanism. It has been reported that the hyperaccumulator plants enhance their metal tolerance by compartmentalizing most of the intracellular metal present in leaves into vacuole (Kramer et al. 2000). Further, the two proton pumps, i.e., vacuolar proton-ATPase (V-ATPase) and vacuolar proton-phosphatase (V-Ppase), strengthen vacuolar uptake of most solutes. The uptake of metal ions can be catalyzed either by channels or by transporters. To date, a wide range of gene families have been identified which are probably involved in transition of metal ions uptake into cell, vacuole sequestration, remobilization of metal from vacuole, xylem loading, and unloading of metals. Several metal transporter proteins have been also reported, *viz*., zinc-regulated transporter (ZRT), iron-regulated transporter (IRT), ATP-binding cassette (ABC) transporters, the P-type metal ATPases, multidrug resistance-associated proteins (MRP), natural resistance-associated macrophage protein (NRAMP) family, ABC transporters of the mitochondria (ATM), cation diffusion facilitator (CDF) family of proteins, copper transporter (COPT) family proteins, yellow-stripe-like (YSL) transporter, Ca^{2+} cation antiporter (CAX), and pleiotropic drug resistance (PDR) transporters (Lee et al. 2005; Kramer et al. 2007; Chiang et al. 2006; Dubey 2011; Hossain et al. 2012a, b).

2.3.4 Metal Chelation by Phytochelatins

To protect themselves from toxicity of heavy metals, chelation of metal ions with high-affinity ligands is one of the prevailing mechanisms of metal detoxification

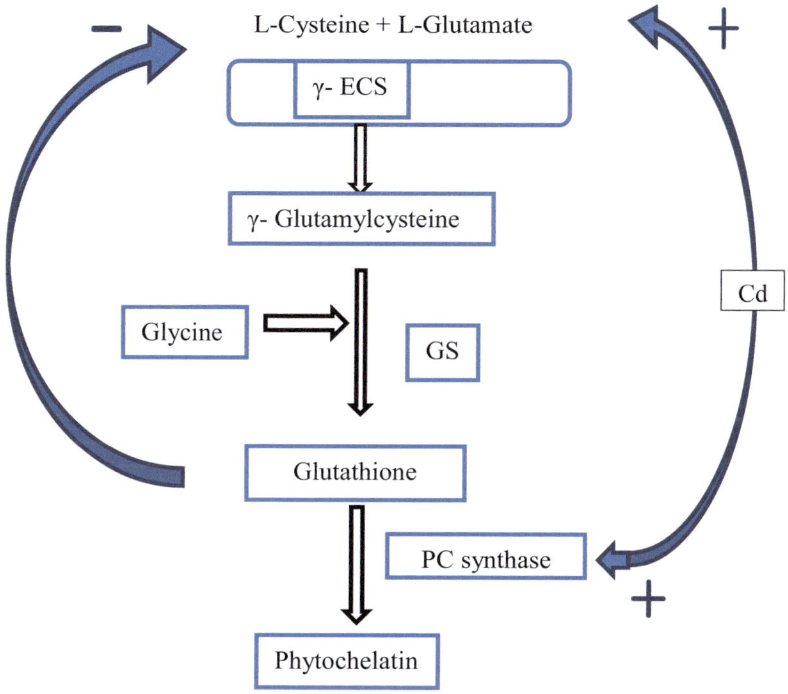

Fig. 2.1 Biosynthesis of phytochelatins in plants (Zenk 1996; Mejare and Bulow 2001)

and tolerance in plants. When a toxic metal enters in a plant cell, it may be scavenged by amino acids, organic acids, and tripeptide GSH or by specific metal-binding ligands. The two classes of peptides or metal-binding ligands are phytochelatins (PCs) and metallothioneins (MTs). The role of phytochelatins in metal detoxification and tolerance has been widely studied in plants (Zenk 1996; Cobbett 2000; Clemens 2001; Mishra et al. 2006). Phytochelatins are not present only in plant cells but have also been reported in fungi and other organism (Grill et al. 1987; Gekeler et al. 1988; Piechalak et al. 2002). Phytochelatins are small, cysteine-rich polypeptides which have potential to form complex with heavy metal ions via thiolate coordination. The general structure of phytochelatin is (γ-Glu-Cys)$_n$X, in which X is Gly, γ-Ala, Ser, or Glu and n is the number of peptides = 2–11. Most of the common forms of PCs have 2–4 peptides. The biosynthesis of phytochelatins is activated in the presence of heavy metals; however, Cd has been reported as the strongest inducer (Grill et al. 1987). PCs are synthesized from glutathione (GSH; (γ-Glu-Cys-Gly)) and related compounds (Fig. 2.1). Their biosynthesis is catalyzed by the enzyme phytochelatin synthase (γ-glutamylcysteine dipeptidyl transpeptidase) which gets activated in the presence of metals (Tomaszewska et al. 1996; Vatamauniuk et al. 2000).

Metal binds to the constitutively PC synthase, thereby activating it to catalyze the conversion of glutathione to phytochelatin. Glutathione is the substrate of the phytochelatin which is synthesized from its constituent amino acids, i.e., L-cystein and

L-glutamate, in two consecutive steps. In the first step, i.e., the formation of γ-glutamylcysteine from L-cystein and L-glutamate is catalyzed by γ-glutamyl-cys synthetase (γ-ECS), while in second step glycine is added to γ-glutamylcysteine by glutathione synthetase (GS). The γ-glutamyl-cys synthetase is dependent on the availability of cysteine and feedback regulated by glutathione (Zenk 1996; Mejare and Bulow 2001).

Further, PCs form complexes with metal ions in cytosol and subsequently transport them into vacuole and protect plant cell from the toxic effects of metals (Salt and Rauser 1995).

2.3.5 Metal Chelation by Metallothioneins

Metallothioneins (MTs) are cysteine-rich (more than 30 % from all amino acids), metal-binding, low-molecular-mass proteins (2–16 kDa) that play a crucial role in detoxification and metabolism of metals. MTs have a unique property of binding d-block metal ions through the 20 cysteinyl groups which are abundant in their structural constituent. MTs were first reported by Margoshes and Vallee in 1957, from a horse renal cortex tissue (Margoshes and Vallee 1957). Metallothioneins have been well reported in bacteria, fungi, and plants (Lerch 1980; Kagi 1991; Murphy and Taiz. 1995; Suzuki et al. 2002; Ryvolova et al. 2011). On the basis of cysteine residue, plant MTs have been subcategorized into three classes, i.e., Cys-Cys, Cys-X-Cys, and Cys-X-X-Cys motifs (in which X denotes an amino acid). The biosynthesis of MTs (gene-encoded polypeptides) is induced by many factors including cytotoxic agents, hormones, and heavy metals (Kagi 1991; Yang et al. 2005; Zhou et al. 2006). Ahn et al. (2012) reported that there are three MT genes, *viz.*, BrMT1, BrMT2, and BrMT3, in *Brassica rapa* which regulates the biosynthesis of MTs under the several metal stress condition. Furthermore, it has also been reported that MTs play an essential role as a Zn donor for several essential metalloproteins comprising matrix metalloproteinases and zinc fingers (Ryvolova et al. 2011).

2.4 Conclusions

Heavy metals/metalloids are important class of inorganic contaminants which enter into the soil and water through various natural and anthropogenic sources. Although some metals like Fe, Cu, Zn, Ni, etc. at required levels are essential for normal growth and metabolism of plants, however, their exposures at high concentration cause several negative impacts on the plant growth. Some metals such as As, Pb, Cr, Cd, etc. are non-essentials, and its contamination in growing medium causes various negative health effects. For terrestrial plants, roots are the primary contact sites exposed directly to the metal contaminants while, in case of aquatic plants, the whole plant body is exposed to metal present in growing medium. Metal contamination in growing environment causes disturbances in the physiological and biochemical processes of plants resulting in altered metabolism, growth reduction, lower biomass, chlorosis, necrosis, wilting, water imbalance, etc.

Heavy metals/metalloids differ in their affinity for O-, N-, and S-containing ligands depending on the physical and chemical properties of the heavy metals/metalloids ions. The metal toxicity is influenced by the binding ability of metals to various ligands present in metal biological system such as carboxylate ion, imidazole, sulfhydryl group, and aliphatic amine. In order to survive, plants have evolved many efficient and specific mechanisms to cope up with the metal stress. Adaptive mechanisms evolved by plants to deal with metal stress are immobilization, plasma membrane exclusion, restriction of uptake and transport, synthesis of specific heavy metal transporters, induction of stress proteins, chelation and sequestration by specific ligands, etc. Increasing research about the natural variation in the potential of plants to accumulate, tolerate, and detoxify heavy metals provides us wealthy information. Therefore, an extensive knowledge from various research domains will further increase our understanding about the fundamental mechanism involved in hyperaccumulation which allows us to find out that plants are more suitable for remediation of heavy metal-contaminated environment.

References

Adrees M, Ali S, Rizwan M, Ibrahim M et al (2015a) The effect of excess copper on growth and physiology of important food crops: a review. Environ Sci Pollut Res 22:8148–8162

Adrees M, Ali S, Rizwan M, Zia-ur-Rehman M et al (2015b) Mechanism of silicon-mediated alleviation of heavy metal toxicity in plants: a review. Ecotoxical Environ Safe 119:186–197

Adriano D (1992) Biogeochemistry of trace metals. Lewis Publishers, Boca Raton

Ahemad M (2015) Enhancing phytoremediation of chromium-stressed soils through plant-growth-promoting bacteria. J Genet Eng Biotechnol 13:51–58

Ahmad MS, Ashraf M (2011) Essential roles and hazardous effects of nickel in plants. Rev Environ Contam Toxicol 214:125–167

Ahn YO, Kim SH, Lee J, Kim HR, Lee H-S, Kwak S-S (2012) Three Brassica rapa metallothionein genes are differentially regulated under various stress conditions. Mol Biol Rep 39(3):2059–2067

Alcantara E, Romera FJ, Canete M, De La Guardia MD (1994) Effects of heavy metals on both induction and function of root Fe(III) reductase in Fe-deficient cucumber (*Cucumis sativus* L.) plants. J Exp Bot 45:1893–1898

Alloway BJ (2008) Micronutrients and crop production: an introduction. In Alloway BJ (ed) Micronutrient deficiency in global crops production. Springer, Dordrecht, pp 1–40

AMAP (2002) Arctic pollution Arctic monitoring and assessment program Oslo, Norvay, XII + 111

Anjum SA, Tanveer M, Hussain S, Shahzad B et al (2016) Osmoregulation and antioxidant production in maize under combined cadmium and arsenic stress. Environ Sci Pollut Res 23:11864–11875

Asgher M, Iqbal M, Khan R, Naser AA, Nafees AK (2015) Minimising toxicity of cadmium in plants—role of plant growth regulators. Protoplasma 252:399–413

Assche VF, Clijsters H (1983) Multiple effects of heavy metals on photosynthesis. In: Marcelle R (ed) Effects of stress on photosynthesis. Nijhoff/Junk The Hague 7: 371–382

Assche VF, Clijsters H (1990) Effect of metals on enzyme activity in plants. Plant Cell Environ 13:195–206

Bachman GR, Miller WB (1995) Iron chelate inducible iron/manganese toxicity in zonal geranium. J Plant Nutr 18:1917–1929

Balestrasse KB, Benavides MP, Gallego SM, Tomaro ML (2003) Effect on cadmium stress on nitrogen metabolism in nodules and roots of soybean plants. Func Plant Biol 30:57–64

Bauddh K, Singh RP (2012) Cadmium tolerance and its phytoremediation by two oil yielding plants *Ricinus communis* (L.) and *Brassica juncea* (L.) from the contaminated soil. Int J Phytorem 14:772–785

Blaylock MJ, Huang JW (2000) Phytoextraction of metals. In: Raskin I, Ensley BD (eds) Phytoremidation of toxic metals-using plants to clean up the environment. Wiley, New York, pp 53–70

Bluskov S, Arocena JM, Omotoso OO, Young JP (2005) Uptake, distribution and speciation of chromium in *Brassica juncea*. Int J Phytorem 7(2):153–155

Bot LJ, Kirkby EA, Beusichem ML (1990a) Manganese toxicity in tomato plants: effects on cation uptake and distribution. J Plant Nutr 13:513–525

Bot LJ, Kirkby EA, Beusichem ML (1990b) Manganese toxicity in tomato plants: effects on cation uptake and distribution. J Plant Nutr 13:5–13

Broadley MR, White PJ, Hammond JP, Zelko I, Lux A (2007) Zinc in plants. New Phytol 173:677–702

Buchanan B, Grusen W, Jones R (2000) Biochemistry and molecular biology of plants. Ame Soc Plant Physiol Maryland 1367

Cakmak I (2000) Possible roles of Zinc in protecting plant cells from damage by reactive oxygen species. New Phytol 146:185–205

Chatterjee C, Gopal R, Dube BK (2006) Physiological and biochemical responses of French bean to excess cobalt. J Plant Nutr 29:127–136

Chiang HC, Lo JC, Yeh KC (2006) Genes associated with heavy metal tolerance and accumulation in Zn/Cd hyperaccumulator *Arabidopsis halleri*: a genomic survey with cDNA microarray. Environ Sci Technol 40(21):6792–6798

Choi JM, Pak CH, Lee CW (1996) Micronutrient toxicity in French marigold. J Plant Nutr 19:901–916

Clarimont KB, Hagar WG, Davis EA (1986) Manganese toxicity to chlorophyll synthesis in tobacco callus. Plant Physiol 80:291–293

Clemens S (2001) Molecular mechanisms of plant metal tolerance and homeostasis. Planta 212(4):475–486

Clemens S (2006) Toxic metal accumulation, responses to exposure and mechanisms of tolerance in plants. Biochimie 88(11):1707–1719

Clijsters H, Van Assche F (1985) Inhibition of photosynthesis by heavy metals. Photosynth Res 7:31–40

Cobbett CS (2000) Phytochelatins and their roles in heavy metal detoxification. Plant Physiol 123(3):825–832

Dalcorso G, Farinati S, Maistri S, Furini A (2008) How plants cope with cadmium: staking all on metabolism and gene expression. J Integr Plant Biol 50(10):1268–1280

Dalcorso G, Farinati S, Furini A (2010) Regulatory networks of cadmium stress in plants. Plant Signal Behav 5(6):1–5

Das P, Samantaray S, Rout GR (1997) Studies on cadmium toxicity in plants: a review. Environ Pollut 98:29–36

Davies BE (1995) Lead and other heavy metals in urban areas and consequences for the health of their inhabitants. In: Majumdar SK, Miller EW, Brenner FJ (eds) Environmental contaminants, ecosystems and human health. The Pennsylvania Academy of Science, Easton, pp 287–307

De DN (2000) Plant cell vacuoles. CSIRO Publishing, Collingwood

Degraeve N (1981) Carcinogenic, teratogenic and mutagenic effects of cadmium. Mutat Res 117:19–27

Delnomdedieu M, Basti MM, Otvos JD, Thomas DJ (1994) Reduction and binding of arsenate and dimethylarsenate by glutathione-a magnetic-resonance study. Chem Biol Interact 90:139–155

Demirevska-kepova K, Simova-Stoilova L, Stoyanova Z, Holzer R, Feller U (2004) Biochemical changes in barely plants after excessive supply of copper and manganese. Environ Exp Bot 52:253–266

Dhankhar R, Sainger PA, Sainger M (2012) Phytoextraction of zinc: physiological and molecular mechanism. Soil Sediment Contam 21:115–133

Dietz K-J, Baier M, Kra¨mer U (1999) Free radicals and reactive oxygen species as mediators of heavy metal toxicity in plants. In: Prasad MNV, Hagemeyer J (eds) Heavy metal stress in plants: from molecules to ecosystems. Springer, Berlin, pp 73–97

Dube BK, Tewari K, Chatterjee J, Chatterjee C (2003) Excess chromium alters uptake and translocation of certain nutrients in citrullus. Chemosphere 53:1147–1153

Dubey RS (2011) Metal toxicity, oxidative stress and antioxidative defense system in plants. In: Gupta SD (ed) Reactive oxygen species and antioxidants in higher plants. CRC Press, Boca Raton, pp 177–203

Ducic T, Polle A (2007) Manganese toxicity in two varieties of Douglas fir (*Pseudotsuga menziesii* var. viridis and glauca) seedlings as affected by phosphorus supply. Funct Plant Biol 34:31–40

Ebbs SD, Kochian LV (1997) Toxicity of zinc and copper to *Brassica* species: implications for phytoremediation. J Environ Qual 26:776–781

Elamin OM, Wilcox GE (1986a) Effect of magnesium and manganese nutrition on musk melon growth and manganese toxicity. J Am Soc Hortic Sci 111:582–587

Elamin OM, Wilcox GE (1986b) Effect of magnesium and manganese nutrition on water melon growth and manganese toxicity. J Am Soc Hortic Sci 111:588–593

Ernst WHO, Verkleij JAC, Schat H (1992) Metal tolerance in plants. Acta Bot Neerlandica 41:229–248

Fageria N, Baligar V, Clark R (2002) Micronutrients in crop production. Adv Agron 77:185–268

Fayiga AO, Ma LQ (2006) Using phosphate rock to immobilize metals in soil and increase arsenic uptake by hyperaccumulator *Pteris vittata*. Sci Total Environ 359:17–25

Fontes RLS, Cox FR (1998) Zinc toxicity in soybean grown at high iron concentration in nutrient solution. J Plant Nutr 21:1723–1730

Foy C, Chaney R, White M (1978) The physiology of metal toxicity in plants. Annu Rev Plant Physiol 29:511–566

Foy CD, Weil RR, Coradetti CA (1995) Differential manganese tolerances of cotton genotypes in nutrient solution. J Plant Nutr 18:685–706

Gajewska E, Sklodowska M, Slaba M, Mazur J (2006) Effect of nickel on antioxidative enzymes activities, proline and chlorophyll contents in wheat shoots. Biol Plant 50:653–659

Gardea-Torresdey JL, de la Rosa G, Peralta-Videa JR, Montes M, Cruz-Jimenez G, Cano-Aguilera I (2005) Differential uptake and transport of trivalent and hexavalent chromium by tumbleweed (*Salsola kali*). Arch Environ Contam Toxicol 48:225–232

Gekeler W, Grill E, Winnacker EL, Zenk MH (1988) Algae sequester heavy metals via synthesis of phytochelatin complexes. Archiv Microbiol 150:197–202

Godzik B (1993) Heavy metal contents in plants from zinc dumps and reference area. Pol Bot Stud 5:113–132

Grill E, Winnacker E-L, Zenk MH (1987) Phytochelatins, a class of heavy-metal-binding peptides from plants are functionally analogous to metallothioneins. Proc Natl Acad Sci U S A 84:439–443

Gruenhage L, Jager IIJ (1985) Effect of heavy metals on growth and heavy metals content of *Allium Porrum* and *Pisum sativum*. Angew Bot 59:11–28

Guest C, Schulze D, Thompson I, Huber D (2002) Correlating manganese X-ray absorption near-edge structure spectra with extractable soil manganese. Soil Sci Soc Am J 66:1172–1181

Gunes A, Pilbeam DJ, Inal A (2009) Effect of arsenic-phosphorous interaction on arsenic-induced oxidative stress in chickpea plants. Plant Soil 314:211–220

Guo J, Dai X, Xu W, Ma M (2008) Overexpressing GSH1 and AsPCS1 simultaneously increase the tolerance and accumulation of cadmium and arsenic in *Arabidopsis thaliana*. Chemosphere 72:1020–1026
Habiba U, Ali S, Farid M, Shakoor MB, Rizwan M, Ibrahim M, Abbasi GH, Hayat T, Ali M (2015) EDTA enhanced plant growth, antioxidant defense system, and phytoextraction of copper by *Brassica napus* L. Environ Sci Pollut Res 22:1534–1544
Haghiri F (1973) Cadmium uptake by plants. J Environ Qual 2:93–96
Hegedüs A, Erdei S, Horváth G (2001) Comparative studies of H_2O_2 detoxifying enzymes in green and greening barley seedlings under cadmium stress. Plant Sci 160:1085–1093
Hernandez LE, Carpena-Ruiz R, Garate A (1996) Alterations in the mineral nutrition of pea seedlings exposed to cadmium. J Plant Nutr 19:1581–1598
Horst WJ (1988) The physiology of manganese toxicity. In: Graham RD, Hannam RJ, Uren NJ (eds) Manganese in soil and plants. Kluwer Academic Publishers, Dordrecht, pp 175–188
Horiguchi T (1988) Mechanism of manganese toxicity and tolerance of plants. IV. Effects of silicon on alleviation of manganese toxicity of rice plants. Soil Sci Plant Nutr 3(4):65–73
Hossain MA, Fujita M (2009) Purification of glyoxalase I from onion bulbs and molecular cloning of its cDNA. Biosci Biotechnol Biochem 73(9):2007–2013
Hossain MA, Hossain MZ, Fujita M (2009) Stress-induced changes of methylglyoxal level and glyoxalase I activity in pumpkin seedlings and cDNA cloning of glyoxalase I gene. Aus J Crop Sci 3(2):53–64
Hossain MA, Hossain MD, Rohman MM, da Silva JAT, Fujita M (2012a) Onion major compounds (flavonoids, organosulfurs) and highly expressed glutathione-related enzymes: possible physiological interaction, gene cloning and abiotic stress response. In: Aguirre CB, Jaramillo LM (eds) Onion consumption and health. Nova, New York
Hossain MA, Piyatida P, Teixeira da Silva JA, Fujita M (2012b) Molecular mechanism of heavy metal toxicity and tolerance in plants: central role of glutathione in detoxification of reactive oxygen species and methylglyoxal and in heavy metal chelation. J Bot. doi:10.1155/2012/872875
Huffman EWD Jr, Allaway HW (1973) Chromium in plants: distribution in tissues, organelles, and extracts and availability of bean leaf Cr to animals. J Agric Food Chem 21:982–986
Humphries J, Stangoulis J, Graham R (2007) Manganese. In: Pilbeam D, Barker A (eds) Handbook of plant nutrition. Taylor and Francis, Boca Raton, pp 351–366
Iwasaki K, Sakurai K, Takahashi E (1990) Copper binding by the root cell walls of Italian ryegrass and red clover. Soil Sci Plant Nutr 36:431–439
James BR, Barlett RJ (1983) Behavior of chromium in soils VII. Adsorption and reduction of hexavalent forms. J Environ Qual 12:177–181
Kagi JHR (1991) Overview of metallothionein. Methods Enzymol 205:613–626
Kara Y (2005) Bioaccumulation of Cu, Zn and Ni from the wastewater by treated *Nasturtium officinal*. Int J Environ Sci Technol 2(1):63–67
Keller C, Rizwan M, Davidian J-C, Pokrovsky OS, Bovet N, Chaurand P, Meunier JD (2015) Effect of silicon on wheat seedlings (Triticum turgidum L.) grown in hydroponics and exposed to 0 to 30 μM Cu. Planta 241:847–860
Khan MIR, Iqbal N, Masood A, Mobin M, Anjum NA, Khan NA (2016) Modulation and significance of nitrogen and sulfur metabolism in cadmium challenged plants. Plant Growth Regul 78:1–11
Kitao M, Lei TT, Koike T (1997a) Effect of manganese toxicity on photosynthesis of white birch (*Betula platyphylla* var. japonica) seedlings. Physiol Plant 101:249–256
Kitao M, Lei TT, Koike T (1997b) Effects of manganese in solution culture on the growth of five deciduous broad-leaved tree species with different successional characters from northern Japan. Photosynth 36:3–14
Kogelmann W, Sharpe W (2006) Soil acidity and manganese in declining and non-declining sugar maple stands in Pennsylvania. J Environ Qual 35:433–441

Kramer U, Pickering IJ, Prince RC, Raskin I, Salt DE (2000) Subcellular localization and speciation of nickel in hyperaccumulator and non-accumulator *Thlaspi* species. Plant Physiol 122:1343–1353

Kramer U, Talke I, Hanikenne M (2007) Transition metal transport. Fed Eur Biochem Soc Lett 581(12):2263–2272

Kumar N, Bauddha K, Kumar S, Dwivedi N, Singh DC, Barman SC (2013) Accumulation of metals in weed species grown on the soil contaminated with industrial waste and their phytoremediation potential. Ecol Eng 61:491–495

Kumar S, Dubey RS, Tripathi RD, Chakrabarty D, Trivedi PK (2015) Omics and biotechnology of arsenic stress and detoxification in plants: current updates and prospective. Environ Int 74:221–230

Lane SD, Martin ES (1977) A histochemical investigation of lead uptake in *Raphanus sativus*. New Phytol 43:231–236

Lee M, Lee K, Lee J, Noh EW, Lee Y (2005) AtPDR12 contributes to lead resistance in Arabidopsis. Plant Physiol 138(2):827–836

Lerch K (1980) Copper metallothionein, a copper binding protein from *Neurospora crassa*. Nature (London) 284:368–370

Lewis S, Donkin ME, Depledge MH (2001) Hsp70 expression in Enteromorpha intestinalis (Chlorophyta) exposed to environmental stressors. Aquat Toxicol 51:277–291

Loneragan JF (1988) Distribution and movement of manganese in plants. In: Hannam RJ, Uren NC, Graham RD (eds) Manganese in soils and plants. Kluwer, Dordrecht, pp 113–124

Lu K, Yang X, Shen J, Robinson B, Huang H, Liu D, Bolan N, Pei J, Wang H (2014) Effect of bamboo and rice straw biochars on the bioavailability of Cd, Cu, Pb and Zn to *Sedum plumbizincicola*. Agric Ecosyst Environ 191:124–132

Mahmood T, Islam KR (2006) Response of rice seedlings to copper toxicity and acidity. J Plant Nutr 29:943–957

Margoshes M, Vallee BL (1957) A cadmium protein from equinr kidney cortex. J Am Chem Soc 79(17):4813–4814

Masion A, Bertsch PM (1997) Aluminium speciation in the presence of wheat root cell walls: a wet chemical study. Plant Cell Environ 20:504–512

McIntyre T (2003) Phytoremediation of heavy metals from soils. In: Springer T (ed) Advances in biochemical engineering/biotechnology, vol 78. Springer, Heidelberg, pp 97–123

Meharg AA, Macnair MR (1992) Suppression of the high affinity phosphate uptake system: a mechanism of arsenate tolerance in *Holcus lanatus* L. J Exp Bot 43:519–524

Mejare M, Bulow L (2001) Metal-binding proteins and peptides in bioremediation and phytoremediation of heavy metals. Trends Biotechnol 19:67–73

Mildvan AS (1970) Metal in enzymes catalysis. In: Boyer DD (ed) The enzymes, vol 11. Academic, London, pp 445–536

Miller RJ, Koeppe DE (1971) Accumulation and physiological effects of lead in corn. In: Proceedings of University of Missouri, Columbia 4: 186–193

Mishra S, Srivastava S, Tripathi RD, Kumar R, Seth CS, Gupta DK (2006) Lead detoxification by coontail (*Ceratophyllum demersum* L.) involves induction of phytochelatins and antioxidant system in response to its accumulation. Chemosphere 65:1027–1039

Miteva E (2002) Accumulation and effect of arsenic in tomatoes. Commun Soil Sci Plant Anal 33(11):1917–1926

Mohanpuria P, Rana NK, Yadav SK (2007) Cadmium induced oxidative stress influence on glutathione metabolic genes of Camellia sinensis (L.) O. Kuntze. Environ Toxicol 22:368–374

Molassiotis A, Sotiropoulos T, Tanou G, Diamantidis G, Therios I (2006) Boron induced oxidative damage and antioxidant and nucleolytic responses in shoot tips culture of the apple rootstock EM9 (Malus domestica Borkh). Environ Exp Bot 56:54–62

Monni S, Salemma M, Millar N (2000) The tolerance of empetrum nigrum to copper and nickel. Environ Pollut 109:221–229

Morel JL, Mench M, Guchert A (1986) Measurement of Pb^{2+}, Cu^{2+} and Cd^{2+} binding with mucilage exudates from maize (*Zea mays* L.) roots. Biol Fertil Soil 2(1):29–34

Moreno-Caselles J, Moral R, Pera-Espinosa A, Marcia MD (2000) Cadmium accumulation and distribution in cucumber plants. J Plant Nutr 23:243–250

Morzck E Jr, Funicclli NA (1982) Effect of lead and on germination of Spartina alterniflora losiel seeds at various salinities. Environ Exp Bot 22:23–32

Murphy A, Taiz L (1995) Comparison of metallothionein gene expression and nonprotein thiols in ten arabidopsis ecotypes correlation with copper tolerance. Plant Physiol 109:945–954

Nagajyoti PC, Lee KD, Sreekanth TVM (2010) Heavy metals occurrence and toxicity for plants: a review. Environ Chem Lett 8(3):199–216

Nakazawa R, Kameda Y, Ito T, Ogita Y, Michihata R, Takenaga H (2004) Selection and characterization of nickel tolerant tobacco cells. Biol Plant 48:497–502

Neelima P, Reddy KJ (2002) Interaction of copper and cadmium with seedlings growth and biochemical responses in *Solanum melongena*. Environ Pol Technol 1:285–290

Nieboer E, Richardson DHS (1980) The replacement of the nondescript term heavy metals by a biologically and chemistry significant classification of metal ions. Environ Poll B 1:3–26

Oehlkers F (1953) Chromosomal breaks influenced by chemicals. Heredity 6:95–105

Pandey N, Sharma CP (2002) Effect of heavy metals Co^{2+}, Ni^{2+} and Cd^{2+} on growth and metabolism of cabbage. Plant Sci 163:753–758

Parida BK, Chhibba IM, Nayyar VK (2003) Influence of nickel contaminated soils on fenugreek (*Trigonella corniculata* L.) growth and mineral composition. Sci Hortic 98:113–119

Parr PD, Taylor FG Jr (1982) Germination and growth effects of hexavalent chromium in Orocol TL (a corrosion inhibitor) on *Phaseolus vulgaris*. Environ Int 7:197–202

Peralta JR, Gardea Torresdey JL, Tiemann KJ, Gomez E, Arteaga S, Rascon E (2001) Uptake and effects of five heavy metals on seed germination and plant growth in alfalfa (*Medicago sativa*) L. Bull Environ Cont Toxicol 66:727–734

Peralta-Videa JR, Lopez ML, Narayana M, Saupea G, Gardea-Torresdeya J (2009) The biochemistry of environmental heavy metal uptake by plants: implications for the food chain. Int J Biochem Cell Biol 41:1665–1677

Piechalak A, Tomaszewska B, Baralkiewicz D, Malecka A (2002) Accumulation and detoxification of lead ions in legumes. Phytochem 60:153–162

Pourrut B, Shahid M, Dumat C, Winterton P, Pinelli E (2011) Lead uptake, toxicity and detoxification in plants. Rev Environ Contam Toxicol 213:113–136

Rao KVM, Sresty TVS (2000) Antioxidative parameters in the seedlings of pigeonpea (*Cajanus cajan* (L). Millspaugh) in response to Zn and Ni stresses. Plant Sci 157:113–128

Reeves RD, Baker AJM (2000) Metal-accumulating plants. In: Raskin I, Ensley BD (eds) Phytoremediation of toxic metals: using plants to clean up the environment. Wiley, New York, pp 193–229

Rizwan M, Ali S, Adrees M et al (2016) Cadmium stress in rice: toxic effects, tolerance mechanisms, and management: a critical review. Environ Sci Pollut Res. doi:10.1007/s11356-016-6436-4

Ros R, Cook DT, Picazo C, Martinez-Cortina I (1992) Nickel and cadmium-related changes in growth, plasma membrane lipid composition, atpase hydrolytic activity and proton pumping of rice (*Oryza sativa* L. cv. Bahia) Shoots. J Exp Bot 43:1475–1481

Rosas I, Carbajal ME, Gomez-Arroyo S, Belmont R, Villalogos-Pietrini R (1984) Cytogenic effects on cadmium accumulation on water hyacinth (*Eichornia crassipes*). Environ Res 33:386–395

Rudakova EV, Karakis KD, Sidorshina ET (1988) The role of plant cell walls in the uptake and accumulation of metal ions. Fiziol Biochim Kult Rast 20:3–12

Ryvolova M, Krizkova S, Adam V, Beklova M, Trnkova L, Hubalek J, Kizek R (2011) Analytical methods for metallothionein detection. Curr Anal Chem 7:243–261

Sakakibara M, Ohmoril Y, Ha NTH, Sano S, Sera K (2011) Phytoremediation of heavy metal-contaminated water and sediment by *Eleocharis acicularis*. Clean Soil Air Water 39(8):735–741

Salt DE, Rauser WE (1995) Mg ATP-dependent transport of phytochelatins across the tonoplast of oat roots. Plant Physiol 107:1293–1301

Samarakoon AB, Rauser WE (1979) Carbohydrate levels and photo-assimilate export from leaves of *Phaseolus vulgaris* exposed to excess cobalt, nickel, and zinc. Plant Physiol 63:1165–1169

Seregin IV, Ivaniov VB (1997) Histochemical investigation of cadmium and lead distribution in plants. Fiziol Rast Plant Physiol 48:525–533

Seregin IV, Ivaniov VB (2001) Physiological aspects of cadmium and lead toxic effects on higher plants. Russ J Plant Physiol 48(4):523–544

Sethy SK, Ghosh S (2013) Effect of heavy metals on germination of seeds. J Nat Sci Biol Med 4(2):272–275

Shanker AK, Cervantes C, Loza-Tavera H, Avudainayagam S (2005) Chromium toxicity in plants. Environ Int 31:739–753

Sharma SS, Dietz KJ (2009) The relationship between metal toxicity and cellular redox imbalance. Trends Plant Sci 14(1):43–50

Sharma P, Dubey RS (2005) Lead toxicity in plants. Braz J Plant Physiol 17:35–52

Silverberg BA (1976) Cadmium-induced ultrastructural changes in mitochondria of freshwater green algea. Phycologia 15:155–159

Singh PK, Tewari SK (2003) Cadmium toxicity induced changes in plant water relations and oxidative metabolism of Brassica juncea L. plants. J Environ Biol 24:107–117

Singh R, Singh DP, Kumar N, Bhargava SK, Barman SC (2010) Accumulation and transcolation of heavy metals in soil and plants from fly ash contamination. Environ Biol 3:421–430

Srivastov RK, Gupta SK, Nýgam KDP, Vasudevan P (1994) Treatment of chromium and nickel in wastewater by using aquatic plants. Water Resour 28(7):1631–1638

Stoeva N, Bineva T (2003) Oxidative changes and photosynthesis in oat plants grown in as-contaminated soil. Bulg J Plant Physiol 29(1–2):87–95

Stoeva N, Berova M, Zlatez Z (2004) Physiological response of maize to arsenic contamination. Biol Plant 47(3):449–452

Suzuki KT, Someya A, Komada Y, Ogra Y (2002) Roles of metallothionein in copper homeostasis: responses to Cu-deficient diets in mice. J Inorg Biochem 88:173–182

Taylor GJ (1991) Current views of the aluminum stress response; the physiological basis of tolerance. Curr Top Plant Biochem Physiol 10:57–93

Thomas F, Malick C, Endreszl EC, Davies KS (1998) Distinct responses to copper stress in the halophyte, *Mesembryanthemum crystallium*. Physiol Plant 102:360–368

Tice KR, Parker DR, DeMason DA (1992) Operationally defined apoplastic and symplastic aluminum fractions in root tips of aluminum-intoxicated wheat. Plant Physiol 100(1):309–318

Tomaszewska B, Tukendorf A, Baralkiewicz D (1996) The synthesis of phytochelatins in lupin roots treated with lead ions. Sci Legum 3:206–217

Tripathi RD, Srivastava S, Mishra S, Singh N, Tuli R, Gupta DK, Maathuis FJM (2007) Arsenic hazards: strategies for tolerance and remediation by plants. Trends Biotechnol 25(4):158–165

Tripathy BC, Bhatia B, Mohanty P (1981) Inactivation of chloroplast photosynthetic electron-transport activity by Ni^{2+}. Biochim Biophys Acta 638:217–224

Vatamauniuk OK, Mari S, Lu YP, Rea PA (2000) AtPCS1, a phytochelatin synthase from Arabidopsis: isolation and in vitro reconstitution. Proc Natl Acad Sci U S A 96:7110–7115

Vazques MD, Poschenrieder C, Barcelo J (1987) Chromium (VI) induced structural changes in bush bean plants. Ann Bot 59:427–438

Venugopal B, Luckey TD (1978) Metal toxicity of mammals. Plenum, New York

Vikram A, Johri T, Tandon PK (2011) Effect of chromium (IV) on growth and metabolism of *Spinacia oleracea* (Spinach) plants. Res Environ Life Sci 4(3):119–124

Von Rosen G (1954) Mutation induced by actions of localization of cadmium and cadmium binding peptides in tobacco leaves. Plant Pathol 92:1086–1093

Wang JR, Zhao FJ, Meharg AA, Raab A, Feldmann J, Mcgrath SP (2002) Mechanisms of arsenic hyperaccumulation in *Pteris vittata* uptake kinetics, interactions with phosphate, and arsenic speciation. Plant Physiol 130:1552–1561

Watmough S, Eimer M, Dillon P (2007) Manganese cycling in central Ontario forests: response to soil acidification. Appl Geochem 22:1241–1247

Wojcik M, Tukiendorf A (2004) Phytochelatin synthesis and cadmium localization in wild type of *Arabidopsis thaliana*. Plant Growth Regul 44:71–80

Wu S (1994) Effect of manganese excess on the soybean plant cultivated under various growth conditions. J Plant Nutr 17:993–1003

Yang XE, Jin XF, Feng Y, Islam E (2005) Molecular mechanisms and genetic basis of heavy metal tolerance/hyperaccumulation in plants. J Integr Plant Biol 47(9):1025–1035

Yang X, Liu J, McGrouther K, Huang H et al (2016) Effect of biochar on the extractability of heavy metals (Cd, Cu, Pb, and Zn) and enzyme activity in soil. Environ Sci Pollut Res 23:974–984

Yang XE, Long XX, Ye HB, He ZL, Calvert DV, Stoffella PJ (2004) Cadmium tolerance and hyperaccumulation in a new Zn-hyperaccumulating plant species (Sedum alfredii Hance). Plant Soil 259:181–189

Zavala YJ, Duxbury JM (2008) Arsenic in rice: estimating normal levels of total arsenic in rice grain. Environ Sci Technol 42(38):56–60

Zayed AM, Terrey N (2003) Chromium in the environment: factors affecting biological remediation. Plant Soil 249(1):139–156

Zenk MH (1996) Heavy metal detoxification in higher plants: a review. Gene 179:21–30

Zhao FJ, Ma JF, Meharg AA, McGrath MP (2009) Arsenic uptake and metabolism in plants. New Phytol 181:777–794

Zhou G, Xu Y, Li J, Yang L, Liu J-L (2006) Molecular analyses of the metallothionein gene family in rice (*Oryza sativa* L.). J Biochem Mol Biol 39(5):595–606

Zhu Y-G, Ralf K, Tong Y-P (2004) vacuolar compartmentalization: a second-generation approach to engineering plants for phytoremediation. Trends Plant Sci 9(1):7–9

Zornoza P, Robles S, Martin N (1999) Alleviation of nickel toxicity by ammonium supply to sunflower plants. Plant Soil 208:221–226

Regulation of Xenobiotics in Higher Plants: Signalling and Detoxification

3

Shikha Singh, Gausiya Bashri, Anita Singh, and Sheo Mohan Prasad

Abstract

Increased anthropogenic activities have aggravated the different chemical pollutants (xenobiotics) in the environment. Xenobiotics are any chemical or other substance that cannot be utilized by plants for their growth and development. Xenobiotics alone and/or in combination can affect the growth and physiology of every organism, which varies species to species. It may also affect the coordinated signalling pathways that alter the gene expression and regulation in higher plants. Therefore, plants have developed the mechanism for the mobilizations of xenobiotics which include three phases, i.e. transformation, conjugation and compartmentation. Further, plants have also evolved various detoxification processes for these xenobiotics. Therefore, in this chapter the different fates of xenobiotics in plant system as well as their signalling and detoxification processes are discussed in detail.

Keywords

Detoxification • Fates • Signalling • Xenobiotics

3.1 Introduction

Xenobiotics (from the Greek word for foreign chemicals) can be defined as any chemical or other substance that cannot be utilized by plants for energy yielding processes and it is not normally found in the ecosystems. Plants, like other organisms in the environment, are continually exposed to natural and synthetic

S. Singh • G. Bashri • A. Singh • S.M. Prasad (✉)
Ranjan Plant Physiology and Biochemistry Laboratory, Department of Botany, University of Allahabad, Allahabad 211002, Uttar Pradesh, India
e-mail: sheomohanp@yahoo.co.in

A. Singh et al. (eds.), *Plant Responses to Xenobiotics*,
DOI 10.1007/978-981-10-2860-1_3

xenobiotics such as heavy metals/metalloids (i.e. lead, cadmium, arsenic, etc.), allelochemicals (cinnamic acid, benzoic acid, etc.), organic pollutants (trinitrotoluene, phenanthrene, etc.), pesticides (atrazine, chlorpyriphos, cypermethrin, endosulfan, etc.) and air pollutants (Zhang et al. 2007; Riechers et al. 2010). These xenobiotics may originate from both natural (fires, volcano eruptions, soil or rock erosion, biodegradation) and anthropogenic (air and soil pollution, herbicides) sources. They may also be characterized according to their environmental targets (air, soil and water) and biological targets (e.g. plants, fungi, mammals and invertebrates). Phytochemical crop protection leads to the release of xenobiotic molecules into the environment by volatilization, spray drift, runoff, leaching and absorption. Such xenobiotics cannot be used for nutrition or as a source of energy, but are nevertheless taken up and accumulate. Exposure of organisms to xenobiotic materials is considered significantly in order to understand the environmental and toxicological chemistry. Combinations of chemical pollution have complex biological effects that are difficult to predict from the effects of single contaminants (Devier et al. 2011; Serra et al. 2013). To detoxify the xenobiotics toxicty, higher plants perform activation or deactivation reactions for their transformation and the rate of xenobiotic transformation depends upon plant species (Saari et al. 1994; Wink 1997). Their mode of action showed that some xenobiotics have an acute (death) or chronic toxicity (e.g. carcinogenesis, mutagenesis) on organisms. Consequently, plants must mount specific and coordinated defence mechanisms for survival under adverse growing conditions through a coordinated network system (Zhang et al. 2007; Ramel et al. 2012). Plants are also capable to metabolize a diverse range of xenobiotics, notably organic pollutants and pesticides, using enzymes that are normally used in the synthesis and processing of endogenous natural products (Cole and Edwards 2000). Herbicides that are used as selective graminicides in wheat, such as the aryloxy phenoxy propionates and phenylureas, are more readily detoxified in the crop than in competing grass weeds (Owen 2000). Significantly, the long-term use of these herbicides has helped to select populations of weeds, especially black-grass, which have developed resistance to multiple classes of herbicides due to an enhanced ability to detoxify graminicides (Hall et al. 1997). Thus, in this chapter we have addressed the fates of xenobiotics, their signalling and detoxification in higher plants.

3.2 Fate of Xenobiotic in Higher Plants

Fate of xenobiotics in plant cells are determined by the uptake, metabolism and compartmentation. The presence and expression of these xenobiotics in cellular systems vary with different plant species, and it is the basis for differences in rates of detoxification and their tolerance. In general, metabolism of an organic pollutant requires its penetration into a living plant cell, and this necessarily involves the passage across the plasma membrane. The majority of xenobiotics are lipophilic organic compounds. Uptake of xenobiotics takes place across the membrane usually passively along the concentration gradient. However, occasionally a specific membrane protein that exists for their transport as well as some endogenous metabolite may transport it. In plants, the leaves have lipophilic wax layer on their surface that collects the pollutants and helps in their penetration into the tissues. In the apoplastic compartment, there are certain enzymes that catalyse the chemical modifications that can affect the uptake and subsequent metabolism of compounds. For example, apoplastic carboxylesterases catalyse the hydrolysis of non-ionized esters to carboxylic acid anions with reduced lower lipophilicity. When any organic compound enters to the plant cell, it can be either transported to the vacuole without any modification or metabolized, and then the final products are transported to the vacuolar compartments. Vacuolar compartmentation is a critical step in the detoxification of organic metabolites because it eliminates conjugated products from vulnerable sites of the cytosol. For the chemical modifications or metabolization of xenobiotics, higher plants involved three major phases known as Phase 1 or conversion, Phase II or conjugation and Phase III or compartmentation (Saari et al. 1994). In Phase I reaction, hydrolysis is catalysed by esterases and amidases, but the major reactions are oxidations catalysed by the cytochrome P-450 system which are responsible for the chemical modification of lipophilic xenobiotics in plants that include non-synthetic processes such as reduction and oxidation. Phase II reaction involves in synthetic reaction (synthesis) in which plants often add sugar moieties with aid of UDP-*O*-glucosyltransferase and UDP-*N*-glucosyltransferase, and the resulting glucosyl derivatives are frequently acylated with malonic acid by –*O*– or –*N*– methyltransferases. In case of plants special importance is attributed to glutathione-*S*-transferases (GSTs) of which different isoforms with varying specificities exist, and these glutathione-*S*-transferases conjugate xenobiotics with reduced glutathione (Timmermann 1989). Although the products of Phase I reactions are more water soluble than the parent xenobiotic, the primary function of this phase is to create reactive sites in the xenobiotic by the addition or exposure of functional group (e.g. hydroxyl or carboxyl) that will prepare the compound for conjugation reactions. In Phase III reaction, the formed xenobiotic conjugates are converted to secondary conjugates or insoluble bound residues and transported from cytosol to the vacuole or other compartments like the cell wall of plant cells for cellular detoxification (Holton and Cornish 1995). A schematic representation of localization of the reactions and enzymatic systems involved in xenobiotics detoxification in the plant cells is shown in Fig. 3.1.

The transport of xenobiotics and sequestration of their conjugates in the vacuole were mediated with the help of specific transporters that belong to the ATP-binding cassette (ABC) transporter superfamily, cytochrome P450 mixed-function oxidases (CYPs), family 1 glucosyltransferases and glutathione transferases (Rishi et al 2004; Baerson et al. 2005; Zhang et al. 2007). The multidrug resistance-associated protein (MRP) is also responsible for the transport of conjugates across the tonoplast in plants. Further, the exploration of genomic databases of *Arabidopsis thaliana* reveals that there are eight members of the AtMRP family. Hitherto, only three of these genes (AtMRP1, AtMRP2 and AtMRP3) and the proteins they encode have been characterized in detail. Analogous to MRP family, other ABC transporters are likely to play an imperative role in the distribution, compartmentation and detoxification of organic xenobiotics (Rea et al. 1998; Davies and Coleman 2000).

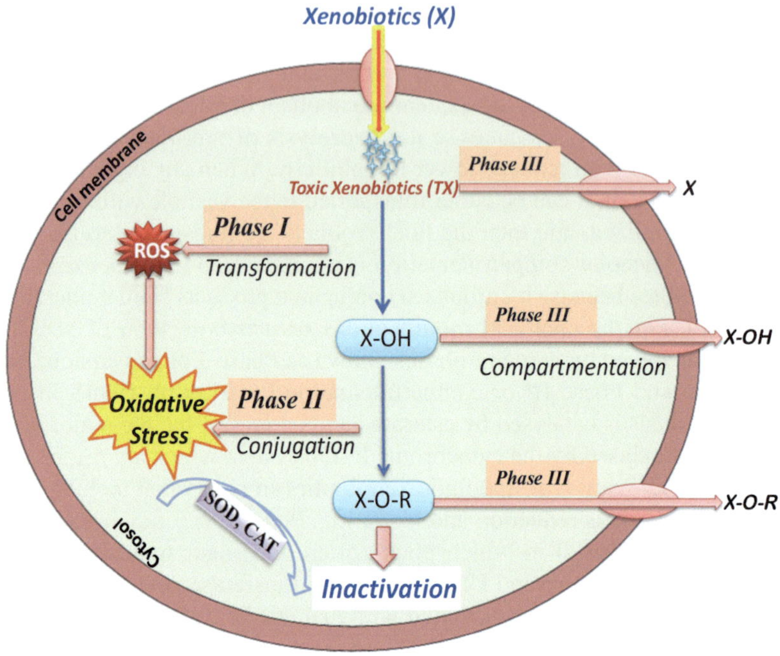

Fig. 3.1 A schematic representation of fate of xenobiotics in the plant cells

Heterologous expression in *Saccharomyces cerevisiae* has shown that all above three proteins catalyse the transport of glutathione conjugates. AtMRP2 and AtMRP3 exhibit transport of malonylated chlorophyll catabolites that are produced during leaf senescence, and AtMRP3 confers cadmium resistance. This difference may reflect the capacity of plants for synthesizing and sequestering large numbers of secondary compounds, and this capability is useful asset for mobilization techniques. Several molecules such as TNT or atrazine may bind to the cell wall polymers (cellulose or lignin) whilst lipophilic molecules will partition into the lipid phases of the cell (membranes). Such immobilization of the xenobiotics will lower translocation and excessive accumulation. Comparatively, organic chemical contaminants' royal demolition explosive (RDX) are readily translocated to the leaves in the transpiration stream and can be accumulated there in amounts, exceeding the safety limit, and can cause damage to the tissue.

3.3 Signalling Response to Xenobiotics

As evident, the xenobiotics are the chemicals compounds that are not internally produced in plants species, and their exposure is continuously increasing in plants, due to enhanced industrial pollutants. To cope with these xenobiotics, plants have evolved different tolerance mechanism and signal perceptions (Mittler et al. 2004;

Ramel et al. 2007, 2012). Sequential analysis of DNA arrays and gene expression have established the transcriptomic profile on the effects of xenobiotic in higher plants (Ekman et al. 2003, 2005; Unver et al. 2010; Weisman et al. 2010; Jin et al. 2011; Peng et al. 2011). The varieties of xenobiotic treatments (trinitrotoluene, atrazine, glyphosate, phenanthrene, polychlorinated biphenyls, naphthalene, fenclorim) provide reliable information for the gene expressions in plants. The effects of xenobiotics in plants were confirmed by the utilization of genome-wide approaches showing xenobiotic-induced gene networks involving induction and repression of hundreds of genes, as in the case for other abiotic stress responses (Desikan et al. 2001; Mittler et al. 2004). In the plant system, several sets of genes such as cytochrome P450s, glutathione-S-transferases (GSTs), thioredoxins and peroxidases are induced for the detoxification and cell protection against xenobiotics toxicity (Ekman et al. 2003; Ramel et al. 2007; Skipsey et al. 2011). They are also able to provoke responses against uncountable natural products like plant hormones, endogenous toxic compounds, allelochemicals, reactive oxygen species (ROS), pathogen-associated molecular patterns, etc. A comparison of phenanthrene-induced responses with a panel of 27 signalling condition has been reported by Weisman et al. (2010), and they have shown significant positive correlations between phenanthrene and reactive oxygen species-induced responses (*Arabidopsis thaliana*). In addition to this, safeners are agrochemicals which are known to selectively protect crop plants from herbicide damage without reducing the activity in target weed species by elevating the expression of xenobiotic detoxifying enzymes, such as glutathione-*S*-transferases (GSTs). Herbicide safener-induced responses are positively correlated with responses of reactive electrophilic species oxylipins in *Arabidopsis thaliana* (Riechers et al. 2010; Skipsey et al. 2011) and also with responses of the allelochemical benzoxazolin-2(3H)-one (Baerson et al. 2005; Riechers et al. 2010). The above discussed correlations may be attributed to common chemical property like the electrophilic strength of these compounds whether exogenous or endogenous. However, it remains difficult to ascribe specific mechanisms to such similarity in the patterns of transcriptomic responses. Interactions between transcription factors or signalling pathways may be involved (Ramel et al. 2007; Ehlting et al. 2008; Riechers et al. 2010). Further, the ZAT12 and bZIP60 transcription factors, which are associated with the atrazine tolerance response of *Arabidopsis thaliana* (Ramel et al. 2007), are involved against oxidative stress (Davletova et al. 2005; Iwata and Koizumi 2005). However, the metabolism of signalling molecules, such as phytohormones, can also be regulated by xenobiotics. Herbicides (Ramel et al. 2007) and herbicide safeners (Riechers et al. 2010; Skipsey et al. 2011) affect genes involved in oxylipin and jasmonate metabolic pathways, and neonicotinoid insecticides induce endogenous synthesis of salicylate (Ford et al. 2010). Various xenobiotics with important biochemical consequences are known to induce ROS production. The ROS can function as cellular second messengers that are likely to modulate many different genes and proteins, thus leading to a various responses (Jaspers and Kangasjarvi 2010). The ROS produced by membrane-bound NAD(P)H oxidases in plant cells activate the calcium channels in plasma membrane which leads to further rise in cytosolic calcium concentration

(Foreman et al. 2003; Mori and Schroeder 2004). This increased cytosolic calcium level activates the downstream signalling that involves MAPK pathway, and with ROS this pathway superficially exerts a direct and central role in induction and stabilization of the MAPK cascade (Kagami et al. 2005; Doczi et al. 2007).

3.4 Detoxification Process of Xenobiotics

Plants have versatile detoxification systems to counter the phytotoxicity of the wide variety of natural and synthetic chemicals (xenobiotics) that are present in the environment. The regular application of protectant for the survival of crop from unwanted threats leads to accumulation of xenobiotic molecules into the environment. The resulting pollution of soil and water frequently consists of mixtures of pesticides (fungicides, insecticides and herbicides) associated with related degradation products and adjuvants. Combinations of these xenobiotics have complex biological effects that are from the effects of a single pollutant (Devier et al. 2011; Serra et al. 2013). As sessile organisms, plant communities are the direct targets of residual agricultural pollution, whether airborne, waterborne or accumulated in soils. These pollutants can affect growth and development of nontarget organisms, with varying toxicity that depends on surrounding physicochemical factors. Such as triazine, herbicides can increase the toxicity of organophosphate insecticides (Hernández et al. 2012). The toxicity of a mixture of pesticides also involves the effects of derived degradation products. The broad-spectrum nonselective herbicide glyphosate and its main degradation product, aminomethylphosphonic acid (AMPA), are often found in freshwater ecosystems (Székács and Darvas 2012). Glyphosate affects plants by blocking the shikimate pathway of synthesis of aromatic amino acids (Steinrücken and Amrhein 1980). Although described as less toxic, AMPA has been reported to be phytotoxic for soybean (*Glycine max*) and canola (*Brassica napus*) through unknown mechanisms (Reddy et al. 2004; Nandula et al. 2007). Serra et al. (2013) have reported that combinations consisting of the herbicide glyphosate, the fungicide tebuconazole, the glyphosate degradation product aminomethylphosphonic acid (AMPA) and the atrazine degradation product hydroxyatrazine negatively affected the growth and physiology of *Arabidopsis thaliana*. Thus, plants have evolved strategy for coping with the potentially negative impacts of xenobiotics on plant growth and development. This is particularly evident in the capability for metabolic detoxification of xenobiotics; plants are able to detoxify herbicides by complex multistep processes that exhibit extraordinary diversity among species (Kreuz et al. 1996). Therefore, under these conditions plants have evolved various detoxification mechanisms for the survival. These detoxification mechanisms have been discussed in this section of this chapter and the summary is presented in Table 3.1.

Table 3.1 Detoxification of xenobiotics by different processes

Detoxifying agent	Xenobiotics	Plant	Response	Reference
(i) Glutathione	Organic xenobiotics	*Phragmites*	Enhanced GST activity	Schröder et al. (2008)
	Chlorothalonil (CHT)	*Solanum lycopersium* Mill.	Increases in GSH content and activities of GSTs and GR enzymes as well as the transcriptional levels of GST1, GST2 and GST3 genes, glutathione synthetase gene (GSH), glutathione reductase gene (GR) and glutathione peroxidase gene (GPX)	Wang et al. (2010)
	Metribuzin and pretilachlor	*Zea mays*	The pesticides increased the glutathione content and activity of GST and GR	Alla et al. (2008)
	Diclofenac, sulfamethoxazole, trimethoprim, 17a-ethinylestradiol	*Medicago sativa*	Increased antioxidant enzymatic activities, as well as upregulated transcript levels of GST7, GST17, H^+-ATPase and CytcOx	Christou et al. (2016)
	Atrazine	*M. sativa*	Upregulation in genes of GST or ABC and glucosyltransferases	Zhang et al. (2016)
(ii) Hormone				
Auxin	Cd	*Hordeum vulgare*	Enhanced GST activity	Bocova et al. (2013)
Salicylic acid				
Brassinosteroids	Cd	*Linum usitatissimum*	Ameliorated the Cd-induced toxicity by increasing NPSH content	Belkadhi et al. (2012)
	Cd and phenanthrene (PHE)	*Solanum lycopersicum* L.	EBR alleviated photosynthetic inhibition and oxidative stress by causing enhancement of the antioxidants activity	Ahammed et al. (2013a)

(continued)

Table 3.1 (continued)

Detoxifying agent	Xenobiotics	Plant	Response	Reference
Brassinosteroids	Polychlorinated biphenyls (PCBs)	*S. lycopersicum* L.	EBR against PCBs stress which may strengthen phytoremediation approaches by enhancing plant tolerance	Ahammed et al. (2013b)
	Phenanthrene (PHE)	*S. lycopersicum* L.	EBR regulates secondary metabolism in tomato which might enhance tolerance to PHE	Ahammed et al. (2013c)
(iii) Sugar				
Sucrose	Atrazine	*Arabidopsis thaliana*	Sucrose-induced acquisition of defence mechanisms was shown by the gene expression pattern of a chloroplastic iron superoxide dismutase and by enhancement of whole-cell glucose-6-phosphate dehydrogenase activity	Sulmon et al. (2006)
	Atrazine	*A. thaliana*	Protected by modifying the gene expression defence mechanisms	Ramel et al. (2007)
	Atrazine	*A. thaliana*	Ameliorated atrazine induced stress and increased photosynthetic activity	Ramel et al. (2009)
(iv) Safeners				
Benoxacor, fenclorim and fluxofenim	Paraquat, metolachlor and glyphosate	*A. thaliana*	Enhanced GST activities toward herbicides	DeRidder et al. (2002)
Isoxadifen-ethyl and mefenpyr-diethyl	Sulfonyl urea	*A. thaliana*	Safeners induce detoxification genes against herbicide	Behringer et al. (2011)

3.4.1 Glutathione-Mediated Detoxification of Xenobiotics

One important detoxification mechanism is chemical modification of the xenobiotic by covalent linkage to the endogenous glutathione. The resulting glutathione conjugates are exported from the cytosol to the vacuole by an ATP-dependent tonoplast transporter. This detoxification pathway shares many features with the pathway used by plants for the synthesis and vacuolar deposition of secondary metabolites, such as anthocyanin. Glutathione (GSH) is a tripeptide having the structure Glu–Cys–Gly, which represents the glutamate, cysteine and glycine molecules, respectively. GSH is abundant in plants, typically exceeding 1 mM in the cytoplasm, and plays a central role in the crucial processes of detoxification and redox buffering. Plants also contain several GSH-dependent detoxifying enzymes, such as glutathione-*S*-transferases (GSTs), which detoxify electrophilic herbicides by catalysing their conjugation with GSH (Fig. 3.2). Indeed, the enhancement of GST expression has become a marker for plant response to stress, although the functional significance of selective GST expression is only just emerging. The expression of specific GSTs vary markedly during plant development (cell division, senescence) and, after exposure to pathogens, changes in environmental conditions and chemical treatment (Marrs 1996).

Individual GST isoenzymes can selectively detoxify specific xenobiotics, with species differences in GST specificity and capacity determining herbicide selectivity (Cole and Edwards 2000). The glutathionylated metabolites are tagged for

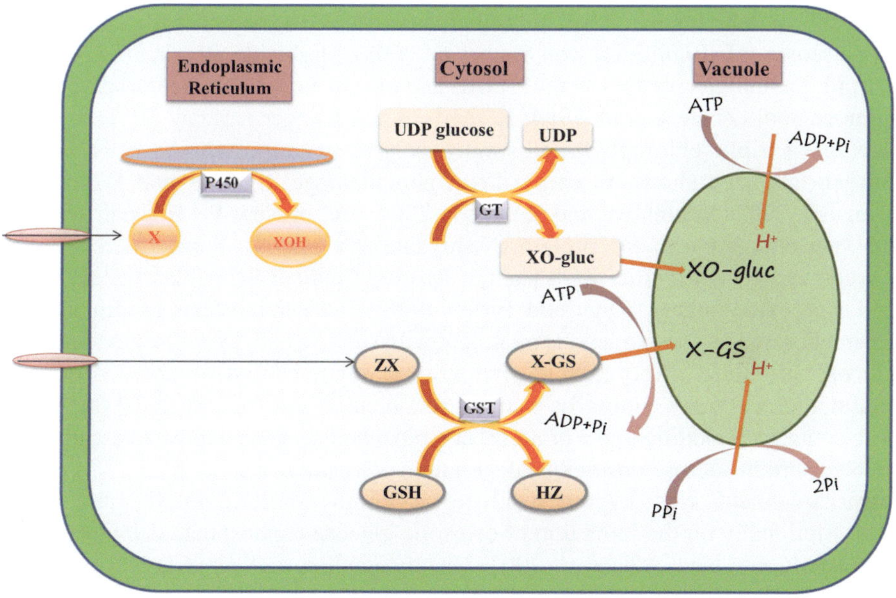

Fig. 3.2 Schematic representation of xenobiotic detoxification in the plant cells. *GT* glucosyltransferase; X and Z-X are the xenobiotics (Source: Kreuz et al. 1965)

vacuolar import by ATP-binding cassette (ABC) transporters, which selectively transport GSH conjugates (Ria 1999). However, the GST mainly functions by (i) catalyse conjugation reactions with natural products similar to those observed with xenobiotics, (ii) function as binding and carrier proteins for phytochemicals between cellular compartments and (iii) catalyse alternative GSH-dependent biotransformation reactions.

3.4.2 Sugar-Mediated Detoxification of Xenobiotics

Stress conditions such as pathogen challenge, drought, salt stress, heavy metals and pesticides cause accumulation of ROS, which may also be associated with soluble sugar accumulation that has been regarded as an adaptive response to the stress condition (Roitsch 1999). Although, the connections between stress and sugar accumulation are difficult to understand because of these situations of stress, the corresponding responses are clearly pleiotropic in terms of targets. Different classes of pesticides act on plants through direct induction of oxidative injury. Ashton and Ziegler (1987) had reported protection of *Chenopodium rubrum* cell suspension cultures by exogenous sucrose which enhanced the oxygen evolution and growth, thus generating less photo-oxidative damage. They also reported psbA mRNA and D1 protein accumulation and suggest that differential glucose/sucrose signalling affects the gene expression. Such carbohydrate-modulated changes of gene expression occur in a number of abiotic and biotic stress situations (Koch 1996; Rolland et al. 2002; Thibaud et al. 2004), and genes involved in antioxidant defence have been shown to be regulated by sugars and hypothesized to be under the control of a sugar sensor (Hauschild and von Schaewen 2003). Herbicide atrazine binds to the D1 protein, inhibits photosystem II (PSII) and blocks electron transfer to the plastoquinone pool (Zheleva et al. 1994; Rutherford and Krieger-Liszkay 2001), by production of triplet chlorophyll and singlet oxygen. In cyanobacterial cells, singlet oxygen has been shown to cause direct photodamage to PS II and D1 protein (Nogushi 2002; Lupinkova and Komenda 2004) and prevent PS II repair by suppressing elongation of D1 protein (Nishiyama et al. 2004). Furthermore, singlet oxygen can generate other ROS such as hydroxyl radical (Rinalducci et al. 2004) and superoxide radical (Ryter and Tyrrell 1998). The lethal effects of atrazine can be ameliorated by the exogenous treatment with sucrose and, to a lesser extent, with glucose that confer a very high level of atrazine tolerance to *Arabidopsis* seedlings (Sulmon et al. 2004). Similarly, sucrose treatment in broccoli enhanced the ascorbate content by modifications of ascorbat e biosynthesis gene expression, and these results were in sucrose dose-dependent manner (Nishikawa et al. 2005). Moreover, sugar treatment was hypothesized to act both transcriptionally and post-transcriptionally on the induction of cytosolic glucose-6-phosphate dehydrogenase (Hauschild and von Schaewen 2003). In mammalian and plant cells, increased glucose-6-phosphate dehydrogenase expression has been related to resistance to oxidative stress (Salvemini et al. 1999; Boada et al. 2000; Debnam et al. 2004).

Similarly, sucrose protected atrazine-treated *Arabidopsis* plantlets and maintained the growth and oxygen evolution (Sulmon et al. 2004), thus suggesting that mechanisms other than phototrophic–photoheterotrophic transitions may be involved in such sucrose-based protection of *Arabidopsis* plantlets against atrazine and singlet oxygen injury. Sucrose-induced attainment of defence mechanisms was shown by the gene expression pattern of a chloroplastic Fe-SOD and by the increase in glucose-6-phosphate dehydrogenase activity. The activation of these defence mechanisms depends on both soluble sugar and atrazine and protected the plantlets. Protection by soluble sugars was concentration-dependent up to 80 mM (Sulmon et al. 2004). The sugar-treated plants are able to maintain PS II activity and phototrophic growth in the presence of atrazine up to 40 μM. Further, Ramel et al. (2009) reported that pretreatment of sugar improved the shoots lengths and photosynthesis under atrazine stress in *A. thaliana.* The specific correlation between pretreatment of sucrose and atrazine tolerance may reflect adaptive mechanisms that link sucrose accumulation, photosynthesis-related stress and sucrose induce stress tolerance.

3.4.3 Hormone-Mediated Detoxification of Xenobiotics

In plants, the expression of specific genes involved in the processes related to detoxification is responsive to xenobiotic compounds such as pesticides (Ramel et al. 2012), suggesting that the intrinsic detoxification mechanisms of higher plants may be exploited to decrease pesticide residues in food produce. Brassinosteroids (BRs) are class of phytohormones involved in the regulation of plant growth, development and stress responses (Gudesblat and Russinova 2011). Previously, it has been reported that BRs-induced stress tolerance is associated with changes in cellular redox homeostasis and expression of a wide range of stress-related genes including those encoding P450 and GST involved in metabolism of xenobiotic compounds (Xia et al. 2009; Jiang et al. 2012). Ahammed et al. (2013a) have reported BRs induce plant tolerance against phenanthrene (PHE) and cadmium (Cd) stress. The foliar application of EBR (0.1 μM) to PHE- and/or Cd-stressed plants alleviated photosynthetic inhibition and oxidative stress by causing enhancement of the activity of the enzymes and related transcript levels of the antioxidant system, secondary metabolism and the xenobiotic detoxification system. Additionally, PHE and/or Cd residues were significantly decreased in both the leaves and roots after application of EBR, more specifically in PHE + Cd-stressed plants when treated with EBR, indicating a possible improvement in detoxification of these pollutants. These results suggest a potential interaction of EBR for PHE and Cd stress alleviation. These results advocate a positive role for EBR in reducing pollutant residues for food safety and also strengthening phytoremediation. Zhou et al. (2015) show that brassinosteroids (BRs) decreased the residues of common organophosphorus, organochlorine and carbamate pesticides by 30–70 % in tomato, rice, tea, broccoli, cucumber, strawberry and other plants when treated externally. Genome-wide microarray analysis showed that fungicide chlorothalonil (CHT) and BR co-upregulated 301 genes, including a set of detoxifying genes encoding cytochrome

P450, oxidoreductase, hydrolase and transferase in tomato plants. The level of BRs was closely related to the respiratory burst oxidase 1 (RBOH1)-encoded NADPH oxide-dependent H_2O_2 production, glutathione biosynthesis and the redox homeostasis and the activity of glutathione-*S*-transferase (GST). Gene silencing treatments showed that BRs decreased pesticide residues in plants likely by promoting their metabolism through a signalling pathway involving BRs-induced H_2O_2 production and cellular redox change. Their study provides a novel approach for minimizing pesticide residues in crops by enhancing pesticide degradation through enhanced glutathione metabolism and glutathione-*S*-transferase (GST) activity via a respiratory burst oxidase homologue1 (RBOH1)-dependent pathway. It has been shown for a number of xenobiotics, especially for some pesticides, that the corresponding conjugates are sequestered in the vacuole (Coupland 1991). For example, the plant growth regulator (2,4-dichlorophenoxy) acetic acid (2,4-D) diffuses into plant cells (*Phaseolus vulgaris*), where it becomes hydroxylated in the 4-position. The hydroxylation occurs via the "NIH shift" (an NIH shift is a chemical rearrangement where a hydrogen atom on an aromatic ring undergoes an intramolecular migration primarily during a hydroxylation reaction) mechanism which involves the displacement of the Cl group (Thomas et al. 1964). The 4-OH-2,5-D is then glucosylated and malonylated; O-(malonyl)glucosyl 4-OH-2,4-D is stored in the vacuole (Schmitt and Sandermann 1982; Sandermann 1987). In the soya bean, 2,4-D forms amide conjugates with various amino acids instead (especially glutamyl and aspartyl derivatives) (Mumma and Davidonis 1983). In addition, the precursor of ethylene is stored as 1-(malonylamino)-cyclopropane-1-carboxylic acid in the vacuole (Bouzayen et al. 1989). The herbicide paraquat slowly accumulates in the vacuoles of root cells of maize seedlings, but is also translocated to the shoot (DiTomaso et al. 1993).

3.4.4 Safener-Mediated Detoxification of Xenobiotics

Another interesting feature of the xenobiotic detoxification system of some monocotyledonous crop species is infusibility by certain synthetic compounds collectively named "safeners". Safeners are the group of chemically diverse compounds with the unique ability to protect grasses from herbicide injury without reducing herbicide activity in target weed species (Davies and Caseley 1999; Hatzios and Burgos 2004). Safeners protection from herbicide injury is accomplished by increasing the expression of genes encoding herbicide-metabolizing enzymes, such as the glutathione-*S*-transferases (GSTs), cytochrome P450 and several others (Hatzios 1989; Farago et al. 1994; Riechers et al. 2005). Cloquintocet mexyl (safener) selectively increased O-glucosyltransferase activities against 4-nitrophenol and 2,4,5-trichlorophenol in wheat plant (Brazier et al. 2002). There are various classes of safeners which have been developed to enhance herbicide tolerance in maize (*Zea mays*), grain sorghum (*Sorghum bicolor*), rice (*Oryza sativa*) and small-grain cereals. Walton and Casida (1995) have characterized the safener-binding protein; along with this Scott-Craig et al. (1998) observed the gene expression patterns in maize seedlings. Safeners induce the expression of genes involved in plant defence and detoxification, such as GSTs and P450s, yet they are not toxic to the target plant and confer protection from herbicide injury. This suggests that safeners have signalling pathway for detoxification of endogenous toxins or xenobiotics (Riechers et al. 2005). Further, it is also hypothesized that safeners may utilize the oxidized lipid, i.e. oxylipins (Mosblech et al. 2009) and cyclopentenone-mediated signalling pathway, which later leads to the expression of GSTs and other proteins involved in detoxification and plant defence (Weber 2002; Loeffler et al. 2005; Mueller et al. 2008; Mueller and Berger 2009). Although the safening activity toward herbicides is only reported in the cereals and some non-domesticated grasses (Skipsey et al. 2011; Zhang et al. 2004), the ability of these compounds to selectively induce xenome (xenomic response) enzymes also extends to most popular dicotyledonous plant, *Arabidopsis thaliana* (Edwards et al. 2005; DeRidder and Goldsbrough 2006).

3.5 Conclusion and Future Perspectives

A characteristic feature of plants and other sessile organisms is the exposure of different xenobiotics, which altered the growth and developments. Therefore, plants have evolved mechanisms for the metabolization and detoxification of these xenobiotics to protect themselves. Many of these xenobiotics are stored in a separate compartment, which is the vacuole in the case of hydrophilic compounds. The storage of vacuolar defence or signal compounds is often tissue specific. Here, the vacuoles function as "defence or signal compartments". It seems likely that some of these functions are exclusive, and not all vacuoles within one cell or within the same tissue must have identical properties. Other than compartmentalization, these xenobiotics are detoxified by different processes, i.e. glutathione, carbohydrates, hormones, safeners, etc. However, the exact mechanism of these processes is still unknown. Thus, for better understanding in the field of xenobiotic action in plants, much more studies are required.

References

Ahammed GJ, Choudhary SP, Chen S et al (2013a) Role of brassinosteroids in alleviation of phenanthrene–cadmium co-contamination induced photosynthetic inhibition and oxidative stress in tomato. J Exp Bot 64:199–213

Ahammed GJ, Ruan YP, Zhou J et al (2013b) Brassinosteroid alleviates polychlorinated biphenyls induced oxidative stress by enhancing antioxidant enzymes activity in tomato. Chemosphere 90:2645–2653

Ahammed GJ, Zhou YH, Xia XJ et al (2013c) Brassinosteroid regulates secondary metabolism in tomato towards enhanced tolerance to phenanthrene. Biol Plant 57(1):154–158

Alla MMN, Badawi AM, Hassan NM et al (2008) Effect of metribuzin, butachlor and chlorimuron-ethyl on amino acid and protein formation in wheat and maize seedlings. Pestic Biochem Physiol 90:8–18

Ashton AR, Ziegler P (1987) Lack of effect of the photosystem II-based herbicides diuron and atrazine on growth of photo- heterotrophic *Chenopodium rubrum* cells at concentrations inhibiting photoautotrophic growth of these cells. Plant Sci 51:269–275

Baerson SR, Sanchez-Moreiras A, Pedrol-Bonjoch N et al (2005) Detoxification and transcriptomr response in *Arabidopsis* seedlings exposed to the allelochemical bzoxazolin-2 (3H)-one (BOA). J Biol Chem 280:21867–21881

Behringer C, Bartsch K, Schaller A (2011) Safeners recruit multiple signalling pathways for the orchestrated induction of the cellular xenobiotic detoxification machinery in *Arabidopsis*. Plant Cell Environ 34:1970–1985

Belkadhi A, Hediji H, Abbes Z et al (2012) Influence of salicylic acid pre-treatment on cadmium tolerance and its relationship with no-protein thiol production in flax root. Afr J Biotechnol 11:9788–9796

Boada J, Roig T, Perez X et al (2000) Cells overexpressing fructose 2, 6- bisphosphatase showed enhanced pentose phosphate pathway flux and resistance to oxidative stress. FEBS Lett 480:251–264

Bocova B, Huttova J, Mistrık I et al (2013) Auxin signalling is involved in cadmium-induced glutathione-*S*-transferase activity in barley root. Acta Physiol Plant 35:2685–2690

Bouzayen M, Latche A, Pech JC et al (1989) Carrier-mediated uptake of 1- (malonylamine)-cyclopropane-1- carboxylic acid in vacuoles isolate from *Catharanthus roseus* cells. Plant Physiol 91:1317–1322

Brazier M, Cole DJ, Edwards R (2002) O-Glucosyl transferase activities toward phenolic natural products and xenobiotics in wheat and herbicide-resistant and herbicide-susceptible black-grass (*Alopecurus myosuroides*). Phytochemistry 59:149–156

Christou A, Antoniou C, Christodoulou C et al (2016) Stress-related phenomena and detoxification mechanisms induced by common pharmaceuticals in alfalfa (*Medicago sativa* L.). Plants Sci Tot Environ 558:652–664

Cole DJ, Edwards R (2000) Secondary metabolism of agrochemicals in plants. In: Roberts TR (ed) Agrochemicals and plant protection. Wiley, Chichester, pp 107–154

Coupland D (1991) Detoxification of herbicides in plants. In: Caseley JC, Cussans GW, Atkin RK (eds) Herbicide resistance in crops. Butterworth-Heineman, Oxford, pp 263–278

Davies J, Caseley JC (1999) Herbicide safeners: a review. Pestic Sci 55:1043–1058

Davies TGE, Coleman JOD (2000) The *Arabidopsis thaliana* ATP-binding cassette proteins : an emerging superfamily. Plant Cell Environ 23(5):431–443

Davletova S, Rizhsky L, Liang H et al (2005) Cytosolic Ascorbate Peroxidase 1 is a central component of the reactive oxygen gene network of *Arabidopsis*. Plant Cell 17:268–681

Debnam PM, Fernie AR, Leisse A et al (2004) Altered activity of the P2 isoform of plastidic glucose 6-phosphate dehydrogenase in tobacco (*Nicotiana tabacum* cv. Samsun) causes changes in carbohydrate metabolism and response to oxidative stress in leaves. Plant J 38:49–59

DeRidder BP, Dixon DP, Beussman DJ et al (2002) Induction of glutathione-S-transferases in *Arabidopsis* by herbicide Safeners. Plant Physiol 130(3):1497–1505

DeRidder BP, Goldsbrough PB (2006) Organ-specific expression of glutathione-*S*-transferases and the efficacy of herbicide safeners in *Arabidopsis*. Plant Physiol 140:167–175

Desikan R, Mackerness SAH, Hancock JT et al (2001) Regulation of the *Arabidopsis* transcriptome by oxidative stress. Plant Physiol 127:159–172

Dévier MH, Mazellier P, Ait-Aissa S et al (2011) New challenges in environmental analytical chemistry: identification of toxic compounds in complex mixtures. Comptes Rendus Chimie 14:766–779

DiTomaso JM, Hart JJ, Kochain LV (1993) Compartmentation analysis of Paraquat fluxes in maize roots as a means of estimating the rate of vacuolar accumulation and translocation to shoots. Plant Physiol 102:467–472

Doczi R, Brader G, Pettko-Szandtner A et al (2007) The *Arabidopsis* mitogen-activated protein kinase kinase MKK3 is upstream of group C mitogen-activated protein kinases and participates in pathogen signaling. Plant Cell 19:3266–3279

Edwards R, Del Buono D, Fordham M et al (2005) Differential induction of glutathione transferases and glucosyltransferases in wheat, maize and *Arabidopsis thaliana* by herbicide safeners. Zeitschrift Naturforsch Biosci C 60:307–316

Ehlting J, Chowrira SG, Mattheus N et al (2008) Comparative transcriptome analysis of *Arabidopsis thaliana* infested by diamondback moth (*Plutella xylostella*) larvae reveals signatures of stress response, secondary metabolism and signalling. BMC Genomics 9:154

Ekman DR, Lorenz WW, Przybyla AE et al (2003) SAGE analysis of transcriptome responses in *Arabidopsis* roots exposed to 2,4,6-trinitrotoluene. Plant Physiol 133:1397–1406

Ekman DR, Wolfe NL, Dean JF (2005) Gene expression changes in *Arabidopsis thaliana* seedling roots exposed to the munition hexahydro-1,3,5-trinitro-1,3,5-triazine. Environ Sci Tech 39:6313–6320

Farago S, Brunold C, Kreuz K (1994) Herbicide safeners and glutathione metabolism. Physiol Plant 91:537–542

Ford KA, Casida JE, Chandran D et al (2010) Neonicotinoid insecticides induce salicylate-associated plant defense responses. Proc Natl Acad Sci USA 107:17527–17532

Foreman J, Demidchik V, Bothwell JH et al (2003) Reactive oxygen species produced by NADPH oxidase regulate plant cell growth. Nature 422:442–446

Gudesblat GE, Russinova E (2011) Plants grow on brassinosteroids. Curr Opin Plant Biol 14:530–537

Hall LM, Moss SR, Powles SB (1997) Mechanisms of resistance to aryloxy phenoxy propionate herbicides in two resistant biotypes of *Alopecurus myosuroides* (black grass): herbicide metabolism as a cross resistance mechanism. Pesticide Biochem Physiol 57:87–98

Hatzios KK (1989) Mechanisms of action of herbicide safeners: an over- view. In: Hatzios KK, Hoagland RE (eds) Crop safeners for herbicides: development, uses and mechanisms of action. Academic, San Diego, pp 65–101

Hatzios KK, Burgos N (2004) Metabolism-based herbicide resistance: regulation by safeners. Weed Sci 52:454–467

Hauschild R, von Schaewen A (2003) Differential regulation of glucose-6-phosphate dehydrogenase isoenzyme activities in potato. Plant Physiol 133:47–62

Hernandez LE, Villasante CO, Montero-Palmero MB et al (2012) Heavy metal perception in a microscale environment: a model system using high doses of pollutants. In: Gupta DK, Sandalio LM (eds) Metal toxicity in plants: perception, signaling and remediation. Springer, Berlin/Heidelberg

Holton T, Cornish EC (1995) Genetics and biochemistry of anthocyanin biosynthesis. Plant Cell 7:1071–1083

Iwata Y, Koizumi N (2005) An Arabidopsis transcription factor, AtbZIP60, regulates the endoplasmic reticulum stress response in a manner unique to plants. Proc Natl Acad Sci USA 102:5280–5285

Jaspers P, Kangasjarvi J (2010) Reactive oxygen species in abiotic stress signaling. Physiologia Plant 138:405–413

Jiang QQ, Yang HQ, Sun X et al (2012) Relation between polyamine metabolism and cell death in roots of *Malus hupehensis* Rehd under cadmium stress. J Int Agri 11:1129–1136

Jin XF, Chen C, Han HJ et al (2011) Microarray analysis of the phytoremediation and phytosensing of occupational toxicant napththalene. J Hazard Mat 189:19–26

Kagami H, Pitzschke A, Hirt H (2005) Emerging MAP kinase pathways in plant stress signalling. Trends Plant Sci 10:339–346

Koch KE (1996) Carbohydrate modulated gene expression in plants. Annu Rev Plant Physiol Plant Mol Biol 47:509–540

Kreuz LE, Levy AH (1965) Physical properties of chick interferon. J Bacteriol 89:462–469

Kreuz KR, Tommasini R, Martinoia E (1996) Old enzymes for a new job: How cells dispose of herbicides. Plant Physiol 111:349–353

Loeffler C, Berger S, Guy A et al (2005) B1-phytoprostanes trigger plant defense and detoxification responses. Plant Physiol 137:328–340

Lupinkova L, Komenda J (2004) Oxidative modifications of the Photosystem II D1 protein by reactive oxygen species: from isolated protein to cyanobacterial cells. Photochem Photobiol 79:152–162

Marrs KA (1996) The function and regulation of glutathione-*S*-transferases in plants. Annu Rev Plant Physiol Plant Mol Biol 47:127–158

Mittler R, Vanderauwera S, Gollery M et al (2004) Reactive oxygen gene network of plants. Trends Plant Sci 9:490–498

Mori IC, Schroeder JI (2004) Reactive oxygen species activation of plant Ca^{2+} channels. A signaling mechanism in polar growth, hormone transduction, stress signaling, and hypothetically mechano transduction. Plant Physiol 135:702–708

Mosblech A, Feussner I, Heilmann I (2009) Oxylipins: structurally diverse metabolites from fatty acid oxidation. Plant Physiol Biochem 47:511–517

Mueller MJ, Berger S (2009) Reactive electrophilic oxylipins: pattern recognition and signalling. Phytochemistry 70:1511–1521

Mueller S, Hilbert B, Dueckershoff K et al (2008) General detoxification and stress responses are mediated by oxidized lipids through TGA transcription factors in *Arabidopsis*. Plant Cell 20:768–785

Mumma RO, Davidonis GH (1983) Plant tissue culture and pesticide metabolism. In: Huston DH, Roberts TR (eds) Prog in pesticide biochem 3. Wiley, Chichester, pp 255–278

Nandula VK, Reddy KN, Rimando AM et al (2007) Glyphosate-resistant and susceptible soybean (*Glycine max*) and canola (*Brassica napus*) dose response and metabolism relationships with glyphosate. J Agric Food Chem 55:3540–3545

Nishikawa F, Kato M, Hyodo H et al (2005) Effect of sucrose on ascorbate level and expression of genes involved in the ascorbate biosynthesis and recycling pathway in harvested broccoli florets. J Exp Bot 56:65–72

Nishiyama Y, Allakhverdiev SI, Yamamoto H (2004) Singlet oxygen inhibits the repair of photosystem II by suppressing the translation elongation of the D1 protein in *Synechocystis* sp. PCC 6803. Biochemistry 43:11321–11330

Nogushi T (2002) Dual role of triplet localization on the accessory chlorophyll in the photosystem II reaction center: photoprotection and photodamage of the D1 protein. Plant Cell Physiol 43:1112–1116

Owen WJ (2000) Herbicide metabolism as a basis for selectivity. In: Roberts T (ed) Metabolism of agrochemicals in plants. Wiley, Chichester, pp 211–258

Peng RH, Xu RR, Fu XY et al (2011) Microarray analysis of the phytoremediation and phytosensing of occupational toxicant napththalene. J Hazard Mat 189:19–26

Ramel F, Sulmon C, Cabello-Hurtado F et al (2007) Genome wide interacting effects of sucrose and herbicide- mediated stress in *Arabidopsis thaliana*: novel insights into atrazine toxicity and sucrose-induced tolerance. BMC Genomics 8:450

Ramel F, Sulmon C, Bogard M et al (2009) Differential dynamics of reactive oxygen species and antioxidative mechanisms during atrazine injury and sucrose-induced tolerance in *Arabidopsis thaliana* plantlets. BMC Plant Biol 9:28

Ramel F, Birtic S, Cuine S et al (2012) Chemical quenching of singlet oxygen by carotenoids in plants. Plant Physiol 158:1267–1278

Rea PA (1999) MRP subfamily ABC transporters from plants and yeasts. J Exp Bot 50:895–913

Rea PA, Martinoia E, Li ZS et al (1998) From vacuolar GS-X pumps to multispecific ABC transporters. Annu Rev Plant Physiol Plant Mol Biol 49:727–760

Reddy KN, Rimando AM, Duke SO (2004) Amino methyl phosphonic acid, a metabolite of glyphosate, causes injury in glyphosate-treated, glyphosate-resistant soybean. J Agric Food Chem 52:5139–5143

Riechers DE, Vaughn KC, Molin WT (2005) The role of plant glutathione-S-transferases in herbicide metabolism. In: Clark JM, Ohkawa H (eds) Environmental fate and safety management of agrochemicals, ACS symposium series 899. American Chemical Society, Washington, DC, pp 216–232

Riechers DE, Kreuz K, Zhang Q (2010) Detoxification without intoxication: Herbicide safeners activate plant defense gene expression. Plant Physiol 153:3–13

Rinalducci S, Pedersen JZ, Zolla L (2004) Formation of radicals from singlet oxygen produced during photoinhibition of isolated light-harvesting proteins of photosystem II. Biochim Biophys Acta 1608:63–73

Rishi A, Muni S, Kapur V et al (2004) Identification and analysis of safener-inducible expressed sequence tags in *Populus* using a cDNA microarray. Planta 220:296–306

Roitsch T (1999) Source-sink regulation by sugar and stress. Curr Opin Plant Biol 2:198–206

Rolland F, Moore B, Sheen J (2002) Sugar sensing and signaling in plants. Plant Cell Suppl 14:S185–S205

Rutherford AW, Krieger-Liszkay A (2001) Herbicide-induced oxidative stress in photosystem II. Trends Biochem Sci 26:648–653

Ryter SW, Tyrrell RM (1998) Singlet molecular oxygen: a possible effector of eukaryotic gene expression. Free Radical Biol Med 24:1520–1534

Saari LL, Cotterman JC, Thill DC (1994) Resistance to aceto lactate synthase inhibiting herbicides. In: Powles SB, Holtum JAM (eds) Herbicide resistance in plants. Lewis Publisher, Boca Raton, pp 83–139

Salvemini F, Franze A, Iervolino A et al (1999) Enhanced glutathione levels and oxidoresistance mediated by increased glucose-6-phosphate dehydrogenase expression. J Biol Chem 274:2750–2757

Sandermann H (1987) Pestizid-Ruckstande in Nahrungspflanzen. Die Rolle des pflanzlichen Metabolismus. Naturwissenschanften 74:573–578

Schmitt R, Sandermann H Jr (1982) Specific localization of beta-D-glucoside conjugates of 2, 4-dichlorophenoxyacetic acid in soy- bean vacuoles [*Glycine max*]. Zeitschriftfuer Naturforschung Section C Biosci 37:772–777

Schröder P, Daubner D, Maier H et al (2008) Phytoremediation of organic xenobiotics– Glutathione dependent detoxification in *Phragmites* plants from European treatment sites. Bioresour Technol 99(15):7183–7191

Scott-Craig JS, Casida JE, Poduje L et al (1998) Herbicide safener binding protein of maize: purification, cloning, and expression of an encoding cDNA. Plant Physiol 116:1083–1089

Serra AA, Nuttens A, Larvor V et al (2013) Low environmentally relevant levels of bioactive xenobiotics and associated degradation products cause cryptic perturbations of metabolism and molecular stress responses in *Arabidopsis thaliana*. J Exp Bot 64:2753–2766

Skipsey M, Knight KM, Brazier-Hicks M et al (2011) Xenobiotic responsiveness of *Arabidopsis thaliana* to a chemical series derived from a herbicide safener. J Biol Chem 286:32268–32276

Steinrücken HC, Amrhein N (1980) The herbicide glyphosate is a potent inhibitor of 5-enol pyruvyl shikimic acid-3-phosphate synthase. Bioch Biophy Res Com 94:1207–1212

Sulmon C, Gouesbet G, Couee I et al (2004) Sugar-induced tolerance to atrazine in *Arabidopsis* seedlings: interacting effects of atrazine and soluble sugars on psbA mRNA and D1 protein levels. Plant Sci 167:913–923

Sulmon C, Gouesbet G, El Amrani A et al (2006) Sucrose-induced tolerance to atrazine in *Arabidopsis* seedlings involves activation of oxidative and xenobiotic stress responses. Plant Cell Rep 25:489–498

Szekacs A, Darvas B (2012) Forty years with glyphosate. In: Hasaneen MNAE-G (ed) Herbicides-properties, synthesis and control of weeds. In Tech, Rijeka, pp 247–284

Thibaud MC, Gineste S, Nussaume L et al (2004) Sucrose increases pathogenesis-related PR-2 expression in *Arabidopsis thaliana* through an SA-dependent but NPR1-independent sig- naling pathway. Plant Physiol Biochem 42:81–88

Thomas EW, Loughman BC, Powell RG (1964) Metabolic rate of 2, 4-D in the stem tissue of *Phaseolus vulgaris*. Nature 204:884–885

Timmermann KP (1989) Molecular characterization of corn glutathione-*S*-transferase isoenzymes involved in herbicide detoxification. Physiol Planta 77:465–471

Unver T, Bakar M, Shearman RC et al (2010) Genome-wide profiling and analysis of *Festuca arundinacea* miRNAs and transcriptomes in response to foliar glyphosate application. Mol Genet Genomics 283:397–413

Walton JD, Casida JE (1995) Specific binding of a dichloroacetamide herbicide safener in maize at a site that also binds thiocarbamate and chloroacetanilide herbicides. Plant Physiol 109:213–219

Wang JT, Jiang YP, Chen SC et al (2010) The different responses of glutathione-dependent detoxification pathway to fungicide chlorothalonil and carbendazim in tomato leaves. Chemosphere 79:958–965

Weber H (2002) Fatty acid-derived signals in plants. Trends Plant Sci 7:217–224

Weisman D, Alkio M, Colon-Carmona A (2010) Transcriptional responses to polycyclic aromatic hydrocarbon-induced stress in *Arabidopsis thaliana* reveal the involvement of hormone and defense signaling pathways. BMC Plant Biol doi. doi:10.1186/1471-2229-10-59

Wink M (1997) Special nitrogen metabolism. In: Dey PM, Harborne J (eds) Plant biochemistry. Academic Press, San Diego/London, pp 439–486

Xia XJ, Zhang Y, Wu JX et al (2009) Brassinosteroids promote metabolism of pesticides in cucumber. J Agric Food Chem 57:8406–8413

Zhang H, Huang Z, Xie B et al (2004) The ethylene, jasmonate, abscisic acid and NaCl-responsive tomato transcription factor JERF1 modulates expression of GCC box-containing genes and salt tolerance in tobacco. Planta 220:262–270

Zhang Q, Xu FX, Lambert KN et al (2007) Safeners coordinately induce the expression of multiple proteins and MRP transcripts involved in herbicide metabolism and detoxification in *Triticum tauschii* seedling tissues. Proteomics 7:1261–1278

Zhang JJ, Lu YC, Zhang SH et al (2016) Identification of transcriptome involved in atrazine detoxification and degradation in alfalfa (*Medicago sativa*) exposed to realistic environmental contamination. Ecotoxicol Environ Saf 130:103–112

Zheleva DT, Tsonev T, Sergiev I et al (1994) Protective effect of exogenous polyamines against atrazine in pea plants. J Plant Growth Regul 13:203–211

Zhou Y, Xia X, Yu G et al (2015) Brassinosteroids play a critical role in the regulation of pesticide metabolism in crop plants. Sci Rep 5:9018. doi:10.1038/srep09018

Metabolic Responses of Pesticides in Plants and Their Ameliorative Processes

4

Gunjan Dubey, Neeraj Mishra, and Sheo Mohan Prasad

Abstract

The major challenge faced by modern agriculture today is to overcome the impending serious food shortage considering the rising population even while battling to maintain crop yields under unfavourable environmental conditions. Pesticides are important chemicals recurrently applied for enhancing crop yields in agriculture and also for amelioration of vector-borne diseases. Rampant use of these chemicals has led to serious health implications to man and his environment. In developing nations (like India), 76 % of the total pesticide used is harmful insecticide as against just 44 % in the developed countries. This difference in the pesticide usage pattern evidently affects the total cost–benefit ratio between these nations. It is commonly believed by the general public that 'if little is good, a lot more will be better'. This unchecked and unwise use of pesticides leads to deposition of 95 % of pesticides in soil, water and atmosphere as well as in non-target organisms. Consequently, pesticide residues have been found in raw as well as processed fruits and vegetables. Pesticides not only kill insects and weeds but also are toxic to a variety of other nontarget organisms like birds, fish, beneficial insects and plants. Soil microflora has also been adversely affected by treatment of soil with pesticides.

G. Dubey
Department of Botany, University of Rajasthan, Jaipur 302004, Rajasthan, India
e-mail: gunjandubey22@gmail.com

N. Mishra
Department of Chemistry, ARSD College, University of Delhi, New Delhi, India
e-mail: neerajmishrarajdhani@gmail.com

S.M. Prasad (✉)
Ranjan Plant Physiology and Biochemistry Laboratory, Department of Botany, University of Allahabad, Allahabad 211002, Uttar Pradesh, India
e-mail: profsmprasad@gmail.com

A. Singh et al. (eds.), *Plant Responses to Xenobiotics*,
DOI 10.1007/978-981-10-2860-1_4

Xenobiotics, such as pesticides, trigger a defence response in plants. The mechanism by which pesticide is perceived and the signalling is transmitted within plant cells still remains elusive. Various xenobiotics induce oxidative stress and often activate cell signalling pathways, leading to production of stress proteins and upregulation of antioxidants. These changes include the evolution of new metabolic pathways, generation/accumulation of low molecular weight metabolites, synthesis of special proteins, detoxification mechanisms and reduction/elevation in phytohormone levels. Comparison of metabolic pathways, detoxification mechanisms and variance in tolerance level of different plants to pesticides can aid us in designing efficient and cheap vegetative treatment systems for decontaminating soil and water containing pesticide residues. Thus, understanding metabolism of pesticides in plants is necessary for developing strategies for amelioration of their harmful effects in nontarget organisms and also for remediation of contaminated soil, water and atmosphere.

Keywords
Metabolic • Pesticide • Toxicity • Uptake • Translocation

4.1 Introduction

Environmental issues have aroused great concern of the general public as well as that of the scientific community in recent years. Many studies have been published related to the subject of pesticide and human health. A cursory glance over the history of pesticide use through the years in the past proves to be an eye opener. Most of the pesticides being used presently did not exist before World War II. Some pesticides being used currently were actually developed during World War II for use in warfare. The organophosphate insecticides were developed as nerve gases, and the phenoxy herbicides (including 2,4-D) were developed to be used as a component of Agent Orange to defoliate large areas of jungle in the Vietnam War. However, post-World War II, these chemicals began to be used as pesticides in agriculture and for environmental spraying to eradicate mosquitoes in homes and gardens. The revolutionary book, *Silent Spring*, authored by Rachel Carson published in 1962, first raised political and public awareness to the hazards posed to wildlife, humans and the ecosystem by the use of pesticides. "The sequel to this book was Our Stolen Future by Colborn et al. (1996a, b). This book documented the adverse effect of endocrine-disrupting chemicals on health." Since then, there have been umpteen scientific studies to determine the relationship between pesticide use and human health hazards. The present chapter is an effort to expand upon the earlier works to present a collection of old and new scientific advances made in the field of biochemistry and xenobiotic chemicals. This chapter is substantive but not exhaustive.

4.2 What Are Pesticides?

Pesticides are important chemicals recurrently applied in agriculture to protect crop plants from harmful pests like fungi, insects, mites and rodents (Freedman 1995) and from competition with abundant but unwanted plant species to:

(i) Obtain a larger yield.
(ii) Obtain crops of high quality.
(iii) Reduce the input of labour and energy into crop production (Ayansina 2009).

They are also used to prevent diseases in livestock and to protect human beings from insect vector-borne diseases. The term pesticide thus includes a wide range of chemical compounds, viz. insecticides, fungicides, herbicides, rodenticides, molluscicides, nematicides, plant growth regulators, synthetic insecticides like organophosphates (OPs), carbamates, pyrethroids, herbicides and fungicides. Herbicides are used to control unwanted plants or 'weeds' in aquatic/agricultural systems to prevent yield losses of agronomically important crops. Organophosphate (OP) pesticides are the most commonly used insecticides (Casida and Quistad 2004). They are widely used for agricultural and public health purposes (Infante-Rivard et al. 1999; Kozawa et al. 2009). OPs account for approximately 50 % of the insecticides used globally and function primarily by inhibiting acetylcholinesterase (AChE) (Kamanyire and Karalliedde 2004; Ogutcu et al. 2009). In the present scenario, insecticides are mostly the only practical way to control the vector-borne diseases of humans and livestock. Pesticides are classified on the basis of:

(i) Their origin/structure
(ii) The pests controlled by them
(iii) Mode/site of action (MoA)

(i) *Classification based on origin/structure of pesticides.* There are two types:
  (A) Chemical pesticides
  (B) Biopesticides
    (A) *Chemical pesticides* have further been classified into four types on the basis of their origin:
      (a) *Organophosphates* – These are the chemical substances which are produced due to reaction between phosphoric acid and alcohols. They cause irreversible blockage of the nervous system as they inhibit the activity of the enzyme acetylcholinesterase. This results in its accumulation causing overstimulation of muscles. These mainly include insecticides, nerve gases, herbicides, etc.
      (b) *Carbamates* – They are carbamic acid esters. Their mode of action is like that of OPs, i.e. by inhibition of acetylcholinesterase, but here the bond formed for inhibition is reversible. These also mainly include insecticides.

(c) *Organochlorines* – These are derivatives of chlorinated hydrocarbons. They are endocrine-disrupting agents which affect the hormonal system of the body, act as duplicates of the normal hormones and thus cause adverse health problems. They persist in environment for a long time and accumulate in the fatty tissues of animals. A well-known example is DDT (dichlorodiphenyltrichloroethane).

(d) *Pyrethroids* – These are potent neuro-poisons and endocrine disruptors and cause paralysis. Pyrethroids are the synthetic form of a natural insecticide pyrethrin, obtained from *Chrysanthemum*. These are ketoalcoholic ester derivatives of chrysanthemic and pyrethroic acids and are more stable in sunlight than pyrethrins. These insecticides are the very most popular insecticides as they can easily pass through the exoskeleton of the insect. Few examples of pyrethroids being currently used are deltamethrin, cypermethrin, etc.

(B) *Biopesticides* are naturally occurring materials derived naturally from living organisms or their metabolites, like bacteria, fungi, plants, etc. These pesticides have microorganisms acting as pest controllers like bacteria, fungi or viruses. The strains of *Bacillus thuringiensis* and its subspecies are commonly used biopesticide. The mode of action of biopesticides generally is by producing a protein that binds to the larval gut receptor which starves the larvae.

(ii) *Classification of pesticides based on the pests they control*:

(a) Insecticides – act especially on insects
(b) Algicides – control/kill the growth of algae
(c) Herbicides – control/kill weeds
(d) Bactericides – act against bacteria
(e) Fungicides – act against fungi.
(f) Rodenticides – kill or prevent rodents
(g) Larvicides – inhibit the growth of larvae
(h) Repellents – repel pests by their taste or smell
(i) Desiccants – act on plants by drying their tissues
(j) Ovicides – inhibit the growth of eggs of insects and mites
(k) Virucides – act against viruses
(l) Molluscicides – inhibit/kill molluscs, i.e. snails usually disturbing growth of plants or crops
(m) Acaricides – kill arachnids like spiders
(n) Nematicides – kill nematodes that act as parasites of plants
(o) Avicides – kill birds
(p) Mothballs – are used to stop any damage to cloths by moth larvae or moulds
(q) Lampricides – target larvae of lampreys which are jawless fish like vertebrates in the river
(r) Piscicides – act against fishes

(iii) *Classification of pesticides based on MoA*:

The Insecticide Resistance Action Committee (IRAC) provides the classification based on mode of action (MoA). Internationally recognised academic and industrial experts in insecticide toxicology have reviewed and approved the classification based on MoA by IRAC (Sparks and Nauen 2015). This classification is essential for the development of insecticide resistance management (IRM) strategies. Currently the MoA classification encompasses more than 25 different MoAs:

1. Acetylcholinesterase (AChE) Inhibitors:
   (i) Carbamates – e.g. aldicarb, bendiocarb, carbaryl, carbofuran, carbosulfan, etc.
   (ii) Organophosphates – e.g. chlorpyrifos, dimethoate, fenitrothion, malathion, monocrotophos, etc.
2. *GABA-Gated Chloride Channel Blockers*:
   (i) Cyclodiene organochlorines – e.g. chlordane, endosulfan, etc.
   (ii) Phenylpyrazoles (fiproles) – e.g. ethiprole, fipronil, etc.
3. Sodium Channel Modulators:
   (i) Pyrethroids and pyrethrins – e.g. Allethrin, cypermethrin, pyrethrins (pyrethrum), α-cypermethrin, β-cypermethrin etc.
   (ii) DDT and methoxychlor
4. Nicotinic Acetylcholine Receptor (NACHR) Competitive Modulators:
   (i) Neonicotinoids – e.g. Acetamiprid, imidacloprid, thiacloprid, etc.
   (ii) Nicotine
   (iii) Sulfoximines – e.g. Sulfoxaflor
   (iv) Butenolides – Flupyradifurone
5. *Nicotinic Acetylcholine Receptor (NACHR) Allosteric Modulators* – e.g. spinetoram and spinosad
6. *Glutamate-Gated Chloride Channel Allosteric Modulators* – e.g. abamectin, lepimectin, milbemectin etc.
7. Juvenile Hormone Mimics:
   (i) Juvenile hormone analogues – e.g. hydroprene, kinoprene, and methoprene
   (ii) Fenoxycarb
   (iii) Pyriproxyfen
8. Miscellaneous Non-specific (Multisite) Inhibitors:
   (i) Alkyl halides – e.g. Methyl bromide and other alkyl halides
   (ii) Chloropicrin
   (iii) Fluorides – e.g. Cryolite and sulfuryl fluoride
   (iv) Borates – e.g. Borax, boric acid, disodium octaborate, sodium borate and sodium metaborate
   (v) Tartaremetic
   (vi) Methyl isothiocyanate generators – e.g. Dazomet and metam
9. Chordotonal Organ TRPV Channel Modulators:
   (i) Pyridine azomethine derivatives – e.g. pymetrozine and pyrifluquinazon

10. Mite Growth Inhibitors:
    (i) Clofentezine, diflovidazin and hexythiazox
    (ii) Etoxazole
11. Microbial Disruptors of Insect Midgut Membranes:
    (i) *Bacillus thuringiensis* and the insecticidal proteins they produce:
    Bacillus thuringiensis var. *aizawai*
    Bacillus thuringiensis var. *israelensis*
    Bacillus thuringiensis var. *kurstaki*
    Bacillus thuringiensis var. *tenebrionis*
    (ii) *Bacillus sphaericus*
12. Inhibitors of Mitochondrial ATP Synthase:
    (i) Diafenthiuron
    (ii) Organotin miticides – e.g. azocyclotin, cyhexatin and fenbutatin oxide
    (iii) Propargite
    (iv) Tetradifon
13. *Uncouplers of Oxidative Phosphorylation via Disruption of the Proton Gradient* – e.g. chlorfenapyr, DNOC and sulfluramid
14. *Nicotinic Acetylcholine Receptor (NACHR) Channel Blockers*: Nereistoxin analogues – e.g. bensultap, cartap hydrochloride, thiocyclam and thiosultap-sodium
15. Inhibitors of Chitin Biosynthesis:
    (i) Type 0: Benzoylureas – e.g. bistrifluron, chlorfluazuron, etc.
    (ii) Type 1: Buprofezin – e.g. buprofezin
16. *Moulting Disruptor* – e.g. cyromazine
17. *Ecdysone Receptor Agonists*: Diacylhydrazines – e.g. chromafenozide, halofenozide, methoxyfenozide and tebufenozide
18. *Octapine Receptor Agonists* – e.g. Amitraz
19. *Mitochondrial Complex III Electron Transport Inhibitors*:
    (i) Hydramethylnon
    (ii) Acequinocyl
    (iii) Fluacrypyrim
    (iv) Bifenazate
20. *Mitochondrial Complex I Electron Transport Inhibitors*:
    (i) METI acaricides and insecticides – e.g. fenazaquin, fenpyroximate, etc.
    (ii) Rotenone
21. Voltage-Dependent Sodium Channel Blockers:
    (i) Oxadiazines – e.g. indoxacarb
    (ii) Semicarbazones – e.g. metaflumizone
22. Inhibitors of Acetyl CoA Carboxylase:
    (i) Tetronic and tetramic acid derivatives – e.g. spirodiclofen, spiromesifen and spirotetramat
23. Mitochondrial Complex IV Electron Transport Inhibitors:
    (i) Phosphides – e.g. aluminium phosphide, calcium phosphide, phosphine and zinc phosphide
    (ii) Cyanides – e.g. calcium cyanide, potassium cyanide and sodium cyanide

24. Mitochondrial Complex II Electron Transport Inhibitors:
    (i) Beta-ketonitrile derivatives – e.g. cyenopyrafen and cyflumetofen
    (ii) Carboxanilides – e.g. pyflubumide
25. Ryanodine Receptor Modulators:
    (i) Diamides – e.g. chlorantraniliprole, cyantraniliprole and flubendiamide
26. Chordotonal Organ Modulators (Undefined Target Site) – e.g. Flonicamid
27. Compounds of Unknown/Uncertain MoA:
    (i) Azadirachtin
    (ii) Benzoximate
    (iii) Bromopropylate
    (iv) Chinomethionat
    (v) Dicofol
    (vi) Lime sulphur
    (vii) Pyridalyl
    (viii) Sulphur

Although pesticides are designed to kill/inhibit organisms that damage crops or animals, yet, they tend to produce harmful effects on other organisms and also to pollute the environment. Pesticides when used in high quantities can sometimes be lethal. According to the estimates of the World Health Organization (WHO), there are about 3,000,000 pesticide poisoning cases each year, with resultant 220,000 deaths worldwide (Jaga and Dharmani 2003, WHO 2007). The use of biopesticides, however, causes reduced negative effects as compared to chemical pesticides (Aktar et al. 2009; Chandler et al. 2011; Kumar and Singh 2015).

4.3 Sources of Pesticide Pollution

The major challenge faced by agriculturists today is to overcome the impending serious food shortage considering the continuously increasing population and to maintain the quantity and quality of crop yields even under unfavourable environmental conditions. According to the 'World Population Prospects: The 2015 Revision' published by the Department of Economic and Social Affairs, Population Division (United Nations), the world population is expected to increase by more than one billion within the next 15 years. It is expected to reach up to 8.5 billion in 2030 and to increase still further to 9.7 billion in 2050 and 11.2 billion by the year 2100. A critical question that needs to be faced is whether the present rate of increase in crop yield would be sufficient for feeding the world population. Food and Agriculture Organization (FAO) projections indicate that the average world cereal yield needs to be improved to 5 t ha^{-1} from its present 3 t ha^{-1} to maintain population growth (Alexandratos and Bruinsma 2012). Consequently this has led to large-scale development of the agrochemical industry during the last few decades. Modern agriculture thus makes use of a large number of pesticides for protection against pests and weeds. As reported, up to 80 % of crop yields could be lost without pesticides (Oerke 2006). These pesticides enter into the atmosphere and terrestrial and aquatic ecosystems by spraying, drifting and leaching, as run-off or sometimes even by accidental spills (Van derWerf 1996). This has led to contamination of agricultural produce as well as the environment in most countries, posing a serious threat to plant, human and animal health worldwide (Giridhar and Indira 1997; Joshi et al. 2005). Exposure of consumers to pesticides commonly occurs via food consumption (González-Rodríguez et al. 2008), causing various adverse health effects (Ló pez-Pérez et al. 2006). The World Health Organization (WHO) estimated that a maximum number of deaths due to pesticide poisoning occur in developing nations (Mbakaya et al. 1994; Smith 2001; Shittu et al. 2012). Here it is important to highlight the fact that the pattern of pesticide usage in developing countries is different from that of the world in general. In the former case out of the total pesticide being used, 76 % is harmful insecticide in contrast to just 44 % used globally (Mathur 1999). Out of the large quantities (several million tons) of pesticides applied annually, the amount of these products that reach the target organism is even less than 5 %. The remaining approximately 95 % is deposited in soil and nontarget organisms and ultimately reaches the atmosphere and water (Pimental and Levitan 1986; de Oliveira et al. 2012; Ahemad and Khan 2010). Further, the general farmers strongly believe in the dictum 'if little is good, a lot more will be better'. This consequently leads to application of these hazardous chemicals to fields in even greater quantities than required.

The large-scale use of pesticides in the recent past has adversely affected both the target biota and the useful nontarget organisms (Shivaramaiah and Kennedy 2006). Thus, in addition to killing insects and weeds, pesticides can be toxic to a host of other organisms including birds, fish, beneficial insects, beneficial soil microorganisms and nontarget plants. Disasters like the endosulfan tragedy of Kerala (India) are a stark reminder of the direct health hazard that pesticide use poses to humans. As reported by Grover and Cessna (1991), most of the pesticides are resistant to chemical and/or photochemical degradation under typical environmental conditions (Grover and Cessna 1991) so they persist in the ecosystem for a longer duration. Thus, the total cost–benefit ratio differs appreciably between developed and developing countries.

4.4 Disadvantages of Pesticides

Plants get exposed to pesticides intentionally or unintentionally. This causes many direct or indirect effects in their survival, health and reproduction bringing about a change in plant community attributes. The surplus amount of added pesticides is absorbed by the plants. This is supported by reports from González-Rodríguez et al. (2011) who found traces of pesticide residues in various raw and processed forms of fruits and vegetables. Consequently selective control of pests by applying pesticides in crop fields has resulted in serious contamination of the environment with greater loss of crop productivity (Vasileva and Ilieva 2007).

Herbicides affect photosynthetic, respiratory growth, cell/nuclear division and synthesis of proteins/carotenoids/lipids (Jaga and Dharmani 2003). 2,4-D is the most commonly used and extensively studied herbicide (WHO 2007). Herbicides persist in soil after application and adversely affect the survival and metabolism of nontarget soil microorganisms which are beneficial for man (Ahemad et al. 2009). Moreover, consistent application of herbicides to soil has resulted in their accumulation in soil which even exceeds the recommended field rates. This alters the architecture of plant roots and reduces the number of root sites used by *Rhizobia* during infection (Eberbach and Douglas 1991).

Organophosphate (OP) insecticides function by resulting into excessive accumulation of acetylcholine (ACh) at the cholinergic receptors in the peripheral and central nervous systems of insects (Kamanyire and Karalliedde 2004). However, reports confirm that toxicity occurs even at doses that do not inhibit acetylcholinesterase enzyme (Pope 1999; Yanai et al. 2002; Gupta 2004; Slotkin 2004, 2005). Therefore, other mechanisms like induction of oxidative stress leading to generation of free radicals and decrease in antioxidants or oxygen free radical scavenging systems have been implicated to explain OP toxicity (Dandapani et al. 2003; Abdollahi et al. 2004; Altuntas et al. 2004; Sharma et al. 2005; Ambali et al. 2007, 2010). For example, chlorpyrifos (CPF; O,O-diethyl-O-3,5,6-trichloro-2-pyridylthiophosphate) is a broad-spectrum OP insecticide and is moderately toxic to mammals (Kenaga et al. 1965). Irrespective of this fact, in the present times, chlorpyrifos is amongst the most widely used OP insecticide for effective crop protection and urban pest control (ATSDR 1997; Steenland et al. 2000, Saulsbutry et al. 2009). Oxidative stress has been reported to be involved in the toxicity of this OP insecticide (Ambali et al. 2010).

4.5 Pesticide Toxicity in Plants

4.5.1 Visible Toxicity Symptoms

Plant communities are often subjected to target as well as nontarget exposure to pesticides, and they often manifest subtle but significant responses. Xenobiotic chemical toxicity is manifested in many ways at low and high concentrations to produce different forms of visible negative effects in plants. These reported symptoms give clear indication of the deleterious effects they have on the health of plants. Visible symptoms like stunting of plant, shoot distortion and reduction in leaf dimensions together with chlorosis are prominent under high levels of these pesticide chemicals coupled with significant decline in photosynthetic pigments (Baruah and Mishra 1986; Yen et al. 2003; Murthy et al. 2005; Kaushik and Inderjit 2006). Complete death or stunted growth of nontarget plants due to pesticides has been observed by Norris and Freed (1966a, b, c) and Dhillion et al. (1968). Sandermann (1992) reported complete phenotype alteration in response to pesticides. According to Radwan (2012), application of clethodim to maize leaves caused yellowing of leaf tip and browning or drying of some leaf parts. Saraf and Sood (2002) reported almost complete inhibition of growth when treated with high concentration of monocrotophos. Similarly Mishra et al. (2008) reported that seedlings of *Vigna unguiculata* treated with dimethoate showed significant reduction in leaf area, shoot and root length and fresh mass and dry mass of the root, shoot and leaf. Xenobiotics especially herbicides have unique affinities for their respective molecular target site within important plant biochemical pathways and/or physiological processes (Dayan et al. 2010; Zhen and Singh 2001). The insecticide dimethoate directly binds to various sites of photosynthetic electron transport chain causing decline in the photosynthetic oxygen yield of *Vigna unguiculata* (Mishra et al. 2008). Similarly studies confirm that diphenyl ethers affect electron transport activities in chloroplasts of higher plants like oxyfluorfen and thus inhibit photosynthesis (Moreland et al. 1970; Pritchard et al. 1980). Thus, decline in photosynthetic process causes substantial decline in accumulation of biomass in plants (Sandermann 1992; Mishra et al. 2008, Radwan et al. 2012). Santi and de Pietri-Tonelli (1959) showed that dimethoate when applied to roots is translocated to other parts of plants making the whole plant insecticidal to certain insects. There is abundant literature illustrating the toxic effects of pesticides and herbicides on target and nontarget plants. Nontarget trees like *Abies* sp. and *Prunus persica* and many others have shown toxicity symptoms such as stunted vegetative growth and fruit size reduction (Obrigawitch et al. 1998). Similarly field crops/grasses (like *Fagopyrum esculentum* Moench, *Brassica napus*, *Cucumis sativus*, *Lens culinaris* Medik., *Allium cepa*, *Solanum tuberosum* L., *Lycopersicon esculentum (Mill.)*, *Helianthus annuus*, etc.) have exhibited marked reduction in seed size, delay in crop maturity and seed yield in response to various sulphonylurea herbicide treatments (Karthikeyan et al. 2004; Obrigawitch et al. 1998).

4.5.2 Indirect Toxicity Effects

Besides the above-mentioned visible toxic effects, scientists have reported various indirect effects caused by the imprudent usage of pesticides in nontarget plants. A common consequence of xenobiotic toxicity is excessive accumulation of reactive oxygen species (ROS). Increased production of ROS during such environmental stresses poses a serious threat to living cells by causing lipid peroxidation, protein oxidation, nucleic acid damage, inhibition of enzymes and activation of programmed cell death pathway which ultimately leads to death of cells (Fig. 4.1).

There is abundant literature to support such indirect toxicity symptoms of pesticides on plants (Shah et al. 2001; Mittler 2002; Meriga et al. 2004; Sharma and Dubey 2005; Mishra et al. 2011):

1. Geoffroy et al. (2002) reported that oxyfluorfen is capable of reducing growth indirectly by generating ROS, viz. singlet oxygen ($^1O_2$), superoxide radical ($O_2^{\bullet -}$), hydrogen peroxide (H_2O_2) and hydroxyl radicals ($^{\bullet}OH$). Increased production of reactive oxygen species (ROS) leads to oxidative stress (Banas´ et al. 1993; 2000; Miteva 2005; Ivanov et al. 2005; Sergiev et al. 2006; Radwan 2012) which eventually might lead to chlorophyll degradation leading to chlorosis. Stress-induced accumulation of ROS is a major cause of loss of crop productivity

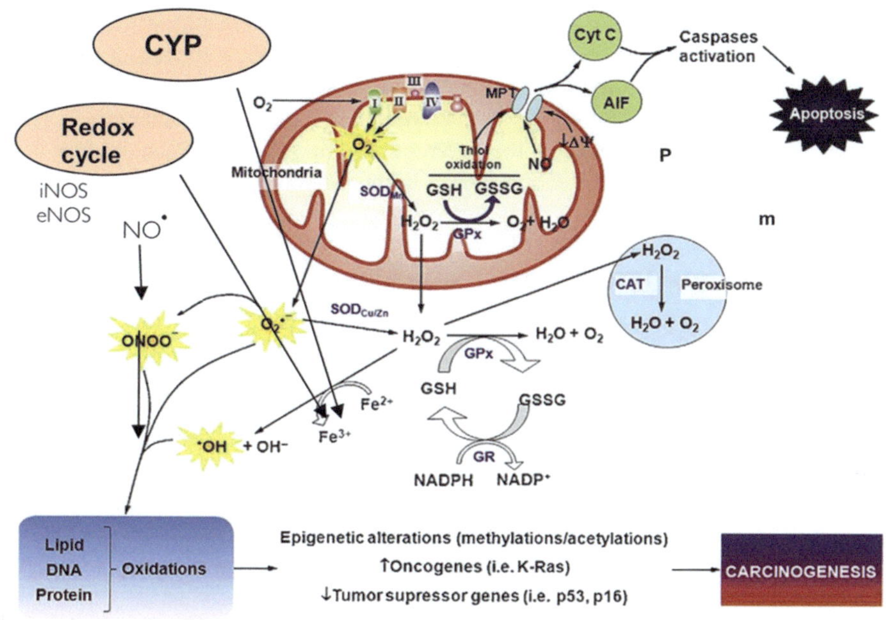

Fig. 4.1 Pesticides promote oxidative stress leading to cell death or procarcinogenic mutations. This figure schematically illustrates the complex molecular network activated by different pesticides. Abbreviations used are the following: AIF, apoptosis-inducing factor; CAT, catalase; CYP, cytochrome P450; Cytc, cytochrome C; eNOS, endothelial nitric oxide synthase; GSH, glutathione; GSSG, glutathione disulphide; iNOS, inducible nitric oxide synthase; MPTP, mitochondrial permeability transition pore; SODCu/Zn, copper-/zinc-type superoxide dismutase; SODMn, mangano-type superoxide dismutase; GPx, glutathione peroxidase; GR, glutathione reductase; ΔΨm, mitochondrial transmembrane potential (Source: Tebourbi et al. 2011 adapted from Mena et al. 2009, with modification)

worldwide (Mittler 2002; Apel and Hirt 2004; Mahajan and Tuteja 2005; Tuteja 2007; Khan and Singh 2008; Gill et al. 2011). Sergiev et al. (2006) observed that glyphosate application in maize plants caused oxidative stress. Reports of elevated amounts of malondialdehyde (MDA) production by Feierabend and Winkelhüsener (1982), Dayan and Watson (2011) and Radwan (2012) also indicate that pesticides either directly or indirectly destabilise the integrity of lipid bilayers and cause disastrous implications leading to cellular death. Enhanced production of ROS by chloroplasts and mitochondria limits carbon dioxide (CO_2) fixation (Mittler et al. 2004) in pesticide-treated plants leading to overproduction of superoxide radicals and H_2O_2 by the photosystems (Foyer and Noctor 2005b).

2. Toxic effects of pesticides also involve biochemical and physiological disruption (Fufezan et al. 2002; Liu et al. 2009; Ramel et al. 2009), as well as disruption of the signalling mechanism (Ramel et al. 2007, 2009; Unver et al. 2010).

3. Another prominent adverse indirect effect of pesticides is that they may alter the competitive interactions amongst species, thereby changing species dominance, richness and distribution, which can lead to further adverse aftereffects in the ecosystem (Kleijn and Snoeijing 1997). Qiu et al. 2002 reported that widespread use of pesticides has limited the productivity of *Nostoc*, which in turn has adversely affected the production of rice in the mountain paddy fields of China.

4.6 Uptake and Transport of Pesticides

Natural sources like water and wind can transport pesticides for long distances. During this transport, pesticides can undergo physical, chemical and/or biological alterations. Xenobiotics carried by wind can be transported to far-off places and finally be deposited onto terrestrial or aquatic vegetation. Transformation or biodegradation of pesticides can also be done by sunlight, microbial action or plant enzymes. Water-transported pesticides are carried upwards in plants during transpiration with the transpiration stream. From there, the volatile pesticides might then enter the atmosphere by evaporation from the large surface area of leaves and stems (Karthikeyan et al. 2004).

The fate of pesticides in a plant after its complete metabolism is a complex process. Uptake of xenobiotics by the roots of plants and their further transport have been reviewed by a number of authors (Briggs et al. 1982; Briggs and Bromilow 1983; Behrendt and Bruggemann 1993; Schroder and Collins 2002). Plants take up pesticides mainly through leaf surfaces, fruits and roots which are then distributed within the plant either from cell to cell or via the plant vascular system. The quantity and method by which a pesticide is absorbed and transported to different parts within the plant are dependent on the physical and chemical properties (such as hydrophilicity, p*K*a/b, etc.) of the pesticide. A pesticide taken up by roots from the soil can alternatively/simultaneously take two pathways to reach the top of the plant via xylem vessels:

1. The apoplastic pathway
2. The symplastic pathway (Karthikeyan et al. 2004)

The *symplastic system* of the plant includes the living plant tissue. It is that part of the tissue system of the plant which remains surrounded by plasmalemma and is interconnected by means of plasmodesmata. It provides a reactive environment by placing chemicals in close proximity to enzymes and other reactants. Movement within the conductive portion of the symplast (phloem) occurs by mass flow and diffusion. The *apoplastic system* on the contrary includes all the nonliving portions of the plant. The cell walls and the xylary elements form a continuous water-permeable column. The transport of solute for both short and long distances occurs through this column by mass flow and diffusion. Certain chemicals remain restricted either to the apoplastic or the symplastic pathways, while some others are *ambimobile*, i.e. they can move in both domains efficiently. The balance between the distribution of pesticides in the apoplastic–symplastic compartments determines the overall transport pattern (Karthikeyan et al. 2004). Uptake and translocation of various xenobiotic pollutants also differ amongst plant species. Uptake may depend on the lipophilicity and dissociation constant of the pesticide (Schroder and Collins 2002). Generally, less lipophilic pesticides take the apoplastic pathway, and more lipophilic pesticides tend to take the symplastic route (Karthikeyan et al. 2004). Movement of pesticides to tips of roots and shoots occurs via active transport in the phloem (Boersma et al. 1988). It also depends largely on species' inherent properties of the root itself and the transport tissues involved. For example, sulphonated anthraquinones are more efficiently translocated and metabolised in those plant species which produce natural anthraquinones like by *Rheum* and *Rumex*, as compared to that by maize or celery which are anthraquinone-nonproducing species (Aubert and Schwitzguebel 2004). Likewise, *Cucurbita pepo* L. shoots accumulate various hydrophobic contaminants like dioxin from soil, but many other plants including *Cucumis sativus* do not accumulate it (Campanella and Paul 2000; Campanella et al. 2002).

4.7 Metabolism of Pesticides in Plants

Pesticides trigger a defence response in plants. Plant responses to the xenobiotic stress are studied under the heads of toxicology, ecotoxicology and stress physiology. Pesticide toxicity is known to alter various physiological functions in target as well as nontarget organisms. Living cells respond and adapt to environmental signals such as toxicants or stressors through multiple mechanisms that involve communication pathways or signal transduction processes (Tebourbi et al. 2011). In this part of the chapter, we focus on the basic ways in which pesticides implicate their toxic effects in plants. According to Tebourbi et al. (2011), pesticides induce oxidative stress in plants in the following ways.

4.7.1 Inhibition of Biochemical Pathways

Pesticides interfere with the normal biological processes of plants. Such alterations/inhibitions of the biochemical pathways induce oxidative stress in plants. Stress-induced enhanced generation/accumulation of ROS by chloroplasts and mitochondria limits carbon dioxide (CO_2) fixation (Mittler et al. 2004) in pesticide-treated plants. This decline in the rate of CO_2 fixation results in overproduction of superoxide radicals and H_2O_2 by the photosystems (Foyer and Noctor 2005b). Mishra et al. (2008) reported that the insecticide dimethoate diminishes photosynthetic oxygen yield due to its direct effect on various sites of photosynthetic electron transport chain. Similarly, diphenyl ethers like oxyfluorfen inhibit photosynthesis by affecting electron transport activities in chloroplasts of higher plants (Moreland et al. 1970; Pritchard et al. 1980). Such site-specific inhibition causes decline in photosynthetic processes ultimately producing substantial reduction of biomass accumulation in plants. On a similar note, herbicides have unique affinities for their respective molecular target site within important plant biochemical pathways and/or physiological processes (Dayan et al. 2010; Zhen and Singh 2001). Herbicides are also reported to interrupt the electron flow by interacting with the plastoquinone-binding site on photosystem II (Oettmeier et al. 1984; Fuerst and Norman 1991). Reports confirm that excess herbicides damage the metabolic and physiological processes in weeds as well as in crop plants (Peixoto et al. 2006; Song et al. 2008; Yin et al. 2008; Jiang et al. 2012).

4.7.2 Induction of Reactive Oxygen Species (ROS)

Induction of oxidative stress is one of the well-known effects of pesticides (Song et al. 2006, 2007). Herbicides are reported to react by overproduction of free radicals and oxidative stress (Banas´ et al. 1993; 2000). Scientific reports confirm that excessive generation/accumulation of superoxide radicals and ROS is accompanied by accumulation of MDA (Feierabend and Winkelhüsener 1982; Sergiev et al. 2006) which causes severe disturbances in the physiological metabolism of plants leading to cell death and accelerated plant senescence (Ogweno et al. 2009; Xu et al. 2006; Li et al. 2004; Guo et al. 2006). The herbicide lactofen, for example, belongs to the diphenyl ether group (Rodrigues Almeida 1998). According to Matringe et al. (1989) and Witkowski and Halling (1989), tissues treated with lactofen accumulate porphyrins in light (Hopf and Whitten 1978). Porphyrins in turn can be the cause of light-dependent peroxidation of membrane lipids which ultimately leads to deleterious herbicidal effects in the form of membrane disruptions (Orr and Hess 1982; Ferreira et al. 2010). As shown in Fig. 4.1, oxidative stress arises if ROS production overrides the detoxification systems. This can happen either when there is excessive generation/accumulation of ROS or when the antioxidative system of the plant is compromised, ultimately leading to the oxidation of lipids, proteins and nucleic acids (Ryter et al. 2007; Franco et al. 2009, 2010; Tebourbi et al. 2011). Oxidative stress induced by xenobiotics often generates numerous secondary entities which act as signalling molecules, such as $^1O_2$ (Wagner et al. 2004) or H_2O_2 (Levine et al. 1994; Desikan et al. 1998; Wang et al. 2010), and numerous forms of denatured or modified proteins and nucleic acids (Teixeira et al. 2007; Xie et al. 2010). These molecules activate cell signalling pathways, production of stress proteins and upregulation of antioxidants. Such changes include processes like evolution of new metabolic pathways, accumulation of low molecular weight metabolites, synthesis of special proteins, changes in phytohormone levels and activation of detoxification mechanisms.

4.7.3 Intrinsic Reduction/Oxidation Cycling Properties

In the redox cycle, the parent chemical compound in the xenobiotic pesticide/herbicide is firstly reduced by the enzyme NADPH-dependent reductase (such as

NADPH–cytochrome P450 reductase) to yield a pesticide radical. This radical donates its unshared electron to O_2, yielding superoxide radical (O_2•-), and the parent compound in turn can undergo another cycle. Therefore, in one cycle, two potentially harmful events occur: a reductant is oxidised and an oxyradical is produced. For example, bipyridylium herbicides become highly reactive free radicals on accepting electrons from PSI. These unstable free radicals undergo autoxidation back to the parent ion. Thus, there is rapid cycling between the parent herbicide ion and its reactive radical and the high flow rate of electrons in PSI. This ultimately is responsible for rapid dehydration of vegetative tissues treated with such inhibitors (Hess 2000).

The mechanism by which pesticide is perceived and the signalling is transmitted within plant cells still remains elusive. However, efforts are on to demystify various mechanisms involved in pesticide toxicity with the application of recent technologies. For example, the herbicide dinoterb has multifaceted mechanism of action which involves inhibition of photosynthesis, uncoupling of oxidative phosphorylation and interference with respiratory oxygen consumption (Belbachir et al. 1980). The mechanism of action of another herbicide monosodium methanearsonate (MSMA) still remains elusive (Dayan and Watson 2011). Senseman (2007) explained that MSMA causes cell membrane destruction. Earlier studies on MSMA by Sachs and Michael (1971) proposed that the possible mechanism of action of MSMA for inhibition could be that it uncouples mitochondrial oxidative phosphorylation. Wauchope (1983) reported that MSMA inhibits photosynthesis and respiration, while Ellis and Camper (1995) suggested that the inhibitory action of MSMA could be attributed to the fact that it forms complexes with sulfhydryl-containing enzymes. Another totally different mechanism of pesticide toxicity as suggested by Luscombe et al. (1995) is that decline in shoot growth due to the pesticide isoxaflutole can be explained on the basis of inhibition in the activity of 4-hydroxyphenylpyruvate dioxygenase (HPPD), an enzyme needed for meristematic tissue in plant. Similarly, the commonly used herbicide glyphosate inhibits biosynthesis of phenylalanine, tyrosine and tryptophan because it interferes with the shikimate pathway (Cole 1985) which ultimately slows down protein synthesis (Jaworski 1972).

4.8 Plant Defence Systems

Xenobiotic chemicals like pesticides, herbicides, insecticides, etc. being widely used in modern agriculture are foreign chemicals for plants. They have contaminated almost every part of our environment. Rampant and irrational use of pesticides in rural, urban and agricultural areas has contributed greatly to the problem. As a result pesticide residues are found everywhere in soil, air as well as surface and groundwater across the countries. Pesticide contamination poses significant risks to the environment and nontarget organisms ranging from beneficial soil microorganisms to insects, plants, fish and birds. Against the common belief, even herbicides can be especially problematic for the environment because they are used in large volumes (Aktar et al. 2009). Plants being immobile cannot escape such environmental stresses. As a result they are subjected to various biotic/abiotic environmental stresses concomitantly. Thus, plants possess intrinsic defence methods to protect them from the adverse effects of small quantities of foreign compounds. There are numerous receptors which sense the presence of foreign compounds in the cell and induce a cascade of events intended to lead to neutralisation and excretion of these compounds (Tebourbi et al. 2011). However, in many cases the metabolism of xenobiotic substances can give rise to metabolites that are even more toxic or can lead to the formation of ROS that can harm the cell further. If plants do not develop a positive detoxification mechanism, they will ultimately die or have other unfavourable effects such as stunted growth. The ability of higher plants to scavenge the toxic effects of ROS is a very important determinant of their tolerance to stresses (Gangwar et al. 2014). The fate of pesticides is dependent on:

(i) Pesticide characteristics (hydrophilicity, p*K*a/b, *K*ow, etc.)
(ii) Abiotic environmental conditions (temperature, moisture, soil pH, etc.)
(iii) Microbial community or plant species or both

It might be assumed that metabolism of xenobiotics in plants is mainly confined to root and leaf tissues and occurs scarcely during transportation in the plant vascular system (Verkleij et al. 2009). Plants possess antioxidative enzymes which scavenge ROS and avoid oxidative damage (Kanazawa et al. 2000). These antioxidant enzymes thus act as critical weapons to maintain optimum health and well-being of plant cells. Plants have mechanisms to degrade or sequestrate most commercial pesticides. There is abundant literature pertaining to plant metabolism of pesticides in target species (Sandermann 1992; Korte et al. 2000; Roberts 2000; Hall et al. 2001). Oxidative stress occurs when the balance between the accumulated/generated ROS of a plant and its antioxidant capacity is disturbed. The antioxidant system comprises of two types of antioxidants (i.e. enzymatic and nonenzymatic). There are several enzymatic antioxidants like superoxide dismutase (SOD), catalase (CAT), ascorbate peroxidase (APX), dehydroascorbate reductase (DHAR), glutathione reductase (GR) and glutathione-S-transferase (GST). Similarly, nonenzymatic antioxidants such as ascorbate and glutathione are effectively involved in the scavenging of ROS in plants (Gondim et al. 2012; Thounaojam et al. 2012). Sandermann (1994) compared the enzymatic cascade responsible for detoxifying, breaking down and finally storing the xenobiotics with a 'green liver' in analogy to human hepatic metabolism (Fig. 4.2).

4.8.1 Phases of Metabolism of Pesticides

According to Jakoby and Ziegler (1990) and Coleman et al. (1997), the detoxification process of pesticides in plants can be divided into three phases (Fig. 4.3).

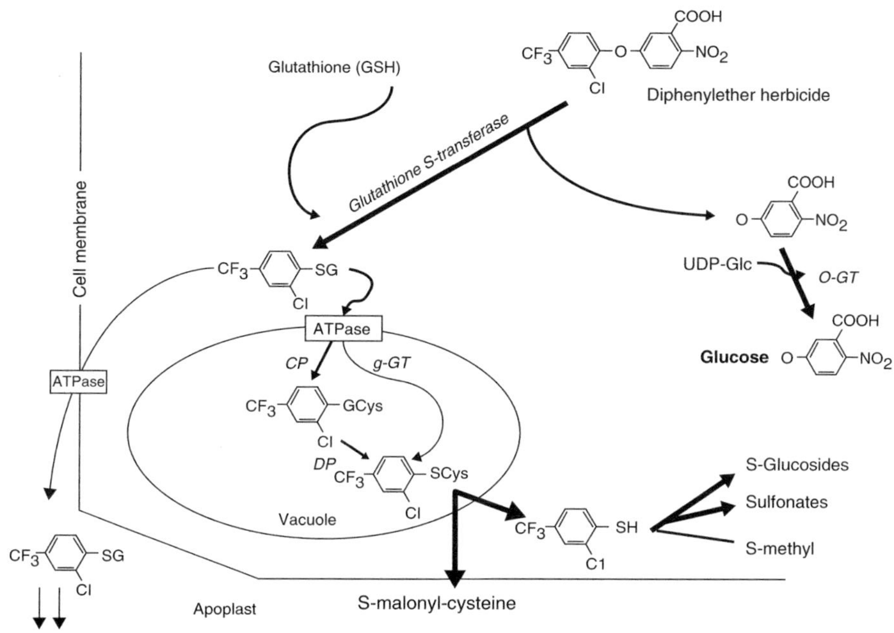

Fig. 4.2 Cascade of enzymes responsible for detoxification, breakdown and final storage of organic xenobiotics (Source: Verkleij et al. (2009) modified after Schroder and Collins 2002). After uptake into the cell, xenobiotics (e.g. a classical diphenyl ether herbicide) are attacked by Phase II enzymes, like the glutathione-S-transferases (GST) or glycosyltransferases (GT). The product conjugates (R-SG) are either sequestered in the vacuole by tonoplast ATPases (Theodoulou 2000; Schroder et al. 2001; Grzam et al. 2007) or there are alternative pathways to leave them in the cytosol or translocate them into the apoplast (Ohkama-Ohtsu et al. 2007). Cleavage of GS-conjugates in vacuoles may occur by a sequence of carboxypeptidase (CP) and dipeptidase (DP) reactions (Wolf et al. 1996) or by γ-glutamyltranspeptidase (Grzam et al. 2007). Normal plant metabolism proceeds towards the formation of S-glucosides, sulphonates or related compounds, to finally yield bound terminal metabolites with the cell wall

4.8.1.1 Phase I Metabolism

Phase I Metabolism – also known as *transformation* phase. Phase I reactions can reduce or modify the phytotoxicity of xenobiotic chemicals and can predispose the molecules to Phase II reactions. It involves oxidation, reduction, hydrolysis, hydroxylation, epoxidation, N-dealkylation, O-dealkylation, desulphurization, dehalogenation, dehydrohalogenation and dehydrogenation reactions (Casida and Lykken 1969; Coleman et al. 1997; Doran 2005; Yuan et al. 2007) to transform the pesticide to produce a more water-soluble and usually a less toxic product than the parent (Karthikeyan et al. 2004). These reactions are primarily catalysed by cytochrome P450 enzymes to generate functional groups in the pesticide structure using different transformation mechanisms (Mougin et al. 2001). A detailed review of the role of plant hydrolytic enzymes in pesticide hydrolysis reactions has been done by

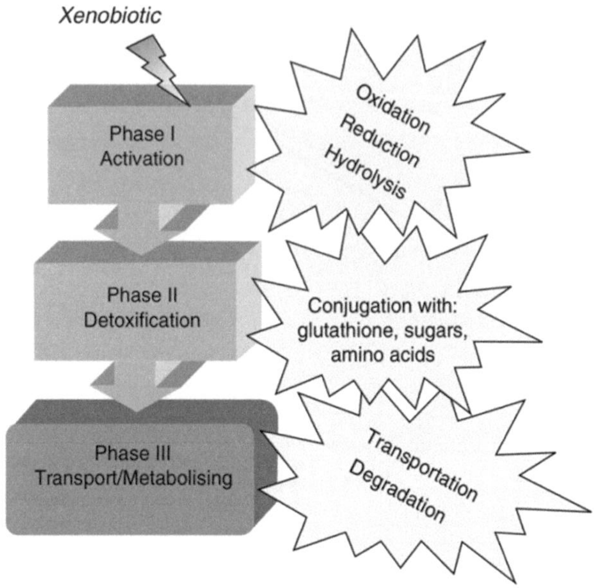

Fig. 4.3 Main phases of the process of detoxification of xenobiotics in plants (Source: Katerova and Miteva (2010) with modifications from Schroder (2001))

Hoagland and Zablotowicz (2001). Several enzymes, viz. cytochrome P450, aryl-acylamidases, amidases, reductases, esterases, monooxygenases, peroxidases (POX), lipases and proteases, participate in pesticide detoxification and degradation (Hall et al. 2001; Zablotowicz et al. 2001). These enzymes are localised in membrane fractions of plant cells, cytochrome P450 and monooxygenases in the apoplast and POX in the cytosol (Werck- Reichhart et al. 2000; Morant et al. 2003; Passardi et al. 2005).

4.8.1.2 Phase II Metabolism

Phase II Metabolism – also known as *conjugation* phase. Conjugation reactions are anabolic processes. These reactions produce compounds which have higher molecular weight together with highly reduced biological activity and mobility and increased water solubility. Phase II involves conjugation of a pesticide/pesticide metabolite to a natural plant constituent like glucose, amino acid, malonic acid or glutathione. This produces an increase in the water solubility and reduction in toxicity compared with the parent pesticide. Detoxification in Phase II is performed by enzymes like glutathione and glucosyl transferases. Glutathione conjugation is an important Phase II transformation in plants (Hatzios 2001). Glutathione in the presence of GST can react with a wide range of substrates like epoxides, aryl and alkyl halides and other electrophilic compounds. Cole and Edwards (2000) published a summary of their research on several agrochemicals metabolised in plants by glutathione conjugation. The products of Phase II metabolism are either slightly or non-phytotoxic and can easily be stored in cellular organelles (Korte et al. 2000; Sandermann 1992).

4.8.1.3 Phase III Metabolism

Phase III Metabolism – also known as *storage* phase. This phase converts Phase II metabolites into nontoxic secondary conjugates (Hatzios 1991). Reactions of Phase III further decrease the water solubility of toxic pesticides and concomitantly reduce their reactivity and toxicity. Theodolou (2000) recently subdivided this last phase into two independent phases:

(a) The first confined to transport and storage in the vacuole.
(b) The second one related to cell wall binding or excretion or copolymerisation with lignin. This results in the formation of non-extractable or bound residue. Sandermann et al. (2001) has reported significant incorporation of pesticide into bound residues. Alternatively, the conjugate in this phase is transferred through active transport by a glutathione pump into the plant cell vacuole.

According to Karthikeyan et al. (2004), the development of detoxification mechanism of pesticides depends on three factors:

1. Type of pesticide together with its present concentration
2. Prevailing environmental conditions
3. Characteristics of the exposed/treated plant

Enhanced ROS production during stress poses threat to cells. However, it is reported that ROS also act as signals for induction of stress response together with defence pathways (Karpinski et al. 1999; Desikan et al. 2001; Knight and Knight 2001; Mittler 2002). Since ROS act as secondary messengers, thus, their level must be tightly regulated by the antioxidative defence system which comprises of both enzymatic and nonenzymatic antioxidants. Under abiotic/biotic stresses, the balance between ROS production and scavenging may get disturbed, and then ROS are generated in excess causing cell death (Apel and Hirt 2004; Mittler et al. 2004; Dietz 2003; Gechev and Hille 2005).

4.8.2 Enzymatic Antioxidants

There are several enzymatic (superoxide dismutase, SOD; catalase, CAT; ascorbate peroxidase, APX; dehydroascorbate reductase, DHAR; glutathione-S-transferase, GST; and glutathione reductase, GR) and nonenzymatic (ascorbate, AsA, and glutathione, GSH) antioxidants which are effectively involved in ROS scavenging in plants (Chen et al. 2010a, b; Elbaz et al. 2010; Gondim et al. 2012; Thounaojam et al. 2012). The above-mentioned antioxidants directly or indirectly participate in scavenging of different ROS and thus help in protecting organisms from

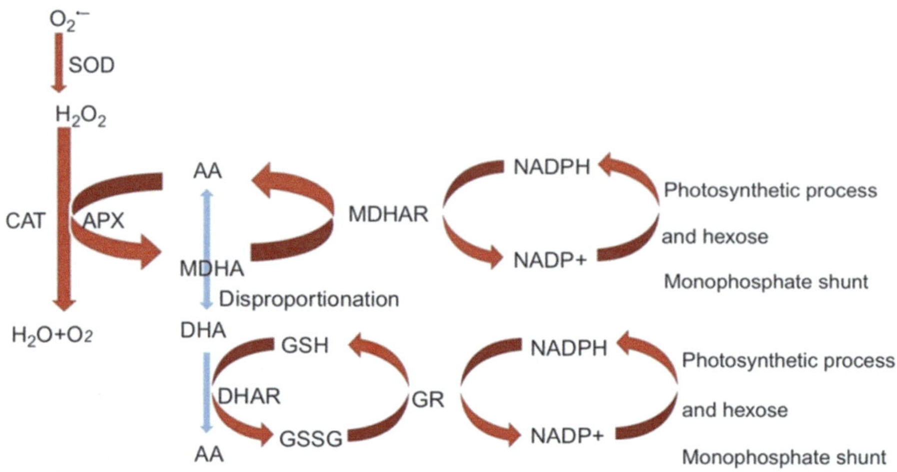

Fig. 4.4 Schematic drawing showing the role of antioxidants in scavenging of reactive oxygen species (Source: Gangwar et al. 2014)

oxidative stress. The importance of enzymatic and nonenzymatic antioxidants in preventing oxidative stress in plants is based on the fact that the level of one or more antioxidants increases under stress; this is generally related to increased stress tolerance as shown in Fig. 4.4.

4.8.2.1 Superoxide Dismutase (SOD; EC 1.15.1.1)

Superoxide dismutase (SOD) is considered to be the first line of defence against ROS. Superoxide radical (SOR; $O_2^{\bullet-}$) produced in the photosynthetic and respiratory chain is very reactive. As shown in Fig. 4.4, it dismutates into comparatively less damaging H_2O_2, and this dismutation is done by the enzyme SOD (Karpinska et al. 2001). SOR are negatively charged and cannot cross the plasma membrane. Thus, they are effectively trapped within the compartments where they are generated. SOD acts as the key enzyme to scavenge $O_2^{\bullet-}$ and provide protection against the toxic effects of oxidative stress (Myouga et al. 2008).

4.8.2.2 Catalase (CAT; EC 1.11.1.6)

Catalase (CAT) is a heme-containing enzyme and is one of the most potent catalysts known (Salin 1988). It is found in the peroxisome. CAT catalyses the conversion of H_2O_2 to water and molecular oxygen (Fig. 4.4).

4.8.2.3 Ascorbate Peroxidase (APX; EC 1.11.1.11)

Ascorbate peroxidase (APX) also scavenges H_2O_2 and is indispensable for the protection of chloroplasts and other cell constituents from damage by H_2O_2 and hydroxyl radicals ($^{\bullet}OH$). APX uses ascorbate (AsA) as its specific electron donor (Fig. 4.4) to reduce H_2O_2 to water with simultaneous production of monodehydroascorbate (MDHA). MDHA spontaneously disproportionates into AsA and dehydroascorbate (DHA).

4.8.2.4 Dehydroascorbate Reductase (DHAR; EC 1.8.5.1)

DHAR is located in the chloroplast (Hossain and Asada 1984) and cytosol (Arrigoni et al. 1981) of higher plants. It plays an important role in the ascorbate–glutathione recycling reaction in higher plants. DHAR uses glutathione (GSH) to reduce dehydroascorbate (DHA) and thereby regenerates reduced ascorbate (AsA) as shown in Fig. 4.4. Ascorbate is oxidised to DHA by means of successive reversible single-electron transfers producing MDHA (monodehydroascorbate) as a free radical intermediate. MDHA spontaneously disproportionates to DHA (Ishikawa et al. 1998) because MDHA radicals have a relatively short lifetime. DHA is then reduced to AsA by DHAR in a reaction requiring reduced GSH. DHAR converts reduced glutathione into oxidised glutathione (GSSG), liberating protons that are incorporated into the recycling reaction of AsA (Washko et al. 1992).

4.8.2.5 Glutathione Reductase (GR; EC 1.6.4.2)

Ascorbate–glutathione cycle also known as *Asada-Halliwell pathway* plays a very important role in the mitigation of ROS. Glutathione reductase (GR) is one of the important enzymes of the ascorbate–glutathione cycle. The enzyme GR is a member of the flavoenzyme family. As shown in Fig. 4.4, it catalyses conversion of oxidised glutathione (GSSG) into its reduced form (GSH), and in turn $NADP^+$ gets converted to NADPH (Rendon et al. 1995). It provides GSH and thus maintains the level of GSH in plant cells for scavenging of ROS by the other enzymes. Thus, GR is involved in defence against oxidative stress in higher plants.

4.8.2.6 Glutathione-S-Transferases (GSTs; EC 2.5.1.18)

Glutathione-*S*-transferases (GSTs) play a key role in cellular detoxification metabolism in different organisms (Tsuchida and Sato 1992; Hayes and Wolf 1988). They catalyse the electrophilic addition of the reduced form of the glutathione (GSH; γ-glutamylcysteine–glycine) to a variety of hydrophobic compounds (Rushmore and Pickett 1993; Mannervik and Danielson 1988). The resulting glutathione-*S*-conjugates are less reactive and more polar than the metabolised xenobiotic chemicals. Plants have to store the soluble glutathione-*S*-conjugates in the vacuole, because of the lack of excretion pathways (Sandermann 1992). GSTs constitute a family of multifunctional enzymes which provide a general enzymatic system of metabolic detoxification in higher plant cells. Plant GSTs metabolise xenobiotics and natural compounds by GSH conjugation. Detoxification by GSH conjugation is a well-documented, highly effective system in higher plants which metabolises various herbicides such as triazines, thiocarbamates and chloroacetanilides (Leavitt and Penner 1979; Ezra and Stephenson 1985). Following the above-mentioned scheme of three-phase detoxification of pesticides, GST can be regarded as Phase II enzyme. Plant GSTs not only inactivate toxic compounds through conjugation but also play an important role in the plant defence system. Stress-inducing agents induce plant

GST gene expression (Thenaken et al. 1995; Ulmasov et al. 1995) through formation of ROS (Levine et al. 1994). The increased GST level in turn protects cellular components. The first reports of herbicide resistance in weeds have been attributed to GSTs (Jensen et al. 1977). Anderson and Gronwald (1991) reported that GSTs are associated with nontarget-site herbicide resistance in the resistant biotype of *Abutilon theophrasti*. They illustrated that increased glutathione conjugation with atrazine correlated with amplified GST activity, which was responsible for herbicide resistance in the weed. Moreover, Marcacci et al. (2006) showed that conjugation to GSH is a major metabolic pathway that detoxifies atrazine in *Chrysopogon zizanioides* Nash, which is species resistant to this herbicide. GST-mediated herbicide resistance can target multiple herbicides also (Hatton et al. 1999; Cummins et al. 1999; Cocker et al. 2001; Letouze and Gasquez 2003; Katerova and Miteva 2010).

4.8.3 Nonenzymatic Antioxidants

4.8.3.1 Ascorbate

Ascorbate (AsA) and glutathione (GSH) are small molecular weight compounds. They detoxify ROS and other such toxic ROS-generated metabolites and thus protect cells from oxidative stress and its incurred damage. This protective action is materialised either directly by scavenging or indirectly through activation of defence mechanisms. Vitamin C (AsA) and GSH are present in very large quantities in majority of plant cells in millimolar concentrations of 10–300 mM and 0.1–25 mM, respectively (Noctor and Foyer 1998; Smirnoff 2000; Noctor et al. 2002; Ruiz and Blumwald 2002; Mou et al. 2003; Ball et al. 2004; Freeman et al. 2004; Gomez et al. 2004; Shao et al. 2008). Hence, they are capable of dealing with very high fluxes of H_2O_2 and other ROS generated/accumulated to maintain the redox homeostasis (Noctor et al. 2002; Latowski et al. 2010).

H_2O_2 scavenging is accomplished by catalases, various peroxidases (POX) and the ascorbate–glutathione pathway, also known as the Asada-Halliwell cycle, which is catalysed by a set of four enzymes (Willekens et al. 1997; Wojtaszek 1997; Kingston-Smith and Foyer 2000; Asada 2006; Almagro et al. 2009; Cosio and Durand 2009). Firstly, H_2O_2 is scavenged via the oxidation of ascorbate by ascorbate peroxidase (APX). This enzyme is involved in the oxidation of AsA to monodehydroascorbate (MDHA), which can be converted back to AsA via monodehydroascorbate reductase (MDHAR). MDHA is further rapidly converted to dehydroascorbate (DHA), which is converted back to AsA by the action of DHAR. DHAR utilises glutathione, which is regenerated by glutathione reductase (GR) from its oxidised form, i.e. glutathione disulphide (GSSG) (Latowski et al. 2010). The ascorbate–glutathione cycle functions in chloroplasts, mitochondria, peroxisomes, cytosol and apoplast (Smirnoff 2000; Mittler 2002, Pignocchi and

Foyer 2003; Asada 2006). AsA and GSH are responsible for maintaining cellular redox homeostasis and concomitantly involved in redox signalling. The total level of AsA and GSH in the cell and the ratio between their reduced and oxidised forms play an important role in activating various defence mechanisms. The important roles of AsA and GSH for tolerance towards environmental stresses have been emphasised in investigations involving mutants and transgenic plant lines by altering the levels of these antioxidants (Pastori et al. 2003; Ball et al. 2004).

4.8.3.2 Glutathione

According to Noctor et al. (2000), GSH is present in almost all living cells where it participates in numerous biochemical reactions. In higher plants, GSH plays an important role within the cell system. It participates in the ascorbate–glutathione cycle, maintains the sulfhydryl groups of cysteine in a reduced form, stores reduced sulphur and also acts as a substrate for GST (Noctor et al. 2000).

Glutathione (GSH) is directly involved in detoxification of xenobiotics. It also acts as an antioxidant against oxidative stress, provoked from herbicides like atrazine, paraquat, etc. The detoxification of xenobiotics occurs via conjugation of GSH with the hazardous chemical molecule. This might either involve the enzyme GST or GSH might accomplish detoxification spontaneously via the three phases explained above: activation (preparation), conjugation and metabolism of conjugates (Fig. 4.5). GSH directly detoxifies ROS like H_2O_2, singlet oxygen, superoxide radical and hydroxyl radical. The process of restoration of GSH from GSSG is important because accumulation of GSSG inhibits protein synthesis during translation. For example, strong oxidative stress following paraquat treatment led to GR inhibition as well (Ananievaa et al. 2004). GSH also protects cells against unfavourable effects of stress through the activation of different defence mechanisms because of its involvement in redox signalling (Foyer et al. 1997; Apel and Hirt 2004; Foyer and Noctor 2005a; Szalai et al. 2009). Culture plants with enhanced levels of GSH have been reported to show greater resistance to similar herbicides (Katerova and Miteva 2010). Millimolar concentrations of GSH in metabolically active tissues act as a redox buffer, blocking protein cysteine groups and ROS (Foyer and Noctor 2005). Glutathione possesses a central cysteine so it undergoes reversible oxidation and reduction. It can exist either in the reduced form (GSH) or in oxidised glutathione disulphide (GSSG) form. Glutathione redox potential depends on the total glutathione concentration and the degree of oxidation (Meyer and Hell 2005) as shown in Fig. 4.6. Changes in GSH redox state are responsible for the protective and regulatory roles of GSH. A high GSH/GSSG ratio is maintained either by increased GSH synthesis or by GSSG reduction. High GSH/GSSG ratio is necessary for efficient protection against ROS accumulation induced by abiotic stress (Szalai et al. 2009). GSH/GSSG ratio can be used for monitoring stress-induced damage (Kranner et al. 2006). Nemat Alla et al. (2008) reported that in *Zea mays*, enhanced tolerance

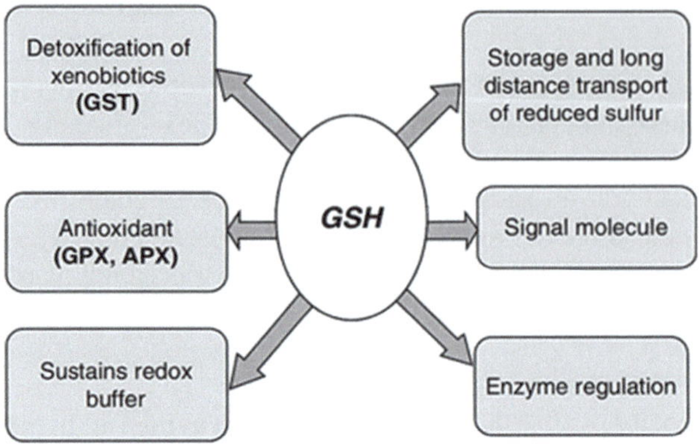

Fig. 4.5 Functions of glutathione in plants. *GPX* guaiacol peroxidase, *APX* ascorbate peroxidase, *GSH* glutathione (Source: Katerova and Miteva 2010)

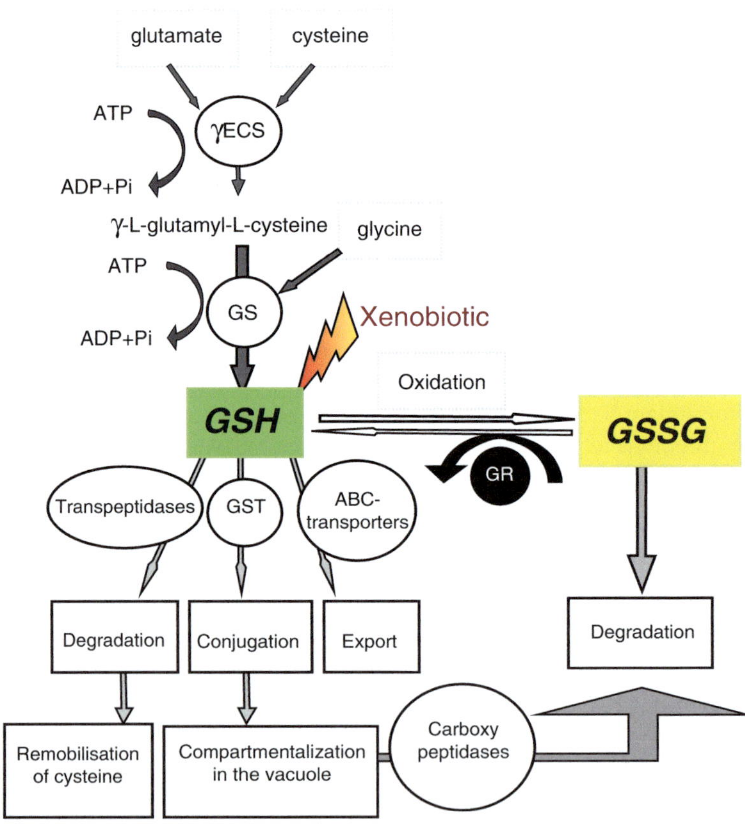

Fig. 4.6 General scheme of glutathione metabolism (Source: Katerova and Miteva 2010 with modifications from Foyer and Noctor 2001). *γECS* γ-glutamylcysteine synthetase, *GS* glutathione synthetase, *GST* glutathione-S-transferase, *GR* glutathione reductase, *GSH* reduced from of glutathione, *GSSG* glutathione disulphide (oxidised form of glutathione). During detoxification of ROS, GSH is oxidised to its dimer form GSSG. The enzyme glutathione reductase (GR) restores GSH

to herbicides is related to increased levels of GSH and the enzymes GSTs, γ-glutamylcysteine synthetase, GSH synthetase and GR.

4.8.4 Amelioration of Toxic Effects of Pesticides in Plants

4.8.4.1 Exogenous Application of Plant Growth Hormones

The plant growth hormone – salicylic acid (SA) – can regulate the adaptive responses of plants to many abiotic stresses (Ding et al. 2002; Yang et al. 2003; Zhou et al. 2009). The role of SA in regulating the antioxidant status and ROS production in plants subjected to biotic/abiotic stresses is well established (Kawano 2003). Exogenous application of SA can influence the antioxidant capacity of plant cells by inhibiting or activating the enzymatic antioxidant system (Ananievaa et al. 2004; Apel and Hirt 2004). Popova et al. (2004) reported that SA is involved in plant response to herbicide stress. Pretreatment of plants with SA in barley before paraquat application protected the plants against paraquat-induced damage due to changes in the antioxidant enzyme activities (Popova et al. 2002). Cui et al. (2010) reported SA-regulated amelioration of napropamide-induced oxidative stress in *Brassica napus* by decreasing ROS production and oxidative damage to the plasma membrane. It prevents herbicides from entering the cells and consequently the cell membrane integrity is improved. Similarly, Radwan (2012) reported SA-induced alleviation of oxidative stress caused by clethodim in maize (*Zea mays* L.).

Brassinosteroids (BRs) are steroidal compounds occurring ubiquitously in almost every part of the plant, such as pollen, flower buds, fruits, seeds, vascular cambium, leaves, shoots and roots. They are required for normal development of plants. BRs have also been reported to be effective in reducing damage caused by pesticides (Bajguz and Hayat 2009). BR treatment-induced alleviation of the damage in rice from treatment of simazine, butachlor or pretilachlor has been confirmed (Sasse 2003). Xia et al. (2009) studied the phytotoxic effect of nine pesticides – three herbicides, three fungicides and three insecticides – on cucumber leaves. They reported that 24 epibrassinolide (24-epiBL) increased the resistance of plants to pesticides probably by enhancing CO_2 assimilation. According to Sharma et al. (2013), exogenous application of 24-epiBL enhanced the tolerance of Pusa Basmati-1 seedlings to stress generated by the insecticide imidacloprid via a stepwise regulation mechanism by:

(i) Enhancing the activity of antioxidative enzymes like SOD, APX, CAT, GR and MDHAR
(ii) Upregulating the expression of genes like Cu/Zn–SOD, Fe–SOD, Mn–SOD, APX, CAT and GR
(iii) Decreasing lipid peroxidation

4.8.4.2 Exogenous Application of Free Radicals

Nitric oxide (NO) is a free radical and a highly reactive molecule. The protective role of NO against oxidative stress is well documented (Beligni and Lamattina

1999a; Hsu and Kao 2004). It scavenges other reactive intermediates and terminates chain-propagated reactions (Kopyra and Gwóźdz 2003). It is reported to counteract oxidative stress in two different ways. It can either directly scavenge ROS (Laspina et al. 2005) or can also act as a signalling molecule in a cascade of events leading to gene expression changes (Lamattina et al. 2003; Laspina et al. 2005). There are many reports which emphasise that NO counteracts the toxic effects of diquat- or paraquat-induced ROS generation in *Solanum tuberosum* (Beligni and Lamattina 1999b) and *Oryza sativa* L. (Hung et al. 2002). Ferreira et al. (2010) also reported that NO reduces oxidative stress generated by lactofen in *Glycine max* L. Merril.

4.8.4.3 Use of Safeners

Safeners are also known as *antagonists/crop protectants/herbicide antidotes*. They provide selective protection to crop plants from herbicide-induced injury without protecting weeds. Many safeners are structurally similar to the herbicides that they antagonise. Safeners promote herbicide metabolism in crop plants. Some safeners protect crop plants like corn, sorghum and rice against damage from soil-applied thiocarbamate and chloracetanilide herbicides in either of the two ways:

(i) By increasing the levels of glutathione
(ii) By stimulating the activity of glutathione-dependent enzymes such as GST

In Phase II reactions, glutathione conjugates with the herbicide to reduce its toxicity. This mechanism provides greater and selective protection to crops, while weeds still remain susceptible. Herbicide safeners regulate the expression of genes associated with herbicide metabolism to ultimately improve crop tolerance to herbicides (Riechers et al. 2010). With this mode of action, safeners provide a relatively inexpensive and flexible method of improving herbicide selectivity without incurring any significant ecological risks.

Soil-Applied Safeners

These safeners are applied with the seed prior to planting or applied to the soil or crop together with the herbicides. They protect large-seeded grass crops (corn, sorghum, rice) against pre-emergence applications of thiocarbamate and chloracetanilide herbicides. Most evidences suggest that they protect crops by stimulating enzymes which are responsible for metabolising the herbicide. They do not affect herbicide absorption/translocation/site of action. For example, they increase the level of glutathione conjugation of chloracetanilide and thiocarbamate herbicides by inducing the enzyme GST (DeRidder et al. 2002). As a result they may also increase the level of glutathione in the plant tissue.

Foliar-Applied Safeners

These safeners protect grass crops by enhancing cytochrome P450 monooxygenase and glucosyltransferase activity from aryloxyphenoxypropionate, sulphonylurea or imidazolinone herbicide injury. An example is the safener called fenchlorazoleethyl used to protect grass crops (i.e. wheat, rye, barley) from fenoxypropethyl injury.

This class of herbicides generally controls grasses only. The crop species (i.e. wheat, rye, barley) possess elevated levels of glutathione as compared to the grassy weeds. The safener fenchlorazoleethyl causes further enhancement in glutathione levels in the crop species, ultimately protecting them from herbicide damage.

4.9 Conclusion and Future Prospects

Environmental stresses, such as xenobiotics, drought, heavy metals, temperature, salts, etc., are major factors that limit agricultural productivity. Since the last few decades, endless efforts have been made to minimise agricultural losses to satisfy the ever-increasing demand of food. This has led to an unwise use of xenobiotic chemicals like pesticides and herbicides by agriculturists and scientists to develop strategies to enhance the quantity and quality of food production under extremely adverse conditions. In recent years, numerous studies illustrate the harmful effects of rampant use of these chemicals on nontarget organisms (plants and animals), soil microflora and aquatic and terrestrial ecosystems on the whole. Moreover, these harmful chemicals undergo biomagnification as they pass down the food chain. As a result pesticide residues have been traced in raw and fresh fruits, vegetables, fish, etc. leading to many reported cases of poisoning. These undesired consequences have urged scientists to understand metabolic responses of pesticides to plants. Furthermore, efforts are also being made to unveil different molecular mechanisms related to pesticide toxicity. Scientists have successfully designed few new and innovative approaches to ameliorate these pesticide toxicity symptoms, and efforts are still on in this direction. So far, there are some excellent studies that support the exogenous application of plant hormones like salicylic acid and brassinosteroids in alleviating xenobiotic toxicity in plants. Comparison of metabolic pathways, detoxification mechanisms, and tolerance of various plants to pesticides may even help us in the future to design vegetative decontamination systems for treating pesticide-contaminated soil and water. Thus, understanding pesticide metabolism in plants is necessary for developing strategies for amelioration of the harmful effects of pesticides in nontarget plants and also for remediation of contaminated soil, water and atmosphere. In plants, the up-/downregulation of the expression of specific genes involved in processes related to detoxification may even be exploited to decrease pesticide residues in food produce. Although there has been much progress, yet, many components of the complex xenobiotic-sensing and signalling network still remain to be identified. Therefore, there is a need to develop an intra- and interdisciplinary approach for elucidation of physiological, biochemical and biotechnological strategies against pesticide stress to increase plant yield and crop productivity for the world's ever-increasing population.

Acknowledgements The authors are thankful to the Head, Department of Botany, University of Rajasthan, Jaipur, and Head, Department of Botany, University of Allahabad, Allahabad, for their kind cooperation. One of the authors (Neeraj Mishra) wishes to thank the Principal, Atma Ram Sanatan Dharma College, University of Delhi, Dhaula Kuan, New Delhi, for his kind cooperation and sympathetic attitude. The authors would also like to thank Mr. Ravi Prakash Dubey, Ms. Pallawi Upadhyaya and Ms. Priya for all sorts of help and cooperation provided during the preparation of this manuscript.

References

Abdollahi M, Mostafalou S, Pournourmohammadi S, Shadnia S (2004) Oxidative stress and cholinesterase inhibition in saliva and plasma of rats following subchronic exposure to malathion. Comp Biochem Physiol C Toxicol Pharmacol 137:29–34

Agency for Toxic Substances and Registry (ATSDR) (1997) Toxicological profile for chlorpyrifos. Available from http://www.atsdr.cdc.gov/toxprofiles/tp84.pdf

Ahemad M, Khan MS (2010) Ameliorative effects of *Mesorhizobium* sp. MRC4 on chickpea yield and yield components under different doses of herbicide stress. Pestic Biochem Physiol 98:183–190

Ahemad M, Khan MS, Zaidi A, Wani PA (2009) Remediation of herbicides contaminated soil using microbes. In: Khan MS, Zaidi A, Musarrat J (eds) Microbes in sustainable agriculture. Nova Publishers, New York, pp 261–284

Aktar MW, Sengupta D, Chowdhury A (2009) Impact of pesticides use in agriculture: their benefits and hazards. Interdiscip Toxicol 2(1):1–12

Alexandratos N, Bruinsma J (2012) World agriculture towards 2030/2050: the 2012 revision. ESA Working paper No. 12–03. FAO, Rome

Almagro L, Ros LVG, Belchi-Navarro S, Bru R, Barceló AR, Pedreño MA (2009) Class III peroxidases in plant defence reactions. J Exp Bot 60:377–390

Altuntas I, KÂlÂnç I, Orhan H, Demirel R, Koylu H, Delibas N (2004) The effects of diazinon on lipid peroxidation and antioxidant enzymes in erythrocytes in vitro. Hum Exp Toxicol 23:9–13

Ambali SF, Abubakar AT, Shittu M, Yaquub LS, Kobo PI, Giwa A (2010) Ameliorative effects of zinc on chlorpyrifos-induced erythrocyte fragility in Wistar rats. N Y Sci J 3:117–122

Ambali SF, Akanbi D, Igbokwe N, Shittu M, Kawu M, Ayo J (2007) Evaluation of subchronic chlorpyrifos poisoning on haematological and serum biochemical changes in mice and protective effect of vitamin C. J Toxicol Sci 32:111–120

Ananievaa EA, Christova KN, Popova LP (2004) Exogenous treatment with salicylic acid leads to increased antioxidant capacity in leaves of barley plants exposed to Paraquat. J Plant Physiol 161(3):319–328

Anderson MP, Gronwald JW (1991) Atrazine resistance in a velvetleaf (*Abutilon theophrasti*) biotype due to enhanced glutathione-S-transferase activity. Plant Physiol 96:104–109

Apel K, Hirt H (2004) Reactive oxygen species: metabolism, oxidative stress, and signal transduction. Annu Rev Plant Biol 55:373–399

Arrigoni O, Dipierro S, Borraccino G (1981) Ascorbate free radical reductase, a key enzyme of the ascorbic acid system. FEBS Lett 125:242–245

Asada K (2006) Production and scavenging of reactive oxygen species in chloroplasts and their functions. Plant Physiol 141:391–396

Aubert S, Schwitzguébel JP (2004) Screening of plant species for the phytotreatment of wastewater containing sulphonated anthraquinones. Water Res 38:3569–3575

Ayansina ADV (2009) Pesticide use in agriculture and microorganisms. In: Khan MS, Zaidi A, Musarrat J (eds) Microbes in sustainable agriculture. Nova, New York, pp 261–284

Bajguz A, Hayat S (2009) Effects of brassinosteroids on the plant responses to environmental stresses. Plant Physiol Biochem 47:1–8

Ball L, Accotto GP, Bechtold U, Creissen G, Funck D, Jimenez A, Kular B, Leyland N, Mejia-Carranza J, Reynolds H, Karpinski S, Mullineaux PM (2004) Evidence for a direct link between glutathione synthesis and stress defence gene expression in Arabidopsis. Plant Cell 16:2448–2462

Banas´ A, Johansson I, Stenlid G, Stymne S (1993) Free radical scavengers and inhibitors of lipoxygenases as antagonists the herbicides haloxyfop and alloxydim. Swed J Agric Res 23:67

Banas´ A, Johansson I, Stenlid G, Stymne S (2000) Selective increase in acyl hydrolase activity by graminicides in wheat. Biochem Soc Trans 28:777

Baruah M, Mishra RR (1986) Effect of herbicides, butachlor, 2,4-D and oxyfluorfen on enzyme activities and CO_2 evolution in submerged paddy field soil. Plant Soil 96:287–291

Behrendt H, Brüggemann R (1993) Modeling the fate of organic-chemicals in the soil–plant environment—model study of root uptake of pesticides. Chemosphere 27:2325–2332

Belbachir O, Matringe M, Tissut M, Chevallier D (1980) Physiological actions of dinoterb, a phenol derivative: 1. Physiological effects on the whole plant and on tissue fragments of pea. Pestic Biochem Physiol 14:303–308

Beligni MV, Lamattina L (1999a) Nitric oxide counteracts cytotoxic processes mediated by reactive oxygen species in plant tissues. Planta 208:337–344

Beligni MV, Lamattina L (1999b) Nitric oxide protects against cellular damage produced by methylviologen herbicides in potato plants. Nitric Oxide Biol Chem 3:199–208

Boersma L, Lindstrom FT, McFarlane C, McCoy EL (1988) Uptake of organic chemicals by plants: a theoretical study. Soil Sci 146:403–417

Briggs GG, Bromilow RH (1983) Relationships between lipophilicity and the distribution of non-ionized chemicals in barley shoots following uptake by the roots. Pestic Sci 14:492–500

Briggs GG, Bromilow RH, Evans AA (1982) Relationships between lipophilicity and root uptake and translocation of non-ionized chemicals by barley. Pestic Sci 13:495–504

Campanella B, Bock C, Schröder P (2002) Phytoremediation to increase the degradation of PCBs and PCDD/Fs—potential and limitations. Environ Sci Pollut Res 9:73–85

Campanella B, Paul R (2000) Presence, in the rhizosphere and leaf extracts of zucchini (*Cucurbita pepo* L.) and melon (*Cucumis melo* L.) of molecules capable of increasing the apparent aqueous solubility of hydrophobic pollutants. Int J Phytorem 2:145–158

Casida JE, Lykken L (1969) Metabolism of organic pesticide chemicals in higher plants. Annu Rev Plant Physiol 20:607–636

Casida JE, Quistad GB (2004) Organophosphate toxicology: safety aspects of on acetylcholinesterase secondary targets. Chem Res Toxicol 17:983–998

Chandler D, Bailey AS, Tatchell GM, Davidson G, Greaves J, Grant WP (2011) The development, regulation and use of biopesticides for integrated pest management. Phil Trans R Soc B 366:1987–1998

Chen F, Wang F, Wu F, Mao W, Zhang G, Zhou M (2010a) Modulation of exogenous glutathione in antioxidant defense system against Cd stress in the two barley genotypes differing in Cd tolerance. Plant Physiol Biochem 48:663–672

Chen WW, Yang JL, Qin C, Jin CW, Mo JH, Ye T et al (2010b) Nitric oxide acts downstream of auxin to trigger root ferric-chelate reductase activity in response to iron deficiency in Arabidopsis. Plant Physiol 154:810–819

Cocker KM, Northcroft DS, Coleman JOD, Moss SR (2001) Resistance to ACCase-inhibiting herbicides and isoproturon in UK populations of *Lolium multiflorum*: mechanisms of resistance and implications for control. Pest Manag Sci 57:587–597

Colborn T, Dumanoski D, Myers JP (1996a) Our stolen future. Dutton, Toronto

Colborn T, Dumanoski D, Myers JP (1996b) Our stolen future: are we threatening our fertility, intelligence, and survival? A scientific detective story. New York: Dutton, p 306, ISBN 0-452-27414-1

Cole DJ (1985) Mode of action of glyphosate—a literature analysis. In: Grossbard E, Atkinson D (eds) The herbicide glyphosate. Butterworths, London, pp 48–74
Cole DJ, Edwards R (2000) Secondary metabolism of agrochemicals in plants. In: Robert TR (ed) Agrochemicals and plant protection. Wiley, Chichester, pp p107–p154
Coleman JOD, Randall RA, Blake-Kalff MMA (1997) Detoxification of xenobiotics by plants: chemical modification and vacuolar compartmentation. TIPS 2:144–151
Cosio C, Durand CH (2009) Specific functions of individual class III peroxidase genes. J Exp Bot 60:391–408
Cui J, Zhang R, Wu GL, Zhu HM, Yang H (2010) Salicylic acid reduces napropamide toxicity by preventing its accumulation in rapeseed (*Brassica napus* L.). Arch Environ Contam Toxicol 59:100–108
Cummins I, Cole DJ, Edwards R (1999) A role for glutathione transferases functioning as glutathione peroxidases in resistance to multiple herbicides in black-grass. Plant J 18:285–292
Dandapani M, Zachariah A, Kavitha MR, Jeyaseelan L, Oommen A (2003) Oxidative damage in intermediate syndrome of acute organophosphorus poisoning. Ind J Med Res 117:253–259
Dayan FE, Duke SO, Grossmann K (2010) Herbicides as probes in plant biology. Weed Sci 58:340–350
Dayan FE, Watson SB (2011) Plant cell membrane as a marker for light-dependent and light-independent herbicide mechanisms of action. Pestic Biochem Physiol 101:182–190
de Oliveira TA, Ronchi-Teles B, da Fonseca CRV, da Silva SLR, Santos PA, Nunez CV (2012) Insecticidal activity of *Vitex cymosa* (Lamiaceae) and *Eschweilera pedicellata* (Lecythidaceae) extracts against *Sitophilus zeamais* adults (Curculionidae). Emir J Food Agric 24:49–56
DeRidder BP, Dixon DP, Beussman DJ, Edwards R, Goldsbrough PB (2002) Induction of glutathione *S*-transferases in Arabidopsis by herbicide safeners. Plant Physiol 130(3):1497–1505
Desikan R, Mackerness SAH, Hancock JT, Neill SJ (2001) Regulation of the *Arabidopsis* transcriptome by oxidative stress. Plant Physiol 127:159–172
Desikan R, Reynolds A, Hancock JT, Neill SJ (1998) Harpin and hydrogen peroxide both initiate programmed cell death but have differential effects on defence gene expression in *Arabidopsis* suspension cultures. Biochem J 330:115–120
Dhillion PS, Byrnes WR, Merritt C (1968) Simazine distribution and degradation in red pine seedlings. Weed Sci 16:374–376
Dietz KJ (2003) Redox control, redox signaling, and redox homeostasis in plant cells. Int Rev Cytol 228:141–193
Ding CK, Wang CY, Gross KC, Smith DL (2002) Jasmonate and salicylate induce the expression of pathogenesis-related-protein genes and increase resistance to chilling injury in tomato fruit. Planta 214:895–901
Doran PM (2005) Application of plant tissue cultures in phytoremediation research: incentives and limitations. Biotechnol Bioeng 103:60–76
Eberbach PL, Douglas LA (1991) Effect of herbicide residues in a sandy loam on the growth, nodulation and nitrogenase activity (C_2H_2/C_2H_4) of *Trifolium subterraneum*. Plant Soil 131:67–76
Elbaz A, Wei YY, Meng Q, Zheng Q, Yang ZM (2010) Mercury-induced oxidative stress and impact on antioxidant enzymes in *Chlamydomonas reinhardtii*. Ecotoxicology 19:1285–1293
Ellis JP, Camper ND (1995) In vitro cultured cocklebur (*Xanthium strumarium* L.) responses to dimercaptopropanesulfonic acid and monosodium methanearsonate. J Plant Growth Reg 14:9–13
Ezra G, Stephenson GR (1985) Comparative metabolism of atrazine and EPTC in proso millet (*Panicum miliaceum* L.) and corn. Pestic Biochem Physiol 24:207–212
Feierabend J, Winkelhüsener T (1982) Nature of photooxidative events in leaves treated with chlorosis-inducing herbicides. Plant Physiol 70(5):1277–1282

Ferreira LC, Cataneo AC, Remaeh LMR, Corniani N, Fumis TF, de Souza YA, Scavroni J, Soares BJA (2010) Nitric oxide reduces oxidative stress generated by lactofen in soybean plants. Pestic Biochem Physiol 97:47–54

Foyer CH, Lopez-Delgado H, Dat JF, Scott IM (1997) Hydrogen peroxide and glutathione-associated mechanisms of acclamatory stress tolerance and signaling. Physiol Plant 100:241–254

Foyer CH, Noctor G (2005a) Redox homeostasis and antioxidant signalling: a metabolic link between stress perception and physiological responses. Plant Cell 17:1866–1875

Foyer CH, Noctor G (2005b) Oxidant and antioxidant signalling in plants: a re-evaluation of the concept of oxidative stress in a physiological context. Plant Cell Environ 28:1056–1071

Foyer C, Noctor G (2001) The molecular biology and metabolism of glutathione. In: Grill D, Tausz M, De Кок L (eds) Significance of glutathione in plant adaptation to the environment, vol 2, Handbook series of plant ecophysiology. Kluwer, Dordrecht, pp 27–56

Franco R, Sanchez-Olea R, Reyes-Reyes EM, Panayiotidis MI (2009) Environmental toxicity, oxidative stress and apoptosis: ménage à trois. Mutat Res 674:3–22

Franco R, Li S, Rodriguez-Rocha H, Burns M, Panayiotidis MI (2010) Molecular mechanisms of pesticide-induced neurotoxicity: relevance to Parkinson's disease. Chem Biol Interact 188:289–300

Freedman B (1995) Environmental ecology. Academic, New York

Freeman JL, Persans MW, Nieman K, Albrecht C, Peer W, Pickering IJ, Salt DE (2004) Increased glutathione biosynthesis plays a role in nickel tolerance in *Thlaspi* nickel hyperaccumulators. Plant Cell 16:2176–2191

Fuerst EP, Norman MA (1991) Interactions of herbicides with photosynthetic electron transport. Weed Sci 39(3):458–464

Fufezan C, Rutherford AW, Krieger-Liszkay A (2002) Singlet oxygen production in herbicide-treated photosystem II. FEBS Lett 532:407–410

Gangwar S, Singh VP, Tripathi DK, Chauhan DK, Prasad SM, Maurya JN (2014) Plant responses to metal stress: the emerging role of plant growth hormones in toxicity alleviation. In: Ahmad P (ed) Emerging technologies and management of crop stress tolerance, 2. Elsevier Academic Press, San Diego, pp 215–248

Gechev TS, Hille J (2005) Hydrogen peroxide as a signal controlling plant programmed cell death. J Cell Biol 168:17–20

Geoffroy L, Teisseire H, Couderchet M, Vernet G (2002) Effect of oxyfluorfen and diurnal one and in mixture on antioxidative enzymes of *Scenedesmus obliquus*. Pestic Biochem Physiol 72:178–185

Gill SS, Khan NA, Anjum NK, Tuteja N (2011) Amelioration of cadmium stress in crop plants by nutrients management: morphological, physiological and biochemical aspects. In: Anjum NA, Lopez-Lauri F (ed) Plant nutrition and abiotic stress tolerance III. Plant Stress 5(1):1–23

Giridhar P, Indira P (1997) Effects of an organophosphorous Nuvan on total lipids and lipase activity of the fresh water fish *Labeo rohita* (Hamilton). Ind J Comp Anim Physiol 15:37–40

Gomez LD, Vanacker H, Buchner P, Noctor G, Foyer CH (2004) Intercellular distribution of glutathione synthesis in maize leaves and its response to short-term chilling. Plant Physiol 134:1662–1671

Gondim FA, Gomes-Filho E, Costa JH, Mendes Alencar NL, Prisco JT (2012) Catalase plays a key role in salt stress acclimation induced by hydrogen peroxide pretreatment in maize. Plant Physiol Biochem 56:62–71

González-Rodríguez RM, Rial-Otero R, Cancho-Grande B, Gonzalez-Barreiro C, Simal-Gándara J (2011) A review on the fate of pesticides during the processes within the food-production chain. Crit Rev Food Sci Nutr 51:99–114

González-Rodríguez RM, Rial-Otero R, Cancho-Grande B, Simal-Gándara J (2008) Occurrence of fungicide and insecticide residues in trade samples of leafy vegetables. Food Chem 107:1342–1347

Grover R, Cessna AJ (eds) (1991) Environmental chemistry of herbicides. CRC Press, Boca Raton
Grzam A, Martin MN, Hell R, Meyer AJ (2007) c-Glutamyl transpeptidase GGT4 initiates vacuolar degradation of glutathione-S-conjugates in *Arabidopsis*. FEBS Lett 581:3131–3138
Guo YP, Zhou HF, Zhang LC (2006) Photosynthetic characteristics and protective mechanisms against photooxidation during high temperature stress in two citrus species. Sci Hortic 108:260–267
Gupta RC (2004) Brain regional heterogeneity and toxicological mechanisms of organophosphates and carbamates. Toxicol Mech Methods 14:103–143
Hall JC, Hoagland RE, Zoblotowicz RM (eds) (2001) Pesticide biotransformation in plants and microorganisms: similarities and divergences, vol 777, ACS symposium series. American Chemical Society, Washington, DC, p 432
Hatton PJ, Cummins I, Cole DJ, Edwards R (1999) Glutathione transferases involved in herbicide detoxification in the leaves of *Setaria faberi* (giant foxtail). Physiol Plant 105:9–16
Hatzios KK (1991) Bio-transformations of herbicides in higher plants. In: Grover R, Cessna AJ (eds) Environmental chemistry of herbicides. CRC Press, Boca Raton, pp 141–185
Hatzios KK (2001) Functions and regulations of glutathione S-transferases. In: Hall JC, Hoagland RE, Zoblotowicz RM (eds) Pesticide biotransformation in plants and microorganisms: similarities and divergences, vol 777, ACS symposium series. American Chemical Society, Washington, DC, pp 218–239
Hayes JD, Wolf CR (1988) Role of glutathione transferase in drug resistance. In: Sies H, Ketterer B (eds) Glutathione conjugation: mechanisms and biological significance. Academic, London, pp 316–356
Hess FD (2000) Light-dependent herbicides: an overview. Weed Sci 48:160–170
Hoagland RE, Zablotowicz RM (2001) The role of plant and microbial hydrolytic enzymes in pesticide metabolism. In: Hall JC, Hoagland RE, Zoblotowicz RM (eds) Pesticide biotransformation in plants and microorganisms: similarities and divergences, vol 777, ACS symposium series. American Chemical Society, Washington, DC, pp 58–88
Hopf FR, Whitten DG (1978) Porphyrins and metalloporphyrins. In: Dolphin D (ed) The porphyrins. Academic, New York, pp p161–p195
Hossain MA, Asada K (1984) Purification of dehydroascorbate reductase from spinach and its characterization as thiol enzyme. Plant Cell Physiol 25:85–92
Hsu YT, Kao CH (2004) Cd toxicity is reduced by nitric oxide in rice leaves. Plant Growth Regul 42:227–238
Hung KT, Chang CJ, Kao CH (2002) Paraquat toxicity is reduced by nitric oxide in rice leaves. J Plant Physiol 159:159–166
Infante-Rivard C, Labuda D, Krajinovi M, Sinnett D (1999) Risk of childhood leukaemia with exposure to pesticides and with gene polymorphisms. Epidemiology 10:481–489
Ishikawa T, Casini AF, Nishikimi M (1998) Molecular cloning and functional expression of rat liver glutathione-dependent dehydroascorbate reductase. J Biol Chem 273:28708–28712
Ivanov SV, Alexieva VS, Karanov EN (2005) Cumulative effect of low and high atrazine concentrations on *Arabidopsis thaliana* plants. Russ J Plant Physiol 52(2):213–219
Jaga K, Dharmani C (2003) Sources of exposure to and public health implications of organophosphate pesticides. Pan Am J Public Health 14:171–185
Jakoby WB, Ziegler DM (1990) The enzymes of detoxication. J Biol Chem 265(34):20715–20718
Jaworski EG (1972) Mode of action of N-phosphonomethylglycine: inhibition of aromatic amino acid biosynthesis. J Agric Food Chem 20:1195–1198
Jensen KIN, Stephenson GR, Hunt LA (1977) Detoxification of atrazine in three Gramineae subfamilies. Weed Sci 25:212–220
Jiang YP et al (2012) Cellular glutathione redox homeostasis plays an important role in the brassinosteroid-induced increase in CO_2 assimilation in *Cucumis sativus*. New Phytol 194:932–943

Verkleij JAC, Golan-Goldhirsh A, Antosiewisz DM, Schwitzguebel J-P, Schroder P (2009) Dualities in plant tolerance to pollutants and their uptake and translocation to the upper plant parts. Environ Exp Bot 67:10–22

Joshi SC, Gulati N, Gajraj A (2005) Evaluation of toxic impacts of mancozeb on testis in rats. Asian J Exp Sci 19:73–83

Kamanyire R, Karalliedde L (2004) Organophosphate toxicity and occupational exposure: in-depth review. Occup Med 54:69–75

Kanazawa S, Sano S, Koshiba T, Ushimaru T (2000) Changes in antioxidative in cucumber cotyledons during natural senescence: comparison with those during dark-induced senescence. Plant Physiol 109:211–216

Karpinska B, Karlsson M, Schinkel H, Streller S, Su¨ss KH, Melzer M et al (2001) A novel superoxide dismutase with a high isoelectric point in higher plants. Expression, regulation, and protein localization. Plant Physiol 126:1668–1677

Karpinski S, Reynolds H, Karpinska B, Wingsle G, Creissen G, Mullineaux P (1999) Systemic signaling and acclimation in response to excess excitation energy in *Arabidopsis*. Science 284:654–657

Karthikeyan R, Davis LC, Erickson LE, Al-Khatib K, Kulakow PA, Barnes PL, Hutchinson SL, Nurzhanova AA (2004) Studies on responses of non-target plants to pesticides: a review. Available at http://www.engg.ksu.edu/HSRC/karthipesticide.pdf

Katerova ZI, Miteva LPE (2010) Glutathione and herbicide resistance in plants In: Anjum NA (eds) Ascorbate-Glutathione pathway and stress tolerance in plants. doi: 10.1007/978-90-481-9404-9_6

Kaushik S, Inderjit S (2006) Phytotoxicity of selected herbicides to mung bean (*Phaseolus aureus* Roxb.). Environ Exp Bot 55:41–48

Kawano T (2003) Roles of the reactive oxygen species-generating peroxidase reactions in plant defense and growth induction. Plant Cell Rep 21:829–837

Kenaga EE, Whitnoy WK, Hardy JL, Doty AE (1965) Laboratory test with dursban insecticide. J Econ Entomol 58:1043–1058

Khan NA, Singh S (eds) (2008) Abiotic stress and plant responses. I K International, New Delhi

Kingston-Smith AH, Foyer CH (2000) Bundle sheath proteins are more sensitive to oxidative damage than those of the mesophyll in maize leaves exposed to paraquat or low temperatures. J Exp Bot 51:123–130

Kleijn G, Snoeijing IJ (1997) Field boundary vegetation and the effects of agrochemical drift: botanical change caused by low levels of herbicide and fertilizer. J Appl Ecol 34(6):1413–1425

Knight H, Knight MR (2001) Abiotic stress signaling pathways: specificity and cross-talk. Trends Plant Sci 6:262–267

Kopyra M, Gwózdz EA (2003) Nitric oxide stimulates seed germination and counteracts the inhibitory effect of heavy metals and salinity on root growth of *Lupinus luteus*. Plant Physiol Biochem 41:1011–1017

Korte F, Kvesitadze G, Ugrekhelidze D, Gordeziani M, Khatisashvili G, Buadze O, Zaalishvili G, Coulston F (2000) Organic toxicants and plants. Ecotoxicol Environ Saf 47:1–26

Kozawa K, Aoyama Y, Mashimo S, Kimura H (2009) Toxicity and actual regulation of organophosphate pesticides. Toxicol Rev 28:245–254

Kranner I, Birtic S, Anderson KM, Pritchard HW (2006) Glutathione half-cell reduction potential: a universal stress marker and modulator of programmed cell death? Free Radic Biol Med 40:2155–2165

Kumar S, Singh A (2015) Biopesticides: present status and the future prospects. J Biofertil Biopestici 6:e129

Lamattina L, Garcia-Mata C, Graziano M, Pagnussat G (2003) Nitric oxide: the versatility of an extensive signal molecule. Annu Rev Plant Biol 54:109–136

Laspina NV, Groppa MD, Tomaro ML, Benavides MP (2005) Nitric oxide protects sunflower leaves against Cd-induced oxidative stress. Plant Sci 169:323–330

Latowski D, Surówka E, Strzaka K (2010) Regulatory role of components of ascorbate–glutathione pathway in plant stress tolerance. In: Anjum NA (ed) Ascorbate-Glutathione pathway and stress tolerance in plants. doi: 10.1007/978-90-481-9404-9_1

Leavitt JRC, Penner D (1979) In-vitro conjugation of glutathione and other thiols with acetanilide herbicides and EPTC sulfoxide and the action of the herbicide antidote R-25788. J Agric Food Chem 27:533–536

Letouze A, Gasquez J (2003) Enhanced activity of several herbicide-degrading enzymes: a suggested mechanism responsible for multiple resistance in blackgrass (*Alopecurus myosuroides* Huds.). Agronomie 23:601–608

Levine A, Tenhaken R, Dixon R, Lamb C (1994) H_2O_2 from the oxidative burst orchestrates the plant hypersensitive disease resistance response. Cell 79:583–593

Li J, Zhao X, Matsui S, Maezawa S (2004) Changes in antioxidative levels in aging cucumber (*Cucumis sativus* L.) leaves. J Jpn Soc Hort Sci 73:491–495

Liu H, Weisman D, Ye YB, Cui B, Huang YH, Colo´ n-Carmona A, Wang ZH (2009) An oxidative stress response to polycyclic aromatic hydrocarbon exposure is rapid and complex in *Arabidopsis thaliana*. Plant Sci 176:375–382

López-Pérez GC, Arias-Estévez M, Ló pez-Periago E, Soto-Gonzalez B, Cancho-Grande B, Simal-Gandara J (2006) Dynamics of pesticides in potato crops. J Agric Food Chem 54:1797–1803

Luscombe BM, Pallett KE, Loubiere P, Millet J, Melgorejo J, Vrabel TE (1995) RPA 201772 a novel herbicide for broad leaf and grass weed control in maize and sugarcane. Proc Brighton Crop Prot Conf Weeds 2:35–42

Mahajan S, Tuteja N (2005) Cold, salinity and drought stresses: an overview. Arch Biochem Biophys 444:139–158

Mannervik B, Danielson UH (1988) Glutathione transferases: structure and catalytic activity. CRC Crit Rev Biochem 23:283–337

Marcacci S, Raveton M, Ravanel P, Schwitzgúebel JP (2006) Conjugation of atrazine in vetiver (*Chrysopogon zizanioides* Nash) grown in hydroponics. Environ Exp Bot 56:205–215

Mathur SC (1999) Future of Indian pesticides industry in next millennium. Pestic Inf 24(4):9–23

Matringe M, Camadro JM, Labbe P, Scalla R (1989) Protoporphyrinogen oxidase as a molecular target for diphenyl ether herbicides. Biochem J 260:231–235

Mbakaya CFL, Ohayo-Mitoko GJA, Ngowi AVF, Mbabazi R, Simwa JM, Maeda DN, Stephens J, Hakuza H (1994) The status of pesticide usage in East Africa. Afr J Health Sci 1:37–41

Mena S, Ortega A, Estrela JM (2009) Oxidative stress in environmental-induced carcinogenesis. Mutat Res 674:36–44

Meriga B, Reddy BK, Rao KR, Reddy LA, Kishor PBK (2004) Aluminium-induced production of oxygen radicals, lipid peroxidation and DNA damage in seedlings of rice (Oryza sativa). J Plant Physiol 161(1):63–68

Meyer A, Hell R (2005) Glutathione homeostasis and redox-regulation by sulfhydryl groups. Photosynth Res 86:435–457

Mishra S, Jha AB, Dubey RS (2011) Arsenite treatment induces oxidative stress, upregulates antioxidant system, and causes phytochelatin synthesis in rice seedlings. Protoplasma 248(3):565–577

Mishra V, Srivastava G, Prasad SM, Abraham G (2008) Growth, photosynthetic pigments and photosynthetic activity during seedling stage of cowpea (*Vigna unguiculata*) in response to UV-B and dimethoate. Pestic Biochem Physiol 92:30–37

Miteva L (2005) Glutathione and glutathione S-transferase – physiological function in plants, grown on normal and stress conditions. Plant Sci 42:483–494

Mittler R (2002) Oxidative stress, antioxidants and stress tolerance. Trends Plant Sci 7:405–410

Mittler R, Vanderauwera S, Gollery M, Van Breusegem F (2004) Reactive oxygen gene network of plants. Trends Plant Sci 9:490–498

Morant M, Bak S, LindbergMoller B, Werck-Reichhart D (2003) Plant cytochromes P450: tools for pharmacology, plant protection and phytoremediation. Curr Opin Biotechnol 14:151–162

Moreland DE, Blackmon WJ, Todd HG, Farmer FS (1970) Effects of diphenylether herbicides on reactions of mitochondria and chloroplasts. Weed Sci 18:636–641

Mou Z, Fan W, Dong X (2003) Inducers of plant systemic acquired resistance regulate NPR1 function through redox changes. Cell 113:935

Mougin CP, Corio-Costet MF, Werck-Reichhart D (2001) Plant and fungal cytochrome P-450s: their role in pesticide transformation. In: Hall JC, Hoagland RE, Zoblotowicz RM (eds) Pesticide biotransformation in plants and microorganisms: similarities and divergences, vol 777, ACS symposium series. American Chemical Society, Washington, DC, pp 166–181

Murthy GP, Prasad GM, Sudarshana MS (2005) Toxicity of different imbibition periods of dimethoate on germination, chlorophyll *a*/*b* and dry matter of *Glycine max* (L.) Merrill. cv. KHSB-2, during early seedling growth. J Physiol Res 18(2):199–201

Myouga F, Hosoda C, Umezaw T, Iizumi H, Kuromori T, Motohashid R et al (2008) A heterocomplex of iron superoxide dismutases defends chloroplast nucleoids against oxidative stress and is essential for chloroplast development in *Arabidopsis*. Plant Cell 20:3148–3162

Nemat Alla M, Badawi A-HM, Hassan NM, El-Bastawisy ZM, Badran EG (2008) Herbicide tolerance in maize is related to increased levels of glutathione and glutathione-associated enzymes. Acta Physiol Plant 30:371–379

Noctor G, Foyer CH (1998) Ascorbate and glutathione: keeping active oxygen under control. Annu Rev Plant Physiol Plant Mol Biol 49:249–279

Noctor G, Gomez L, Vanacker H, Foyer CH (2002) Interactions between biosynthesis, compartmentation and transport in the control of glutathione homeostasis and signalling. J Exp Bot 53:1283–1304

Noctor G, Veljovic-Jovanovic S, Foyer CH (2000) Peroxide processing in photosynthesis: antioxidant coupling and redox signalling. Philos Trans R Soc B Biol Sci 355:1465–1475

Norris LA, Freed VH (1966a) The absorption and translocation characteristics of several phenoxyalkyl acid herbicides in bigleaf maple. Weed Res 6(3):203–211

Norris LA, Freed VH (1966b) The metabolism of a series of chlorophenoxyalkyl acid herbicides in bigleaf maple. Weed Res 6(3):212–220

Norris LA, Freed VH (1966c) The absorption, translocation, and metabolism characteristics of 4-(2,4-dichlorophenoxy)butyric acid in big leaf maple. Weed Res 6(4):283–291

Obrigawitch TT, Cook G, Wetherington J (1998) Assessment of effects on non-target plants from sulfonylurea herbicides using field approaches. Pestic Sci 52:199–217

Oerke EC (2006) Crop losses to pests. J Agri Sci 144:31

Oettmeier W, Soil H-J, Neumann E (1984) Herbicide and plastoquinone binding to photosystem II. Z Naturforsch 39(c):393–396

Ogutcu A, Suludere Z, Kalender Y (2009) Dichlorvos-induced hepatotoxicity in rats and the protective effects of vitamins C and E. Environ Toxicol Pharmacol 26:355–361

Ogweno JO, Song XS, Hu WH, Shi K, Zhou YH, Yu JQ (2009) Detached leaves of tomato differ in their photosynthetic physiological response to moderate high and low temperature stress. Sci Hortic 123:17–22

Ohkama-Ohtsu N, Radwan S, Peterson A, Zhao P, Badr AF, Xiang C, Oliver DJ (2007) Characterization of the extracellular c-glutamyl transpeptidases, GGT1 and GGT2, in *Arabidopsis*. Plant J 49:865–877

Orr GL, Hess FD (1982) Mechanism of action of the diphenyl ether herbicide acifluorfen-methyl in excised cucumber (*Cucumis sativus* L.) cotyledons: light activation and the subsequent formation of lipophilic free radicals. Plant Physiol 69:502–507

Passardi F, Cosio C, Penel C, Dunand C (2005) Peroxidases have more functions than a Swiss army knife. Plant Cell Rep 24:255–265

Pastori GM, Kiddle G, Antoniw J, Bernard S, Veljovic-Jovanovic S, Verrier PJ, Noctor G, Foyer CH (2003) Leaf vitamin C contents modulate plant defense transcripts and regulate genes controlling development through hormone signaling. Plant Cell 15:1212–1226

Peixoto F, Alves-Fernandes D, Santos D, Fontaínhas-Fernandes A (2006) Toxicological effects of oxyfluorfen on oxidative stress enzymes in tilapia *Oreochromis niloticus*. Pest Biochem Physiol 85:91–96

Pignocchi C, Foyer CH (2003) Apoplastic ascorbate metabolism and its role in the regulation of cell signaling. Curr Opin Plant Biol 6:379–389

Pimental D, Levitan L (1986) Pesticides: amounts applied and amounts reaching pests. Biosciences 36:86–91

Pope CN (1999) Organophosphorus pesticides: do they all have the same mechanism of toxicity? J Toxicol Environ Health 2:161–181

Popova LP, Ananieva EA, Alexieva VS (2002) Treatment with salicylic acid decreases the effects of paraquat on photosynthesis. J Plant Physiol 159:685–693

Popova LP, Ananieva EA, Christov KN (2004) Exogenous treatment with salicylic acid leads to increased antioxidant capacity in leaves of barley plants exposed to Paraquat. J Plant Physiol 161:319–328

Pritchard MK, Warren GF, Dilley RA (1980) Site of action of oxyfluorfen. Weed Sci 28(6):640–645

Qiu BS, Liu JY, Liu ZL, Liu SX (2002) Distribution and ecology of the edible cyanobacterium Ge-Xian-Mi (*Nostoc*) in rice fields of Hefeng county in China. J Appl Phycol 14:423–429

Radwan DEM (2012) Salicylic acid induced alleviation of oxidative stress caused by clethodim in maize (*Zea mays* L.) leaves. Pestic Biochem Physiol 102:182–188

Ramel F, Sulmon C, Bogard M, Coue´e I, Gouesbet G (2009) Differential dynamics of reactive oxygen species and antioxidative mechanisms during atrazine injury and sucrose-induced tolerance in *Arabidopsis thaliana* plantlets. BMC Plant Biol 9:28

Ramel F, Sulmon C, Cabello-Hurtado F, Taconnat L, Martin-Magniette ML, Renou JP, El Amrani A, Coue´e I, Gouesbet G (2007) Genome-wide interacting effects of sucrose and herbicide-mediated stress in *Arabidopsis thaliana*: novel insights into atrazine toxicity and sucrose-induced tolerance. BMC Genomics 8:450

Rendon JL, Pardo JP, Mendoza-Hernandez G, Rojo-Dominguez A, Hernandez-Arana A (1995) Denaturing behavior of glutathione reductase from cyanobacterium *Spirulina maxima* in guanidine hydrochloride. Arch Biochem Biophys 318:264–270

Riechers DE, Kreuz K, Zhang Q (2010) Detoxification without intoxication: herbicide safeners activate plant defense gene expression. Plant Physiol 153:3–13

Roberts T (ed) (2000) Metabolism of agrochemicals in plants. Wiley, New York, p 300

Rodrigues BN, Almeida FLS (1998) Guia de Herbicidas. IAPAR, Londrina, p 648

Ruiz JM, Blumwald E (2002) Salinity-induced glutathione synthesis in *Brassica napus*. Planta 214:965–969

Rushmore TH, Pickett CB (1993) Glutathione *S*-transferase: structure, regulation, and therapeutic implications. J BiolChem 268:11475–11478

Ryter SW, Kim HP, Hoetzel A, Park JW, Nakahira K, Wang X, Choi AM (2007) Mechanisms of cell death in oxidative stress. Antioxid Redox Signal 9:49–89

Sachs RM, Michael JL (1971) Comparative phytotoxicity among four arsenical herbicides. Weed Sci 19:558–564

Salin ML (1988) Toxic oxygen species and protective systems of the chloroplast. Physiol Plant 72:681–689

Sandermann H (1994) Higher plant metabolism of xenobiotics: the green liver concept. Pharmacogenetics 4:225–241

Sandermann H Jr (1992) Plant metabolism of xenobiotics. Trends Biochem Sci 17:82–84

Sandermann H Jr, Hertkorn N, May RG, Lange BM (2001) Bound pesticidal residues in crop plants: chemistry, bioavailability, and toxicology. In: Hall JC, Hoagland RE, Zoblotowicz RM (eds) Pesticide biotransformation in plants and microorganisms: similarities and divergences, vol 777, ACS symposium series. American Chemical Society, Washington, DC, pp 119–128

Santi R, de Pietri-Tonelli P (1959) Mode of action and biological properties of the S-(Methylcarbanyl) methyl O, O-dimethyldithiophosphate. Nature 183:398–399

Saraf M, Sood N (2002) Influence of monocrotophos on growth oxygen uptake and exopolysaccharide production of *Rhizobium* NCIM 2771 on chickpea. J Indian Bot Soc 82:157–164
Sasse JM (2003) Physiological actions of brassinosteroids: an update. J Plant Growth Regul 22:276–288
Saulsbury MD, Heyliger SO, Wang K, Johnson DJ (2009) Chlorpyrifos induces oxidative stress in oligodendrocyte progenitor cells. Toxicology 259:1–9
Schroder P, Collins CJ (2002) Conjugating enzymes involved in xenobiotic metabolism of organic xenobiotics in plants. Int J Phytorem 4:247–265
Schroder P, Scheer C, Belford EJD (2001) Metabolism of organic xenobiotics in plants: conjugating enzymes and metabolic endpoints. Minerva Biotechnol 13:85–91
Senseman SA (2007) Herbicide handbook, 9th edn. Weed Science Society of America, Lawrence
Sergiev IG, Alexieva VS, Ivanov SV, Moskova II, Karanov EN (2006) The phenylurea cytokinin 4PU-30 protects maize plants against glyphosate action. Pest Biochem Physiol 85:139–146
Shah K, Kumar RG, Verma S, Dubey RS (2001) Effect of cadmium on lipid peroxidation, superoxide anion generation and activities of antioxidant enzymes in growing rice seedlings. Plant Sci 161(6):1135–1144
Shao HB, Chu LY, Lu ZH, Kang CM (2008) Primary antioxidant free radical scavenging and redox signaling pathways in higher plant cells. Int J Biol Sci 4:8–14
Sharma I, Bhardwaj R, Pati PK (2013) Stress modulation response of 24-epibrassinolide against imidacloprid in an elite indica rice variety Pusa Basmati-1. Pestic Biochem Physiol 105:144–153
Sharma P, Dubey RS (2005) Drought induces oxidative stress and enhances the activities of antioxidant enzymes in growing rice seedlings. Plant Growth Regul 46(3):209–221
SharmA Y, Bashir S, Irshad M, Gupta SD, Dogra TD (2005) Effects of acute dimethoate on antioxidant status of liver and brain of experimental rats. Toxicology 206:49–57
Shittu M, Ayo JO, Ambali SF, Fatihu MY, Onyeanusi BI, Kaw MU (2012) Chronic chlorpyrifos-induced oxidative changes in the testes and pituitary gland of Wistar rats: Ameliorative effects of vitamin C. Pestic Biochem Physiol 102:79–85
Shivaramaiah HM, Kennedy IR (2006) Biodegradation of endosulfan by a soil bacterium. J Environ Sci Health B 41:895–905
Slotkin TA (2004) Cholinergic systems in brain development and disruption by neurotoxicants: nicotine, environmental tobacco smoke, organophosphates. Toxicol Appl Pharmacol 198:132–151
Slotkin TA (2005) Developmental neurotoxicity of organophosphates: a case study of chlorpyrifos. In: Gupta RC (ed) Toxicity of organophosphate and carbamate pesticides. Elsevier Academic Press, San Diego, pp p293–p314
Smirnoff N (2000) Ascorbic acid: metabolism and functions of a multi-facetted molecule. Curr Opin Plant Biol 3:229–235
Smith C (2001) Pesticide exports from US ports, 1997–2000. Int J Occup Environ Health 74:266–274
Song NH, Chen L, Yang H (2008) Effect of dissolved organic matter on mobility and activation of chlorotoluron in soil and wheat. Geoderma 146:344–352
Song NH, Yang ZM, Zhou LX, Wu X, Yang H (2006) Effect of dissolved organic matter on the toxicity of chlorotoluron to *Triticum aestivum*. J Environ Sci 18:101–108
Song NH, Yin X, Chen GF, Yang H (2007) Biological responses of wheat (*Triticum aestivum*) plants to the herbicide chlorotoluron in soils. Chemosphere 69:1779–1787
Sparks TC, Nauen R (2015) IRAC: mode of action classification and insecticide resistance management. Pestic Biochem Physiol 121:122–128
Steenland K, Dich RB, Howell RJ, Chrislip DW, Hines CJ, Reid TM, Lehman E, Laber P, Krieg EF, Knot C (2000) Neurologic function among termiticide applicators exposed to chlorpyrifos. Environ Health Perspect 108:293–300
Szalai G, Kellős T, Kocsy G (2009) Glutathione as an antioxidant and regulatory molecule in plants under abiotic stress condition. Plant Growth Regul 28:66–80

Tebourbi O, Sakly M, Rhouma KB (2011) Molecular mechanisms of pesticide toxicity. In: Stoytcheva M (ed) Pesticides in the modern world – Pests control and pesticides exposure and toxicity assessment. ISBN: 978-953-307-457-3

Teixeira MC, Duque P, Sa´-Correia I (2007) Environmental genomics: mechanistic insights into toxicity of and resistance to the herbicide 2,4-D. Trends Biotechnol 25:363–370

Thenaken R, Levine A, Brisson LF, Dixon RA, Lamb C (1995) Function of the oxidative burst in hypersensitive disease resistance. Proc Natl Acad Sci U S A 92:4158–4163

Theodolou F (2000) Plant ABC transporters. Biochim Biophys Acta 1465:79–103

Thounaojam TC, Panda P, Mazumdar P, Kumar D, Sharma GD, Sahoo L et al (2012) Excess copper induced oxidative stress and response of antioxidants in rice. Plant Physiol Biochem 53:33–39

Tsuchida S, Sato K (1992) Glutathione *S*-transferases and cancer. CRC Crit Rev Biochem Mol Biol 27:337–384

Tuteja N (2007) Mechanisms of high salinity tolerance in plants. Meth Enzymol: Osmosens Osmosignal 428:419–438

Ulmasov T, Ohimiya A, Hagen G, Guilfoyle T (1995) The soybean GH2/4 gene that encodes a glutathione -*S*-transferase has a promoter that is activated by a wide range of chemical agents. Plant Physiol 108:919–998

United Nations, Department of Economic and Social Affairs, Population Division (2015) World population prospects: The 2015 revision, key findings and advance tables. Working Paper No. ESA/P/WP.241

Unver T, Bakar M, Shearman RC, Budak H (2010) Genome-wide profiling and analysis of *Festuca arundinacea* miRNAs and transcriptomes in response to foliar glyphosate application. Mol Gen Genomics 283:397–413

Van derWerf HMG (1996) Assessing the impact of pesticides on the environment. Agric Ecosys Environ 60:81–96

Vasileva V, Ilieva A (2007) Effect of presowing treatment of seeds with insecticides on nodulating ability, nitrate reductase activity and plastid pigments content of lucerne (*Medicago sativa* L.). Agron Res 5:87–92

Wagner D, Przybyla D, Op den Camp R, Kim C, Langraf F, Lee KP, Wursch M, Laloi C, Nater M, Hideg E, Apel K (2004) The genetic basis of singlet oxygen-induced stress responses of *Arabidopsis thaliana*. Science 306:1183–1185

Wang P, Du Y, Li Y, Ren D, Song CP (2010) Hydrogen peroxide-mediated activation of MAP kinase 6 modulates nitric oxide biosynthesis and signal transduction in *Arabidopsis*. Plant Cell 22:2981–2998

Washko PW, Welch RW, Dhariwal KR, Wang Y, Levine M (1992) Ascorbic acid and dehydroascorbic acid analyses in biological samples. Anal Biochem 204:114

Wauchope RD (1983) Uptake, translocation and phytotoxicity of arsenic in plants. In: Lederer WH, Fensterhein RJ (eds) Arsenic: industrial biomedical and environmental perspective. Van Nostrand Reinhold, New York, pp 348–375

Werck-Reichhart D, Hehn A, Didierjean L (2000) Cytochromes P450 for engineering herbicide tolerance. Trends Plant Sci 5:116–122

Willekens H, Chamnongpol S, Davey M, Schraudner M, Langebartels C, Van Montagu M, Inzé D, Van Camp W (1997) Catalase is a sink for H_2O_2 and is indispensable for stress defence in C_3 plants. EMBO J 16:4806–4816

Witkowski DA, Halling BP (1989) Inhibition of plant protoporphyrinogen oxidase by the herbicide acifluorfen-methyl. Plant Physiol 90:1239–1242

Wojtaszek P (1997) Oxidative burst: an early plant response to pathogen. Biochem J 322:681–692

Wolf AE, Dietz KJ, Schroder FP (1996) A carboxypeptidase degrades glutathione conjugates in the vacuoles of higher plants. FEBS Lett 384:31–34

World Health Organization (2007) The impact of pesticides on health preventing intentional and unintentional deaths from pesticide poisoning. Available at:http://www.who.int/mentalhealth/prevention/suicide/en/pesticidesHealth2.pdf

Xia XJ et al (2009) Reactive oxygen species are involved in brassinosteroid-induced stress tolerance in cucumber. Plant Physiol 150:801–814

Xie X, Zhou Q, He Z, Bao Y (2010) Physiological and potential genetic toxicity of chlortetracycline as an emerging pollutant in wheat (*Triticum aestivum* L.). Environ Toxicol Chem 29:922–928

Xu S, Li J, Zhang X, Wei H, Cui L (2006) Effects of heat acclimation pretreatment on changes of membrane lipid peroxidation, antioxidant metabolites, and ultrastructure of chloroplasts in two cool-season turfgrass species under heat stress. Environ Exp Bot 56:274–285

Yanai J, Vatury O, Slotkin TA (2002) Cell signalling as a target and underlying mechanism for neurobehavioural teratogenesis. Ann NY Acad Sci 965:473–478

Yang ZM, Wang J, Wang SW, Xu LL (2003) Salicylic acid-induced aluminum tolerance by modulation of citrate efflux from roots of *Cassia tora* L. Planta 217:168–174

Yen JH, Sheu WS, Wang YS (2003) Dissipation of herbicide oxyfluorfen in subtropical soils and its potential to contaminate groundwater. Ecotoxicol Environ Saf 54:151–156

Yin XL, Jiang L, Song NH, Yang H (2008) Toxic reactivity of wheat (*Triticum aestivum*) plants to herbicide isoproturon. J Agric Food Chem 56:4825–4831

Yuan JS, Tranel PJ, Stewart CN Jr (2007) Non-target-site herbicide resistance: a family business. Trends Plant Sci 12:6–13

Zablotowicz RM, Hoagland RE, Lee H, Alber T, Trevors JT, Hall JC, Locke MA (2001) Transformation of nitroaromatic pesticides and related xenobiotics by microorganisms and plants. In: Hall JC, Hoagland RE, Zoblotowicz RM (eds) Pesticide biotransformation in plants and microorganisms: similarities and divergences, vol 777, ACS symposium series. American Chemical Society, Washington, DC, pp 194–217

Zhen RG, Singh BJ (2001) From inhibitors to target site genes and beyond—herbicidal inhibitors as powerful tools for functional genomics. Weed Sci 49:266–272

Zhou ZS, Guo K, Elbaz AA, Yang ZM (2009) Salicylic acid alleviates mercury toxicity by preventing oxidative stress in roots of *Medicago sativa*. Environ Exp Bot 65:27–34

Assessment of Antioxidant Potential of Plants in Response to Heavy Metals

5

Namira Arif, Vaishali Yadav, Shweta Singh, Bishwajit Kumar Kushwaha, Swati Singh, Durgesh Kumar Tripathi, Kanchan Vishwakarma, Shivesh Sharma, N.K. Dubey, and D.K. Chauhan

Abstract

Heavy metals (HMs) are consequential environmental contaminant, and their prodigious bioaccumulation in the surroundings has become an enigma for all living organisms including plants. Heavy metal has the potential to react with various indispensable cellular components like DNA, protein, and enzymes and in turn induce several stress responses in plants like oxidative stress which is the root cause for the progression of cell death in the plant. Stress responses inflicted by oxidative stress include severe morphological, metabolic, and physiological amendments in plants like DNA strand breakage, defragmentation of proteins, and damage of pho-

N. Arif • V. Yadav • S. Singh • B.K. Kushwaha • S. Singh • D.K. Chauhan (✉)
D D Pant Interdisciplinary Research Laboratory, Department of Botany,
University of Allahabad, Allahabad 211002, Uttar Pradesh, India
e-mail: dkchauhanau@yahoo.com

D.K. Tripathi
Centre of Advanced Study in Botany, Banaras Hindu University,
Varanasi 221005, Uttar Pradesh, India

Centre for Medical Diagnostic and Research, Motilal Nehru National Institute of Technology,
Allahabad, Uttar Pradesh, India
e-mail: dktripathiau@gmail.com

K. Vishwakarma
Department of Biotechnology, Motilal Nehru National Institute of Technology,
Allahabad, Uttar Pradesh, India

S. Sharma
Department of Biotechnology, Motilal Nehru National Institute of Technology, Allahabad,
Uttar Pradesh, India

Centre for Medical Diagnostic and Research, Motilal Nehru National Institute of Technology,
Allahabad, Uttar Pradesh, India

N.K. Dubey
Centre of Advanced Study in Botany, Banaras Hindu University,
Varanasi 221005, Uttar Pradesh, India

A. Singh et al. (eds.), *Plant Responses to Xenobiotics*,
DOI 10.1007/978-981-10-2860-1_5

tosynthetic pigment, which may stimulate cell death. In reaction, plants have a range of mechanisms to minimize the heavy metal toxicity. Plants are endowed with antioxidant defense mechanism, which can be divided into two groups such as enzymatic antioxidants and nonenzymatic antioxidants, for instance, SOD, CAT, APX, GPX, GR and AsA, GSH, carotenoids, alkaloids, tocopherols, proline, and phenolic compounds, respectively, that together act as the scavengers for free radicals to mitigate the damaging impacts of heavy metal agglomeration in the cells. These antioxidant potentials could be assessed by different in vivo and in vitro methods such as hydrogen atom transfer and electron transfer through which we can evaluate the ROS detrimental action of antioxidant enzymes. Therefore, the present chapter attempts to provide the contemporary knowledge regarding the metal-influenced antioxidant status in plants and also provides the precise pathway that should follow for the future research in the area of antioxidant potentials.

Keywords
Antioxidant • Oxidative stress • Heavy metal • Detoxification

5.1 Introduction

Being restricted in distribution, plants are inevitably exposed to several environmental factors (abiotic and biotic), which constitute their macro- and microenvironment. Any digression in these factors from the optimum level is harmful and eventually leads to stress in plants (Kumar et al. 2008; Parvaiz and Satyawati 2008; Sharma et al. 2016). Momentous abiotic factors such as heavy metals (HMs) are imperative environmental pollutants, and their toxicity is a problem for environmental grounds (Nagajyoti et al. 2010). Industry and mining have escort to a relocation of heavy metals, which further resulted in a soil and water pollution. Heavy metals that occur in nature are mainly in two forms: essential and nonessential. Crucial HMs, like copper, iron, zinc, or nickel, are micronutrients, causing toxicity when present at higher concentrations, while nonessential heavy metals, like lead, cadmium, and mercury, are not recognized to have any physiological functions (Nowicka et al. 2016). Increased amount of metals in available soil fractions led an increased bioaccumulation in various parts of the plants (Kabata-Pendias 2004), which potentially induces several functional disorders at multiple level in plants, possibly from the oxidative action of metals (Sun et al. 2007; Shamsi et al. 2008; Kafel et al. 2010). Plants are often susceptible both to the shortage and to the glut accessibility of some HM ions as the increased accumulation of several vital HMs induced plausible changes in the plant (Nagajyoti et al. 2010). Zn, Cu, and Pb are acknowledged as prooxidants, and responsible for the production of the ROS at the higher concentration (Ferrat et al. 2003; Fatima and Ahmad 2005; Drążkiewicz et al. 2004; Caregnato et al. 2008). However, as a consequence of higher net production of reactive oxygen species, there occurs a photooxidative disintegration of DNA, proteins, and lipids that eventually causes cell fatality in plants (Tripathy and Oelmüller 2012). In view of the fact that the stimulation of oxidative stress is a significant process of HM lethality (Nagajyoti et al. 2010; Yadav 2010) likewise, the ability to detoxify

ROS is also a significant factor for excessive concentration of metal tolerance. That is why to ensure continued existence, plants have developed proficient antioxidant mechanism that possesses two arms: (i) enzymatic components such as superoxide dismutase (SOD), catalase (CAT), ascorbate peroxidase (APX), guaiacol peroxidase (GPX), glutathione reductase (GR), monodehydroascorbate reductase (MDHAR), and dehydroascorbate reductase (DHAR) and (ii) nonenzymatic antioxidants such as ascorbic acid (AA), reduced glutathione, α-tocopherol, carotenoids, flavonoids, and the osmolyte proline (Das and Roychoudhury 2014). Chromanols and prenylquinones (isoprenoid antioxidants) are thought to involve in response to heavy metal-stimulated stress (Nowicka et al. 2016). This antioxidant system marks the essentiality of ROS detoxification for the cellular existence (Gill et al. 2011; Das and Roychoudhury 2014). Plants are the source of dietary antioxidants; approximately all plants possess antioxidant prospects in retort to generated stress (Krishnaiah et al. 2011; Kasote et al. 2015). The secondary metabolites also participate significantly in therapeutic properties of plants (Abeysinghe et al. 2014). Gill and Tuteja (2010) in their article propounded that the antioxidant resistance mechanism works in recital to manage the rush of uncontrolled oxidation and defend plant cells from oxidative damage through the escaping of free radical. Thus, the efficacy of its antioxidant defenses is very decisive for a plant's resistance to metals (Kafel et al. 2010).

5.2 Occurrence, Accumulation, and Transport of Heavy Metals (HMs)

HMs are characterized as metals with the atomic mass over 20 and the density higher than 5 $g \cdot cm^{-3}$ (Emamverdian et al. 2015). Heavy metals are regarded as trace elements because of their trace concentration (less than 10 ppm) in the plant (Kabata and Pendias 2001; Tchounwou et al. 2012). Most of the HMs are positively charged, nondegradable, and persistent in the environment (Eshagberi 2012). Naturally HMs are present abundantly into the outermost layer of the earth (Tchounwou et al. 2012). High degree of HM pollution can be observed in the surroundings (Hajar et al. 2014) and these heavy metals cause toxicity even at very low concentration (Lenntech Water Treatment and Air Purification 2014; Nagajyoti et al. 2010). Different anthropogenic activities such as industrial, agricultural, domestic medical, and technological uses have led to their extensive allocation in the environment (Tchounwou et al. 2012). HMs include lead (Pb), cadmium (Cd), nickel (Ni), cobalt (Co), iron (Fe), zinc (Zn), chromium (Cr), iron (Fe), arsenic (As), silver (Ag), and the platinum group elements (Nagajyoti et al. 2010), among which Cd, Cr, Cu, Hg, Pb, and Zn are the major toxic elements present in the environment (Lasat 2000; Tangahu et al. 2011). These contaminations occur through the weathering of rock, volcanic eruptions, and many anthropogenic activities (He et al. 2005). Anthropogenic sources of HMs are the differential industrial activities such as waste from metal processing refineries; contamination from the nuclear power stations; coal and petroleum combustion power plants; wood preservation; waste from the plastic, paper, and textile manufacturing plants; microelectronics; and high-tension electrical lines (Arruti et al. 2010; Tchounwou et al. 2012).

Many varieties of plants successfully absorb hazardous contaminants like Pb, Cd, Cr, As, and an assortment of radionuclides from soils, as they enter into the food web and show progressive bioaccumulation at successive trophic levels. The accessibility of metals is an active process in soil that depends on precise combinations of chemical, biological, and environmental constraints (Peijnenburg and Jager 2003; Hajar et al. 2014). Absorption, movement, and transportation of these metals within the plant tissue are largely reliant on plant species, type of concentrations, and also the oxidation state of HMs (Tangahu et al. 2011). The pH, reduction capability, and soil organic matters (SOM) influence the HMs to exist in ionic form for easy availability to the plant (Fritioff and Greger 2003). The plant performs as "hyperaccumulators" as well as "excluders." Accumulators continue to be present in spite of concerted pollutants in the shoots. The excluders confined pollutant uptake (Sinha et al. 2007). Basic HM tolerance is present in all plant species. Thus, they scamper a compound organization, including absorption, transportation, and chelation; these imperative metals are concerned firmly in homeostasis of essential metal micronutrients. The traits of these HM elements separate the plant kingdom into two categories: hyperaccumulating and non-accumulating plants (Viehweger 2014). "Hyperaccumulator" plants could thrive in toxic environments, require little maintenance, and produce high biomass, whereas non-accumulating plants (typically have a shoot-to-root ratio considerably less than one) can accumulate toxic ion at higher concentration approximately thousands ppm level (Salido et al. 2003; Singh et al. 2015). Hyperaccumulator plants can accommodate heavy metals 1000 times more than excluder plants (Tangahu et al. 2011). Different types of transport mechanism such as intrinsic protein, proton pumps, and co- and antitransporters implicated in ion uptake and transportation, after absorption transportation of these ions into shoots, are desirable (Fernández et al. 2015). Different types of heavy metal transporters such as IRT1, ZnT1, heavy metal ATPase-HMA2, and HMA4 are able to transport Zn, Cu, Cd, Pb, Ni, and Fe (Viehweger 2014). Contaminants are translocate from the root to shoot in the plant by two regulatory mechanism that is evaporation and transpiration (Tangahu et al. 2011).

5.3 Heavy Metal-Induced Oxidative Stress and Antioxidant Potential of Plant System

There are two sorts of metals that reside in the earth's crust that correspond as an imperative micronutrients for plant development such as iron, manganese, zinc, copper, magnesium, molybdenum, and nickel and nonessential elements such as cadmium, antimony, chromium, lead, arsenic, selenium, and mercury. Plants entail them in petite quantities for their growth, metabolism, and development, though the concentration of essential and inessential metals is a significant aspect in the plant development and growth, but their surplus concentration can restrict the plant growth (Zengin and Munzuroglu 2005; Emamverdian et al. 2015; Tripathi et al. 2016). All plant species, either sensitive or tolerant, could tolerate a minimal amount of metal stress. Heavy metals, irrespective of their redox-associated mode of action, are capable of disturbing antioxidant equilibrium in plant cells, inducing ROS, and

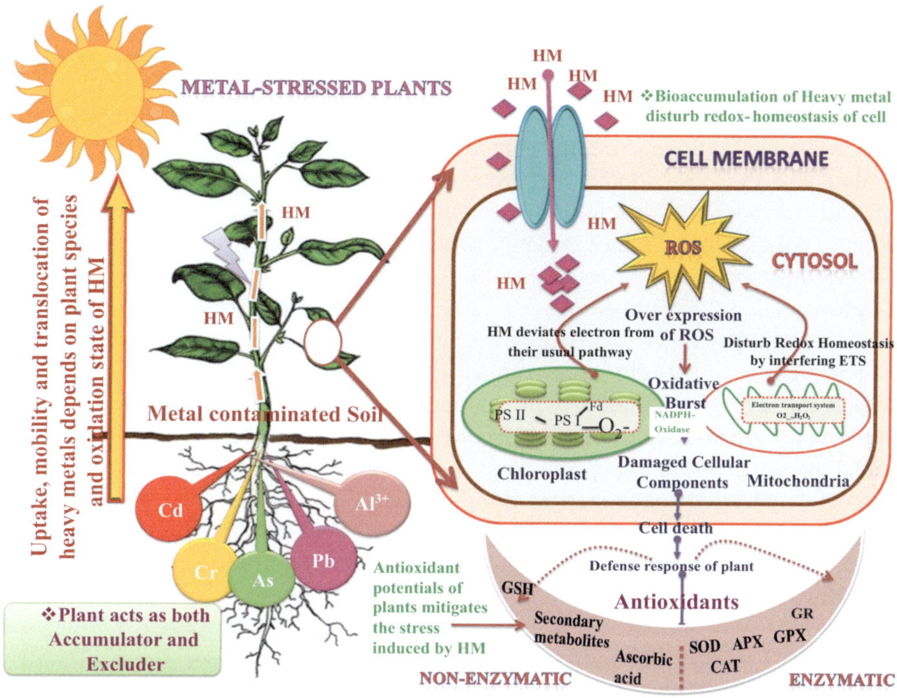

Fig. 5.1 Integrated response of plant toward metal-induced oxidative damage and activation of antioxidant potential of plant

directly reacting with functioning cellular macromolecules and organelles. Likewise, substitution of these crucial cations with the toxic HMs can disturb the equilibrium between cations and enzymatic cofactors (Tangahu et al. 2011). Some redox-active HMs like iron, copper, and chromium can exist in various oxidative states which could produce reactive oxygen species through the Fenton-type reactions and Haber–Weiss cycling, whereas non-redox metals like cadmium, lead, and mercury produce ROS indirectly, mostly by causing depletion of glutathione and through distracting the ETC (electron transport chain) (Pinto et al. 2003; Yadav 2010; Nowicka et al. 2016). However, non-oxido-reducing metals such as zinc and lead induced indirectly oxidative stress as a result of toxicity to metabolic pathways and membrane-coupled ETC (Verma and Dubey 2003; Caregnato et al. 2008). The generation of ROS is a usual process in HM stress treatment. Reactive oxygen species like $O_2^{\cdot\bullet}$, H_2O_2, and OH$^\bullet$ are usually produced due to stress; further they bear strong oxidizing activities that can react with different biomolecules (Fig. 5.1).

Plants in contact with several HM ions move the poise of free radical metabolism toward an accommodation of hydrogen peroxide (Mithöfer et al. 2004). Elevated free radical concentrations exert an inhibitory impact on cell molecules like DNA, proteins, and lipids, for instance, nonenzymatic lipid peroxidation, consequently escort to the accommodation of oxidative burst in various cell sites (Schrader and Fahimi 2006). Hg^{2+} ions restrain the functions of antioxidative enzymes particularly of glutathione reductase and also elevate an ephemeral depletion of GSH (Schützendübel and Polle 2002; Mithöfer et al. 2004). Weihong et al. (2009) through the experiment, studied the effect of HMs like Cd and Zn on plant *Vetiveria zizanioides* and illustrated that Cd and Zn was found involved in plant growth inhibition. The level of antioxidants became enhanced such as SOD, POD and CAT, MDA and proline whereas GSH content and water-soluble proteins decreased as the level of Zn and Cd increased to a toxic level. Plants occupy various innate and extrinsic defense policies for tolerance or detoxification whenever confronted with the stressful circumstance, which occurred through the higher concentrations of HMs (Viehweger 2014; Emamverdian et al. 2015). To study the oxidative stress and antioxidant response under Cu toxicity on nodules of *white lupin* and soybean plant Sánchez-Pardo (2012) did an investigation and revealed that Cu in excess concentration cause severe damages in ultrastructures due to emerged oxidative stress in the White lupin nodules, such damages were reported as the breakage of peribacteroidal membrane with rising numbers of vesicles in the cytosol. While in the nodules of soybean damage appeared in the form of degradation of bacteroidal membrane, and precipitation in vacuoles cells. Although white lupin was proved as more sensitive to Cu stress, the antioxidative effect (total thiol content and APX activity) was found less effective in white lupin than soybean.

5.4 Delineating the Complete Outline of Free Radical Production in Plants

The source of production of reactive oxygen species in plants is mainly the chloroplast, mitochondria, peroxisomes and over and above ER, cell membrane, cell wall, and the apoplast (Das and Roychoudhury 2014). ROS generated in these cell organelles due to stress induced signalling and enzymes like peroxidase, amine oxidase and NADPH oxidase present in cell walls and plasma membrane (Tripathy and Oelmüller 2012). Reactive oxygen species are very fatal which induce a broad injury to protein, DNA, and lipids and disturb the normal cellular pathways (Apel and Hirt 2004).

Furthermore, Gill and Tuteja (2010) have demonstrated that generally the production of ROS in plant tissue occurs mainly in photosystem I and photosystem II of the chloroplast and plasma membrane and also in complex I (ubiquinone) and complex III of the mitochondrial ETC. In a regular physiological activity of the plant, the electron moves from PSI and PSII of the chloroplasts, mitochondrial membrane, ETC, and peroxisome (Kasote et al. 2015). These negative ions (electron) react with molecular oxygen and form superoxide radical ($O_2^{-\bullet}$) (Fig. 5.1; Table 5.1). The superoxide radical is subsequently converted to hydroperoxyl radical ($HO_2^{\bullet}$) and finally to H_2O_2 (Zhao et al. 2005; Kasote et al. 2015). The ROS comprise of highly reactive free radicals (containing unpaired electrons) like $O_2^{-\bullet}$ (superoxide radical) and $OH^{\bullet}$ (hydroxyl radical), the most highly reactive and toxic form of oxygen, and non-radicals (has no unpaired electrons) like H_2O_2 (hydrogen peroxide) and $^1O_2$ (singlet oxygen) (Gill and Tuteja 2010). Environmental

Table 5.1 Effect of different heavy metals on the plant growth and elevating antioxidant potential of different plant spp.

Metals	Plant species	Concentration	Effect	Antioxidant enzymes	References
Co (cobalt)	Duckweed (*Lemna minor*)	1 mM	Impaired function of oxygen-evolving complex, decreases plant growth and biomass, chlorophyll content, starch accumulation, water potential, and transpiration rate	SOD and POD	Begović et al. (2016)
			Increase ROS production		
	Wilczek (*Vigna radiata*)	100–250 mg/kg	Decreased root and shoot length of plant	Catalase, peroxidase, and polyphenol oxidase	Jaleel et al. (2009)
			Inhibition of cell division and elongation		
			Dry weight of root and shoot declined		
	Soybean (*Glycine max*)	250 mg/kg	Decreased root nodule formation and leg hemoglobin content	CAT, SOD, POX, and APX	Jayakumar et al. (2008)
			Increase ROS production		
Pb (lead)	Bean (*Vicia faba*)	5 μm	Decreased nutrient uptake by plant	SOD, GPOX, APX, GR, and CAT	Shahid et al. (2014)
	Thale cress (*Arabidopsis thaliana*)	100 μm	Induces increase of H_2O_2 and lipid hydroperoxide	SOD, CAT, GR, GPX, and POD	Phang et al. (2011)
	Water hyacinths (*Eichhornia crassipes*)	1000 mg/L Pb	Higher accumulation in roots, petiole, and leaf tissue	SOD and CAT	Malar et al. (2014)
			Decreased chlorophyll content	APX and POX	
			MDA content increased in leaf and root tissues		
			Biomass reduction		
			Chlorosis and drying at edges in seedlings		

(continued)

Table 5.1 (continued)

Metals	Plant species	Concentration	Effect	Antioxidant enzymes	References
Cr (chromium)	Pea (*Pisum sativum*)	100 μm	Reduce photosynthetic process and nutrient uptake by plant	SOD and APX	Tripathi et al. (2015)
	Green gram (*Vigna radiata*)	50 μM	Increased lipid peroxidation and H_2O_2 generation	SOD and APX	Shanker et al. (2004)
			Decreased dry weight, root, and shoot		
	Tomato plant (*Lycopersicum esculentum*)	50 mg/L	Higher concentration of Cr accumulated in roots	SOD, CAT, and POX	Mangabeira et al. (2006)
			Yellowing of leaves and complete wilting		
			Root and shoot dry biomass decreased		
Cd (cadmium)	Mung bean (*Vigna radiata*)	1.5 mM	Severe oxidative stress, decreased plant height and root length, and reduced chlorophyll content	SOD,CAT, and GPX	Nahar et al. (2016)
	Mutant tobacco plant (*Nicotiana tabacum*)	500 μm	Increased H_2O_2 and superoxide production, plant growth reduced, and chlorophyll content declined	SOD, proline, and glutathione	Iannone et al. (2015)
	Pea (*Pisum sativum*)	68 μmol	Reduction in plant growth and photosynthetic pigment, cause oxidative injury by enhancing the production of ROS	SOD, CAT, POD, and GR	Agrawal and Mishra (2009)
	Raddish (*Raphanus sativus*)	50 μM	Oxidative stress and accelerated cell senescence in mesophyll area of leaf blade	Phenols, catalase, and glutathione reductase	Vitoria et al. (2001)
Ni (nickel)	Watercress (*Nasturtium officinale*)	25 mg/l	Weak plant growth, chlorosis, metabolic disorder, and ROS production	SOD, APX, and CAT	Duman and Ozturk (2010)
	Wheat (*Triticum aestivum*)	50 μg/l	Growth reduction, decreased chlorophyll content, and increased ROS production	Proline and SOD	Parlak (2016)
	Pigeon pea (*Cajanus cajan*)	1.5 mM	Reduction in seedling growth, decreased dry weight, increased lipid peroxidation, and elevated ROS generation	SOD, POD, and GR	Rao and Sresty (2000)
	Rapeseed (*Brassica napus*)	0.5 mM	Stunted plant growth, brownish roots, chlorosis, and induced ROS generation	APX, GPX, and CAT	Kazemi et al. (2010)

Hg (mercury)	Rattlebush (*Sesbania drummondii*)	100 mg/l	Reduction in biomass, photosynthetic activity declined, and increased oxidative stress	GSH, SOD, APX, and GR	Israr et al. (2006)
	Alfalfa (*Medicago sativa*)	40 μm	Physiological disorder such as stomatal closure, water flow, and trigger oxidative stress	SOD, POD, CAT,	Zhou et al. (2008)
				APX and GR	
	Cucumber seedlings (*Cucumis sativus*)	250 μm	Decreased chlorophyll content and root, shoot, length induced oxidative stress	CAT and APX	Cargnelutti et al. (2006)
	Saltbush (*Atriplex codonocarpa*)	1 mg/l	Increased solute leakage, growth inhibition, and elevated ROS production	Ascorbate and GR	Lomonte et al. (2010)
	Wheat (*Triticum aestivum*)	30 mg/kg	Oxidative damage, reduced enzymatic activity	SOD, CAT, and APX	Li et al. (2013)
Zn (zinc)	Maize plant (*Zea mays*)	50 μM	Inhibited biomass production	SOD, CAT, GPX, and POD	Islam et al. (2014)
			Decreased chlorophyll content		
			Total soluble protein		
			Elevate the ROS production		
	Alfalfa (*Medicago sativa*)	900 μM	Induced production of H_2O_2	SOD, CAT, APX, and GR	Dai et al. (2015)
			Reducing growth, caused leaf chlorosis and nutritional disturbances		
	Vetiver (*Vetiveria zizanioides*)	80 mg/kg	Total biomass decreased, proline content increased, total soluble protein decreased	SOD, POD, and CAT	Weihong et al. (2009)
	Tomato (*Lycopersicon esculentum*)	500 μM	Decreased chlorophyll content, reduced growth and ROS generation	CAT, GST, and APX	Sbartai et al. (2012)

(continued)

Table 5.1 (continued)

Metals	Plant species	Concentration	Effect	Antioxidant enzymes	References
Cu (copper)	Mustard plant (*Brassica juncea*)	150 mg/kg	Reduced growth of plant, activities of nitrate reductase, and carbonic anhydrase. Decreased chlorophyll content and proline increased	CAT, POD, and SOD	Fariduddin et al. (2009)
	Mustard plant (*Brassica junica*)	150 mg/kg	Decreased shoot–root length, biomass, decreased chlorophyll content	CAT,POX, and SOD	Yusuf et al. (2016)
			Reduced activities of RuBisCO and carbonic anhydrase		
	White lupin (*Lupinus albus*)	192 μM	Reduced nitrogen fixation, ROS production, chlorosis, necrosis, and abnormal root morphology	APX and CAT	Sanchez-Pardo et al. (2012)
	Soybean plant (*Glycine max*)				
	Ashwagandha (*Withania somnifera*)	200 μM	Inhibition of cell elongation and division, reduction of biomass, reducing membrane fluidity, and decreased concentration of carotenoid and photosynthetic pigment, ROS generation	APX, MDHAR, DHAR, GST and G-POD	Khatun et al. (2008)
As (arsenic)	Tobacco (*Nicotiana tabacum*)	5 mg/l	Impedes the photosynthesis, reduced essential nutrient content, ROS generation	SOD, POD, and GSH	Han et al. (2015)
	Rice plant (*Oryza sativa*)	50 μM	Hindered plant length and weight, oxidative stress enhanced	SOD, APX, GPX, and CAT	Dixit et al. (2016)
	Watercress (*Nasturtium officinale*)	100 μM	Decreased dry weight of roots and shoots and chlorophyll content of leaves, oxidative damage	CAT,APX, and SOD	Namdjoyan and kermanian (2013)
	Mung bean (*Phaseolus aureus*)	10 μM	Inhibited growth caused physiological disorders such as membrane damage, oxidative damage	SOD, CAT, APX, GR, ASC, GHS	Malik et al. (2012)

fluctuations such as increased salt concentration, low water availability, and elevated HM concentration result in closure of stomata which further leads to inadequate intracellular carbon dioxide level and induced ROS formation which induce rigorous injury in the photosystem (Das and Roychoudhury 2014).

5.5 Why Does All Plant Possess Antioxidant Potential?

Metals cause phytotoxicity when it is transported to the plant from the earth's crust. The most prominent consequence of HMs in plant cells is on the growth productivity (Kumar et al. 2013). HM stress declines the capability of the plant to assimilate carbon and elevate the photosynthetic electron flow toward oxygen from which the formation of $O_2^{-\bullet}$, H_2O_2, and $OH^{\bullet}$ radical increases (Gill and Tuteja 2010). As mentioned above, in plants, ROS are constantly generated chiefly in chloroplasts, mitochondria, and peroxisomes. Therefore, generation and elimination of reactive oxygen species should be regulated by the antioxidative defense system in restricted manner (Apel and Hirt 2004), but in the stress condition, the production of ROS elevates and destructs the whole cell metabolism (Sharma et al. 2012). These destructive properties of ROS generate the complex range of nonenzymatic and enzymatic detoxification device in plants (Apel and Hirt 2004). Antioxidants are reducing agents which restrain the oxidation of other molecules, because oxidation reactions generate free radicals which create cell damage (Sies 1997; Bansal and Kaushal 2014). Plants generate antioxidants like glutathione and ascorbic acid (AA) in the chloroplast, stroma, and cytosol with the help of NADPH (Alscher et al. 1997). These antioxidants interact with numerous cellular molecules and affect the growth productivity and development of the plant by interfering in cell division and cell elongation (Foyer and Noctor 2005). These antioxidants also influence gene expression to elevate the defense mechanism in the plant cell. The key reason for the stimulation of these antioxidant mechanisms might be the genetic structure of plants which have innate capacity to produce phytochemicals to execute their continuous physiological task (Kasote et al. 2015). Plants produce secondary metabolites which also illuminate the reactive oxygen species because these metabolites play a significant role in adjustment of plants against environmental fluctuations (Baier and Dietz 2005). ROS can cause many disorders in the cell by affecting many physiological reactions (Ragavendran et al. 2012). Stress damages the cell by increasing the production of ROS (Rahman 2007). So for the inhibition of these reactive species, detoxification system evolves such as enzymatic and nonenzymatic antioxidant. These systems include catalase, peroxidase, SOD, ascorbic acid tocopherol, GSH, etc. (Prakash and Sharma 2014; Gout et al. 2001) (Fig. 5.1). SOD enzyme scavenges the superoxide radical and forms hydrogen peroxide which is also highly toxic for the cell (Kusvuran 2012). SOD destroys superoxide anion by converting it to peroxide (Cannon et al. 1987). Catalase breaks the H_2O_2 into H_2O and oxygen (Mittler 2002). Polyphenol oxidase is an antioxidant enzyme which scavenges H_2O_2 in chloroplasts and plays a significant function in lignin biosynthesis (Mittler 2002). Ascorbic oxidase regulates the reduced glutathione and NADPH. Vitamin C is a water-soluble antioxidant which scavenges the peroxy radicals (Sies 2007).

5.6 Enzymatic Antioxidant

5.6.1 Superoxide Dismutase (SOD)

SOD is considered as the essential defensive antioxidant against oxygen free radicals. SOD is a metalloenzyme which converts superoxide anion ($O_2^{-\bullet}$) to H_2O_2. SOD has been present in all aerobes that work against toxic oxygen species which are usually produced as the by-products of many biological oxidation reactions (Imlay 2008). Begović et al. (2016)) reported increased concentration of SOD in duckweed (*Lemna minor*) in retort to toxicity of cobalt. SOD is localized in mitochondria, chloroplast, cytosol, and peroxisomes (Mittler 2002), and the amount of SOD escalates in accordance with the level of stress condition. Superoxide is the initial product of the monovalent reduction of oxygen and also the first free radical in the plant cell. SOD catalyzes the dismutation reaction by metal ion like Cu, Mn, and Fe at the active site. Based on metal ion, superoxide dismutase is categorized mainly in three isozymes: Mn-SOD, Fe-SOD, and Cu/Zn-SOD (Mittler 2002). The effect of the Cr toxicity on SOD transcription has been demonstrated on the green gram and black gram resulted in a substantial elevation in the production of ROS due to reduced SOD synthesis (Karuppanapandian et al. 2006; Karuppanapandian and Manoharan 2008).

5.6.2 Catalase

Catalase is the foremost discovered and characterized enzyme, which possesses antioxidant activity, and it is a Fe-containing enzyme present in diverse organisms, including prokaryotes (Zamocky et al. 2008). It consists of polypeptides of 50–70 kDa which are arranged in tetramers and each monomer encloses a heme prosthetic group (Regelsberger et al. 2002). It catalyzes the dismutation reaction of H_2O_2 into H_2O and O_2. Catalase obliterates the H_2O_2 generate in peroxisome by β-oxidation of fatty acids, photorespiration, and purine catabolism (Mittler 2002; Vellosillo et al. 2010) and prevents the diffusion of H_2O_2 from cytosol (Lopez-Huertas et al. 2000). There is elevated level of catalase in a bean (*Vicia faba*) for the destruction of ROS produced due to lead toxicity (Shahid et al. 2014). However during stress condition like salinity, drought, and HMs, the enzyme production is found to be reduced (Karuppanapandian et al. 2006; Karuppanapandian and Manoharan 2008) which limits the plant's tolerance to environmental stress. Li et al. (2013) conducted their experiment on *Triticum aestivum* (wheat plant) with mercury (Hg)-contaminated soil and found the increased intense activity of catalase (CAT) antioxidant enzyme in a wheat plant grown in a highly polluted soil.

5.6.3 Ascorbate Peroxidase and Guaiacol Peroxidase

Ascorbate peroxidase (APX) is a heme peroxidase present in higher eukaryotes (Takeda et al. 1998). In chloroplast and cytosol, the level of H_2O_2 is illuminated by the APX. It uses ascorbic acid for the breakdown of H_2O_2 and yields water and monodehydroascorbate (Asada 2000). APX isoforms are classified on the basis of subcellular localization, such as chloroplasts, mitochondria, peroxisome, and cytosol (Caverzan et al. 2012). APX activity frequently increases with the function of other enzymes, like CAT, SOD, and GSH reductase (Shigeoka et al. 2002). Sharma et al. (2016) reported in their article about the significant increase in ascorbate peroxidase (APX) activity with chromium (Cr)-stressed *Oryza sativa* (rice) seedling under the influence of EBL (epibrassinolide).

Guaiacol peroxidase (GPX) is a significant member of peroxidase enzyme. GPXs are usually acknowledged as "stress enzymes" and found in the cellular cytoplasm and apoplasm (Sharma et al. 2012). GPX is reported to involve in many processes such as growth of plants and its development. It also takes part in ROS scavenging. GPX is an iron-enclosing protein and oxidizes certain substrates at the expenditure of H_2O_2. It relieves the cell from excess peroxide which generates in stress condition (Sharma et al. 2012). GPX deteriorate indole-3-acetic acid and also play a significant part in the biosynthesis of lignin (Karuppanapandian et al. 2011). GPX scavenges H_2O_2 produced due to stress from the cytosol, vacuole, and cell wall and in the extracellular space (Koji et al. 2009). The study reported on *Arabidopsis thaliana* seedlings exposed to lethal lead (Pb) level represented the increased activity of GPX antioxidant enzyme (Phang et al. 2011). The study on *Avicennia marina* (gray mangrove) relevant to glutathione antioxidant system for the evaluation of HM stress showed the incensement of GPX activity in a dose-dependent fashion in response to accumulated leaf metals (Zn, Cu, or Pb) (Caregnato et al. 2008). Similarly, in another study on *Vicia faba* plant showed the increased activity of APX and GPX in relation to lead stress (Shahid et al. 2014).

5.6.4 Monodehydroascorbate Reductase (MDHAR) and Dehydroascorbate Reductase (DHAR)

MDHAR is a FAD enzyme and important constituent of the glutathione–ascorbate cycle which is the major antioxidant system of plant tissue (Yoon et al. 2004). MDHAR catalyzes the ascorbate production through the MDA radical. Ascorbate is used to detoxify H_2O_2 via APX (Mittler 2002). MDHAR regenerate the ascorbate with the help of NAD(P)H. The monodehydroascorbate reductase functionality has been seen in many cell organelles such as chloroplast, cytosol, mitochondria, glyoxysomes, and peroxisomes (Leterrier et al. 2005).

DHAR is assessed as a chloroplast enzyme and contains thiol group. It plays an active role in the protection against oxidative stress (Noctor and Foyer 1998). DHAR also catalyzes the revival of ascorbic acid. Ascorbate regenerates from the DHA by the thiol enzyme DHAR, but the MDHAR produce more ascorbate than DHAR (Asada 2006; Minkov et al. 1999). DHAR overproduction in tobacco and *Arabidopsis* had been shown under environmental stress (Chen and Gallie 2006; Eltayeb et al. 2007). An investigation on *Raphanus sativus* (radish) to cadmium (Cd) treatment showed increased concentration of antioxidant enzyme MDHAR and

DHAR via the activation of ascorbate–glutathione cycle for the removal of H_2O_2 (Vitória et al. 2001).

5.6.5 Glutathione Reductase (GR)

Glutathione reductase is a flavoprotein present in all organisms (Romero-Puertas et al. 2006). Glutathione reductase (GR) is also recognized as glutathione disulfide reductase (GSR) (Kotapati et al. 2014). Glutathione reductase is a homodimeric and oxidoreductase enzyme which is NADPH dependent. It is an imperative enzyme of the ASH–GSH cycle which scavenges hydrogen peroxide with the united exploit of some antioxidant enzymes such as ascorbate peroxidase, monodehydroascorbate reductase, dehydroascorbate reductase GSH, and ascorbic acid (Noctor and Foyer 1998; Gutteridge and Halliwell 2000). It plays an essential function in scavenging of ROS by catalyzing reduction of glutathione disulfide to the sulfhydryl form GSH (Zitka et al. 2012). GR is localized mainly in chloroplasts and also little quantity of GR has been found in mitochondria and cytosol (Ding et al. 2012). Agrawal and Mishra (2009) reported increased concentration of glutathione reductase in *Pisum sativum* under cadmium stress.

5.7 Nonenzymatic Antioxidant

5.7.1 Ascorbic Acid

Ascorbic acid (AA) is the plentiful, influential, and water-soluble antioxidant present in chloroplast and all cell organelles including the cell wall. Ascorbic acid takes active role in plant growth and development in stress condition (Sharma et al. 2012). Ascorbate also works as a cofactor for some hydroxylase enzymes like prolyl hydroxylase (Kuiper and Vissers 2013). AA acts to avoid or minimize the harmful effect caused by ROS in plants (Smirnoff 2005; Ahmad et al. 2000). Hg-stimulated oxidative burst in saltbush (*Atriplex codonocarpa*) is found to be decreased by ascorbate (Lomonte et al. 2010). It destroys the several forms of ROS including singlet oxygen, superoxide, and hydroxyl radicals (Padh 1990) and shields the membranes from oxidative damage. Ascorbic acid also maintains α-tocopherol in the reduced state (Traber and Stevens 2011) and indirectly scavenges H_2O_2 through the AsA peroxidase (Chugh et al. 2011). The study on *Phaseolus vulgaris* (bean) seedling with heavy metal (Pb, Cu, Cd, and Hg) showed the significant increase in ascorbic acid content in primary leaves after 10 days of metal exposure (Zengin and Munzuroglu 2005).

5.7.2 Tocopherols

α-Tocopherol is a lipophilic antioxidant and generates only by photosynthetic organism. Tocopherol has a chromanol head group attached to the phytyl tail (Wang and Quinn 2000). Tocopherols are proficient denominator of free radicals (Kiffin et al. 2006). Tocopherols are crucial component of biological membrane and act as antioxidant in higher plants (Kiffin et al. 2006). They protect the chlorophyll membrane by physical quenching and it also undergoes reaction with oxygen (O_2) in chloroplast and shielding the photosynthetic pigment (Igamberdiev et al. 2004). The study conducted on *Arabidopsis thaliana* plant in the existence of Cd and Cu shows marked increase in α-tocopherol (Collin et al. 2008).

5.7.3 Glutathione (GSH)

Glutathione is a thiol tripeptide, a low molecular weight enzyme which presents in cytosol, ER, mitochondria, peroxisomes, vacuoles, and apoplast (Das and Roychoudhury 2014). GSH is the most significant endogenous antioxidant enzyme which is active in the neutralization of ROS directly and also maintains the exogenous antioxidants like ascorbate and tocopherol in their reduced forms (Ahmad et al. 2012). GSH has a high reductive capacity due to nucleophilic character (Halliwell 2006). GSH scavenges H_2O_2, $OH^{\bullet}$, and $O_2^{\bullet-}$ and prevents the reduction of different biomolecules. GSH also act as an imperative function in the regeneration of ascorbic acid (Ahmad et al. 2012). Glutathione occurs in the cell in two states: reduced and oxidized, the reduced form is GSH and oxidized form is GSSG. As reported in an article, conducted on *Pisum sativum* plant, glutathione (GSH) is found to be increased under cadmium (Cd) stress (Metwally et al. 2005).

5.8 Secondary Metabolites

Plants generate an ample of secondary metabolites such as flavonoids, phenolic acids, alkaloids, etc. (Hartmann et al. 1995). These secondary metabolites have no contribution in the photosynthetic mechanism, substrate oxidation, solute transportation, translocation, nutrient absorption, and differentiation (Mazid et al. 2011), but these metabolites play a significant function in ROS disintegration (Fini et al. 2011). These secondary metabolites are also important for plants to survive under stress condition. Their liberation differs from plant to plant and species to species on exposure to stress (Korkina 2007). These metabolites are formed by basic pathways like glycolysis or shikimic acid pathways (Kasote et al. 2015). Phenolics have shown the most prominent antioxidant functionality between all secondary metabolites (Kasote et al. 2015)

5.8.1 Secondary Metabolites as Antioxidant

Plant metabolites are chiefly differentiated into primary and secondary forms. Primary metabolites are those compounds which produced through primary metabolism, like sugars, amino acids, fatty acids, etc. Primary metabolites are indispensable for cell maintenance (Kliebenstein and Osbourn 2012), whereas secondary metabolites are requisite for the normal cell growth and development. Secondary metabolites also take part in the defense system of the plant (Korkina 2007). Secondary metabolites constantly remain in the plant cell. Secondary metabolites also occur in passive and active forms. In passive form, metabolites already exist in tissue, while active forms of secondary metabolites are generated in response to stress (Korkina 2007), and these metabolites are synthesized by basic pathways like glycolysis or shikimic acid pathways (Aharoni et al. 2005).

These secondary metabolites may also be of two types: one is nitrogen containing such as alkaloids containing terpenoid indole alkaloids, tropane alkaloids, and purine alkaloids (Ziegler and Facchini 2008) and the other is nitrogen deficient like terpenoids and phenolics (Kasote et al. 2015). Phenolics have shown the most prominent antioxidant reactivity between all secondary metabolites.

5.8.2 Phenolics

Plant phenolics are chiefly categorized into different groups, such as phenolic acids, flavonoids, lignins, stilbenes, and tannins (Myburgh 2014). Phenolic compounds usually have more than one aromatic ring with hydroxyl groups. The antioxidant capability of phenolics elevated with increase in hydroxyl group numbers and its conjugation with the side chain of aromatic rings (Flora 2009). Between all these phenolics, flavonoids are the chief active plant's secondary metabolite and act as an antioxidant under stress condition (Hernández et al. 2009). Posmyk et al. (2009) have observed increased level of phenolic compound in red cabbage seedling exposed to copper. Flavonoids occur broadly within the plant tissue and are usually found in leaves, floral parts, and pollens. Flavonoids generally concentrate in the plant vacuole as glycosides. Flavonoids act as a secondary ROS scavenger and get activated on the loss of photosynthetic system, because of the more excitation energy (Fini et al. 2011). Flavonoids perform as an ROS scavenger in the plant tissue by neutralizing the free radicals before they injured the cell (Løvdal et al. 2010).

Flavonoids are also capable to modify peroxidation reaction by altering the lipid packing arrangement (Sharma et al. 2012). They stabilize membranes by diminishing membrane fluidity. Most of the plant root exudates elevate the amount of phenolics on exposure to heavy metals (Winkel-Shirley 2002). Many flavonoid biosynthetic genes are activated under stress conditions. In many stress conditions like wounding, drought, metal toxicity, and nutrient deficiency, it has been seen that flavonoid concentration increases in response to these stresses (Winkel-Shirley 2002). Anthocyanins, a derivative of flavonoids, gather in the vacuoles and possess an antioxidative capability (Kähkönen and Heinonen 2003), but its location prevents them to contact directly with ROS generation sites. However, its level is found to be increased under Cd stress (Mobin and Khan 2007). Keilig and Ludwig-Müller (2009) propounded in their article about the potential role of flavonoids with response to cadmium (Cd) in tolerant *Arabidopsis thaliana* seedling.

5.8.3 Terpenoids

Terpenoids are a huge class of secondary metabolites containing more than 40,0000 different compounds (Aharoni et al. 2005), ranging in structure from linear to polycyclic. Terpenoids are organic compounds derived from the isoprene unit which also have an antioxidative role in plants (Grassmann et al. 2002). Based on the different compositions, it is classified into monoterpenes, diterpenes, triterpenes, and tetraterpenes (Rabi and Bishayee 2009). Monoterpenes, sesquiterpenes, and diterpenes are acquired remarkable antioxidant activity in different in vitro analyses (Baratta et al. 1998). Tetraterpenes possess strong antioxidant activity within both in vivo and in vitro studies (Palozza and Krinsky 1992; Kasote et al. 2015).

5.8.4 Alkaloids

Alkaloids are nitrogen-containing most plentiful secondary metabolites present at 10–15 % concentration, in nearly all plant tissues (Schardl et al. 2006). Alkaloids are heterocyclic compounds containing negatively charged nitrogen due to which it possesses antioxidant properties. Caffeine obtained from the *Thea sinensis* leaves and *Coffea arabica* also shows antioxidant activity. Alkaloids are frequently accommodated in the plant tissue in response to several stresses (Ali and Alqurainy 2006). Several alkaloids have been established as effective inhibitors of $^1O_2$ such as indole alkaloids like strychnine and brucine that have a basic nitrogen atom in a rigid, cage-like conformation. These alkaloids are physical quenchers and not smashed chemically by the course of quenching. Thus, in principle, they could destroy singlet oxygen. Srivastava and Srivastava (2010) reported in his article about the increased alkaloid content in the root of *Catharanthus roseus* in response to cadmium and nickel stress.

5.8.5 Carotenoids

Carotenoids are lipid-soluble molecules and beta carotene is the main precursor of vitamin A. Carotenoids defend the plant from oxidative stress (Britton et al. 2009). Carotenoids are present in photosynthetic organisms as a light-harvesting pigment, expanding the light spectrum range, which utilize in the photosynthetic mechanism. Carotenoids also quench the $^1O_2$ within the photosynthetic machinery (Li et al. 2012). They absorb light in the region from 450 to 570 nm and pass the confined energy to chlorophyll pigment and also serve as an antioxidant scavenging superoxide anion produced by quenching of the triplet state of the chlorophyll molecules (Young and Lowe 2001). Andrianos et al. (2016) described the increased concentration of carotenoids in *Solanum tuberosum* and *Daucus carota* cultivated in a greenhouse and irrigated with a water solution including different concentrations of chromium and nickel.

5.9 In Vitro and In Vivo Strategies for ROS and Plant Antioxidant Potential Measurement

There is a rising curiosity among the scientific world and ingenuities with regard to the measurement of ROS and antioxidant prospective in plant tissue. In plant tissue, reactive oxygen species detect mainly with the help of histochemical method. Because of their highly reactive nature and extremely short lifetimes, the studies of free radical generation in plants are very difficult. The quantitative biochemical analysis does not make available exact information for the localization of reactive oxygen species in plants (Cheeseman 2006). The histochemical localization of ROS provides the opportunity to identify the specific sites of their in situ production that greatly helps to detect the distribution and accommodation of reactive oxygen species in the cell. Histochemical revealing of ROS is mainly done by the use of 3,3′-diaminobenzidine (DAB) and nitro blue tetrazolium (NBT) stain (Kuźniak et al. 2014). ROS detection could also be done by the use of fluorescent probes which is the simplest, greatest, and accessible method. Dihydroethidium (DHE), MitoSOX Red, and 5-(and 6)-chloromethyl-2,7-dichlorodihydrofluorescein diacetate (CM-H2DCFDA) are used to measure superoxide anion and hydrogen peroxide production in the cell (Fryer et al. 2003), while for the evaluation of antioxidant, a credible number of methods are available through which an easy evaluation could be carried for the measuring of reactive oxygen species scavenging activity. Approximately 19 in vitro and 10 in vivo methods are applied for the measurement of antioxidant ability (Alam et al. 2013).

There are copious in vitro assays that are available to fully elucidate the antioxidant behavior of plants conversely, and every method has its own margins concerning its applicability. In vitro methods are usually used to confirm the antioxidant ability of the plant particularly on the basis of certain reaction like reduction, quenching, or metal chelation, and on that basis they are further classified as primary and secondary antioxidants (Kasote et al. 2015) (Fig. 5.2; Table 5.2).

The primary antioxidant works by donating a proton, whereas secondary metabolites work by binding of metal ion which is able to catalyze oxidative reactions and UV absorbance and impeding hydroperoxide activities (Kasote et al. 2015). The efficiency of antioxidant mechanism mainly depends on bond dissociation energy and ionization potential (Karadag et al. 2009). Based on the inactivation mechanism involved, a basic classification of antioxidant assays falls under two categories:

(i) Hydrogen atom transfer (HAT)-based assays
(ii) Electron transfer (ET)-based assays

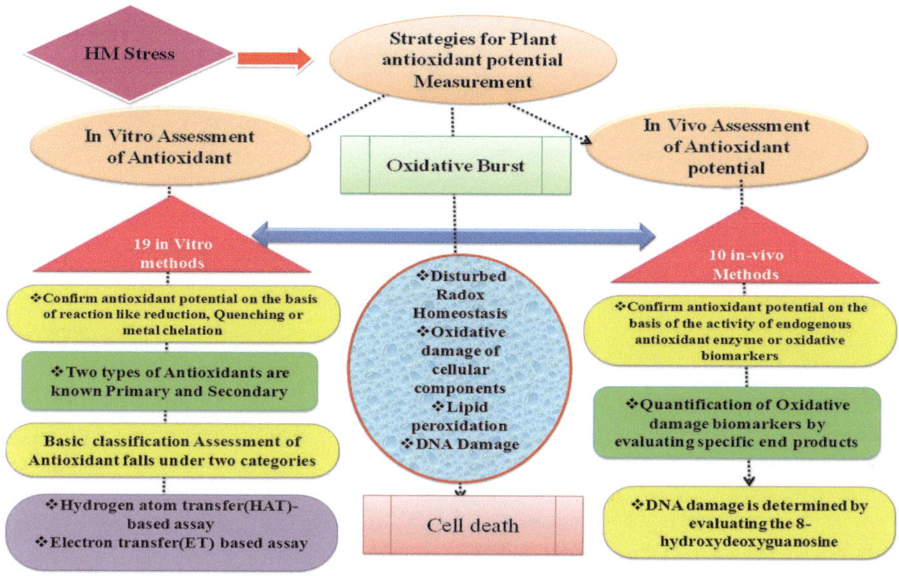

Fig. 5.2 Theoretical model illustrating the most probable strategy adopted by plant against antioxidant potential measurement. Model representing the overall strategies adopted for the measurement of antioxidant potentials that generally falls under two categories: in vitro assessment method (hydrogen atom transfer (HAT)-based assays and electron transfer (ET)-based assays) and in vivo assessment method

The HAT-based methods are used to determine the potential of an antioxidant to destroy the ROS and in the formation of stable compound. Antioxidant potential evaluation primarily depends on the competition kinetics. HAT assay reaction is fast and completed in minutes and the reactions are pH dependent.

HAT-based assays include oxygen radical absorbance capacity (ORAC) method, lipid peroxidation inhibition capacity (LPIC) assay, total radical trapping antioxidant parameter (TRAP), inhibited oxygen uptake (IOC), crocin bleaching nitric oxide radical inhibition activity, hydroxyl radical scavenging activity by p-NDA (p-butrisidunethyl aniline), scavenging of H_2O_2 radicals, ABTS radical scavenging method, and scavenging of superoxide radical formation by alkaline (SASA) (Badarinath et al. 2010).

ET-based methods calculate the potential of an antioxidant. The color of oxidant gets changed on the reduction (Fig. 5.2). The extent of color change is interconnected to the concentration of antioxidants in the sample. Electron transfer reactions are usually slow and require longer times to attain a final point, so antioxidant potential calculations are mainly based on percent decline in the product rather than kinetics. ET reactions depend upon the pH (Prior et al. 2005) (Table 5.2).

ET-based assay includes Trolox equivalent antioxidant capacity (TEAC) decolorization, ferric reducing antioxidant power (FRAP), DPPH free radical scavenging assay, copper (II) reduction capacity total phenols by Folin–Ciocalteu, and

Table 5.2 Assessment of antioxidant potential in plants

In vitro assay	Methods	References
β-carotene or crocin bleaching assay	HAT	Ordoudi and Tsimidou (2006)
ORAC (oxygen radical absorbance capacity)	HAT	Haytowitz and Bhagwat (2010)
IOU (inhibited oxygen uptake)	HAT	Filippenko et al. (2009)
LPIC (lipid peroxidation inhibiting capacity) assay	HAT	Shalaby and Shanab (2013)
TRAP (total radical trapping antioxidant parameter)	HAT	Sies (2007)
Copper reduction assay	ET	Campos et al. (2009)
FRAP (ferric reducing antioxidant power) assay	ET	Ou et al. (2002)
Total phenolic content assay by Folin–Ciocalteu reagent	ET	Ainsworth and Gillespie (2007)
TEAC (Trolox equivalent antioxidant capacity)	ET	Gliszczyńska-Świgło (2006)
DMPD (N,N-dimethyl-p-phenylenediamine) assay	ET	Çekiç et al. (2015)
ABTS [(2,2'-azinobis-(3-ethylbenzothiazolin-6-sulfonic acid)] assay	HAT and ET	Johnston et al. (2006)
DPPH (2,2-diphenyl-1-picrylhydrazyl) assay	HAT and ET	Ozgen et al. (2006)
SASA (scavenging of superoxide radical formation by alkaline)	HAT	Badarinath et al. (2010)

N,N-dimethyl-p-phenylenediamine (DMPD) assay (Badarinath et al. 2010), whereas in the in vivo assay, plants' antioxidant potential is usually examined on the basis of the reactivity of endogenous antioxidant enzymes or oxidative biomarkers prior and subsequent stimulation of oxidative stress (Kasote et al. 2015). In these techniques the action of antioxidant enzyme like superoxide dismutase, catalase, GPX, and GR is directly estimated. While several other techniques are made by the evaluation of oxidative damage biomarkers and definite yield formed by the interaction of ROS and biologically significant macromolecules like DNA, lipids, and protein-like DNA, damage is determined by evaluating the 8-hydroxydeoxyguanosine (Kasote et al. 2015) (Table 5.2).

5.10 Conclusion and Future Outlook

The planet's inhabitants are burgeoning exponentially and stretching the earth's limited resources; as the population is increasing, food consumption follows the same upward trend (FAO 2009). Based on the UN report (2015), the world population reached 7.3 billion as of mid-2015, whereas the Indian population reached 1.3 billion (World Population Prospects: The 2015 Revision. New York: United Nations. 2015). Population detonation by diverse human activities upshot the quantity of waste production and pollution is on the rise. The environmental collision of various activities affects abiotic and biotic factors, such as water quality, soil and sediment quality, air quality, noise, and vibration generated beyond the permissible limits and various types of waste generated. Among them, heavy metal pollution is the major pollution. Besides natural source, anthropogenic activities such as flawed disposal of waste from different industries (nanoparticles manufacturing factories, smelters, power plants, electroplating, and mines), conflagration by-product, and automobile discharges are the major sources of HM pollution. HMs in limited quantity are significant for the healthy growth of plants, but their accumulation in productive soil in excess leads to phytotoxicity which declines the physical and biochemical activities, germination and growth retardation, structural breakage, and reduced yield.

In these contexts, plants produce and accommodate numerous enzymatic and nonenzymatic antioxidants like AA, glutathione, and phenolics. In response to heavy metal stress, plants trigger increased ROS level through the Fenton-type reactions and Haber–Weiss cycling. These ROS species scavenge by the erection of enzymes and nonenzymatic antioxidants. Significant scientific information has been gathered in the form of plant redox biology and the antioxidant resistance device possessed by it. Therefore, it becomes a prerequisite to delineate the different activities that are generating heavy metal saddle on the environment. However, this chapter, though, covers largely the discernible detrimental impacts induced by HMs in plants with the integrated response adopted by plants toward metal stress, particularly in the form of antioxidant ability and also assessment strategies adopted toward the measurement of antioxidant ability; further research is still required for cultivating plant species with improved antioxidant potentials that could be able to feed the ever-growing world population. Furthermore, there is a need to produce such transgenic plant varieties or genetically modified (GM) plants that have the potential to resist against the weed, pest, diseases, soil salinity, and also heavy metal-induced phytotoxicity.

References

Abeysinghe DC, Wijerathne SMNK, Dharmadasa RM (2014) Secondary metabolites contents and antioxidant capacities of *acmella oleraceae* grown under different growing systems. World J Agric Res 2(4):163–167

Agrawal SB, Mishra S (2009) Effects of supplemental ultraviolet-B and cadmium on growth, antioxidants and yield of *Pisum sativum* L. Ecotoxicol Environ Safe 72:610–618

Aharoni A, Jongsma MA, Bouwmeester HJ (2005) Volatile science? Metabolic engineering of terpenoids in plants. Trends Plant Sci 10:594–602

Ahmad I, Hamid T, Fatima M, Chand HS, Jain SK, Athar M, Raisuddin S (2000) Induction of hepatic antioxidants in freshwater catfish (*Channa punctatus* Bloch) is a biomarker of paper mill effluent exposure. Biochim Biophys Acta Gen Subj 1523:37–48

Ahmad P, Azooz MM, Prasad MNV (eds) (2012) Ecophysiology and responses of plants under salt stress. Springer Science & Business Media

Ainsworth EA, Gillespie KM (2007) Estimation of total phenolic content and other oxidation substrates in plant tissues using Folin–Ciocalteu reagent. Nat Protoc 2:875–877

Alam MN, Bristi NJ, Rafiquzzaman M (2013) Review on in vivo and in vitro methods evaluation of antioxidant activity. Saudi Pharm J 21:143–152

Ali AA, Alqurainy F (2006) Activities of antioxidants in plants under environmental stress. The lutein-prevention and treatment for diseases. In: Motohashi N (ed) The lutein prevention and treatment for disease. Kerala, Transworld Research Network, pp 187–256

Alscher RG, Donahue JL, Cramer CL (1997) Reactive oxygen species and antioxidants: relationships in green cells. Physiol Plant 100:224–233

Andrianos V, Stoikou V, Tsikrika K, Lamprou D, Stasinos S, Proestos C, Zabetakis I (2016) Carotenoids and antioxidant enzymes as biomarkers of the impact of heavy metals in food chain. Curr Res Nutr Food Sci 4:15–24

Apel K, Hirt H (2004) Reactive oxygen species: metabolism, oxidative stress, and signal transduction. Annu Rev Plant Biol 55:373–399

Arruti A, Fernández-Olmo I, Irabien Á (2010) Evaluation of the contribution of local sources to trace metals levels in urban PM2. 5 and PM10 in the Cantabria region (Northern Spain). J Environ Monit 12:1451–1458

Asada K (2000) The water–water cycle as alternative photon and electron sinks. Philos Trans R Soc Lond B Biol Sci 355:1419–1431

Asada K (2006) Production and scavenging of reactive oxygen species in chloroplasts and their functions. Plant Physiol 141:391–396

Badarinath AV, Rao KM, Chetty CMS, Ramkanth S, Rajan TVS, Gnanaprakash K (2010) A review on in-vitro antioxidant methods: comparisons, correlations and considerations. Int J PharmTech Res 2:1276–1285

Baier M, Dietz KJ (2005) Chloroplasts as source and target of cellular redox regulation: a discussion on chloroplast redox signals in the context of plant physiology. J Exp Bot 56:1449–1462

Bansal M, Kaushal N (2014) Oxidative stress mechanisms and their modulation. Springer, New Delhi

Baratta MT, Dorman HJ, Deans SG, Figueiredo AC, Barroso JG, Ruberto G (1998) Antimicrobial and antioxidant properties of some commercial essential oils. Flavour Frag J 13:235–244

Begović L, Mlinarić S, Dunić JA, Katanić Z, Lončarić Z, Lepeduš H, Cesar V (2016) Response of *Lemna minor* L. to short-term cobalt exposure: the effect on photosynthetic electron transport chain and induction of oxidative damage. Aquat Toxicol 175:117–126

Britton G, Liaaen-Jensen S, Pfander H (eds) (2009) Carotenoids volume 5: nutrition and health (Vol. 5) SSBM

Cannon RE, White JA, Scandalios JG (1987) Cloning of cDNA for maize superoxide dismutase 2 (SOD2). Proc Nalt Acad Sci 84:179–183

Campos C, Guzmán R, López-Fernández E, Casado Á (2009) Evaluation of the copper (II) reduction assay using bathocuproinedisulfonic acid disodium salt for the total antioxidant capacity assessment: The CUPRAC–BCS assay. Anal Biochem 392(1):37–44

Caregnato FF, Koller CE, MacFarlane GR, Moreira JC (2008) The glutathione antioxidant system as a biomarker suite for the assessment of heavy metal exposure and effect in the grey mangrove, *Avicennia marina* (Forsk.) Vierh. Mar Pollut Bull 56:1119–1127

Cargnelutti D, Tabaldi LA, Spanevello RM, de Oliveira JG, Battisti V, Redin M, Morsch VM (2006) Mercury toxicity induces oxidative stress in growing cucumber seedlings. Chemosphere 65:999–1006

Caverzan A, Passaia G, Rosa SB, Ribeiro CW, Lazzarotto F, Margis-Pinheiro M (2012) Plant responses to stresses: role of ascorbate peroxidase in the antioxidant protection. Genet Mol Biol 35:1011–1019

Çekiç SD, Avan AN, Uzunboy S, Apak R (2015) A colourimetric sensor for the simultaneous determination of oxidative status and antioxidant activity on the same membrane: N, N-Dimethyl-p-phenylene diamine (DMPD) on Nafion. Anal Chim Acta 865:60–70

Cheeseman JM (2006) Hydrogen peroxide concentrations in leaves under natural conditions. J Exp Bot 57:2435–2444

Chen Z, Gallie DR (2006) Dehydroascorbate reductase affects leaf growth, development, and function. Plant Physiol 142:775–787

Chugh V, Kaur N, Gupta AK (2011) Evaluation of oxidative stress tolerance in maize (*Zea mays* L.) seedlings in response to drought. Indian J Biochem Biophys 48:47–53
Collin VC, Eymery F, Genty B, Rey P, Havaux M (2008) Vitamin E is essential for the tolerance of *Arabidopsis thaliana* to metal induced oxidative stress. Plant Cell Environ 31:244–257
Dai HP, Shan CJ, Zhao H, Li JC, Jia GL, Jiang H, Wang Q (2015) The difference in antioxidant capacity of four alfalfa cultivars in response to Zn. Ecotoxicol Environ Safe 114:312–317
Das K, Roychoudhury A (2014) Reactive oxygen species (ROS) and response of antioxidants as ROS-scavengers during environmental stress in plants. Front Environ Sci 2:53
Ding S, Lei M, Lu Q, Zhang A, Yin Y, Wen X, Lu C (2012) Enhanced sensitivity and characterization of photosystem II in transgenic tobacco plants with decreased chloroplast glutathione reductase under chilling stress. BBA Bioenerg 1817:1979–1991
Dixit G, Singh AP, Kumar A, Mishra S, Dwivedi S, Kumar S, Tripathi RD (2016) Reduced arsenic accumulation in rice (*Oryza sativa* L.) shoot involves sulfur mediated improved thiol metabolism, antioxidant system and altered arsenic transporters. Plant Physiol Biochem 99:86–96
Drążkiewicz M, Skórzyńska-Polit E, Krupa Z (2004) Copper-induced oxidative stress and antioxidant defence in *Arabidopsis thaliana*. Biometals 17:379–387
Duman F, Ozturk F (2010) Nickel accumulation and its effect on biomass, protein content and antioxidative enzymes in roots and leaves of watercress (*Nasturtium officinale* R. Br.). J Environ Sci 22:526–532
Eltayeb AE, Kawano N, Badawi GH, Kaminaka H, Sanekata T, Shibahara T, Tanaka K (2007) Overexpression of monodehydroascorbate reductase in transgenic tobacco confers enhanced tolerance to ozone, salt and polyethylene glycol stresses. Planta 225:1255–1264
Emamverdian A, Ding Y, Mokhberdoran F, Xie Y (2015) Heavy metal stress and some mechanisms of plant defense response. Sci World J 2015 Article ID 756120: 18
Eshagberi GO (2012) Toxic effects of heavy metal on crop plants. Multidiscip J Emp Res 10:1–10
Fariduddin Q, Yusuf M, Hayat S, Ahmad A (2009) Effect of 28-homobrassinolide on antioxidant capacity and photosynthesis in *Brassica juncea* plants exposed to different levels of copper. Environ Exp Bot 66:418–424
Fatima RA, Ahmad M (2005) Certain antioxidant enzymes of *Allium cepa* as biomarkers for the detection of toxic heavy metals in wastewater. Sci Total Environ 346:256–273
FAO (2009) Press release, 19 June 2009. http://www.fao.org/news/story/en/item/20568/icode/
Fernández LG, Fernández-Pascual M, Mañero FJG, García JAL (2015) Phytoremediation of contaminated waters to improve water quality. In: Phytoremediation. Springer International Publishing, pp 11–26
Ferrat L, Pergent-Martini C, Roméo M (2003) Assessment of the use of biomarkers in aquatic plants for the evaluation of environmental quality: application to seagrasses. Aquat Toxicol 65:187–204
Filippenko V, Frenette M, Scaiano JC (2009) Solvent-independent antioxidant activity from thermally generated carbon-centered radical antioxidants. Org Lett 11:3634–3637
Fini A, Brunetti C, Di Ferdinando M, Ferrini F, Tattini M (2011) Stress-induced flavonoid biosynthesis and the antioxidant machinery of plants. Plant Signal Behav 6:709–711
Flora SJ (2009) Structural, chemical and biological aspects of antioxidants for strategies against metal and metalloid exposure. Oxid Med Cell Longev 2:191–206
Foyer CH, Noctor G (2005) Redox homeostasis and antioxidant signaling: a metabolic interface between stress perception and physiological responses. Plant Cell 17:1866–1875
Fritioff Å, Greger M (2003) Aquatic and terrestrial plant species with potential to remove heavy metals from stormwater. Int J Phytorem 5:211–224
Fryer MJ, Ball L, Oxborough K, Karpinski S, Mullineaux PM, Baker NR (2003) Control of ascorbate peroxidase 2 expression by hydrogen peroxide and leaf water status during excess light stress reveals a functional organisation of Arabidopsis leaves. Plant J 33:691–705
Gill SS, Tuteja N (2010) Reactive oxygen species and antioxidant machinery in abiotic stress tolerance in crop plants. Plant Physiol Biochem 48:909–930

Gill SS, Khan NA, Anjum NA, Tuteja N (2011) Amelioration of cadmium stress in crop plants by nutrients management: morphological, physiological and biochemical aspects. Plant Stress 5:1–23

Gliszczyńska-Świgło A (2006) Antioxidant activity of water soluble vitamins in the TEAC (trolox equivalent antioxidant capacity) and the FRAP (ferric reducing antioxidant power) assays. Food Chem 96:131–136

Gout E, Boisson AM, Aubert S, Douce R, Bligny R (2001) Origin of the cytoplasmic pH changes during anaerobic stress in higher plant cells. Carbon-13 and phosphorous-31 nuclear magnetic resonance studies. Plant Physiol 125:912–925

Grassmann J, Hippeli S, Elstner EF (2002) Plant's defence and its benefits for animals and medicine: role of phenolics and terpenoids in avoiding oxygen stress. Plant Physiol Biochem 40:471–478

Gutteridge J, Halliwell B (2000) Free radicals and antioxidants in the year 2000: a historical look to the future. Ann N Y Acad Sci 899:136–147

Hajar EWI, Sulaiman AZB, Sakinah AM (2014) Assessment of heavy metals tolerance in leaves, stems and flowers of *Stevia rebaudiana* plant. Proc Environ Sci 20:386–393

Halliwell B (2006) Oxidative stress and neurodegeneration: where are we now? J Neurochem 97:1634–1658

Han D, Xiong S, Tu S, Liu J, Chen C (2015) Interactive effects of selenium and arsenic on growth, antioxidant system, arsenic and selenium species of *Nicotiana tabacum* L. Environ Exp Bot 117:12–19

Hartmann A, Nieβ AM, Grünert-Fuchs M, Poch B, Speit G (1995) Vitamin E prevents exercise-induced DNA damage. Mut Res Lett 346:195–202

Haytowitz DB, Bhagwat S (2010) USDA database for the oxygen radical absorbance capacity (ORAC) of selected foods, Release 2. US Department of Agriculture

He ZL, Yang XE, Stoffella PJ (2005) Trace elements in agroecosystems and impacts on the environment. J Trace Elem Med Biol 19:125–140

Hernández I, Alegre L, Van Breusegem F, Munné-Bosch S (2009) How relevant are flavonoids as antioxidants in plants? Trends Plant Sci 14:125–132

Iannone MF, Groppa MD, Benavides MP (2015) Cadmium induces different biochemical responses in wild type and catalase-deficient tobacco plants. Environ Exp Bot 109:201–211

Igamberdiev AU, Mikkelsen TN, Ambus P, Bauwe H, Lea PJ, Gardeström P (2004) Photorespiration contributes to stomatal regulation and carbon isotope fractionation: a study with barley, potato and Arabidopsis plants deficient in glycine decarboxylase. Photosynth Res 81:139–152

Imlay JA (2008) Cellular defenses against superoxide and hydrogen peroxide. Annu Rev Biochem 755

Islam F, Yasmeen T, Riaz M, Arif MS, Ali S, Raza SH (2014) *Proteus mirabilis* alleviates zinc toxicity by preventing oxidative stress in maize (*Zea mays*) plants. Ecotoxical Environ Safe 110:143–152

Israr M, Sahi S, Datta R, Sarkar D (2006) Bioaccumulation and physiological effects of mercury in *Sesbania drummondii*. Chemosphere 65:591–598

Jaleel CA, Jayakumar K, Chang-Xing Z, Azooz MM (2009) Antioxidant potentials protect *Vigna radiata* (L.) Wilczek plants from soil cobalt stress and improve growth and pigment composition. Plant Omics 2:120

Jayakumar K, Vijayarengan P, Changxing Z, Gomathinayagam M, Jaleel CA (2008) Soil applied cobalt alters the nodulation, leg-haemoglobin content and antioxidant status of *glycine max* (L.). Merr Collid Surf B 67:272–275

Johnston JW, Dussert S, Gale S, Nadarajan J, Harding K, Benson EE (2006) Optimisation of the azinobis-3-ethyl-benzothiazoline-6-sulphonic acid radical scavenging assay for physiological studies of total antioxidant activity in woody plant germplasm. Plant Physiol Biochem 44:193–201

Kabata A, Pendias H (2001) Trace elements in soils and plants. CRC, Washington

Kabata-Pendias A (2004) Soil–plant transfer of trace elements—an environmental issue. Geoderma 122:143–149

Kafel A, Nadgórska-Socha A, Gospodarek J, Babczyńska A, Skowronek M, Kandziora M, Rozpędek K (2010) The effects of *Aphis fabae* infestation on the antioxidant response and heavy metal content in field grown *Philadelphus coronarius* plants. Sci Total Environ 408:1111–1119

Kähkönen MP, Heinonen M (2003) Antioxidant activity of anthocyanins and their aglycons. J Agric Food Chem 51:628–633

Karadag A, Ozcelik B, Saner S (2009) Review of methods to determine antioxidant capacities. Food Anal Method 2:41–60

Karuppanapandian T, Manoharan K (2008) Uptake and translocation of tri- and hexa-valent chromium and their effects on black gram (*Vigna mungo* L. Hepper cv. Co4) roots. J Plant Biol 51:192–201

Karuppanapandian T, Sinha PB, Haniya AMK, Mamoharan K (2006) Differential antioxidative responses of ascorbate-glutathione cycle enzymes and metabolites to chromium stress in green gram (*Vigna radiata* L. Wilczek) leaves. J Plant Biol 49:440–447

Karuppanapandian T, Moon JC, Kim C, Manoharan K, Kim W (2011) Reactive oxygen species in plants: their generation, signal transduction, and scavenging mechanisms. Aust J Crop Sci 709

Kasote DM, Katyare SS, Hegde MV, Bae H (2015) Significance of antioxidant potential of plants and its relevance to therapeutic applications. Int J Biol Sci 982

Kazemi N, Khavari-Nejad RA, Fahimi H, Saadatmand S, Nejad-Sattari T (2010) Effects of exogenous salicylic acid and nitric oxide on lipid peroxidation and antioxidant enzyme activities in leaves of *Brassica napus* L. under nickel stress. Sci Hortic –Amsterdam 126:402–407

Keilig K, Ludwig-Müller J (2009) Effect of flavonoids on heavy metal tolerance in *Arabidopsis thaliana* seedlings. Bot Stud 50:311–318

Khatun S, Ali MB, Hahn EJ, Paek KY (2008) Copper toxicity in *Withania somnifera:* growth and antioxidant enzymes responses of in vitro grown plants. Environ Exp Bot 64:279–285

Kiffin R, Bandyopadhyay U, Cuervo AM (2006) Oxidative stress and autophagy. Antioxid Redox Signal 8:152–162

Kliebenstein DJ, Osbourn A (2012) Making new molecules–evolution of pathways for novel metabolites in plants. Curr Opin Plant Biol 15:415–423

Koji Y, Shiro M, Michio K, Mitsutaka T, Hiroshi M (2009) Antioxidant capacity and damages caused by salinity stress in apical and basal regions of rice leaf. Plant Prod Sci 12:319–326

Korkina LG (2007) Phenylpropanoids as naturally occurring antioxidants: from plant defense to human health. Cell Mol Biol 53:15–25

Kotapati KV, Palaka BK, Kandukuri A, Reddy R (2014) Response of antioxidative enzymes and lipoxygenase to drought stress in finger millet leaves (*Eleusine Coracana* (L.) Gaertn)

Krishnaiah D, Sarbatly R, Nithyanandam R (2011) A review of the antioxidant potential of medicinal plant species. Food Bioprod Process 89:217–233

Kuiper C, Vissers MC (2013) Ascorbate as a co-factor for Fe-and 2-oxoglutarate dependent dioxygenases: physiological activity in tumor growth and progression. Front Oncol 4:359–359

Kumar JN, Soni H, Kumar RN, Bhatt I (2008) Assessing heavy metal hyper-accumulation and mobility in selected vegetable crops: a case study of organic farm, Gujarat, India. Nat Environ Pollut Technol 7(2):203–210

Kumar RR, Goswami S, SinghK GK, Rai GK, Rai RD (2013) Modulation of redox signal transduction in plant system through induction of free radical/ROS scavenging redox-sensitive enzymes and metabolites. Aust J Crop Sci 1744

Kusvuran S (2012) Influence of drought stress on growth, ion accumulation and antioxidative enzymes in okra genotypes. Int J Agric Biol 14:401–406

Kuźniak E, Głowacki R, Chwatko G, Kopczewski T, Wielanek M, Gajewska E, Skłodowska M (2014) Involvement of ascorbate, glutathione, protein S-thiolation and salicylic acid in benzothiadiazole-inducible defence response of cucumber against *Pseudomonas syringae pv lachrymans*. Physiol Mol Plant 86:89–97

Lasat MM (2000) Phytoextraction of metals from contaminated soil: a review of plant/soil/metal interaction and assessment of pertinent agronomic issues. J Hazard Sub Res 2:1–25

Lenntech Water Treatment and Air Purification (2004) Water treatment. Lenntech, Rotterdamseweg, Netherlands (http://www.excelwater.com/thp/filters/Water-Purification.htm)

Leterrier M, Corpas FJ, Barroso JB, Sandalio LM, Luis A (2005) Peroxisomal monodehydroascorbate reductase. Genomic clone characterization and functional analysis under environmental stress conditions. Plant Physiol 138:2111–2123

Li Z, Keasling JD, Niyogi KK (2012) Overlapping photoprotective function of vitamin E and carotenoids in *Chlamydomonas*. Plant Physiol 158:313–323

Li X, Yang Y, Jia L, Chen H, Wei X (2013) Zinc-induced oxidative damage, antioxidant enzyme response and proline metabolism in roots and leaves of wheat plants. Ecotoxicol Environ Saf 89:150–157

Lomonte C, Sgherri C, Baker AJ, Kolev SD, Navari-Izzo F (2010) Antioxidative response of *Atriplex codonocarpa* to mercury. Environ Exp Bot 69:9–16

Lopez-Huertas E, Charlton WL, Johnson B, Graham IA, Baker A (2000) Stress induces peroxisome biogenesis genes. EMBO J 19:6770–6777

Løvdal T, Olsen KM, Slimestad R, Verheul M, Lillo C (2010) Synergetic effects of nitrogen depletion, temperature, and light on the content of phenolic compounds and gene expression in leaves of tomato. Phytochemistry 71:605–613

Malar S, Vikram SS, Favas PJ, Perumal V (2014) Lead heavy metal toxicity induced changes on growth and antioxidative enzymes level in water hyacinths [*Eichhornia crassipes* (Mart.)]. Bot Stud 54

Malik JA, Goel S, Kaur N, Sharma S, Singh I, Nayyar H (2012) Selenium antagonises the toxic effects of arsenic on mungbean (*Phaseolus aureus* Roxb.) plants by restricting its uptake and enhancing the antioxidative and detoxification mechanisms. Environ Exp Bot 77:242–248

Mangabeira PA, Gavrilov KL, De Almeida AAF, Oliveira AH, Severo MI, Rosa TS, Mielke MS (2006) Chromium localization in plant tissues of *Lycopersicum esculentum* Mill using ICP-MS and ion microscopy (SIMS). Appl Surf Sci 252:3488–3501

Mazid M, Khan TA, Mohammad F (2011) Role of secondary metabolites in defense mechanisms of plants. Biol Med 3:232–249

Metwally A, Safronova VI, Belimov AA, Dietz KJ (2005) Genotypic variation of the response to cadmium toxicity in *Pisum sativum* L. J Exp Bot 56:167–178

Minkov IN, Jahoubjan GT, Denev ID, Toneva VT (1999) Photooxidative stress in higher plants. Handb Plant Crop Stress 2:499–525

Mithöfer A, Schulze B, Boland W (2004) Biotic and heavy metal stress response in plants: evidence for common signals. Febs Lett 566:1–5

Mittler R (2002) Oxidative stress, antioxidants and stress tolerance. Trends Plant Sci 7:405–410

Mobin M, Khan NA (2007) Photosynthetic activity, pigment composition and antioxidative response of two mustard (*Brassica juncea*) cultivars differing in photosynthetic capacity subjected to cadmium stress. J Plant Physiol 164:601–610

Myburgh KH (2014) Polyphenol supplementation: benefits for exercise performance or oxidative stress? Sports Med 44:57–70

Nagajyoti PC, Lee KD, Sreekanth TVM (2010) Heavy metals, occurrence and toxicity for plants: a review. Environ Chem Lett 8:199–216

Nahar K, Hasanuzzaman M, Alam MM, Rahman A, Suzuki T, Fujita M (2016) Polyamine and nitric oxide crosstalk: antagonistic effects on cadmium toxicity in mung bean plants through upregulating the metal detoxification, antioxidant defense and methylglyoxal detoxification systems. Ecotoxicol Environ Safe 126:245–255

Namdjoyan S, Kermanian H (2013) Exogenous nitric oxide (as sodium nitroprusside) ameliorates arsenic-induced oxidative stress in watercress (*Nasturtium officinale* R. Br.) plants. Sci Hortic-Amsterdam 350–356

Noctor G, Foyer CH (1998) Simultaneous measurement of foliar glutathione, γ-glutamylcysteine, and amino acids by high-performance liquid chromatography: comparison with two other assay methods for glutathione. Anal Biochem 264:98–110

Nowicka B, Pluciński B, Kuczyńska P, Kruk J (2016) Prenyllipid antioxidants participate in response to acute stress induced by heavy metals in green microalga *Chlamydomonas reinhardtii*. Environ Exp Bot 123:98–107

Ordoudi SA, Tsimidou MZ (2006) Crocin bleaching assay step by step: observations and suggestions for an alternative validated protocol. J Agric Food Chem 54:1663–1671

Ou B, Huang D, Hampsch-Woodill M, Flanagan JA, Deemer EK (2002) Analysis of antioxidant activities of common vegetables employing oxygen radical absorbance capacity (ORAC) and ferric reducing antioxidant power (FRAP) assays: a comparative study. J Agric Food Chem 50:3122–3128

Ozgen M, Reese RN, Tulio AZ, Scheerens JC, Miller AR (2006) Modified 2, 2-azino-bis-3-ethylbenzothiazoline-6-sulfonic acid (ABTS) method to measure antioxidant capacity of selected small fruits and comparison to ferric reducing antioxidant power (FRAP) and 2, 2′-diphenyl-1-picrylhydrazyl (DPPH) methods. J Agric Food Chem 54:1151–1157

Padh H (1990) Cellular functions of ascorbic acid. Biochem Cell Biol 68:1166–1173

Palozza P, Krinsky NI (1992) β-Carotene and α-tocopherol are synergistic antioxidants. Arch Biochem Biophys 297:184–187

Parlak KU (2016) Effect of nickel on growth and biochemical characteristics of wheat (*Triticum aestivum* L.) seedlings. NJAS-Wagen J Life Sci 76:1–5

Parvaiz A, Satyawati S (2008) Salt stress and phyto-biochemical responses of plants-a review. Plant Soil Environ 54(3):89–99

Peijnenburg WJGM, Jager T (2003) Monitoring approaches to assess bioaccessibility and bioavailability of metals: matrix issues. Ecotoxicol Environ Safe 56:63–77

Phang C, Leung DW, Taylor HH, Burritt DJ (2011) The protective effect of sodium nitroprusside (SNP) treatment on *Arabidopsis thaliana* seedlings exposed to toxic level of Pb is not linked to avoidance of Pb uptake. Ecotoxicol Environ Safe 74:1310–1315

Pinto E, Sigaud-kutner T, Leitao MA, Okamoto OK, Morse D, Colepicolo P (2003) Heavy metal–induced oxidative stress in algae1. J Phycol 39:1008–1018

Posmyk MM, Kontek R, Janas KM (2009) Antioxidant enzymes activity and phenolic compounds content in red cabbage seedlings exposed to copper stress. Ecotoxicol Environ Safe 72:596–602

Prakash D, Sharma G (eds) (2014) Phytochemicals of nutraceutical importance. CABI

Prior RL, Wu X, Schaich K (2005) Standardized methods for the determination of antioxidant capacity and phenolics in foods and dietary supplements. J Agric Food Chem 53:4290–4302

Rabi T, Bishayee A (2009) Terpenoids and breast cancer chemoprevention. Breast Cancer Res Treat 115:223–239

Ragavendran P, Sophia D, Arul Raj C, Starlin T, Gopalakrishnan VK (2012) Phytochemical screening, antioxidant activity of *Aerva lanata (L)*–an in vitro study. Asian J Pharm Clin Res 5:77–81

Rahman K (2007) Studies on free radicals, antioxidants, and co-factors. Clin Interv Aging 219

Rao KM, Sresty TVS (2000) Antioxidative parameters in the seedlings of pigeonpea (*Cajanus cajan* (L.) Millspaugh) in response to Zn and Ni stresses. Plant Sci 157:113–128

Regelsberger G, Atzenhofer W, Rüker F, Peschek GA, Jakopitsch C, Paumann M, Obinger C (2002) Biochemical characterization of a membrane-bound manganese-containing superoxide dismutase from the cyanobacterium *Anabaena* PCC 7120. J Biol Chem 277:43615–43622

Romero-Puertas MC, Corpas FJ, Sandalio LM, Leterrier M, Rodríguez-Serrano M, Del Río LA, Palma JM (2006) Glutathione reductase from pea leaves: response to abiotic stress and characterization of the peroxisomal isozyme. New Phytol 170:43–52

Salido AL, Hasty KL, Lim JM, Butcher DJ (2003) Phytoremediation of arsenic and lead in contaminated soil using Chinese brake ferns (*Pteris vittata*) and Indian mustard (*Brassica juncea*). Int J Phytorem 5:89–103

Sánchez-Pardo B, Fernández-Pascual M, Zornoza P (2012) Copper microlocalisation, ultrastructural alterations and antioxidant responses in the nodules of white lupin and soybean plants grown under conditions of copper excess. Environ Exp Bot 84:52–60

Sbartai H, Djebar MR, Sbartai I, Berrabbah H (2012) Bioaccumulation of cadmium and zinc in tomato (*Lycopersicon esculentum* L.). C R Biol 335:585–593
Schardl CL, Panaccione DG, Tudzynski P (2006) Ergot alkaloids–biology and molecular biology. Alkaloids Chem Boil 63:45–86
Schrader M, Fahimi HD (2006) Peroxisomes and oxidative stress. Biochim Biophys Acta Mol Cell Res 1763:1755–1766
Schützendübel A, Polle A (2002) Plant responses to abiotic stresses: heavy metal-induced oxidative stress and protection by mycorrhization. J Exp Bot 53:1351–1365
Shahid M, Pinelli E, Pourrut B, Dumat C (2014) Effect of organic ligands on lead-induced oxidative damage and enhanced antioxidant defense in the leaves of *Vicia faba* plants. J Geochem Explor 144:282–289
Shalaby E, Shanab S (2013) Antioxidant compounds, assays of determination and mode of action. Afr J Pharm Pharmacol 7:528–539
Shamsi IH, Wei K, Zhang GP, Jilani GH, Hassan MJ (2008) Interactive effects of cadmium and aluminum on growth and antioxidative enzymes in soybean. Biol Plant 52:165–169
Shanker AK, Djanaguiraman M, Sudhagar R, Chandrashekar CN, Pathmanabhan G (2004) Differential antioxidative response of ascorbate glutathione pathway enzymes and metabolites to chromium speciation stress in green gram (*Vigna radiata* (L.) R. Wilczek. cv CO 4) roots. Plant Sci 166:1035–1043
Sharma P, Jha AB, Dubey RS, Pessarakli M (2012) Reactive oxygen species, oxidative damage, and antioxidative defense mechanism in plants under stressful conditions. J Bot 2012: Article ID 217037, 26 pages
Sharma P, Kumar A, Bhardwaj R (2016) Plant steroidal hormone epibrassinolide regulate–Heavy metal stress tolerance in *Oryza sativa* L. by modulating antioxidant defense expression. Environ Exp Bot 1–9
Shigeoka S, Ishikawa T, Tamoi M, Miyagawa Y, Takeda T, Yabuta Y, Yoshimura K (2002) Regulation and function of ascorbate peroxidase isoenzymes. J Exp Bot 53:1305–1319
Sies H (1997) Oxidative stress: oxidants and antioxidants. Exp Physiol 82:291–295
Sies H (2007) Total antioxidant capacity: appraisal of a concept. J Nutr 137:1493–1495
Sinha RK, Herat S, Tandon PK (2007) Phytoremediation: role of plants in contaminated site management. In: Environmental Bioremediation Technologies, pp 315–330
Singh S, Srivastava PK, Kumar D, Tripathi DK, Chauhan DK, Prasad SM (2015) Morpho-anatomical and biochemical adapting strategies of maize (*Zea mays* L.) seedlings against lead and chromium stresses. Biocatal Agric Biotechnol 4(3):286–295
Smirnoff N (2005) Ascorbate, tocopherol and carotenoids: metabolism, pathway engineering and functions. Antioxidants and Reactive Oxygen Species in Plants 53–86
Srivastava NK, Srivastava AK (2010) Influence of some heavy metals on growth, alkaloid content and composition in *Catharanthus roseus* L. Indian J Pharm Sci 775
Sun RL, Zhou QX, Sun FH, Jin CX (2007) Antioxidative defense and proline/phytochelatin accumulation in a newly discovered Cd-hyperaccumulator, Solanum nigrum L. Environ Exp Bot 60:468–476
Takeda T, Yoshimura K, Ishikawa T, Shigeoka S (1998) Purification and characterization of ascorbate peroxidase in *Chlorella vulgaris*. Biochimie 80:295–301
Tangahu BV, Sheikh Abdullah SR, Basri H, Idris M, Anuar N, Mukhlisin M (2011) A review on heavy metals (As, Pb, and Hg) uptake by plants through phytoremediation. Int J Chem Eng 2011: Article ID 939161, 31 pages
Tchounwou PB, Yedjou CG, Patlolla AK, Sutton DJ (2012) Heavy metal toxicity and the environment. In: Molecular, clinical and environmental toxicology, pp 133–164 EXS. 101:133–164
Traber MG, Stevens JF (2011) Vitamins C and E: beneficial effects from a mechanistic perspective. Free Radic Biol Med 51(5):1000–1013
Tripathi DK, Singh VP, Prasad SM, Chauhan DK, Dubey NK (2015) Silicon nanoparticles (SiNp) alleviate chromium (VI) phytotoxicity in *Pisum sativum* (L.) seedlings. Plant Physiol Biochem 96:189–198

Tripathi A, Tripathi DK, Chauhan DK, Kumar N (2016) Chromium (VI)-induced phytotoxicity in river catchment agriculture: evidence from physiological, biochemical and anatomical alterations in Cucumis sativus (L.) used as model species. Chem Ecol 32(1):12–33

Tripathy BC, Oelmüller R (2012) Reactive oxygen species generation and signaling in plants. Plant Signal Behav 7:1621–1633

Vellosillo T, Vicente J, Kulasekaran S, Hamberg M, Castresana C (2010) Emerging complexity in reactive oxygen species production and signaling during the response of plants to pathogens. Plant Physiol 154:444–448

Verma S, Dubey RS (2003) Lead toxicity induces lipid peroxidation and alters the activities of antioxidant enzymes in growing rice plants. Plant Sci 164:645–655

Viehweger K (2014) How plants cope with heavy metals. Bot Stud 35

Vitória AP, Lea PJ, Azevedo RA (2001) Antioxidant enzymes responses to cadmium in radish tissues. Phytochemistry 57:701–710

Wang X, Quinn PJ (2000) The location and function of vitamin E in membranes (Review). Mol Membr Biol 17:143–156

Weihong XU, Wenyi LI, Jianping HE, Singh B, Xiong Z (2009) Effects of insoluble Zn, Cd, and EDTA on the growth, activities of antioxidant enzymes and uptake of Zn and Cd in Vetiveria zizanioides. J Environ Sci 21:186–192

Winkel-Shirley B (2002) Biosynthesis of flavonoids and effects of stress. Curr Opin Plant Biol 5:218–223

Yadav SK (2010) Heavy metals toxicity in plants: an overview on the role of glutathione and phytochelatins in heavy metal stress tolerance of plants. S Afr J Bot 76:167–179

Yoon HS, Lee H, Lee IA, Kim KY, Jo J (2004) Molecular cloning of the monodehydroascorbate reductase gene from *Brassica campestris* and analysis of its mRNA level in response to oxidative stress. BBA Bioenerg 1658:181–186

Young AJ, Lowe GM (2001) Antioxidant and prooxidant properties of carotenoids. Arch Biochem Biophys 385:20–27

Yusuf M, Khan TA, Fariduddin Q (2016) Interaction of epibrassinolide and selenium ameliorates the excess copper in *Brassica juncea* through altered proline metabolism and antioxidants. Ecotoxical Environ Safe 129:25–34

Zamocky M, Furtmüller PG, Obinger C (2008) Evolution of catalases from bacteria to humans. Antioxid Redox Signal 10:1527–1548

Zengin FK, Munzuroglu O (2005) Effects of some heavy metals on content of chlorophyll, proline and some antioxidant chemicals in bean (*Phaseolus vulgaris* L.) seedlings. Acta Biol Cracov Bot 47:157–164

Zhao J, Davis LC, Verpoorte R (2005) Elicitor signal transduction leading to production of plant secondary metabolites. Biotechnol Adv 23:283–333

Zhou ZS, Wang SJ, Yang ZM (2008) Biological detection and analysis of mercury toxicity to alfalfa (*Medicago sativa*) plants. Chemosphere 70:1500–1509

Ziegler J, Facchini PJ (2008) Alkaloid biosynthesis: metabolism and trafficking. Annu Rev Plant Biol 59:735–769

Zitka O, Skalickova S, Gumulec J, Masarik M, Adam V, Hubalek J, Kizek R (2012) Redox status expressed as GSH: GSSG ratio as a marker for oxidative stress in paediatric tumour patients. Oncol Lett 4:1247–1253

Impact of Heavy Metals on Physiological Processes of Plants: With Special Reference to Photosynthetic System

6

Richa Rai, Madhoolika Agrawal, and S.B. Agrawal

Abstract

Heavy metal pollution is one of the major environmental problems which is caused by unchecked and uncontrolled discharge of hazardous chemicals consisting of heavy metals. Heavy metals affect agricultural land and plant's productivity by affecting its most vital process, i.e., photosynthesis. Exposure of plants to heavy metals leads to generation of ROS, and few heavy metals are directly involved in electron transport. It also leads to lipid peroxidation destroying plant's cell membrane and its associated organelles. Chlorophyll biosynthesis is affected by heavy metals as it substitutes Mg ion from chlorophyll. Few heavy metals also affect enzyme involved in chlorophyll biosynthesis as well as disturb electron transport in light reactions and affect various enzymes in dark reactions. This detailed information helps us to understand the mechanism linked with the impact of heavy metals on physiological responses affecting plant biomass and productivity which is of concern for agriculture. Therefore, the present chapter consists of whole information regarding the impact of heavy metals on photosynthetic pigments, photosynthetic apparatus, and light and dark reactions.

Keywords

Heavy metals • Lipid peroxidation • ROS • Photosynthetic pigments • PS I • PS II • Calvin cycle

R. Rai • M. Agrawal (✉) • S.B. Agrawal
Laboratory of Air Pollution and Global Climate Change, Department of Botany, Banaras Hindu University, Varanasi 221005, Uttar Pradesh, India
e-mail: madhoo58@yahoo.com

A. Singh et al. (eds.), *Plant Responses to Xenobiotics*,
DOI 10.1007/978-981-10-2860-1_6

6.1 Introduction

Urbanization and industrialization have led to several disturbances in the environment and among them heavy metal pollution is one of the serious problems. Heavy metal applies to the group of metal and metalloids with the specific density greater than 5 g cm^{-3} or greater than water (Michalak 2006). Heavy metals include lead (Pb), cadmium (Cd), nickel (Ni), cobalt (Co), iron (Fe), zinc (Zn), chromium (Cr), iron (Fe), arsenic (As), silver (Ag), and the platinum group elements. Sources of heavy metals are metalliferous mining and smelting, use of chemicals in agricultural practices, waste disposal, and discharge of metals (Michalak 2006; Lin and Mark 2012). There are different sources of heavy metals in the environment such as natural, agricultural, industrial, domestic effluent, atmospheric, and other sources. Natural sources of heavy metals include heavy metals originated within the Earth's crust through weathering process. The composition and concentration of heavy metals mainly depended on the rock type and environmental conditions which activated the weathering process. The geologic plant materials normally have high concentrations of Cr, Mn, Co, Ni, Cu, Zn, Cd, Sn, Hg, and Pb. While igneous rocks such as olivine, augite, and hornblende are rich in Mn, Co, Ni, Cu, and Zn (Pandey et al. 2014), sedimentary rocks have highest concentrations of Cr, Mn, Co, Ni, Cu, Zn, Cd, Sn, Hg, and Pb (Canon et al. 1998). Eruption of volcanoes along with toxic gases also emits high levels of Al, Zn, Mn, Pb, Ni, Cu, and Hg (Barghiani et al. 1987).

In Sahara region, wind dust has high levels of Fe and lesser concentrations of Mn, Zn, Cr, Ni, and Pb (Pacyna 1986). While the long-range transport of dust, particularly from the Sahara, has received considerable recent attention, factors such as marine aerosols and forest fires are responsible for transport of some heavy metals in the environment. In agriculture, sources of heavy metals are the use of inorganic and organic fertilizers, and apart from fertilizers, liming, sewage sludge, irrigation waters, and pesticides are also other sources of heavy metals in the agricultural soils. Fungicides, inorganic fertilizers, and phosphates have variable levels of Cd, Cr, Ni, Pb, and Zn depending on their sources. Normally, the levels of heavy metals in agricultural soil are very small, but constant use of phosphate fertilizer may lead to high accumulation of some metals (Atafar et al. 2010). Industrial sources of heavy metals include mining operation which emits different heavy metals depending on the type of mining, like As, Cd, Fe, etc., and lead to the enrichment of the soil around the coalfield directly or indirectly (Pandey et al. 2014). The utilization of Hg in gold mining and the mobilization of significantly high amounts of Hg led to the significant source of Hg in the environment (Lacerda 1997). Heavy metals affect crop, animals, and human health via food chain (Dziubanek et al. 2015; Augustsson et al. 2015).

It is reported that heavy metals unfavorably affect about 12 % of the world agricultural land and crop productivity (Dziubanek et al. 2015). Heavy metal initiates different responses in plants which ranged from physiological, biochemical to crop yield (Dubey 2011; Villiers et al. 2011). Variation in heavy metal toxicity depends on plant species, specific metal, concentration, chemical form and soil composition, and pH, as many heavy metals are considered to be essential for plant growth like Cu and Zn which either serve as cofactor and activators of enzyme reactions, e.g., in forming enzymes/substrate metal complex (Lin and Mark 2012), or exert a catalytic property such as prosthetic group in metalloproteins. Essential trace metal nutrients take part in redox reactions, electron transfer, and structural functions in nucleic acid metabolism. Heavy metals influence the function of photosynthetic apparatus at various level of organization. Heavy metals posses direct effect on plants by affecting PS II and PS I and indirect effect on photosynthesis, growth, and yield. Cd, Hg, and As are phytotoxic to metal-sensitive enzymes, resulting in growth inhibition and death of organisms.

The present review focuses on impact of heavy metals on physiological processes of plant from its uptake, ROS formation, and its effect on photosynthetic pigments, photosynthetic apparatus, and photosynthetic reactions.

6.2 Heavy Metals Induced ROS Generation

The toxicity of heavy metals is interpreted in many ways after accumulation in plant cells. Heavy metals can be divided into two groups: redox active (Fe, Cu, Cr, Co) and redox inactive (Cd, Zn, Ni, Al, etc.). The redox-active heavy metals such as Fe, Cu, and Cd are directly involved in the redox reaction in cells and result in the formation of $O_2^{\bullet -}$ and subsequently in H_2O_2 and •OH production via the Haber-Weiss and Fenton reactions (Schutzendubel and Polle 2002). Exposure of plants to redox-inactive heavy metals also results in oxidative stress through indirect mechanisms such as interaction with the antioxidant defense system, disruption of the electron transport chain, and induction of lipid peroxidation. Źróbek et al. (2009) showed induction of ROS in *Nicotiana tabacum* L. in Cd and Zn treatment. Weckx and Clijsters (1997) detected increased levels of H_2O_2 in zinc-stressed bean leaves.

Heavy metals like Fe, Cu, Zn, Cd, Hg, and Pb are involved in the direct or indirect generation of free radicals and ROS in the following ways: (1) direct transfer of electrons in single-electron reactions, (2) disturbance of metabolic pathways results in an increase in the rate of free radicals and ROS formation, (3) inactivation and downregulation of enzymes of the antioxidative defense system, and (4) depletion of low molecular weight antioxidants.

Heavy metals are also involved in chemical reactions, biochemical pathways, and physiological processes which generate and release free radicals in plants (Fraustro da Silva and Williams 1991). Within biological systems, oxygen-, sulfur-, nitrogen-, and carbon-centered radicals generate ROS such as hydroxyl, lipoxy, thiyl, phenyl, and nitroxide radicals which occur as a result of radical chain reactions and direct electron transfer (Kalyanaram 1996). Among the most prominent and most damaging highly reactive oxygen-centered radicals are superoxide radical anions, hydroxyl radicals, and lipoxy radicals.

Plants exposed to heavy metal stress also show alteration in the lipid composition of thylakoid membranes. Lipid peroxidation process involves the peroxidative degradation of polyunsaturated fatty acids of membrane lipids and thus brings about membrane deterioration (De Vos et al. 1993). Lipid peroxidation capacity gets enhanced markedly in Cd^{2+}-treated wheat seedlings compared to control seedlings suggesting that Cd^{2+} induced the alteration in the structural composition of the membrane. According to Skórzńska et al. (1998), enhanced activity of galactolipase in Cd^{2+}-treated plants was recorded which led to an increased release of fatty acids due to degradation of galactolipids (MGDG). Somashekaraiah et al. (1992) have reported the importance of lipoxygenase in induction of the lipid peroxidation (through the production of free radicals from deoxygenation of membrane lipids and unsaturated fatty acids) in Cd^{2+}-treated plants. In plants exposed to heavy metals, chlorophyll bleaching occurs along with malondialdehyde (MDA) production (Somashekaraiah et al. 1992). Hence, heavy metal stress-induced lipid peroxidation also leads to the reduction in the photosynthetic oxygen evolution (Alia et al. 1997). Within the plant cell, plasma membranes are the primary target of heavy metal-induced ROS (Emamuerdian et al. 2015). The phenomenon of lipid peroxidation is most common in polyunsaturated fatty acids and involves three distinct stages: (i) initiation, (ii) progression, and (iii) termination. Under normal conditions, lipid peroxidation in green tissues is generally initiated by $O_2^{\bullet-}$ by-product of light capture in photosystem II (Pandey et al. 2009). Heavy metals are known to inhibit PS II and thus increase $O_2^{\bullet-}$ production in leaves which led to increased lipid peroxidation (Farmer and Mueller 2013). In an aerobic environment, oxygen reacts with the fatty acids, thereby creating another unstable peroxyl-fatty acid radical. Once initiated, ROS groups are capable to continue the peroxidation chain reaction by receiving a hydrogen atom from neighboring polyunsaturated fatty acids (Karuppanapandian et al. 2011). The resulting lipid hydroperoxide is a highly unstable molecule and decays into several reactive species such as lipid epoxides, aldehydes (malonylaldehydes), lipid alkoxyl radicals, alkanes, and alcohols (Bhattacharjee 2012). Lipid peroxidation led to increased membrane leakiness, decreased membrane fluidity, and damage to membrane proteins. Recent studies reported that heavy metal toxicity to different physiological processes occurs via ROS-induced lipid peroxidation and even by-products of lipid peroxidation like acrolein, linolenic acid −13- ketotriene, and 12-oxo-phytodienoic acid may strongly affect PS II and photosynthetic rate (Farmer and Mueller 2013; Shahid et al. 2014).

6.3 Effect of Heavy Metals on Photosynthetic Pigments

Chlorophylls (Chl) and carotenoids are important photosynthetic pigments and play a prime role in photosynthesis where solar energy converts to chemical energy. Particularly, heavy metal affects the biosynthesis of photosynthetic pigments. Chlorosis and retardation of plant growth are frequently observed in metal-polluted environments that indicate an impairment of photosynthetic pigment biosynthetic pathways, hence affecting plastid development, photosynthetic efficiency, and general metabolism. Heavy metal also reduces accumulation of photosynthetic pigments.

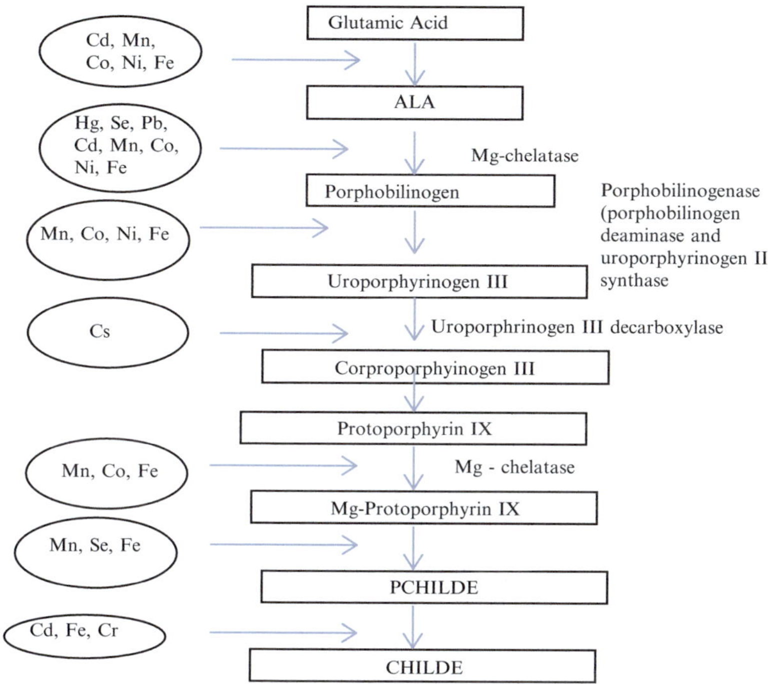

Fig. 6.1 Effect of heavy metals on chlorophyll biosynthesis (Modified: Kurdziel and Strazalka 1999)

6.3.1 Biosynthesis of Chlorophyll and Heavy Metals

Chlorophyll plays a fundamental role in the process of photosynthesis because of its ability to absorb light. In the branch of the pathway leading to the protochlorophyllide (Pchilde) synthesis, Mg is incorporated into protoporphyrin IX by Mg-chelatase (Fig. 6.1). The Mg porphyrin IX is methylated ester which is followed by the formation of the isocyclic ring characteristic to Chl-related pigment. Under heavy metal stress, chlorosis is the most common symptom. Metals interfere with pigment and influence the Chl content in plants as Mg^{2+} in chlorophyll is substituted by Cu^{2+}, Zn^{2+}, Cd^{2+}, Hg^{2+}, Pb^{2+}, or Ni^{2+} (Kupper et al. 1996) (Table 6.1).

There is a direct proportional relationship between metal toxicity and the rate of Mg substitution in chlorophyll ring, and the magnitude of this toxic effect could be ordered as Hg^{2+}, Cu^{2+}, $Cd^{2+} > Zn^{2+} > Ni^{2+} > Pb^{2+}$ (Kupper et al. 1996). Magnesium insertion into protoporphyrin IX is a specific reaction for Chl biosynthesis which is catalyzed by Mg-chelatase, and the inhibition of this reaction is caused by Cd^{2+} in cyanobacteria (*Anacystis nidulans*) which lead to accumulation of protoporphyrinogen IX which delayed Chl biosynthesis. Higher Cd^{2+} concentration caused cell growth retardation and prevention of protoporphyrinogen IX transformation (Csatorday et al. 1984). On the other hand, Mn^{2+} inhibited Chl biosynthesis and

Table 6.1 Effects of heavy metals on different processes of photosynthesis

Heavy metals	Effects
Cd	1. Reduction in chlorophyll content and Chl a/Chl b ratio
	2. Inhibits chlorophyll formation
	3. Decreased Rubisco activity
	4. Inhibits both PS I and PS II
	5. Increased lipoxygenase activity
Pb	1. Changes lipid composition of thylakoid membranes
	2. Influences PS I and PS II
Ni	1. Reduces chlorophyll concentration
	2. Affects both PS I and PS II
	3. Alters organization of oxygen-evolving complex
Cu	1. Disturbs architecture of thylakoid membranes and alter overall chloroplast ultrastructure
	2. Inhibits photosynthetic electron transport of both PS I and PS II
	3. Inhibit RuBP carboxylase activity
Mn	1. Inhibits chlorophyll biosynthesis
	2. Decreases chlorophyll a and b levels
	3. Reduces net photosynthesis rate
Zn	1. Decrease total chlorophyll content and chla/chl b ratio
	2. Inhibits CO_2 assimilation
	3. Interferes activity of oxygen-evolving complex (OEC)
Fe	1. Impairs photosynthetic electron transport
	2. Induces oxidative stress

(Modified from Prasad and Strzalka 1999)

enhanced the accumulation of protoporphyrin IX and Mg-protoporphyrin IX monoethyl ester in green leaves of etiolated barley.

ALA (δ-aminolevulinic acid) is the first common precursor of all tetrapyrroles. Cadmium inhibited the synthesis of ALA in etiolated bean seedlings (Padmaja et al. 1990) mainly due to the interaction of Cd^{2+} with active thiol groups in ALA synthase. The conversion of ALA to porphobilinogen is catalyzed by ALA dehydratase (ALAD, porphobilinogen synthase). ALAD is a regulatory enzyme in Chl biosynthesis and is a metal-sensitive enzyme which requires Mg^{2+} or Zn^{2+} for its activity. The toxic effect of Hg, Se, and Pb on the activity of ALAD was demonstrated in mung bean and bajra, and the result showed an increase in ALAD activity up to the fourth day, and later on it decreased both in metal-treated plants compared to control (Prasad and Prasad 1987). The possible mechanism of ALAD inhibition was suggested to be the interaction of the heavy metals with SH groups of the enzyme at its active sites.

Heavy metals cause ultrastructural changes in the chloroplast of plants growing in a heavy metal-contaminated areas. Another possible target of heavy metals is thylakoid in chloroplast membrane. Wheat plants grown in copper-contaminated soils showed reduced number of chloroplasts, starch, grains, plastoglobuli per chloroplast, surface area of chloroplasts, volume fraction of internal membrane system, and starch grains (Eleftheriou and Karataglis 1989). Cu modified the morphology and structure of primary leaves and their chloroplast ultrastructure in runner beans (*Phaseolus coccineus* L.) (Maksymiec et al. 1995). In radish cotyledons, Cd^{2+} induced changes in the composition and structure of the light-harvesting Chl a/b protein complex II (Krupa et al. 1988). Ahmed and Tajmir-Riahi (1993) observed the interaction of Cd^{2+}, Hg^{2+}, and Pb^{2+} with the light-harvesting proteins (LHC II) in spinach thylakoid membranes using Fourier transform infrared (FTIR) spectroscopy. The structural damage to the structure of chloroplast was found highest for Hg^{+}, lower for Cd^{2+}, and lowest for Pb^{2+} (Ouzounidou 1997).

It was also shown that the inhibition of the photosynthetic electron transport was caused mainly by changes in the structure and composition of the thylakoid membranes as observed in Cu-sensitive spinach (*Spinacia oleracea* L. cv. Matador) which preceded the loss of photochemical activity of PS I and PS II. Cu was also shown to cause slower Chl incorporation into PS I and PS II in greening barley seedlings (Capsi et al. 1999).

6.4 Role of Heavy Metals on Light Reactions and Dark Reactions

Photosynthesis mainly consists of light and dark reactions. Light reaction depends on the interplay of the photosystems which are linked by common intermediates. Photosystem I (PS I) responds to light with wavelengths shorter than 700 nm, and PS II responds to the wavelength shorter than 680 nm. Electrons flow first through photosystem II, then through cytochrome bf, a membrane-bound complex, and then to PS I. The electrons are derived from splitting of water where two molecules of H_2O are oxidized to generate a molecule of O_2 and four electrons sent through this electron transport chain, and the electrons end up by reducing $NADP^{+}$ to NADPH. These processes generate a proton gradient across the thylakoid membrane that drives the formation of ATP (Siedlecka and Baszynski 1993).

6.4.1 Impact of Heavy Metals on Photosystem I (PS I)

PS I is a membrane-bound protein complex which catalyzes the oxidation of plastocyanin and reduction of ferredoxin under light conditions. A site of Cd^{2+} toxicity was found on the reducing side of PS I (Fig. 6.2). Siedlecka and Baszynski (1993) compared electron transport activities (DCIP-MV and DCIP-$NADP^{+}$) in isolated chloroplast of 21-day-old maize, and results showed that site of Cd^{2+} inhibition is between primary electron acceptor and $NADP^{+}$. Cd^{2+} treatment led to Fe deficiency indicating that the light phase of photosynthesis was affected in the treated plants due to Cd-induced Fe deficiency. Ferredoxin was found to be the place of Hg^{2+} and Cu^{2+} action (Ŝeršeň and Kráľová 2013) (Table 6.1). Cu^{2+} interrupted the electron transport from ferredoxin to $NADP^{+}$ in *Chlorella vulgaris* (Šeršeň et al. 1996).

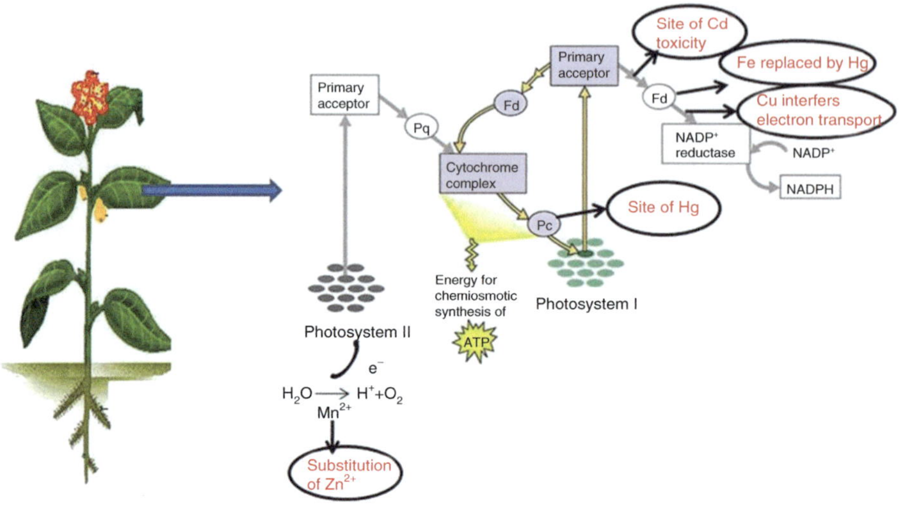

Fig. 6.2 Effect of heavy metals on different sites of light reactions

Mercury induced the process of P700 oxidation in darkness, but it did not inhibit the electron transport around PS I (Šeršeň et al. 1996). Mercury interacted with the donor side of PS I and with plastocyanin (Myśliwa-Kurdziel et al. 2013). Lead inhibited PS I activity in isolated bundle sheath and mesophyll chloroplasts of maize (Wong and Govindjee 1976).

6.4.2 Impact of Heavy Metal on Photosystem II (PS II)

PS II is a multisubunit pigment-protein complex with the enzymatic activity of light-dependent water-oxidizing plastoquinone reductase, leading to the release of electrons, protons, and molecular oxygen, and most of the heavy metals inhibit PS II activity (Fodor 2013). The majority of studies dealing with the effect of heavy metals on the light phase of photosynthesis are devoted to reactions related to PS II. PS II is located in the grana or appressed lamellae. This multi-protein complex consists of at least 25 different subunits (Aro et al. 1993). The photosynthetic electron transport at the level of PS II both at oxidizing (donor) and reducing (acceptor) sides is effectively inhibited by different heavy metals. Heavy metals may impair the functions of PS II directly via the plastoquinone pool or indirectly via feedback regulation by inhibition of the photosynthetic carbon reduction cycle enzymes and changes in ATP level.

Zinc is reported to interact with the donor side of PS II (Prasad and Strzalka 1999). The oxygen-evolving complex (OEC), located on the luminal side of the thylakoids which donates electron from water to the PS II reaction center, was reported to be the primary target of heavy metal toxicity. The primary effect is the inhibition of photosynthetic O_2 evolution, reduction of NADP, and photophosphorylation. Secondary effects include delayed Chl degradation, a decreased Chl a/Chl b ratio, and changes in nitrogen metabolism (Prasad and Strzalka 1999). In studies with *Nicotiana plumbaginifolia*, it was found that Fe excess decreased photosynthetic rate by 40 % due to the reduction of PS II activity and an increase in thylakoid energization (Kampfenkel et al. 1995). Iron due to its participation in oxidation-reduction reactions within the cells is believed to generate oxidative stress in plants when taken in excess, thereby leading to increased activities of antioxidative enzymes. Fe toxicity also induces photorespiration (Keunen et al. 2011).

Zinc affects water-oxidizing complex due to the local competition between Zn^{2+} and Mn^{2+} on the water splitting of PS II and substitution of Mn^{2+} by Zn^{2+} (Dasgupta et al. 2008). Under Zn^{2+} toxicity, alteration of the inner structure and composition of the thylakoid membrane occurred led to the decline in photochemical activities (Dasgupta et al. 2008). In rice plants, a significant inhibition in Hill reaction activity was noticed under Zn toxicity and also on the quantum yield of PS II. Pb toxicity also influences both PS II and PS I activity. Rashid et al. (1994) showed that Pb competes for binding sites near the calcium and chloride binding in the water-oxidizing complex. Studies have shown that loss of the extrinsic polypeptides of 17 and 24 kDa in lead treated PS II sub-membrane fraction.

Cd toxicity binds both sites of donor and acceptor sides of PS II (Sigfridsson et al. 2004). On the donor side, the presence of Cd^{2+} toxicity led to inhibition of photosynthetic oxygen-evolving center by exchange of Cd with high-affinity Ca^{2+} cofactor in the Ca/Mn cluster that constitutes the oxygen-evolving center (Faller et al. 2005), which results in inhibition of photosynthetic oxygen evolution. Cd also inhibits electron transfer from redox-active tyrosine residues D1-161 (Wang et al. 2009).

6.4.3 Impact of Heavy Metals on Dark Reactions

Photosynthesis is very important to plant physiological reactions, and heavy metals inhibit not only light reactions, but dark reactions are also affected (Gill et al. 2011) and reported in many plants (Table 6.2). The first step of enzymes catalyzing carbon metabolism in both C_3 and C_4 photosynthesis is Rubisco and phosphoenol-pyruvate carboxylase (PEPC), and they are the most sensitive targets of heavy metal stress. Heavy metal toxicity in plants has been attributed to the binding of these metal ions to enzymes resulting in the alteration of their catalytic functions (Clijsters and Van Assche 1985). Decreased activity of key enzymes of CO_2 fixation (RuBPcase, PEP-carboxylase) has been observed in various crops under heavy metal treatments (Sheoran et al. 1990). Decreased Ribulose Bisphosphate carboxylase/oxygenase (RUBPCase) activity has been observed under Zn^{2+} treatment (Van Assche and Clijsters 1983). Metal substitution might be the possible reason for this effect (Table 6.1). Uptake of heavy metals induces metal substitution in the metalloproteins, impairing enzyme activity substitution of Zn^{2+} and Mn^{2+} by Cd^{2+} toxicity for carbonic anhydrase (Ernst 1980). In vitro substitution of Mg^{2+} by Mn^{2+} or Co^{2+} and Ni^{2+}

Table 6.2 Effect of heavy metals toxicity on photosynthetic pigments, light reactions, and gas exchange parameters in different plants

Heavy metal	Plant	Parameters	Percent decrease	References
Cd	Pea	*Chl* a	50.6	Hattab et al. (2009)
		Chl b	51.89	
		Carotenoids	45.53	
Cd	Pea	Ps	16.7	Januškaitienė (2010)
	Barley		12.8	
Cd2+	Mustard cv. Varuna	Ps	39	Gill et al. (2011)
		g_s	41	
		CA	40	
		Total chl	46.6	
Zn	Phaseolus	Ps	39	Vassilev et al. (2011)
		g_s	31	
		Chl a	52	
		Chl b	41	
		Carotenoids	31	
Zn	Rice	Total chl	32.9	Song et al. (2014)
		Chl a/*Chl* b	17.4	
		Ps	49.9	
		g_s	64.7	
		Ci	32.1	
Cd	Bermuda grass	Total chl	14.8	Xie et al. (2014)
Hg	Maize	O_2 evolution	82	Srinivasulu and Murthy (2015)
		Electron transport activity PS II	71	
Ni	Maize	O_2 evolution	77	Srinivasulu and Murthy (2015)
		Electron transport activity PS II	78	
Hg	Turf grass	Ps	59.6	Guo et al. (2015)
		g_s	62.5	

Ps photosynthetic rate, *gs* stomatal conductance, *chl a* cholorophyll a, *chl b* chlorophyll b, *Total chl* total chlorophyll, *Ci* intercellular CO_2, *CA* carbonic anhydrase

decreased the RuBPcase/oxygenase activity ratio. The changed ratio showed an increased CO_2 compensation and accumulation of glycolate. A detailed study in pigeon pea has shown that the activity of various enzymes of Calvin cycle, namely, RuBPcase, 3-PGA kinase, NAD and NADP-glyceraldehyde phosphate dehydrogenases, aldolases, and fructose-1,6-bisphosphatase, gets reduced only by 10–40 %. Cd^{2+} drastically affects the enzyme functioning related to carbon metabolism. The activity of ADPG pyrophosphorylase was most sensitive to this treatment. Levels of various Calvin cycle intermediates, namely, 3-phosphoglyceric acid (PGA), DHAP (dihydroxyacetone phosphate), fructose-1,6-biphosphate, fructose-6-phosphate, ribulose-1,5-bisphosphate, and glyceraldehyde 6-phosphate, were reduced under heavy metal stress, and reductions ranged from 31 to 80 % (Mourato et al. 2015). Phosphoglyceric acid kinase (PGK) and glyceraldehyde 3-phosphate dehydrogenase (GAPDH) enzymes are inhibited by Cu and Cd treatment in cucumber cotyledons (Burzynski and Zurek 2007). Metals additionally decreased the potential activity of PGK which suggests that they can influence not only the activity of this protein. The negative effect of Cd on PGK and GAPDH was observed by Sheoran et al. (1990) and Stirborova et al. (1986) in the pigeon pea and maize leaves. However, reductions recorded in various enzymes and metabolites could not clearly explain the main reason of reduction in photosynthesis at the whole plant level.

6.5 Conclusions

Heavy metals influence the functions of photosynthetic apparatus. Heavy metals interact directly with the photosynthetic apparatus at various levels of organization and architecture affecting chlorophyll synthesis, the organization of chlorophyll by interfering with the organization of pigment-protein complexes that are essential for optimal function of PS II. Magnesium insertion during chlorophyll biosynthesis is substituted by Cu, Zn, Hg, Pb, and Ni. Most of the heavy metals react with thiol groups or influence enzymes involved in oxidation-reduction reactions by modulating its donor site. Heavy metals also affect the enzymes involved in Calvin cycle. So, overall high concentration of metal can hinder the photosynthetic pathways that consequently lead to reduction in the productivity of the plants.

Acknowledgments The authors are thankful to the Head of the Department of Botany for all the laboratory facilities and to the Department of Science and Technology, New Delhi, and Council of Scientific and Industrial Research, New Delhi, for providing financial support to the work. Richa Rai is grateful to the Department of Science and Technology, New Delhi, for awarding Fast Track Young Scientist.

References

Ahmed A, Tajmir-Riahi HA (1993) Interaction of toxic metal ions Cd^{2+}, Hg^{2+} and Pb^{2+} with light-harvesting proteins of chloroplast thylakoid membranes. An FITR spectroscopic study. J Inorg Biochem 50:235–243

Alia P, Saradhi P, Mohanty P (1997) Involvement of proline in protecting thylakoid membranes against free radical-induced photodamage. J Photochem Photobiol B: Biol 38:253–257

Aro EM, Virgin I, Andersson B (1993) Photoinhibition of photosynthesis II. Inactivation, protein damage and turn-over. Biochim Biophys Acta 11443:113–134

Atafar Z, Mesdaqhinia A, Nouri J, Homaee M, Yunesian M, Ahmadimoqhaddam M, Mahvi AH (2010) Effect of fertilizer application on soil heavy metal. Environ Monit Assess 160:83–89

Augustsson AL, Uddh-Söderberg TE, Hogmalm KJ, Filipsson ME (2015) Metal uptake by home grown vegetables – the relative importance in human health risk assessment sat contaminated sites. Environ Res 138:181–190

Barghiani C, Gloffre D, Bargali R (1987) Mercury content in Pinus Sp of the Mt. Etna volcanic area. In: Lindberg JE, Hutchinson TC (eds) Heavy metals in the environment, vol 2, New Orleans.51

Bhattacharjee S (2012) The language of reactive oxygen species signaling in plants. J Bot 2012:1–22

Burzynski M, Zurek A (2007) Effects of copper and cadmium on photosynthesis in cucumber cotyledons. Photosynthetica 45:239–244

Canon HL, Connally GG, Epstein JB, Parker JG, Thornton I, Wixson G (1998) Rocks: geological sources of most trace elements. In: Report to the workshop at south scas plantation Captiva Island, FL, US. Geochem Environ 3:17–31

Capsi V, Droppa M, Horváth G, Malkin S, Marder JB, Raskin VI (1999) The effect of copper on chlorophyll organization during greening of barley leaves. Photosynth Res 62:165–174

Clijsters H, Van Assche F (1985) Inhibition of photosynthesis by heavy metals. Photosynth Res 7:31–40

Csatorday K, Gombos Z, Szalontai C (1984) Mn^{2+} and Co^{2+} toxicity in chlorophyll biosynthesis. Proc Natl Acad Sci U S A 81:476–478

Dasgupta J, Ananyev GM, Dismukes DC (2008) Photoassembly of the water-oxidizing complex in photosystem II. Coord Chem Rev 252:347–360

De Vos CHR, Ten Bookum W, Vooijs R, Schat H, De Kok IJ (1993) Effect of copper on fatty acid composition and peroxidation of lipids in the roots of copper-tolerant and sensitive *Silene cucubalus*. Plant Physiol Biochem 31:151–158

Dubey RS (2011) Metal toxicity, oxidative stress and antioxidative defense system in plants. In: Gupta SD (ed) Reactive oxygen species and antioxidants in higher plants. CRC Press, Boca Raton, pp 177–203

Dziubanek G, Piekut A, Rusin M, Baranowska R, Hajok I (2015) Contamination of food crops grown on soils with elevated heavy metals content. Ecotoxicol Environ Safe 118:183–189

Ernst WHO (1980) Biochemical aspects of cadmium in plants. In: Nriagu JO (ed) Cadmium in the Environment Part I. Wiley, New York, pp 639–653

Eleftheriou EP, Karataglis S (1989) Ultrastructural and morphological characteristics of cultivated wheat growing on copper polluted fields. Bot Acta 102:134–140

Emamuerdian A, Ding Y, Mokhberdoran F, Xie Y (2015) Heavy metal stress ad some mechanisms of plant response. Sci World, 756120

Faller P, Kienzler K, Krieger-Liszkay A (2005) Mechanism of Cd^{2+} toxicity: Cd^{2+} inhibits photoactivation of Photosystem II by competitive binding to the essential Ca^{2+} site. Biochim Biophy Acta Bioenerg 1706:158–164

Farmer EE, Mueller MJ (2013) ROS-Mediated lipid peroxidation and RES-activated signaling. Annu Rev Plant Biol 64:429–450

Fodor F (2013) Physiological responses of vascular plants to heavy metals. In: Prasad MNV, Strzalka K (eds) Physiology and biochemistry of metal toxicity and tolerance in plants, Springer- Verlag

Fraustro da Silva JJR, Williams RJP (1991) The biological chemistry of the elements. Clarendon, Oxford, p 114

Gill SS, Khan NA, Tuteja N (2011) Differential cadmium stress tolerance in five indian mustard (*Brassica juncea* L.) cultivars An evaluation of the role of antioxidant machinery. Plant Signal Behav 6:293–300

Guo YC, Liu YY, Wang RY, Wang S, Lu XP, Wang B (2015) Effect of mercury stress on photosynthetic characteristics of two kinds of warm season turf grass. Int J Environ Monit Anal 3:293–297

Hattab S, Dridi B, Chouba L, Kheder MB, Bousetta H (2009) Photosynthesis and growth responses of pea *Pisum sativum* L. under heavy metals stress. J Environ Sci 21:1552–1556

Januškaitienė I (2010) Impact of low concentration of cadmium on photosynthesis and growth of pea and barley. Environ Res Eng Manag 3:24–29

Kalyanaram B (1996) Thiol radicals in biological systems: significant or trivial? Biochem Soc Symp 61:55–63

Kampfenkel K, Montagu MV, Inze D (1995) Effects of iron excess on *Nicotiana plumbaginifolia* plants (Implications to oxidative stress). Plant Physiol 107:725–735

Karuppanapandian T, Moon J, Kim C, Manoharan K, Kim W (2011) Reactive oxygen species in plants: their generation, signal transduction, and scavenging mechanisms. Aust J Crop Sci 5:709–725

Keunen E, Remans T, Bohler S, Vangronsveld J, Cuypers A (2011) Metal-induced oxidative stress and plant mitochondria. Int J Mol Sci 12:6894–6918

Krupa Z, Skorzynska E, Maksymiec W, Baszynski T (1988) Effect of cadmium treatment on the photosynthetic apparatus and its photochemical activities in greening radish seedlings. Photosynthetica 21:156–164

Kupper H, Kupper F, Spiller M (1996) Environmental relevance of heavy metal-substituted chlorophylls using the example of water plants. J Exp Bot 47:259–266

Kurdziel BM, Strazalka K (1999) Influence of metals on biosynthesis of photosynthetic pigments In: Prasad MNV, Strazalka M (eds) Physiology and biochemistry of metal toxicity and tolerance in plants, Springer, pp 201–228

Lacerda LD (1997) Global mercury emissions from gold and silver mining. Water Air Soil Pollut 97:209–221

Lin YF, Mark GMA (2012) The molecular mechanism of zinc and cadmium stress response in plants. Cell Mol Life Sci 69:3187–3206

Maksymiec W, Bednara J, Baszynski T (1995) Responses of runner bean plants to excess copper as a function of plant growth stages: lffects on morphology and structure of primary leaves and their chloroplast ultrastructure. Photosynthetica 31:427–435

Michalak A (2006) Phenolic compounds and their antioxidant activity in plants growing under heavy metal stress. Pol J Environ Stud 15:523–530

Mourato MP, Moreira IN, Leitão I, Pinto FR, Sales JR, Martins LL (2015) Effect of heavy metals in plants of the genus *Brassica*. Int J Mol Sci 16:17975–17998

Myśliwa-Kurdziel B, Prasad MNV, Strzalke K (2013) Photosynthesis in heavy metals stressed plants. In: Prasad MNV (ed) Heavy metal stress in plants: from biomolecules to ecosystems. Springer, Berlin

Ouzounidou G (1997) Sites of copper in the photosynthetic apparatus of maize leaves: kinetic analysis of chlorophyll fluorescence, oxygen evolution, absorption changes and thermal dissipation as monitored by photoacoustic signals. Aust J Plant Physiol 24:81–90

Pandey B, Agrawal M, Singh S (2014) Assessment of air pollution around coal mining area: emphasizing on spatial distributions, seasonal variations and heavy metals, using cluster and principal component analysis. Atmos Pollut Res 5:79–86

Pandey N, Pathak GC, PAndey DK, Pandey R (2009) Heavy metals Co, Ni, Cu, Zn and Cd produce oxidative damage and evoke differential antioxidant responses in spinach. Bra J Plant Physiol 21:103–111

Pacyna JM (1986) Atmospheric trace elements from natural and anthropogenic sources. In: Nriagu JO, Davidson CI (eds) Toxic metals in the atmosphere, Chap 2. Wiley, New York

Padmaja K, Prasad DDK, Prasad ARK (1990) Inhibition of chlorophyll synthesis in *Phaseolus vulgaris* L. seedlings by cadmium acetate. Photosynthetica 24:399–405

Prasad DDK, Prasad ARK (1987) Altered delta-aminolevulinic acid metabolism by lead and mercury in germinating seedlings of bajra (*Pennisetum typhoideum*). J Plant Physiol 127:241–249

Prasad MNV, Strzalka K (1999) Impact of heavy metals on photosynthesis. In: Prasad MNV J, Hagemeyer J (eds) Heavy metal stress in plants: from molecules to ecosystems. Springer, Berlin, pp 117–138

Rashid A, Camin EL, Ekramoddoulah AKM (1994) Molecular mechanism of action of Pb^{2+} and Zn^{2+} on water oxidizing complex of photosystem II. FEBS Lett 350:296–298

Schützendübel A, Polle A (2002) Plant responses to abiotic stresses: heavy metal-induced oxidative stress and protection by mycorrhization. J Exp Bot 53:1351–1365

Somashekaraiah BV, Padmaja K, Prasad ARK (1992) Phytotoxicity of cadmium ions on germinating seedlings of mung bean (*Phaseolus vulgaris*): involvement of lipid peroxides in chlorophyll degradation. Physiol Plant 5:85–89

Shahid M, Powrut B, Dumat C, Nadeem M, Aslam M, Pinelli E (2014) Heavy metals induced reactive oxygen species: phytotoxicity and physicochemical changes in plant. Rev Environ Contam Toxicol 232:1–44

Siedlecka A, Baszynski T (1993) Inhibition of electron flow around photosystem I in chloroplasts of Cd-treated maize plants is due to Cd-induced iron deficiency. Physiol Plant 87:199–202

Šeršeň F, Kráľová K, Blahová M (1996) Photosynthesis of chlorella vulgaris as affected by 4-chloro-2- methylphenoxyacetato)copper(II) complex. Biol Plant 38(1):71–75

Šeršeň F, Kráľová K (2013) Photosynthesis EPR spectroscopy — a valuable tool to study photosynthesizing organisms exposed to abiotic stresses. In: Dubinsky Z (ed) Photosynthesis, Intech, pp 247–283

Sheoran IS, Singal HR, Singh R (1990) Effect of cadmium and nickel on photosynthesis and the enzymes of photosynthetic carbon reduction cycle in pigeon pea. Photosynth Res 23:345–351

Sigfridsson KGV, Bernát G, Mamedov F, Styring S (2004) Molecular interference of Cd^{2+} with photosystem II. Biochim Biophys Acta 1659:19–31

Skórzńska -Polit E, Tukendorf A, Selstam E, Baszynski T (1998) Calcium modifies Cd effect on runner bean plants. Environ Exp Bot 40:275–286

Song A, Li P, Fan F, Li Z, Liang Y (2014) The effect of silicon on photosynthesis and expression of its relevant genes in rice (*Oryza sativa* L.) under high-zinc stress. PLoS ONE 9(11):e113782

Srinivasulu P, Murthy SDS (2015) Action of selected heavy metals on photosynthetic electron transport activities of maize thylakoid membranes. Int J Plant Anim Environ Sci 5:90–93

Stirborova M, Doubravorva M, Leblova S (1986) A comparative study of the effect of heavy metal ions on ribulose 1-5- bisphosphate carboxylase and phosphenol pyruvate carboxylase. Biochem Physiol Pflanz 181:373–379

Van-Assche F, Clijsters H (1983) Multiple effects of heavy metal toxicity on photosynthesis. In: Marcelle R, Clijsters H, Van Pouckem (eds) Effects of stress on photosynthesis, Martinus-Nijhoff, The Hague, pp 371–382

Vassilev A, Nikolova A, Koleva L, Lidon F (2011) Effects of excess Zn on growth and photosynthetic performance of young bean plants. J Phytol 3:58–62

Villiers F, Ducruix C, Hugouvieux V (2011) Investigating the plant response to cadmium exposure by proteomic and metabolomic approaches. Proteomics 11:1650–1663

Wang H, Zhao SC, Liu RL, Zhou W, Jin JY (2009) Changes of photosynthetic activities of maize (*Zea mays* L.) seedlings in response to cadmium stress. Photosynthetica 47:72–78

Weckx JEJ, Clijsters H (1997) Zn phytotoxicity induces oxidative stress in primary leaves of *Phaseolus vulgaris*. Plant Physiol Biochem 35:405–410

Wong D, Govindjee (1976) Effects of lead ions on photosystem I in isolated chloroplasts: studies on the reaction centre P700. Photosynthetica 10:241–254

Xie Y, Hu L, Du Z, Sun X, Amombo E (2014) Effects of cadmium exposure on growth and metabolic profile of bermudagrass [*Cynodon dactylon* (L.) Pers.]. PLoS ONE 9 e115279. doi:10.1371/journal.pone.0115279

Źróbek AS, Asard H, Górska KK, Górecki RJ (2009) Cadmium and zinc mediated oxidative burst in tobacco BY-2 cell suspension cultures. Acta Physiol Plant 31:43–49

Impact, Metabolism, and Toxicity of Heavy Metals in Plants

7

Anil Kumar and N.C. Aery

Abstract

Plants are sessile organisms, and in order to survive they have to combat with the surrounding environment. Due to the numerous anthropogenic activities, an excessive level of different heavy metals accumulates in the soil system. The uptake of eminent concentration of these metals is toxic for the living organisms existing in that region. The troubling impact of heavy metals on plants is associated with the deformed growth and development, ionic imbalance, reduced photosynthetic rate, degradation of photosynthetic pigments and chloroplast, alteration in elemental composition, and disturbed plant water relation. The influence of metal ion is more complex by interaction between different ions because a high level of one metal ion may interfere with the uptake and transport of others and disturb the nutritional composition of plants and induce toxic symptoms. Several mechanisms have been evolved by the plants to sustain suitable physiological concentrations of metal ions and to minimize exposure of cellular processes to toxic heavy metals. Plants exposed to toxic concentrations of heavy metals attempt to prevent entry of these metal ions in roots as well as translocation from root to aerial parts by restricting metal ions to the apoplastic region, binding them to the cell wall, extracellular chelation with root exudates, or by reducing long-distance transport. Once metals enter in the cell, several storage and detoxification strategies including metal transport, chelation, and sequestration into the vacuole take place to diminish the toxic effects. The entry and

A. Kumar (✉)
Laboratory of Plant Physiology and Biochemistry, Department of Botany, University of Rajasthan, Jaipur 302004, Rajasthan, India
e-mail: rathoreanilk@yahoo.com

N.C. Aery
Laboratory of Geobotany and Biogeochemistry, Department of Botany, Mohanlal Sukhadia University, Udaipur 313001, Rajasthan, India

A. Singh et al. (eds.), *Plant Responses to Xenobiotics*,
DOI 10.1007/978-981-10-2860-1_7

transport of heavy metal in plants, strategies of plants to overcome the lethal consequences, and the specific toxic effects of heavy metals on plants when exposed to metal-enriched environment are emphasized in this chapter.

Keywords
Heavy metals • Uptake • Accumulation • Detoxification • Toxicity • Plants

7.1 Introduction

The group of metals and metalloids with atomic density higher than 4 g/cm^3 or 5 times or more, greater than water, have been called heavy metals (Hawkes 1997). However, the toxic properties are more relevant to the chemical properties of the heavy metals as compared to the density. Heavy metal includes lead (Pb), cadmium (Cd), nickel (Ni), cobalt (Co), zinc (Zn), chromium (Cr), iron (Fe), arsenic (As), silver (Ag), etc.

Heavy metals are mostly found in dispersed form in rock formations. These metals are present mainly in soil as well as aquatic ecosystems and also present in relatively low proportion as particulates or vapors in atmosphere. Heavy metals which are usually confined in the Earth's crust and are benign to living organisms have been significantly exploited and discharged into the biosphere in the last few decades.

Nieboer and Richardson (1980) proposed a classification of metals to avoid the use of the term "heavy metal." They have classified the elements in three classes on the basis of equilibrium constant that describes the formation of metal ion/ligand complexes. These are:

(i) Class A elements which show preferences for ligands containing oxygen
(ii) Class B elements which show preferences for ligands with nitrogen or sulfur
(iii) Borderline elements of intermediate character between Class A and B

Heavy metals such as Hg, Cd, Pb, and Ni are highly toxic to different organisms and come under Class B elements (Nieboer and Richardson 1980). Some of these heavy metals are able to persist in the environment for a long duration. However, these metals are difficult to metabolize, but they are bioaccumulative and accumulate in the ecological food chain through uptake at primary producer level and transfer to different trophic levels.

Hossain et al. (2012) classified the heavy metals into two groups: (i) redox active (Fe, Cu, Cr, Co) and (ii) redox inactive (Cd, Zn, Ni, Al, etc.). The redox-active heavy metals are directly involved in the redox reactions of cells whereas redox inactive heavy metals indirectly induce oxidative stress by interacting with the antioxidant defense system, disrupting the electron transport chain or induction of lipid peroxidation (Dietz et al. 1999; Schützendübel and Polle 2002; Kumar and Aery 2011; Hossain et al. 2012).

Heavy metal contamination is a severe problem worldwide. These metals are the major inorganic contaminants of soil, and a considerable large area of land is contaminated with them due to anthropogenic activities like extensive application of fertilizers and pesticides, sludge or municipal composts, emissions from municipal waste incinerators, vehicle exhaust, residues of metalliferous mines and smelting industries, etc. (Garbisu and Alkorta 2003; Halim et al. 2003). Contamination of agricultural soil by heavy metals has become a critical environmental concern due to their long-term persistent nature and potential harmful ecological effects.

There is a very complex mechanism used by plants to metabolize heavy metals. A complex network of uptake, transport, compartmentalization, chelation, and sequestration processes are found in the plants to maintain the concentrations of essential metal ions in cytosol within a narrow physiological range and minimize the damage caused by nonessential metal ions (O'Halloran and Cullota 2000; Clemens 2001).

7.2 Impact of Heavy Metals on Plants

Plants are stationary, and roots of a plant are the main contact site for heavy metal ions. Due to this contact, plants absorb heavy metals mainly by roots and also by the leaf surfaces due to deposition of particles containing these metals. Those plants which are grown in aquatic systems face more toxic influence as in this type of plants whole plant body is exposed to toxic ions.

Some heavy metals (Fe, Cu, Mo, and Zn) are considered as essential elements for plants. Involvement in the redox reactions and being an integral part of enzymes are two key functions of essential heavy metals in cells. Heavy metals like Cu, Mo, Zn, etc. serve as cofactor and activator of different enzyme reactions and play a vital role in enzymes/substrate metal complex formation (Mildvan 1970) or show a catalytic property as prosthetic group in metalloenzymes take part in electron transport and structural functions in nucleic acid metabolism (Nagajyoti et al. 2010).

The heavy metal toxicity causes serious threats to living organisms because these metals are persistent and present for a long period in the environment (Gisbert et al. 2003). Heavy metals such as Cd, Hg, and Pb are strong inhibitors of metal-sensitive enzymes, and exposure to these metals results in growth inhibition and even death of organisms. However, heavy metal toxicity in plants depends upon plant species, specific metal, chemical form and soil composition, soil pH, and concentration in tissues.

In some instances, low concentrations of toxic metals have been observed to be promoting the growth parameters such as length, dry weight, nodule number, and weight of plants (Sarkar and Aery 1988; Matsumoto et al. 1976; Marschner 2012). This phenomenon is known as hormesis (Cedergreen et al. 2007; Poschenrieder et al. 2013).

For metals such as Cu, Zn, Fe, and Mo which plays an essential role in plant growth, adverse effects have been recorded at high concentrations as compared to metals (Pb, Cd, Hg, As) which do not play an essential role in plant growth; harmful effects have been recorded at very low concentrations in the growth medium.

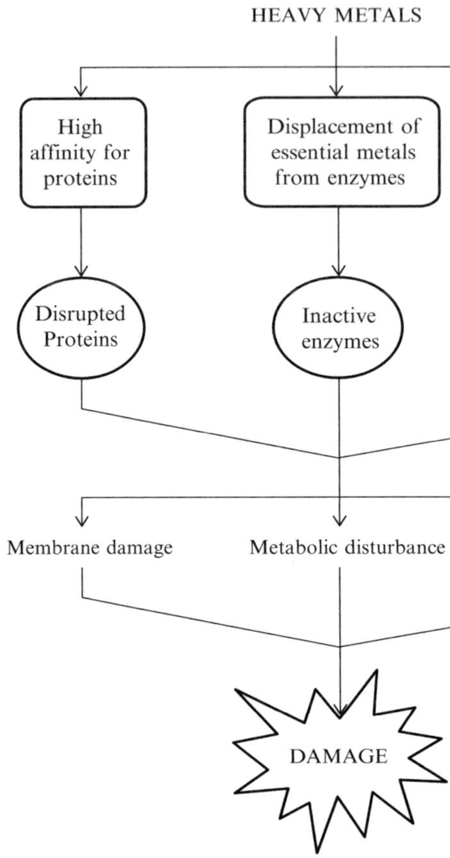

Fig. 7.1 Summary of heavy metal-induced toxicity mechanism in plants

Essential heavy metals improve the plant growth at low concentrations in the substrate. However, at higher concentrations of these metals, reductions in plant growth have been reported (Hagemeyer 1999; Kumar and Aery 2010; Kumar and Aery 2012b).

Heavy metals directly or indirectly affect the growth and development of plants (Fig. 7.1). The direct effects include decline in physiological and biological activities, inhibition of cytoplasmic enzymes, and damage of cellular structure and functions of plants by causing oxidative stress (Assche and Clijsters 1990; Aery 1998, 2012). In indirect manner, these heavy metals replace the essential nutrients at cation exchange site of plants (Taiz and Zeiger 2002). The adverse effect of heavy metals on the growth and activities of soil microbes may also indirectly affect the plant growth (Kumar 2013).

The mechanism of heavy metal toxicity in plants is summarized in Fig. 7.1. Metal toxicity is attributed to three most important reasons:

(i) Direct interaction with proteins due to high affinities for thiol, histidyl, and carboxyl groups, directing the metals to target structural, catalytic, and transport sites of the cell;

Table 7.1 Toxic effect of some heavy metals in plants

Heavy metal	Toxic effect	References
Cd	Growth retardation, stunted root, chlorosis of leaves, and red-brown coloration of leaf margins or veins	Aery and Sarkar (2012a, b) and Baszyński (2014)
Hg	Stunted growth, reduced root development, inhibition of photosynthesis and respiration, reduced chlorophyll synthesis, low exchange of gases	Patra and Sharma (2000)
Pb	Destruction of plasma membrane, disturbance of electron transfer reactions, reduced elasticity and plasticity of cell wall	Zimdahl (1976), Lane et al. (1978), and Woźny (1998)
Cr	Wilting of top, root injury, chlorosis, reduced biomass production	Baszyński (1981) and Mukhopadhyay and Aery (2000)
Al	Reduced root development, disruption of microtubules and microfilament, low chlorophyll synthesis, reduced rate of photosynthesis	Foy (1988) and Rana and Aery (1999)

(ii) Induce generation of reactive oxygen species (ROS) that alters the antioxidant defense system and results in oxidative stress;
(iii) Displacement of essential metal ions from the binding sites, causing functions to collapse (Sharma and Dietz 2009).

The interference of heavy metals with ionic homeostasis and enzyme activity affects physiological processes involving single organ (like nutrient uptake by roots) followed by several processes such as germination, photosynthesis, respiration, plant water balance, metabolism, and reproduction. Indeed, visible symptoms of heavy metal toxicity include chlorosis, necrosis, senescence, wilting, stunted growth, low biomass production, limited numbers of seeds, and eventually death. The plants growing under heavy metal stress have to expend more energy for their survival, which otherwise would have been available for their other process. This deficiency of the required amount of energy may result in the overall decrease in the plant's growth in such hostile metal-stressed environment (Aery and Sarkar 1991). Toxic effects of some heavy metals are given in Table 7.1.

7.3 Metabolism of Heavy Metals

Heavy metals, especially nonessential heavy metals, at higher accumulation generate an unease situation to the functioning of plant cells. As the heavy metal enters in the plant system, several different types of strategies are employed by the plants to combat the situation. Response of plants to heavy metal depends mainly on efficiency of metal uptake, translocation, and sequestration of heavy metal in specialized tissues or in trichomes and organelles such as vacuoles. Several studies have been carried out to understand the mechanism of metabolic processes in plants at an elevated metal concentration. The possible mechanisms undertaken by plants in response to heavy metals include exclusion, inclusion and accumulation of heavy metals, binding to the cell wall, reduced transport across the plasma membrane, active efflux, compartmentalization, and chelation (Prasad 1995; Aery 1998, 2012; Raskin and Ensley 2000; Hall 2002).

7.3.1 Uptake and Accumulation

Plants are able to solubilize and take up different nutrients from the soil by producing chelating agents and changing in pH and redox reactions. They also have highly specific mechanisms for translocation and accumulation of different nutrients. These mechanisms also accomplish the uptake and translocation of heavy metals, whose physical and chemical properties are similar to those of essential elements (Salisbury and Ross 1978; Tangahu et al. 2011).

The uptake of elements by a plant is much affected by a number of factors, viz., the organ of plant, age of plant organ, health of plant, the pH of the soil, depth of root system, temperature, oxygen partial pressure, carbohydrate level, respiration rate, interaction between elements, presence of microorganisms, etc. (Aery and Tyagi 1988). It has been reported that administration of glucose or citric acid significantly improved the extractable Cu concentration in soil (Chen et al. 2006).

The uptake of heavy metals by plants depends upon its availability in the substrate. Those heavy metals which are present as soluble component in the soil solution or easily solubilized by root exudates are available for plants (Chibuike and Obiora 2014). It has been reported that the plants liberate organic acids such as malic acid from radicle apex on exposure to heavy metals (Delhaize and Ryan 1995; Larsen et al. 1998). These organic acids inhibit metal uptake by forming a complex with the metal at the rhizosphere. Citrate-inhibited Al uptake and tolerance in several plant species is an example of this mechanism (de la Fuente et al. 1997; Pineros and Kochian 2001; Papernik et al. 2001).

Once a heavy metal is bioavailable to the plant, the entry of metal ions into the plant is either through the symplastic or through the apoplastic process and depends on the type of heavy metal. The entry of heavy metal ions in plant cells is an energy-dependent process and completed by the help of some specific ion carriers or channels (Bubb and Lester 1991).

Generally plants adopt an avoidance strategy to avoid excess entry of heavy metal and prevent the toxic effects of heavy metals. The maximum fraction of heavy metals is accumulated in peripheral regions of roots such as the root hairs, epidermis and endodermal cell layers, and some parts in the cortex and excluded from the vascular cylinder (Kumar and Aery 2011; Hajiboland and Poschenrieder 2015). The entry of heavy metals in plants is restricted either by precipitating or by making complex in the root environment. Plants are able to precipitate heavy metals by changing the pH of the rhizosphere or by excreting anions such as phosphate (Hossain et al. 2012). The surface of the root is able to bind many heavy metals during the process of adsorption. These heavy metals (Cd, Ni, Pb, Sr) rapidly concentrate into root tissues of plants (Salt et al. 1995, 1999).

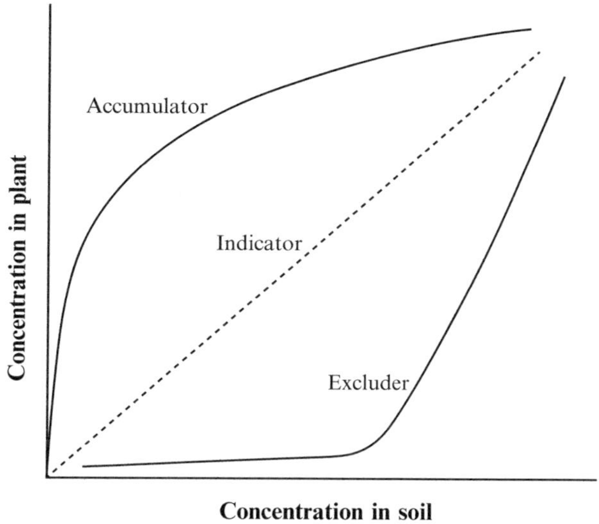

Fig. 7.2 Patterns of metal accumulation in plants (after Baker 1981)

There are some proteins of plasma membrane which participate in ion uptake and translocation. These are (1) proton pumps (ATPases that require energy and generate electrochemical gradients), (2) transporters (co and anti) (use the electrochemical gradients generated by ATPases to drive the active uptake of ions), and (3) channels (proteins that facilitate the transport of ions into the cell). These proteins are liable to take up a range of ions, but the interaction of other ionic species influences the uptake of heavy metals. After uptake by roots, these metal ions are translocated to the shoot. The forms in which metal ions are translocated from the roots to the shoots are not well studied.

Nonessential metal ions enter in plant cells via the uptake system of essential ions. It is reported for wheat that protein LCT1, when expressed heterologously in *Saccharomyces cerevisiae*, renders yeast cells Cd^{2+} hypersensitive because of LCT1-mediated Ca^{2+} influx (Clemens et al. 1998). It shows that Cd^{2+} ions enter in the yeast cells using protein for Ca^{2+} influx. In transgenic *Brassica juncea*, overproduction of either γ-glutamylcysteine synthetase or glutathione synthetase leads to improved accumulation and tolerance of different metals such as Cd, Cr, and As (Reisinger et al. 2008). Overproduction of ferritin by genetic alteration leads to increased uptake of Fe, Cd, Mn, and Zn (Sappin-Didier et al. 2005). This may be due to Fe deficiency at high pH, which stimulates metal uptake and translocation in shoots by an increase in ferric reductase and H^+-ATPase activities in the roots (Revathi and Venugopal 2013).

In addition to uptake from the soil, elemental Hg can be absorbed through leaves. Total mercury content of corn and wheat plants exposed to Hg vapors reflects total Hg concentrations in the air rather than the soil (Niu et al. 2011). The exact mechanism of metal uptake from leaves is not known, but likely involves gas exchange through the stomata.

Baker (1981) suggested that plants could be classified into three categories (Fig. 7.2). These are

1. Excluders: those plants which grow in metal-contaminated soil and maintain the shoot concentration at low level up to a critical soil value, and above which relatively unrestricted root-to-shoot transport is found.
2. Accumulators: those plants which accumulate metals in the above-ground part at low as well as high soil levels.
3. Indicators: In this type of plants, the uptake and transport of metals to the shoot are regulated so that internal concentration reflects external levels.

7.3.1.1 Hyperaccumulation

Hyperaccumulation is the process by which plants accumulate metals more than 0.1–1 % of the dry weight. This term was coined by Baker and Brooks (1989) to define plants that contain higher than 1000 mg g^{-1} of nickel in leaves. According to Baker et al. (2000), plants that accumulate more than 100 mg Cd kg^{-1} (0.01 %) or more than 500 mg Cr kg^{1} (0.05 %) in dry leaf tissue can be considered as hyperaccumulators.

A hyperaccumulator plant can accumulate and tolerate high levels of heavy metals. Some plant species have the capacity to grow in the soil, contaminated with heavy metals and have the ability to accumulate elevated amount of heavy metals as an ecological adaptation in metalliferous soil (Maywald and Weighl 1997).

The hyperaccumulator plants have an unusual capability to absorb metals from the soil even under low and high metal concentrations in the substrate and accumulate them in the shoots (Ma et al. 2001; Yang et al. 2002). More than 400 plant species have been reported so far that hyperaccumulate metals (Baker et al. 2000; McIntyre 2003).

The major processes involved in hyperaccumulation of metals in plants include bioactivation of metals in the rhizosphere through root-microbe interaction, enhanced activity of metal transporter proteins in the cell membranes, detoxification of metals by limiting to the apoplasts, chelation of metals in the cytoplasm by different ligands, and sequestration of metals into the vacuole by transporters present on tonoplasts (Lombi et al. 2002; Yang et al. 2005).

Ferns have been studied for their ability to accumulate As in the fronds. The Chinese brake fern (*Pteris vittata*) is able to take up As (V), reduce it to As (III), translocate it through the xylem with water and minerals as an As (III)-S compound, and store it as As (III) in the fronds (Ma et al. 2001). Tripathi et al. (2007) reported that the plants take up and mobilize this As (V) through the phosphate transport channels.

Generally plants absorb heavy metals with respect to concentration of metals present in the substrate. Tiagi and Aery (1986) observed that some plants such as *Talinum portulacifolium*, *Tephrosia villosa*, *Rhus mysorensis*, and *Bouchea marrubifolia* accumulate Cu in their leaves at very high concentrations (394, 288, 244, and 147 μg g^{-1} on dry weight basis, respectively). Moreover, *Tephrosia villosa* has very high biological absorption coefficient (concentration of the element in plant tissue divided by the concentration of same element in substrate) value. However, at very high concentration, plants limit to accumulate the element. On attainment of the threshold value, a gradual decrease in Cu contents of *Rhus mysorensis* and *Talinum portulacifolium* is reported with any further increase in the Cu content of the soil (Tiagi and Aery 1986).

7.3.2 Transport of Heavy Metals

Once a heavy metal is taken up by plants from the soil, it is then transported to the aerial part. Non-hyperaccumulator plants retain most of the heavy metal taken up from the soil in root cells, detoxify them by chelation in the cytoplasm or sequestrating them into vacuoles, whereas hyperaccumulator plants rapidly and efficiently translocate these elements to the shoot via the xylem (Monferrán and Wunderlin 2013). The transport of heavy metal depends upon the availability of heavy metal for xylem loading, which derives from a reduced sequestration and efflux of the vacuoles, possibly due to specific features of root cell tonoplast (Lasat et al. 2000).

There are four classes of membrane transporters that have been implicated in transport of heavy metals in different organisms and could serve such role in plants. These are heavy metal ATPases, the natural resistance-associated macrophage protein (Nramp) family, members of the cation diffusion facilitator (CDF) family, and the ZIP family (Williams et al. 2000; Guerinot 2000; Williams and Hall 2000).

Heavy metal ATPases (also known as P1B-type ATPases) operate heavy metal transport and play an important role in metal homeostasis and tolerance (Axelsen and Palmgren 1998). It belongs to P-type ATPase superfamily and uses energy from ATP hydrolysis to efflux various metal cations across the plasma membrane (Axelsen and Palmgren 2001). Heavy metal ATPases are internal transporters, responsible for the loading of Cd and Zn into the xylem from the nearby tissues, and act as an efflux pump. These transporters are able to transfer monovalent as well as divalent cation and are more selective than the transporters involved in metal uptake (HMA2, HMA3, and HMA4) (Baxter et al. 2003; Krämer et al. 2007). It has been reported in *Arabidopsis thaliana* that AtHMA3 transporter contributes in sequestration of several heavy metals such as Cd, Pb, Co, and Zn (Morel et al. 2009; Manara 2012).

Nramp transporters (AtNRAMP3 and AtNRAMP4) are present on the tonoplast and help in the transportation of Fe from the vacuole (Thomine et al. 2003; Lanquar et al. 2005). Further, it has been reported that Cd sensitivity in plants increases on overexpression of AtNRAMP3 (Thomine et al. 2000).

Metal tolerance proteins (MTPs) are metal efflux transporters in plants that belong to the CDF transporter family involved in transport of divalent metal cations such as Zn, Cd, Co, Fe, Ni, and Mn and help in transportation from the cytoplasm to the vacuole (Nies 1992; Krämer et al. 2007; Montanini et al. 2007; Manara 2012).

The ZIP family transporters are well known for uptake of divalent metal. It consists of eight transmembrane domains and a histidine-rich domain believed to involve in specific metal binding (Guerinot 2000; Nishida et al. 2008). IRT1 was the first reported transporter of the ZIP family in root cells of *Arabidopsis thaliana*. It plays an important role in Fe^{2+} uptake from the soil (Vert et al. 2002) and also transport of Mn^{2+}, Zn^{2+}, and Cd^{2+} (Korshunova et al. 1999). AtZIP4 proteins are involved in Zn transport and also help in Cd uptake from soil into the root cells and Cd transport from root to aerial parts (Krämer et al. 2007).

The multidrug and toxin efflux (MATE) is a family of small organic molecule transporters that are active in heavy metal translocation in hyperaccumulator plants. The FDR3 protein, which is localized at root pericycle plasma membranes, usually operates the xylem influx of citrate, which is a ligand for Fe homeostasis and transport (Durrett et al. 2007). Furthermore, FDR3, a gene encoding a member of this family, is constitutively overexpressed in the roots of *Thlaspi caerulescens* and *Arabidopsis halleri* (Talke et al. 2006; van de Mortel et al. 2006).

Heavy metal transporter proteins of plants can be classified as metal uptake proteins and metal efflux proteins. The metal uptake proteins are responsible for the transportation of essential heavy metals into the cytoplasm and, in the absence of these essential heavy metals or due to the competition between ions, also able to transport toxic heavy metals. The metal efflux proteins are proteins involved in detoxification, which can efflux excess and toxic heavy metals from the cytoplasm or transfer these metals into the vacuole (Jin et al. 2010).

The translocation of heavy metals is mainly dependent upon the metal-chelate complex. Generally heavy metals are chelated to organic acids such as histidine (Krämer et al. 1996), nicotianamine, citrate, and malate or oxalate (Senden et al. 1995). Application of citric acid stimulates the transportation of Cd from the roots to the shoots by converting the metal into a more easily transported form and decreases its toxicity in radish (Chen et al. 2003). It has been reported that histidine acts as a ligand and responsible for the long-distance (root to shoot) transport of Ni in Ni hyperaccumulator plants such as *Alyssum lesbiacum* and *Thlaspi caerulescens* (Krämer et al. 1996; Morel et al. 2009).

Citrate is transported into the xylem by FRD3 transporter, a member of the MATE family. FRD3 protein is essential for efficient iron translocation by vascular system (Durrett et al. 2007). Citrate is also involved in Zn translocation as FRD3 is much higher expressed in roots of *T. caerulescens* (Monferrán and Wunderlin 2013) than those of *A. thaliana* (van de Mortel et al. 2006). Some divalent metal ions like Fe, Zn, Ni, Cd, and Cu are also chelated and transported in plants by nicotianamine (Ling et al. 1999; Pich et al. 2001; Takahashi et al. 2003).

Rapid and efficient transport of heavy metals from root to aerial parts of hyperaccumulator plants depends on improved xylem loading of these metals by a constitutive overexpression of gene coding for transport systems common to non-hyperaccumulators. The genes encoding bivalent cation transporters belonging to heavy metal transporting ATPases are overexpressed in the roots and shoots of Zn/Cd hyperaccumulators *Thlaspi caerulescens* and *Arabidopsis halleri* (Papoyan and Kochian 2004; Talke et al. 2006; Hanikenne et al. 2008). Further, the overexpression of HMA4 gene supports Cd and Zn efflux from the root symplasm into the xylem vessels. In fact, the increased expression of HMA4 enhances the expression of genes belonging to the ZIP family, implicated in heavy metal uptake (Monferrán and Wunderlin 2013). This suggests that the root-to-shoot translocation acts as a driving force of the hyperaccumulation, by creating a permanent metal deficiency in roots (Hanikenne et al. 2008).

The movement of As from root to shoot is controlled by the external As concentration (Singh and Ma 2006). The higher arsenic translocation of As to the shoot in hyperaccumulator *Pteris vittata*, as compared with non-hyperaccumulator ferns, occurs mainly as arsenite, which accounts for over 90 % of the As in the xylem sap (Su et al. 2008). This is because in the root of hyperaccumulator ferns, most of arsenate (AsV) is quickly reduced to arsenite (AsIII) by the activity of glutathione-dependent arsenate reductase (Duan et al. 2005).

7.4 Role of Cell Wall

The cell wall of the root is in direct contact with heavy metals present in the soil, and adsorption of these metals on the cell wall is a key step in the activity of metal ions on the surface of the cell membrane as well as on other organelles. Heavy metals primarily get accumulated in the root apoplast and then translocated to other tissues. Plant cell wall is rich in compounds able to bind divalent and trivalent metal cations. Polysaccharides play an essential role in binding of heavy metals in the cell wall, although other components such as proteins and amino acids are also able to bind metal ions. The pectin matrix of the cell wall has some negatively charged carboxylic groups in the apoplast and is responsible for binding of metal ions. These compounds bound the maximum fraction of heavy metals entered in the plant system and result in the altered structural and mechanical properties of the cell wall that also led to a reduction in the mechanical extensibility of the cell wall essential for normal cell expansion in the root elongation zone (Kochian et al. 2005; Aery 2012). Absorption of metal ions on the cell wall reduces the movement of water as well as solutes through the apoplasm which results in the decreased acquisition of nutrients by the root.

Low-methylesterified pectin level in the cell wall of different plant species such as *Funaria hygrometrica*, *Populus tremula*, and *Lemna trisulca* is known to increase under heavy metal stress (KrzesBowska et al. 2013). The absorption of heavy metals at the cell wall represents a protective action over these metal-susceptible cellular organelles like ribosome, mitochondria, etc. (Aery and Sarkar 1988).

It has been reported that a large proportion of heavy metals taken up by the plants accumulates at the cell wall. Hg^{2+} from the soil is readily taken up and accumulates in root systems with up to 80 % bound to the cell wall (Wang and Greger 2004). About 40–50 % of Al binds to the cell wall without any differentiation among leaves and roots of different ages (Hajiboland and Poschenrieder 2015). Furthermore, heavy metal fixed to the cell wall can be removed as a complex using citric acid (Varga et al. 1997).

7.5 Heavy Metal Efflux Through Plasma Membrane

The plasma membrane plays a crucial role in plants to reduce the accumulation of heavy metals into the cell by active efflux pumping of these metals outside the cell. In some instances it has been observed that ion exclusion and reduced accumulation play a sole protective mechanism in plants under heavy metal stress. Active efflux system is the most common mechanism used to reduce the accumulation of heavy metals in cells. This mechanism is reported in bacteria (Silver 1996) and animals (Palmiter and Findley 1995). However, there are only few evidences indicating the plasma membrane efflux transporters in plants. The presence of MTP1 transporter in plasma membrane suggests that it can also operate in Zn and Ni efflux from the cytoplasm to the cell wall (Kim et al. 2004). Compartmental flux analysis using ^{65}Zn suggested that *Deschampsia cespitosa* is capable of actively pumping Zn across the plasma membrane into the external medium (Brookes et al. 1981). It has also been reported that Cd in the environment of plants stimulates H^+-coupled and Mg ATP-driven efflux of Cd across the cell membrane (Migocka et al. 2011).

7.6 Vacuolar Compartmentalization/Sequestration

Vacuole is considered as the most important storage place of heavy metals in plant cells, and vacuolar compartmentalization of heavy metals is quite effective in controlling the distribution and concentration of metal ions. This organelle "captures and detains" the metal ions into a restricted location so that other parts of the cell have no access to those hazardous metal ions. Vacuolar compartmentalization depends on two vacuolar pumps (V-ATPase and V-PPase) and a set of transporter proteins of tonoplast and primary ATP-dependent pumps (Sharma et al. 2016). It has been reported that the vacuole is the site for the accumulation of many heavy metals such as Zn and Cd (Ernst et al. 1992; De 2000). Non-hyperaccumulator plants largely sequester toxic heavy metals in the vacuoles of root cells, whereas hyperaccumulators usually sequester them in vacuoles of leaf cells using efficient long-distance translocation (Sharma et al. 2016).

The role of vacuole in Cd detoxification and tolerance is well studied (Salt et al. 1995). Root-selective expression of AtCAX4 and AtCAX2 (calcium exchangers) plays a role in plant (*Nicotiana tabacum* L.) tolerance possibly due to the vacuolar Cd sequestration (Korenkov et al. 2007). The toxicity of Cd in plants induces the synthesis of phytochelatins. Interaction between phytochelatins and Cd forms a Cd-PC complex molecule, which is subsequently transferred to the vacuole by a Cd/H antiport and an ATP-dependent PC transporter (Dräger et al. 2004; Hammond et al. 2006). Apart from the Cd-PC accumulation, the best evidence for the role of vacuolar accumulation of heavy metal is of Zn (Hall 2002). Guo et al. (2012) reported that heavy metal chelation by thiols and vacuolar compartmentalization increased tolerance and accumulation of Cd and As in transgenic *Arabidopsis thaliana*. The meristematic cells of *Festuca rubra* roots show increased vacuolation on exposure with Zn, while uptake analysis using Zn^{65} with barley leaves suggests that rapid compartmentalization of Zn into the vacuole is a key mechanism to deal with elevated concentrations of Zn in plants (Davies et al. 1995; Brune et al. 1994). However, Cd, Zn, and Mo accumulate mainly in the vacuole in barley leaves, but Ni primarily accumulates in the cytosol (Brune et al. 1995).

Hyperaccumulators have great efficiency to detoxify and sequester heavy metals. This feature allows heavy metals to concentrate in huge amounts in their body without suffering any phytotoxic effect. These plants evolved their own mechanism to protect themselves from heavy metal stress. The detoxifying mechanisms found in hyperaccumulators include heavy metal complexation by ligands and/or in their removal from metabolically active cytoplasm by moving them into inactive compartments, mainly the vacuole and cell wall.

7.7 Chelation

Chelation of heavy metals is the main strategy adopted by plants to detoxify and tolerate high concentrations of heavy metals. When a ligand binds to a heavy metal ion through donor atoms, chelation occurs. Ligands are substances that interact with the electronic orbitals of the central metal ion and form secondary valence bonds resulting in a complex molecule. Overproduction of metal chelating molecules like organic acids, phytochelatins, metallothioneins (MTs), phytosiderophores, and ferritin or overexpression of genes coding metal transporter proteins enhances tolerance and accumulation of heavy metals in plants.

Upon exposure to heavy metals, plants often synthesize various metabolites that accumulate in millimolar range, mainly specific amino acids such as proline and histidine and the amines spermine, spermidine, putrescine, nicotianamine, and mugineic acids (Sharma and Dietz 2006).

The process of chelation starts, as the heavy metals approaches to the rhizosphere. It occurs both outside the plant body and within the plant cell. Extracellular chelation of heavy metals occurs by different organic acids present in roots exudate, whereas inside the plant cell, different organic acids, amino acids, and peptides are responsible for intracellular heavy metal chelation.

Metal-binding proteins and peptides are preferentially metal specific so that only toxic metals (e.g., Cd, Hg, Pb, and Cr) are sequestered excluding essential metals such as Zn, Cu, etc. (Ryu et al. 2003). Metal-peptide complex is often transported to the plant vacuole to prevent interaction with other components of cellular metabolism.

On the basis of their origin, chelators can be classified as natural and synthetic chelators. Among natural chelators, phytochelatins and metallothionein are widely studied. Metallothioneins and phytochelatins are the best-characterized S-containing metal-binding ligands that contribute to heavy metal homeostasis, detoxification, and/or tolerance by reducing the metal ion concentrations (Verkleij et al. 2003; Aery 2012). Plants produce a number of ligands including organic acids like citric acid and malic acid, phytin, amino acids, and S-containing compounds which form complexes with heavy metals (Rauser 1999). These ligands have a major role as detoxifying factors and are used in preventing the persistence of heavy metals as free ions in the cytoplasm and their entrapment in vacuoles where the metal-organic acid chelates are primarily located. Citrate is the chief ligand of Ni in leaves of *Thlaspi goesingense* (Krämer et al. 2000), while citrate and acetate are reported to bind Cd in leaves of *Solanum nigrum* (Sun et al. 2006). Moreover, a large proportion of Zn in *Arabidopsis halleri* and Cd in *Thlaspi caerulescens* is complexed with malate (Salt et al. 1999; Sarret et al. 2002).

Enhancement in the release of the phytosiderophore 2′-deoxymugineic acid (a chelating agent of Cd) from the roots under Fe deficiency conditions is reported in *Zea mays* on exposure to Cd. It has been reported that a fraction of Cd is accumulated in root cell vacuoles by phytochelatins in wheat (Cobbett 2000; Stolt et al. 2003).

The major detoxification strategy in Se hyperaccumulator plants is to protect from selenoamino acids, mainly selenocysteine (Se- Cys), derived from selenate assimilation in leaf chloroplasts (Sors et al. 2009). It has been shown that Al induced the release of organic anions such as malate, oxalate, or citrate, which chelate Al^{3+} in the rhizosphere and present its entry into the root of a number of plant species (wheat, maize, buckwheat, rye, taro, snap bean) (Ma et al. 2001).

Metallothioneins are classified into three types:

Type I metallothioneins are the polypeptides related to mammals that consist of 61 amino acids and lack aromatic amino acids or histidines.

Type II MTs are from *Candida albicans*, a yeast, or cyanobacteria (Winge et al. 1985); a similar chelator belonging to this class is *Saccharomyces cerevisiae* metallothioneins, contributing to high copper tolerance in plants (Kagi 1991).

Type III MTs are typical polypeptides composed of γ-glutamylcysteinyl units. Phytochelatins belong to the Third Type metallothioneins.

Genes responsible for production of metallothioneins have been isolated from some plant species including maize, soybean, rice, wheat, tobacco, and rapeseed. Type I MT genes are expressed mainly in the roots, whereas type II MT genes are expressed mainly in the leaves (Nedkorska and Atanassov 1995). According to Kagi (1991), heavy metals such as Cd, Zn, Hg, Ag, and Pb induce the production of metallothionein especially in animal and plant species. Cd is a best activator of metallothionein production, followed by other metals like Ag, Bi, Pb, Zn, Cu, Hg, and Au (Krämer et al. 1996). Cd accumulation increased significantly in tobacco plants having the transgene coding for the polyhistidine cluster combined with yeast metallothionein (Macek et al. 2002). Introduction of a metallothionein gene in tobacco also enhances the tolerance to some heavy metals such as Cd, Zn, and Ni (de Borne et al. 1998; Pavlíková et al. 2004).

Phytochelatins are the well-known heavy metal chelators in plants, especially in the context of Cd tolerance (Cobbett 2000). These are a family of metal-binding peptides with the general structure (γ-Glu-Cys)nGly ($n = 2$–11) (Cobbett and Goldsbrough 2002). Phytochelatins are glutathione-derived peptides synthesized in the cytosol and are reported in a variety of plant species including monocots, dicots, gymnosperms, and algae (Gekeler et al. 1989). Plants exposed to Cd stress had 2.7–3 times more total phytochelatins rather than the plants of same lines grown without Cd (Guo et al. 2012). Phytochelatins form PC-metal(loid) complexes that are transported into vacuoles and thus help in removal of toxic metals from the cytosol (Clemens 2006).

Proline functions as an osmolyte, radical scavenger, electron sink, and stabilizer of macromolecules and a component of the cell wall (Matysik et al. 2002). It is a widely studied molecule in context of plant responses to abiotic stresses such as water deficit (Paleg and Aspinall 1981), salinity (Ashraf and Harris 2004), low temperature (Naidu et al. 1991), high temperature (Kuznetsov and Shevyakova 1997), and high concentration of heavy metal (Panchal and Aery 2008; Kumar and Aery 2011, 2012a; Aery 2012). An increase in the level of proline may provide protection by chelating the metals in the cytoplasm and maintaining the water balance which is often disturbed by heavy metals (Xu et al. 2009; Kumar 2013). Panchal and Aery (2008) observed increased proline contents in plants grown in manganese waste. Kumar and Aery (2012) and Jayakumar et al. (2010) have also observed the elevated accumulation of proline in plants with the application of tungsten, manganese, and cobalt.

Polyamines are low-molecular-weight aliphatic polycations, highly charged and universally present in all organisms. They are cations due to protonation at cytoplasmic pH, i.e., putrescine^{2+}, spermidine^{3+}, and spermine^{4+} (Sharma and Dietz 2006). It has been reported that tobacco BY-2 cells exposed to 0.05 mM $CdCl_2$ produced a distinct accumulation of total polyamines (Kuthanová et al. 2004). Choudhary and Singh (2000) have also reported accumulation of polyamines in mung bean seedlings treated with 0.1–1.5 mM $CdCl_2$. A specific mechanism of action of polyamines in plants under heavy metal stress is yet unknown. However, there is a strong possibility that polyamines can efficiently stabilize and guard the membrane systems of plants against the toxic effects of metallic ions specially the redox-active metals (Sharma and Dietz 2006).

Beyond the above-mentioned natural chelators, some synthetic chelators are also known. Ethylenediaminetetraacetate (EDTA) has been practiced more intensively among synthetic chelators (Blaylock et al. 1997; Huang et al. 1997; Grcman et al. 2001). EDTA binds to heavy metal ions and makes them uncharged. An uncharged ion is of high mobility and much easier to pass through plasma membrane. Application of EDTA improved the uptake of Pb in *Brassica juncea* (L.) Czern, 1000–10,000 times greater than control plants (Blaylock et al. 1997). In some cases ethylenediamine disuccinate (EDDS) (a structural isomer of EDTA) has recently been used to enhance metal uptake through phytoextraction (Grcman et al. 2003; Luo et al. 2005).

7.8 Some Toxic Heavy Metals

7.8.1 Cadmium

The average Cd content in the Earth's crust and soil is 0.1 and 0.41 mg kg^{-1}, respectively. Cd is mainly used in battery production (Ni–Cd and Ag–Cd). It is also used as pigments (yellow), for coatings, and as stabilizers at relatively high amount. Due to some unique physical and chemical properties, Cd is also added to alloys and used as stabilizer for various plastics.

Generally Cd is considered as one of the most toxic heavy metal that adversely affects all biological processes of living organisms including human beings, animals, and plants. Although it is considered to be a nonessential element for plants, it is efficiently absorbed by the root and leaves of plants. Cd^{2+} ions present in the mobile fraction of the soil are apparently taken up by the root of plants mainly via transporters of essential metals (Fe^{2+}, Zn^{2+}, and Mn^{2+}) and probably by Ca^{2+} channels (Lay and Levina 2014). There are evidences that a considerable portion of Cd is taken up passively by roots, but it is also absorbed actively (Smeyers-Verbeke et al. 1978).

Though Cd is widely reported as a toxic metal for plants, a general toxic concentration is not established. Kloke et al. (1984) reported the toxic concentrations of Cd to be 5–10 mg kg^{-1}, whereas Macnicol and Beckett (1985) estimated 10–20 mg Cd kg^{-1} as critical concentration of Cd for plants.

Elevated accumulation of Cd affects cell biology (viability, proliferation), cell structure (microtubules) (Fusconi et al. 2007), and physiology (photosynthesis, disturbs transpiration, reduces water transport, alters the permeability of cell membranes, and disturbs stomatal conductance and electron transport systems) (Seregin and Ivanov 2001; Prasad 2005). Cd exerts an inhibitory effect on seed germination of fenugreek (Aery and Sarkar 1990).

Huang et al. (1974) observed a decrement in the nodule weight and activity of enzyme nitrogenase in soybean under high concentration of Cd and Pb. Toxicity symptoms induced by the higher accumulation of Cd in plants are growth retardation, stunted root, chlorosis of leaves, and red-brown coloration of leaf margins or veins (Aery and Sarkar 2012a, b). It has been reported that Cd concentration above 5 μg g^{-1} decreases growth parameters such as root-shoot length, dry weight, leaf area, seed number, seed weight, as well as nodule number and nodule weight of soybean at seedling, vegetative, flowering, and fruiting stage (Aery and Sarkar 1991). Inhibition of growth by Cd is accompanied by the responses like suppression of chlorophyll synthesis and recurvature of the embryo (Imai and Siegel 1973).

The toxicity of Cd in plants mainly lies in its tendency to disturbing the activity of different enzymes, resulting from substitution of other metal ions (like Zn^{2+}, Cu^{2+}, and Ca^{2+}) and its affinity to biological structures containing –SH groups such as proteins, enzymes, and nucleic acids (Jacobson and Turner 1980; Stohs and Bagchi 1995).

The response of plants to Cd present in growth media depends on the age of plants and time of Cd exposure. Sensitivity to Cd increases with the aging in plants and is more related to functioning of the photosynthetic apparatus than to growth parameters. Cd significantly reduced the photochemical activities of chloroplasts at 50–75 μM L^{-1} concentrations in nutrient media (Tukendorf and Baszynki 1991). The amount of chlorophyll has been known to be a function of the Cd concentration in plant tissues and proposed as an indicator of the upper critical Cd level in plants (Burton et al. 1986). Aery (1994) studied Cd-Zn-induced chlorosis in soybean. Lower concentrations of Zn (10 μg g^{-1}) as well as Cd (5 μg g^{-1}) increase the contents of chlorophyll "a" and chlorophyll "b." Beyond the above-mentioned concentration of Zn and Cd, a decrement in chlorophyll "a" and "b" and intraveinal chlorosis were reported. Cunningham et al. (1975) and Baszyński (2014) also reported inhibition of the synthesis of anthocyanin and chlorophyll pigments in plants treated with Cd. Excess of Cd may affect cell acidosis, due to an imbalance in H^+ production associated with changes in anaplerotic reactions (Nocito et al. 2008). Cd is also reported as an efficient and specific inhibitor of the biological reduction of NO_2^- to NO (McKenney and Vriesacker 1985).

Mutagenic activity and degradation of DNA (Bertin and Averbeck 2006) and cytological abnormalities (Wang et al. 2016) have been reported under Cd stress. The elevated concentration of Cd has an inhibitory impact on the Calvin's cycle by disrupting the function of key enzyme, especially ribulose, diphosphate, and carboxyhydrazine (Kabata-Pendias 2011). Elevated concentrations of Cd in the rooting medium of barley resulted in an increase in shoot thionein mRNA (Krämer et al. 1999). Cd is also known to inhibit the DNA-mediated transformation in microorganisms and interfere with symbiotic relationship of microorganisms and plants, as well as increases predisposition of plants to fungal invasion (Kabata-Pendias 2011).

Metal toxicity may result from competition between Cd and ferrous ions at the enzyme site (Aery 1994). Cd can interact with the uptake and biochemical functions of several other elements. These interactions take place at different stages of absorption, partitioning, as well as on the biological functions of essential elements.

The interactions between Cd and Zn are commonly observed, and both depressing and enhancing effects have been reported (McKenna et al. 1993; Hart et al. 2002; Aery and Rana 2007). It is known that Zn antagonistically affects the uptake of Cd. When the Cd/Zn ratio in plant tissues is limited to 1, the Cd content is restricted to below 5 mg kg^{-1}, thus below its phytotoxic level (Chaney and Hornick 1978). According to Nan et al. (2002), Cd-Zn interaction is synergistic under field condition, and increasing the concentration of both metals in soils resulted in increased accumulation in plants. Turner (1973) observed that the effect of Cd on Zn accumulation in leaves depends on plant species, i.e., synergistic effect on some plant species (e.g., lettuce) and antagonistic effect on others. McKenna et al. (1993) observed a synergistic effect of Cd application on Zn accumulation in the leaves of lettuce at low as well as high Zn treatments.

Aery (1994) found that leaf iron contents decreased with increase in the leaf Cd concentration. Cd-Fe interactions are related to disturbances in the photosynthetic apparatus. Only a moderate excess of Fe has a detoxificatory influence on Cd-treated plants, although a higher dose is toxic. Cd uptake can also be modified by Fe nutrition (Lombi et al. 2002). At low Fe supply, Cd can be transferred from the soil solution into the roots by Fe transporters of the IRT1 family (Nakanishi et al. 2006).

Ca^{2+} ions are able to replace Cd^{2+} in carrier mechanisms, and thus Cd absorption by plants may be inhibited by an excess of Ca cations. Ca in substrate affects the uptake of Cd because Cd competes with Ca for Ca channels (Wojas et al. 2007). Low Ca in substrate enhances the uptake of Cd and Pb and thus induces the toxicity in many plant species (Suzuki 2005; Wojas et al. 2007). The antagonistic interactions between Cd and Se are also observed in certain plants. Seleno-urea forms a complex with Cd^{2+} of limited solubility, which might decrease the availability of Cd ions (Feroci et al. 2005). Phosphorus affects the uptake of Cd in the roots of plants.

Both increased and decreased contents of Cd under phosphate treatment are known. An antagonistic impact of Cd on the P absorption by plants is significant for plant growth and agricultural practices. The P uptake by plants can be reduced up to 40 % by Cd application (Kabata-Pendias 2011).

It is well known that a higher concentration of essential cations such as Zn^{2+} and Ca^{2+} plays a protective role against Cd^{2+} toxicity (Antonovics et al. 1971). The absorption of Zn and Cu by plants is influenced in the presence of Cd and Pb in the soil (Luo and Rimmer 1995). Interactions of Cd with several essential elements (e.g., Mg and K) have been reported and may be related with an impaired effect of Cd on the plasma membrane, which disturbs the uptake and transport of these elements within different organs.

The concentration of Cl and S in the shoot of alfalfa increased by the addition of heavy metals in the soil (Peralta-Videa et al. 2002). This may be due to the complexation of Cl and S to the heavy metals, which results in their common translocation with the metals in the plant (McLaughlin et al. 1998).

Wu et al. (2003) studied the interactions of Cd and uptake and translocation of Zn, Cu, Mn, and Fe in different barley genotypes. Cd addition to the substrate not only decreased Zn concentrations in tissues but also inhibited the translocation from root to aerial parts, leading to high root/shoot Zn ratio. Cadmium addition also reduced the accumulation of Mn, Cu, and Fe in different plant parts. Significant negative correlation between Zn, Cu, and Mn contents and Cd contents in different plant parts suggests the possibility of increasing the accumulation of Cd in barley by administration of these elements on the Cd-polluted soils (Wu et al. 2003).

However, it has also been known that an increase in the concentration of Cd in the substrate results in an improved uptake and translocation of Mn in lettuce. This is in contrast to the response of the other essential micronutrients (Fe, Cu, and Zn). Further, an increase in the concentration of Mn in the chloroplasts suggests relations between Cd and Mn at the chloroplast (Ramos et al. 2002).

Li et al. (2016) reported that the concentrations of different nutrients in the tissues of Welsh onion are influenced by Cd and varied by the type of element and plant tissue. The concentrations of the nutrient in the roots decreased at the 1 mg Cd kg^{-1} level, although significant differences were only reported for P, Ca, Mg, Fe, Cu, and Mn in some tissues. At Cd concentration of 2.5 mg kg^{-1}, the contents of K and Zn increased significantly in all tissues. The concentrations of Ca, Mg, and Fe were similarly influenced by Cd treatment. Except for Fe in the pseudostems, the concentrations of nutrients decreased under low Cd treatment (1 mg kg^{-1}), but increased under 2.5–5 mg kg^{-1}. A significant increase in Ca and Fe contents in roots and leaves, and Mg in leaves at 5 mg kg^{-1} of Cd, has been reported (Li et al. 2016).

7.8.2 Mercury

The average content of mercury (Hg) in the Earth's crust is 0.07 mg kg^{-1}; whereas in soils of different groups all over the world, it ranges between 0.58 and 1.8 mg kg^{-1}, and the worldwide mean content is estimated as 1.1 mg kg^{-1} (Kabata-Pendias 2011).

Mercury is mainly used in chlor-alkali production, vinyl chloride monomer production, gold mining, batteries, measuring instruments, florescent lamps, wood impregnation, paints, pesticides, control instruments, and electrical lightings production. Due to its vast usage, this metal is accumulated at different sites, and as in now it is considered as global pollutant.

The uptake of Hg by plants is directly dependent on its concentration in the soil. As the concentrations of Hg increased in soil, an increase in the Hg content of plants was reported. However, a maximum fraction of Hg is found in roots, but leaves and grains also accumulated a significant fraction of Hg (Hogg et al. 1978; Lipsey 1975; Sorterberg 1980). Hg is not only absorbed by plants from the soil, but it is also absorbed from Hg vapor gradually released from soils.

Israr et al. (2006) reported that *Sesbania drummondii* accumulates high concentrations of Hg. The accumulation of Hg was 998 and 41,403 mg kg^{-1} in shoots and roots, respectively, at a nutrient solution concentration of 100 mg L^{-1}. It has also been studied that transgenic *Arabidopsis thaliana* with the genes for mercuric ion reductase (reduce toxic Hg^{2+} to the relatively inert Hg^0) exhibits resistance to high Hg concentrations (McGrath 1998).

Toxic effects of Hg in plants were reported at a low concentration even at 1 μg L^{-1} Hg concentration in the nutrient solution (Mhathre and Chaphekar 1984). Kloke et al. (1984) observed retarded growth of plants at Hg content ranging from 0.5 to 1 mg kg^{-1}, whereas Macnicol and Beckett (1985) established the critical level of Hg toxicity from 1 to 8 mg kg^{-1}.

The most common symptoms of Hg toxicity are stunted growth, reduced root development, and inhibition of photosynthetic processes. It is also responsible for failure of various metabolic processes like photosynthesis, chlorophyll synthesis, exchange of gases, and respiration. The higher accumulation of Hg in the root inhibits K^+ uptake by plants. However, it is also observed that lower Hg concentrations stimulate uptake of K^+ (Hendrix and Higinbotham 1974).

The toxicity of volatilized elemental Hg is known to be the most serious for plants. Hg vapor induces senescence-related processes by increasing ethylene production, and the most active toxicant is elemental Hg, not its ionic form. Further, young plants are more sensitive to Hg-saturated air than mature plants (Siegel et al. 1984).

Hg has strong affinity to amino acids of several proteins and enzymes. Its binding nature to sulfhydryl groups is apparently the key reaction in disrupting metabolism of plants (McNear et al. 2012). Yathavakilla and Ceruso (2007) observed an association of Hg with Se in high-molecular-weight molecules in soybean root. Further, in several cases enhanced activity of antioxidative enzymes is reported on application of Hg in growth media (Israr and Sahi 2006; Zhou et al. 2008).

7.8.3 Lead

The average lead (Pb) content in the Earth's crust is estimated as 15 mg kg^{-1}. In the terrestrial environment, two types of Pb are known, i.e., primary and secondary.

Primary Pb is of geogenic origin and incorporated into minerals at the time of their formation, whereas secondary Pb is of radiogenic origin from the decay of U and Th. The largest use of Pb is in production of lead acid batteries. It is also used in solders, alloys, cables, and chemicals.

Among the total soil Pb, the amount of available Pb to plants is unknown. In some cases, only about 0.005–0.13 % of Pb in the soil solution is available to plants (Davies 1995), whereas on the other hand, only 0.003–0.005 % of the total Pb in soils is available to plants (Wilson and Cline 1966). The uptake of Pb, however, varies significantly over the concentration ranges and with different forms of Pb present in the soil.

Along with many other nonessential elements, Pb is taken up by plants. All plants are able to absorb Pb from soil and transport it to different parts (Aery and Jagetiya 1997). In addition, seasonal variation of Pb accumulation in plants (like *Acacia leucophloea*, *Azadirachta indica*, *Cassia siamea*, etc.) growing on the tailings dam region of Zawar Mines (India) is also reported (Singh and Aery 1993).

The toxicity of organic Pb not only exceeds by far the toxicity of its inorganic forms but the effects caused by both types of agents also differ in quality (Roderer 1984). This difference is presumably due to various physical and chemical properties of organic and inorganic Pb compounds. Although even a very low Pb concentration may inhibit vital plant processes, Pb poisoning has seldom been observed in plants growing under field conditions. This is perhaps due to a relatively low Pb concentration in the soil and/or low Pb availability even under contaminated soil (Zimdahl 1976).

Pb accumulates in various parts of cells and affects their structures. The most deteriorating effect of Pb is the destruction of the plasma membrane which subsequently disturbs the permeability for water and leads to impaired plant growth (Woźny 1998). Reports have described the inhibitory effects of low Pb levels on plant metabolism (Aery and Jagetiya 1997; Patra et al. 2004; Sharma and Dubey 2005). Owing to the interactions of Pb with other elements and with many environmental factors, it has not been simple to establish toxic Pb concentrations to vital plant processes. There are several reports on toxic effects of Pb on the fundamental biological processes such as photosynthesis, respiration, mitosis, and water absorption; however, the toxic symptoms in plants are not very specific (Koeppe 1981; Sharma and Dubey 2005; Islam et al. 2008).

Subcellular effects of Pb on plant tissues are associated to the inhibition of respiration and photosynthesis due to the disturbance of the electron transfer reactions. These reactions are inhibited by Pb concentrations as low as 1 mg kg^{-1} in mitochondria of corn (Zimdahl 1976). Photosynthesis processes in sunflower leaves were also reduced by half at a Pb concentration of about 1 μM g^{-1} (Carlson et al. 1975).

According to Lane et al. (1978), Pb becomes strongly bound to cell walls and that pectic acid is most active in the Pb sorption. Thus, Pb has a marked influence on the elasticity and plasticity of cell wall, resulting in an increase in tissue wall rigidity.

Pb interferes with the uptake of some other ions such as Zn and Cd. It stimulates the uptake of Cd by plant root. This stimulatory effect may be due to a secondary effect of the disturbance of transmembrane ion transport. An antagonistic relation is also known between Pb and Zn. Zn-Pb adversely affects the translocation of each element from root to shoot. Pb can mimic the physiological behavior of Ca; therefore, interference of Pb with Ca is of metabolic importance, and thus Pb inhibits the activity of some enzymes. Cd and Pb inhibit the accumulation Zn and Mn in the root and shoot of plants (Motesharezadeh and Savaghebi 2012). Recently, it has been reported that S application decreases the accumulation of Pb in straw and grains of rice. This may be due to the development of Fe plaque formation, decrement in Pb availability, and increased contents of reduced glutathione in rice leaves (Yang et al. 2016).

7.8.4 Chromium

Chromium (Cr) is the seventh most abundant element on the earth and 21st in the Earth's crust, with an average concentration of 100 mg kg^{-1}. It is mainly used for stainless steel, pigments, metal finishing, wood preservatives, chemicals, and chromate plating. Cr is commonly used in the production of green tints for paints, varnishes, glazes, inks, and paper. Leather tanning also utilizes substantial amounts of Cr compounds. The main source of Cr pollution in the environment is dyestuffs and leather tanning when wastes are released directly into waste streams, either in liquid or solid form. Sewage treatment plants from industrial as well as residential sources also discharge a considerable quantity of Cr.

Chromium is not considered as an essential element for the plants. It is slightly available and not easily translocated in different parts of the plants and mainly concentrated in roots, apparently because of the tendency of Cr to bind to the cell wall (Zayed et al. 1998). The most available form of chromium to plants is Cr^{6+}. It is a very unstable form under normal soil conditions, and its availability depends on soil properties especially on soil texture and soil pH. However, Cr^{3+} and some anions (e.g., CrO_4^{2-}) are also available to plants.

Cr is taken up by plant roots (through sulfate channels) predominantly in the form of highly soluble $[CrO_4]^{2-}$ (Lay and Levina 2014). A very low Cr content (1–2 mg kg^{-1}) inhibits the growth of plants (Kloke et al. 1984). The elevated Cr content in soil is known to be responsible for the poor growth of forest trees (Wedepohl 1974).

Symptoms of Cr toxicity appear as wilting of tops and root injury, chlorosis in young leaves, chlorotic bands on cereals, and brownish-red leaves. Increased levels of Cr in the nutrient solution (up to 104 μM) disorganize the fine structure of chloroplasts and the chloroplast membranes of *Lemna minor* (Baszyński et al. 1981). There is evidence that Cr^{6+} is transformed into Cr^{3+} form in plant cells which readily interacts with DNA and protein compounds (Zayed and Terry 2003).

An increased level of Cr^{6+} in nutrient solution decreased CO_2 assimilation of *Lolium perenne* leaves (Vernay et al. 2007). Chatterjee and Chatterjee (2000) reported that a higher application of Cr to plants leads to decrease in protein contents by disrupting N metabolism. Cr-induced competition in assimilation and translocation of mineral elements indirectly affect water status and plant growth (Vernay et al. 2007). In some instances improved activities of antioxidative enzymes such as superoxide dismutase and peroxidase also are reported in the plants grown in Cr-treated soil (Dong et al. 2007).

The toxicity of Cr is also related to readily available forms of chromate ion. Cr_2O_7 addition to bush bean at the 0.0001 N concentration level decreased plant growth by about 25 %, and Cr uptake of 2.2 mg kg^{-1} but at the same level $Cr_2(SO_4)_3$ did not affect the plant growth and Cr content was 1.3 mg kg^{-1} (Wallace et al. 1977).

The phytotoxic concentrations of Cr in shoots of barley seedlings were reported to be 18–24 mg kg^{-1} in tobacco, 4–8 mg kg^{-1} in corn, and 10–100 mg kg^{-1} in barley (Davis et al. 1978; Gough et al. 1979; Kitagishi and Yamane 1981). The toxic level of Cr reported by Macnicol and Beckett (1985) is much broader and ranges from 1 to 10 mg kg^{-1}. At the same dose (750 mg kg^{-1} soil) of Cr^{6+} and Cr^{3+}, an increased uptake of Cr and reduced oat yield were observed; however, harmful effects were significantly greater in the case of Cr^{6+} addition (McGrath 1982). Similarly, Mukhopadhyay and Aery (2000) reported that toxic effects of Cr(VI) on seed germination, seedling growth, and fresh and dry weight of wheat were higher than Cr(III).

Moreira et al. (2005) explained that the reduced uptake of essential nutrients in Cr-treated plants is due to the inhibition of the activity of the plasma membrane-localized proton-ATPase pump. The decrease in proton-ATPase activity causes a decrease in proton extrusion. This results in a decreased transport of the root cell membranes and reduces the uptake of elements (Fernandes et al. 2002).

In some cases, synergistic interactions between Cr and some essential nutrients (Ca, Mg, Fe, Mn, Cu) have also been observed (Dong et al. 2007; Vernay et al. 2007). Excess Cr exposure of plants leads to substantial changes in elemental contents of roots as well as leaves that may cause ions deficiency with visible symptoms and could be detrimental to plants (Vernay et al. 2007). Turner and Rust (1971) studied the initial symptoms of Cr toxicity with the addition of low Cr application (0.5 mg kg^{-1} Cr to the nutrient culture and 60 mg kg^{-1} to the soil culture). These Cr additions resulted in decreased concentrations of almost all major nutrients in shoots and K, P, Fe, and Mg in roots.

Cr is competitive for P and Fe on surface root sites, in binding and transport (Chatterjee and Chatterjee 2000). Schiavon et al. (2008) reported a significant decrease of the S uptake by plants under Cr stress. Chromium causes severe decrease in Ca contents in the plants (Marschner 1999). Mn deficiency in *Brassica oleracea* and *Vigna radiata* has been reported in response to Cr stress (Chatterjee and Chatterjee 2000; Sinha et al. 2006).

7.8.5 Aluminum

Aluminum (Al) is the third most abundant element in the Earth's crust, occurring at about 8 %. It has some versatile properties that allow us to use it in different industrial sectors like metallurgical, packaging, transportation, electrical, and chemical productions. Different compounds of Al are also used in paper manufacturing, sugar refining, water purification, wood preservation, leather tanning, and waterproofing textiles.

Aluminum is present in soil as insoluble aluminosilicates and oxides. It is easily absorbed by roots and translocated within the plant at the initial stages of plant growth, but then sharply declines with advancing maturity. The absorption of Al is affected by the concentration of other elements (P and Ca) in the rhizosphere.

Al in soil can inhibit growth of plants at a level as low as 1 ppm (Foy 1974). A decrement in growth and biomass of wheat is also reported at low Al (3 $\mu g\ g^{-1}$) in soil (Rana and Aery 2000). The earliest symptoms of Al toxicity appear on the root tip, which turns brown. The damage is limited to actively growing tissues of the root tip (Fleming and Foy 1968).

Root growth inhibition on the exposure to Al has been widely used as a measurement of Al toxicity (Foy 1988). Inhibition of root growth is the initial symptom of Al toxicity in plants (Ryan et al. 1993; Sivaguru et al. 1999; Kollmeier et al. 2000). More specifically the distal part of the transition zone of root apex is very sensitive to Al toxicity (Sivaguru and Horst 1998). This widespread damage in the structure of root results in reduced and damaged root system and limited water as well as mineral elements uptake (Jones and Kochian 1995; Barcelo and Poschenrieder 2002) which subsequently results in deformed growth.

Al can enter in the cytosol within a short time after the exposure (Vazquez et al. 1999). A maximum part of Al enters in the plants and reacts with apoplastic binding sites; a small fraction enters the symplasm to interact with symplastic binding sites. Disrupted organization of microtubules and microfilaments in root cells are also reported on Al exposure (Blancaflor et al. 1998; Frantzios et al. 2000, 2001; Alessa and Oliveira 2001; Schwarzerová et al. 2002; Sivaguru et al. 2003).

Al interferes with a large range of physical and cellular processes. It causes morphological damage as well as decreases in chlorophyll contents and rate of photosynthesis (Rana and Aery 1999). Decrement in chlorophyll "a" and "b" and total chlorophyll under different treatments of $AlCl_3$, $Al_2(SO_4)_3$, and $Al(NO_3)_3$ is also reported (Rana and Aery 1999). A drastic reduction in chlorophyll contents of wheat is reported with increasing concentration of Al in soil (Rana and Aery 2000).

Al exposure induces the peroxidative damage of membrane lipids by producing ROS (Cakmak and Horst 1991). However, lipid peroxidation is enhanced after a long exposure to Al. Thus, lipid peroxidation does not appear to be a primary mechanism of Al toxicity (Horst et al. 1992; Yamamoto et al. 2001). Al-induced ROS generation and associated mitochondrial dysfunction could play a more general role in Al inhibition of root growth (Yamamoto et al. 2002). It has also been reported that plants synthesize phenolic compounds under Al stress (Rana and Aery 1999) to neutralize ROS in order to survive and prevent molecular damage (Dicko et al. 2006).

Prolonged exposures of Al cause structural abnormalities within the nucleus, detrimentally affecting DNA composition, chromatin structure, and cell division (Matsumoto 1991; Silva et al. 2000). Al can also inhibit the activity of the enzyme phospholipase C of the phosphoinositide pathway related to Ca^{2+} signaling

(Jones and Kochian 1995, 1997). The Al-induced disruption of ion fluxes could directly lead to changes in cytosolic ion activities (e.g., Ca^{2+} homeostasis) as well as signaling pathways (e.g., inhibition of Ca^{2+}-dependent enzymes such as phospholipase C) which would ultimately reflect in any of the physiological and morphological changes (Ramos-Díaz et al. 2007).

Exposure of plants to Al induces the synthesis of callose (b-l,3-glucan) on plasma membrane. Callose accumulation disturbs intercellular transport through plasmodesmatal connections and resulted in cellular damage (Sivaguru et al. 2000). A significant fraction of Al in root is associated with the apoplastic region, predominantly in cell walls of the root periphery (Vazquez et al. 1999). Among the different components of the cell wall, pectins have been known to be a critical site for Al action (Blamey et al. 1993). This results in displacement of other cations (e.g., Ca^{2+}) required for cell wall stability (Schmohl and Horst 2000; Tabuchi and Matsumoto 2001). Consequently, aluminum alters the structural and mechanical properties of the cell wall by strong and rapid binding, making it highly rigid, leading to reduction in the mechanical extensibility of the cell wall.

Aluminum is known to affect the uptake, transport, and functions of other nutrients like Ca, Mg, P, and K (Foy et al. 1978), and reduced nutrient uptake leads to nutrient deficiencies. Application of Al to plants can block the uptake of many cations such as Ca^{2+}, Mg^{2+}, K^{+}, and NH^{4+} by interacting directly with several different channel proteins (Rengel and Elliott 1992; Ryan and Kochian 1993; Gassmann and Schroeder 1994; Pineros and Kochian 2001; Pineros and Tester 1995).

7.8.6 Silver

The concentration of silver (Ag) in the Earth's crust and soil averages around 0.06 and 0.13 mg/kg, respectively. It is mainly used in photographic industries and also useful in other sectors like batteries, coins, jewelry, silverware, catalysts, and electronics brazing and soldering.

Silver toxicity depends on the concentration of active free silver ions (Ag^{+}), therefore, mainly found in the aqueous stage. Several processes in medium and water characteristics reduce silver toxicity by preventing the formation of free Ag^{+} or avoiding binding of Ag^{+} to the reactive surfaces of organisms (Ratte 1999).

The toxic effects of Ag on plants grown on soil have not been reported for a long period of time. Wallace et al. (1977) reported that about 5 mg kg^{-1} Ag in the shoots and about 1500 mg kg^{-1} in the roots of bush beans greatly reduced yields without showing any toxicity symptoms. Ozoliniya (1986) reported that Ag at a very low concentration (10 μg L^{-1}) in the nutrient medium stimulated the growth of roots of grass. He speculated that some cations (e.g., Ag, Co, and Cu) can indirectly change cell metabolism and subsequently result in a higher growth rate of cells. Hendrix and Higinbotham (1974) observed that Ag substitutes K^{+} sites in membranes and inhibits the absorption of other cations by the roots. Recently, Krizkova et al. (2008) observed that high concentrations of Ag (up to 1 μM/L) significantly decrease growth and protein contents, whereas they increase the activity of enzyme urease in sunflower.

7.9 Conclusion

Heavy metal contamination of soil is a global issue as these metals are the chief group of inorganic pollutants. The problem is of more concern because anthropogenic activities are continuously increasing the levels of these contaminants in agricultural soils. It is evident from the previous reports that heavy metals impose several toxic effects on the plant and adversely affect the growth as well as development of plants. However, various cellular, physiological, and molecular strategies are adopted by plants to tolerate heavy metal stress. There are some key components like metal transporters, hyperaccumulation, organic compounds, phytochelatins, and metallothionein proteins that help plant to thrive under heavy metal stress condition. In order to perceive further, the heavy metal toxicity on plants and their potential to uptake the metal, it is essential to explore more research toward this area.

References

Aery NC (1994) Cadmium-Zinc induced chlorosis in soybean and its relation to iron stress. Asian J Plant Sci 6(1&2):27–33

Aery NC (1998) Accumulation, toxicity and detoxification of heavy metals in plants. In: Sood PP, Prakash R (eds) Heavy metal pollution, toxication and chelation. M.D. Publications Pvt. Ltd., New Delhi, pp 277–306

Aery NC (2012) Plant defence against heavy metal stress. In: Merillon JM, Ramawat KG (eds) Defence: biological control. Springer, Dordrechts, pp 241–269

Aery NC, Jagetiya BL (1997) Relative toxicity of cadmium, lead, and zinc on barley. Commun Soil Sci Plant Anal 28(11–12):949–960

Aery NC, Rana D (2007) Interactive effects of Zn, Pb and Cd in barley. J Environ Sci Eng 49(1):71–76

Aery NC, Sarkar S (1988) Histochemical localization of accumulated zinc in soybean and fenugreek. Bionature 8(2):109–112

Aery NC, Sarkar S (1990) Heavy metal toxicity and temperature interactions in fenugreek. Bionature 10(1&2):15–18

Aery NC, Sarkar S (1991) Studies on the effect of heavy metal stress on growth parameters of soybean. J Environ Biol 12(1):15–24

Aery NC, Sarkar S (2012a) Metal species *vis-à-vis* seed germination and early seedling growth responses in soybean. J Chem Bio Phy Sci Sec B 2(2):763–769

Aery NC, Sarkar S (2012b) Responses of Zn and Cd treatment in soybean and fenugreek. NBU J Plant Sci 6(1):71–76

Aery NC, Tyagi S (1988) Cu/Zn ratio: an indicator of mineralization in biogeochemistry. Bionature 8(2):142–144

Alessa L, Oliveira L (2001) Aluminum toxicity studies in *Vaucheria longicaulis* var. macounii (Xanthophyta, Tribophyceae). II. Effects on the F-actin array. Environ Exp Bot 45(3):223–237

Antonovics J, Bradshaw AD, Turner RG (1971) Heavy metal tolerance in plants. Adv Ecol Res 7:1–85

Ashraf MP, Harris PJ (2004) Potential biochemical indicators of salinity tolerance in plants. Plant Sci 166(1):3–16

Assche FV, Clijsters H (1990) Effects of metals on enzyme activity in plants. Plant Cell Environ 13(3):195–206

Axelsen KB, Palmgren MG (1998) Evolution of substrate specificities in the P-type ATPase superfamily. J Mol Evol 46(1):84–101
Axelsen KB, Palmgren MG (2001) Inventory of the superfamily of P-type ion pumps in Arabidopsis. Plant Physiol 126(2):696–706
Baker AJ (1981) Accumulators and excluders-strategies in the response of plants to heavy metals. J Plant Nutr 3(1–4):643–654
Baker AJ, Brooks R (1989) Terrestrial higher plants which hyperaccumulate metallic elements: a review of their distribution, ecology and phytochemistry. Biorecovery 1(2):81–126
Baker AJM, McGrath SP, Reeves RD et al (2000) Metal hyperaccumulator plants: a review of the ecology and physiology of a biological resource for phytoremediation of metal-polluted soils. In: Terry N, Bañuelos G (eds) Phytoremediation of contamined soil and water. CRC Press LLC, Boca Raton, pp 85–107
Barcelo J, Poschenrieder C (2002) Fast root growth responses, root exudates, and internal detoxification as clues to the mechanisms of aluminium toxicity and resistance: a review. Environ Exp Bot 48(1):75–92
Baszyński T (2014) Interference of Cd^{2+} in functioning of the photosynthetic apparatus of higher plants. Acta Soc Bot Pol 55(2):291–304
Baszyński T, Krol M, Wolinka D (1981) Effect of chromate on photosynthetic apparatus of *Lemna minor* L. In: Akoynoglou G (ed) Photosynthesis II. Electron transport and photophosphorylation, Balbon International Science Services, Philadelphia, pp 245–246
Baxter I, Tchieu J, Sussman MR et al (2003) Genomic comparison of P-type ATPase ion pumps in *Arabidopsis* and rice. Plant Physiol 132(2):618–628
Bertin G, Averbeck D (2006) Cadmium: cellular effects, modifications of biomolecules, modulation of DNA repair and genotoxic consequences (a review). Biochimie 88(11):1549–1559
Blamey FP, Asher CJ, Edwards DC et al (1993) In vitro evidence of aluminum effects on solution movement through root cell walls. J Plant Nutr 16(4):555–562
Blancaflor EB, Jones DL, Gilroy S (1998) Alterations in the cytoskeleton accompany aluminum-induced growth inhibition and morphological changes in primary roots of maize. Plant Physiol 118(1):159–172
Blaylock MJ, Salt DE, Dushenkov S et al (1997) Enhanced accumulation of Pb in Indian mustard by soil-applied chelating agents. Environ Sci Technol 31(3):860–865
Brookes A, Collins JC, Thurman DA (1981) The mechanism of zinc tolerance in grasses. J Plant Nutr 3(1–4):695–705
Brune A, Urbach W, Dietz KJ (1994) Compartmentation and transport of zinc in barley primary leaves as basic mechanisms involved in zinc tolerance. Plant Cell Environ 17(2):153–162
Brune A, Urbach W, Dietz KJ (1995) Differential toxicity of heavy metals is partly related to a loss of preferential extraplasmic compartmentation: a comparison of Cd-, Mo-, Ni- and Zn stress. New Phytol 129(3):403–409
Bubb JM, Lester JN (1991) The impact of heavy metals on lowland rivers and the implications for man and the environment. Sci Total Environ 100:207–233
Burton KW, Morgan E, Roig A (1986) Interactive effects of cadmium, copper and nickel on the growth of Sitka spruce and studies of metal uptake from nutrient solutions. New Phytol 1:549–557
Cakmak I, Horst WJ (1991) Effect of aluminium on lipid peroxidation, superoxide dismutase, catalase, and peroxidase activities in root tips of soybean (*Glycine max*). Physiol Plant 83(3):463–468
Carlson RW, Bazzaz FA, Rolfe GL (1975) The effect of heavy metals on plants: II. Net photosynthesis and transpiration of whole corn and sunflower plants treated with Pb, Cd, Ni, and Tl. Environ Res 10(1):113–120
Cedergreen N, Streibig JC, Kudsk P et al (2007) The occurrence of hormesis in plants and algae. Dose-Response 5:150–162
Chaney RL, Hornick SB (1978) Accumulation and effects of cadmium on crops. In: Proceedings of the First International Cadmium Conference, San Francisco. Metals Bulletin Ltd., London, pp 125–140

Chatterjee J, Chatterjee C (2000) Phytotoxicity of cobalt, chromium and copper in cauliflower. Environ Pollut 109(1):69–74
Chen YX, Lin Q, Luo YM et al (2003) The role of citric acid on the phytoremediation of heavy metal contaminated soil. Chemosphere 50(6):807–811
Chen Y, Wang Y, Wu W et al (2006) Impacts of chelate-assisted phytoremediation on microbial community composition in the rhizosphere of a copper accumulator and non-accumulator. Sci Total Environ 356(1):247–255
Chibuike GU, Obiora SC (2014) Heavy metal polluted soils: effect on plants and bioremediation methods. Appl Environ Soil Sci. doi:10.1155/2014/752708
Choudhary A, Singh RP (2000) Cadmium-induced changes in diamine oxidase activity and polyamine levels in *Vigna radiata* Wilczek seedlings. J Plant Physiol 156(5):704–710
Clemens S (2001) Molecular mechanisms of plant metal tolerance and homeostasis. Planta 212(4):475–486
Clemens S (2006) Toxic metal accumulation, responses to exposure and mechanisms of tolerance in plants. Biochimie 88(11):1707–1719
Clemens S, Antosiewicz DM, Ward JM et al (1998) The plant cDNA LCT1 mediates the uptake of calcium and cadmium in yeast. Proc Natl Acad Sci U S A 95(20):12043–12048
Cobbett CS (2000) Phytochelatins and their roles in heavy metal detoxification. Plant Physiol 123(3):825–832
Cobbett C, Goldsbrough P (2002) Phytochelatins and metallothioneins: roles in heavy metal detoxification and homeostasis. Annu Rev Plant Bio 53(1):159–182
Cunningham JD, Keeney DR, Ryan JA (1975) Phytotoxicity and uptake of metals added to soils as inorganic salts or in sewage sludge. J Environ Qual 4(4):460–462
Davies BE (1995) Lead and other heavy metals in urban areas and consequences for the health of their inhabitants. In: Majumdar SK, Miller EW, Brenner FJ (eds) Environmental contaminants, ecosystems and human health. The Pennsylvania Academy of Science, Easton, pp 287–307
Davies KL, Davies MS, Francis D (1995) The effects of zinc on cell viability and on mitochondrial structure in contrasting cultivars of *Festuca rubra* L. – a rapid test for zinc tolerance. Environ Pollut 88:109–113
Davis RD, Beckett PHT, Wollan E (1978) Critical levels of twenty potentially toxic elements in young spring barley. Plant Soil 49:395–408
De DN (2000) Plant cell vacuoles. CSIRO Publishing, Collingwood
de Borne FD, Elmayan T, de Roton C et al (1998) Cadmium partitioning in transgenic tobacco plants expressing a mammalian metallothionein gene. Mol Breed 4(2):83–90
de la Fuente JM, Ramirez-Rodriguez V, Cabrera-Ponce JL et al (1997) Aluminium tolerance in transgenic plants by alteration of citrate synthesis. Science 276(5318):1566–1568
Delhaize E, Ryan PR (1995) Aluminum toxicity and tolerance in plants. Plant Physiol 107:315–321
Dicko MH, Gruppen H, Traore AS et al (2006) Phenolic compounds and related enzymes as determinants of sorghum for food use. Biotechnol Mol Biol Rev 1:21–38
Dietz KJ, Baier M, Krämer U (1999) Free radicals and reactive oxygen species as mediators of heavy metal toxicity in plants. In: Prasad MNV, Hagemeyer J (eds) Heavy metal stress in plants: from biomolecule to ecosystem. Springer, Berlin/Heidelberg, pp 73–97
Dong J, Wu F, Huang R et al (2007) A chromium-tolerant plant growing in Cr-contaminated land. Int J Phytorem 9(3):167–179
Dräger DB, Desbrosses-Fonrouge AG, Krach C et al (2004) Two genes encoding *Arabidopsis halleri* MTP1 metal transport proteins co-segregate with zinc tolerance and account for high MTP1 transcript levels. Plant J 39(3):425–439
Duan GL, Zhu YG, Tong YP et al (2005) Characterization of arsenate reductase in the extract of roots and fronds of Chinese brake fern, an arsenic hyperaccumulator. Plant Physiol 138(1):461–469
Durrett TP, Gassmann W, Rogers EE (2007) The FRD3-mediated efflux of citrate into the root vasculature is necessary for efficient iron translocation. Plant Physiol 144(1):197–205
Ernst WH, Verkleij JA, Schat H (1992) Metal tolerance in plants. Acta Bot Neerl 41(3):229–248

Fernandes MA, Santos MS, Alpoim MC et al (2002) Chromium(VI) interaction with plant and animal mitochondrial bioenergetics: a comparative study. J Biochem Mol Toxicol 16:53–63
Feroci G, Badiello R, Fini A (2005) Interactions between different selenium compounds and zinc, cadmium and mercury. J Trace Elem Med Bio 18(3):227–234
Fleming AL, Foy CD (1968) Root structure reflects differential aluminium tolerance in wheat varieties. Agron J 60:172–176
Foy CD (1974) Effects of aluminium on plant growth. In: Carson EW (ed) The plant root and its environment. University Press of Virginia, CharlottesvIlle, pp 601–642
Foy CD (1988) Plant adaptation to acid, aluminum-toxic soils. Commun Soil Sci Plant Anal 19:959–987
Foy CD, Chaney RT, White MC (1978) The physiology of metal toxicity in plants. Annu Rev Plant Physiol 29(1):511–566
Frantzios G, Galatis B, Apostolakos P (2000) Aluminum effects on microtubule organization in dividing root-tip cells of *Triticum turgidum.* I. Mitotic cells. New Phytol 145(2):211–224
Frantzios G, Galatis B, Apostolakos P (2001) Aluminum effects on microtubule organization in dividing root-tip cells of *Triticum turgidum.* II. Cytokinetic cells. J Plant Res 114(2):157–170
Fusconi A, Gallo C, Camusso W (2007) Effects of cadmium on root apical meristems of *Pisum sativum* L.: cell viability, cell proliferation and microtubule pattern as suitable markers for assessment of stress pollution. Mutat Res Genet Toxicol Environ Mutagen 632:9–19
Garbisu C, Alkorta I (2003) Basic concepts on heavy metal soil bioremediation. Eur J Miner Process Environ Prot 3(1):58–66
Gassmann W, Schroeder JI (1994) Inward-rectifying K^+ channels in root hairs of wheat (a mechanism for aluminium sensitive low-affinity K^+ uptake and membrane potential control). Plant Physiol 105:1399–1408
Gekeler W, Grill E, Winnacker EL et al (1989) Survey of the plant kingdom for the ability to bind heavy metals through phytochelatins. Z Naturforsch C 44:361–369
Gisbert C, Ros R, De Haro A et al (2003) A plant genetically modified that accumulates Pb is especially promising for phytoremediation. Biochem Biophys Res Commun 303(2):440–445
Gough LP, Shacklette HT, Case AA (1979) Element concentrations toxic to plants, animals, and man. US Geol Surv Bull 1466:80
Grcman H, Velikonja-Bolta Š, Vodnik D et al (2001) EDTA enhanced heavy metal phytoextraction: metal accumulation, leaching and toxicity. Plant Soil 235(1):105–114
Grcman H, Vodnik D, Velikonja-Bolta Š et al (2003) Ethylenediaminedissuccinate as a new chelate for environmentally safe enhanced lead phytoextraction. J Environ Qual 32(2):500–506
Guerinot ML (2000) The ZIP family of metal transporters. Biochim Biophys Acta 1465:190–198
Guo J, Xu W, Ma M (2012) The assembly of metals chelation by thiols and vacuolar compartmentalization conferred increased tolerance to and accumulation of cadmium and arsenic in transgenic *Arabidopsis thaliana.* J Hazard Mater 15(199):309–313
Hagemeyer J (1999) Ecophysiology of plant growth under heavy metal stress. In: Prasad MNV, Hagemeyer J (eds) Heavy metal stress in plants: from molecules to ecosystems. Springer, Berlin, pp 157–182
Hajiboland R, Poschenrieder C (2015) Localization and compartmentation of Al in the leaves and roots of tea plants. Phyton 84:86–100
Halim M, Conte P, Piccolo A (2003) Potential availability of heavy metals to phytoextraction from contaminated soils induced by exogenous humic substances. Chemosphere 52(1):265–275
Hall JL (2002) Cellular mechanisms for heavy metal detoxification and tolerance. J Exp Bot 53(366):1–11
Hammond JP, Bowen HC, White PJ et al (2006) A comparison of the *Thlaspi caerulescens* and *Thlaspi arvense* shoot transcriptomes. New Phytol 170(2):239–260
Hanikenne M, Talke IN, Haydon MJ et al (2008) Evolution of metal hyperaccumulation required cis-regulatory changes and triplication of HMA4. Nature 453:391–396
Hart JJ, Welch RM, Norvell WA et al (2002) Transport interactions between cadmium and zinc in roots of bread and durum wheat seedlings. Physiol Plant 116(1):73–78
Hawkes SJ (1997) What is a "heavy metal"? J Chem Edu 74(11):1374–1378

Hendrix DL, Higinbotham N (1974) Heavy metals and sulphhydryl reagents as probes of ion uptake in pea stem. In: Zimmermann PU, Dainty J (eds) Membrane transport in plants. Springer, Berlin/Heidelberg, pp 412–417

Hogg TJ, Stewart JWB, Bettany JR (1978) Influence of the chemical form of mercury on its adsorption and ability to leach through soils. J Environ Qual 7(3):440–445

Horst WJ, Asher CJ, Cakmak I et al (1992) Short-term responses of soybean roots to aluminium. J Plant Physiol 140:174–178

Hossain MA, Piyatida P, da Silva JA et al (2012) Molecular mechanism of heavy metal toxicity and tolerance in plants: central role of glutathione in detoxification of reactive oxygen species and methylglyoxal and in heavy metal chelation. J Bot 2:2012

Huang CY, Bazzaz FA, Vanderhoeff LN (1974) The inhibition of soybean metabolism by cadmium and lead. Plant Physiol 54:122–124

Huang JW, Chen JJ, Berti WR et al (1997) Phytoremediation of lead-contaminated soils: role of synthetic chelates in lead phytoextraction. Environ Sci Technol 31:800–805

Imai I, Siegel SM (1973) A specific response to toxic cadmium levels in red kidney bean embryos. Physiol Plant 29(1):118–120

Islam E, Liu D, Li T et al (2008) Effect of Pb toxicity on leaf growth, physiology and ultrastructure in the two ecotypes of *Elsholtzia argyi*. J Hazard Mater 154(1):914–926

Israr M, Sahi SV (2006) Antioxidative responses to mercury in the cell cultures of *Sesbania drummondii*. Plant Physiol Biochem 44(10):590–595

Israr M, Sahi S, Datta R et al (2006) Bioaccumulation and physiological effects of mercury in *Sesbania drummondii*. Chemosphere 65(4):591–598

Jacobson KB, Turner JE (1980) The interaction of cadmium and certain other metal ions with proteins and nucleic acids. Toxicology 16(1):1–37

Jayakumar K, Azooz MM, Vijayarengan P et al (2010) Biochemical changes with exogenous cobalt application in soybean. J Phytol 2(1):7–12

Jin F, Wang C, Lin HJ et al (2010) Heavy metal-transport proteins in plants: a review. J Appl Ecol 21(7):1875–1882

Jones D, Kochian L (1995) Aluminum inhibition of the inositol 1,4,5-triphosphate signal transduction pathway in wheat roots; a role in aluminum toxicity? Plant Cell 7:1913–1922

Jones D, Kochian L (1997) Aluminum interaction with plasma membrane lipids and enzyme metal binding sites and its potential role in Al cytotoxicity. FEBS Lett 400:51–57

Kabata-pendias A (2011) Trace elements in soils and plants. CRC Press, London

Kagi JHR (1991) Overview of metallothionein. Methods Enzymol 205:613–626

Kim D, Gustin JL, Lahner B et al (2004) The plant CDF family member TgMTP1 from the Ni/Zn hyperaccumulator *Thlaspi goesingense* acts to enhance efflux of Zn at the plasma membrane when expressed in *Saccharomyces cerevisiae*. Plant J 39(2):237–251

Kitagishi K, Yamane I (1981) Heavy metal pollution in soils of Japan. Japan Scientific Societies Press, Tokyo

Kloke A, Sauerbeck DR, Vetter H (1984) The contamination of plants and soils with heavy metals and the transport of metals in terrestrial food chains. In: Nriagu JO (ed) Changing metal cycles and human health. Springer, Berlin/Heidelberg, pp 113–141

Kochian LV, Pineros MA, Hoekenga OA (2005) The physiology, genetics and molecular biology of plant aluminum resistance and toxicity. Plant Soil 274(1):175–195

Koeppe DE (1981) Lead: understanding the minimal toxicity of lead in plants. In: Lepp NW (ed) Effect of heavy metal pollution on plants. Springer, Dordrecht, pp 55–76

Kollmeier M, Felle HH, Horst WJ (2000) Genotypical differences in aluminum resistance of maize are expressed in the distal part of the transition zone. Is reduced basipetal auxin flow involved in inhibition of root elongation by aluminum? Plant Physiol 122:945–956

Korenkov V, Hirschi K, Crutchfield JD et al (2007) Enhancing tonoplast Cd/H antiport activity increases Cd, Zn, and Mn tolerance, and impacts root/shoot Cd partitioning in *Nicotiana tabacum* L. Planta 226(6):1379–1387

Korshunova YO, Eide D, Clark WG et al (1999) The IRT1 protein from *Arabidopsis thaliana* is a metal transporter with a broad substrate range. Plant Mol Biol 40(1):37–44

Krämer U, Cotter-Howells JD, Charnock JM et al (1996) Free histidine as a metal chelator in plants that accumulate nickel. Nature 379:635
Krämer U, Golldack D, Kiriazidou G et al (1999) Induction of plant pathogenesis-related proteins by heavy metals. In: Proceedings of 5th international conference on the biogeochemistry trace elements, Vienna, July 11–15, 1158
Krämer U, Pickering IJ, Prince RC et al (2000) Subcellular localization and speciation of nickel in hyperaccumulator and non-accumulator *Thlaspi* species. Plant Physiol 122:1343–1354
Krämer U, Talke IN, Hanikenne M (2007) Transition metal transport. FEBS Lett 581:2263–2272
Krizkova S, Ryant P, Krystofova O et al (2008) Multi-instrumental analysis of tissues of sunflower plants treated with silver (I) ions – plants as bioindicators of environmental pollution. Sensors 8(1):445–463
KrzesBowska M, Rabęda I, Lewandowski M et al (2013) Pb induces plant cell wall modifications-in particular-the increase of pectins able to bind metal ions level. In E3S web of conferences 2013, 1:26008
Kumar A (2013) Studies on the effect of tungsten and molybdenum on soil quality parameters and growth and metabolic activities of certain plant species. PhD thesis, Mohanlal Sukhadia University, Udaipur
Kumar A, Aery NC (2010) Studies on the effect of tungsten on seed germination and early seedling growth of cowpea. In: Masih MR, Singh B, Pareek DK et al (eds) Natural resource management for sustainable agriculture. Aavishkar Publishers, Jaipur, pp 274–280
Kumar A, Aery NC (2011) Effect of tungsten on growth, biochemical constituents, molybdenum and tungsten contents in wheat. Plant Soil Environ 57(11):519–525
Kumar A, Aery NC (2012a) Effect of tungsten on the growth, dry-matter production, and biochemical constituents of cowpea. Commun Soil Sci Plant Anal 43(7):1098–1107
Kumar A, Aery NC (2012b) Influence of tungsten and molybdenum on seed germination and early seedling growth of wheat – a comparative study. NBU J Plant Sci 6(1):1–5
Kuthanová A, Gemperlová L, Zelenková S et al (2004) Cytological changes and alterations in polyamine contents induced by cadmium in tobacco BY-2 cells. Plant Physiol Biochem 42(2):149–156
Kuznetsov VV, Shevyakova NI (1997) Stress responses of tobacco cells to high temperature and salinity. Proline accumulation and phosphorylation of polypeptides. Physiol Plant 100(2):320–326
Lane SD, Martin ES, Garrod JF (1978) Lead toxicity effects on indole-3-acetic acid-induced cell elongation. Planta 144(1):79–84
Lanquar V, Lelièvre F, Bolte S et al (2005) Mobilization of vacuolar iron by AtNRAMP3 and AtNRAMP4 is essential for seed germination on low iron. EMBO J 24(23):4041–4051
Larsen PB, Degenhardt J, Stenzler LM et al (1998) Aluminium-resistant *Arabidopsis* mutant that exhibit altered patterns of aluminum accumulation and organic acid release from roots. Plant Physiol 117:9–18
Lasat MM, Pence NS, Garvin DF et al (2000) Molecular physiology of zinc transport in the Zn hyperaccumulator *Thlaspi caerulescens*. J Exp Bot 342:71–79
Lay PA, Levina A (2014) Chromium. In: Maret W, Wedd A (eds) Binding, transport and storage of metal ions in biological cells. Royal Society of Chemistry, Cambridge, pp 188–222
Li X, Zhou Q, Sun X et al (2016) Effects of cadmium on uptake and translocation of nutrient elements in different welsh onion (*Allium fistulosum* L.) cultivars. Food Chem 1(194):101–110
Ling HQ, Koch G, Baumlein H et al (1999) Map-based cloning of chloronerva, a gene involved in iron uptake of higher plants encoding nicotianamine synthase. Proc Natl Acad Sci U S A 96(12):7098–7103
Lipsey RL (1975) Accumulation and physiological effects of methyl mercury hydroxide on maize seedlings. Environ Pollut 8(2):149–155
Lombi E, Zhao FJ, Wieshammer G et al (2002) In situ fixation of metals in soils using bauxite residue: biological effects. Environ Pollut 118(3):445–452
Luo Y, Rimmer DL (1995) Zinc-copper interaction affecting plant growth on a metal-contaminated soil. Environ Pollut 88(1):79–83

Luo C, Shen Z, Li X (2005) Enhanced phytoextraction of Cu, Pb, Zn and Cd with EDTA and EDDS. Chemosphere 59(1):1–11

Ma LQ, Komar KM, Tu C et al (2001) A fern that hyperaccumulates arsenic. Nature 409(6820):579

Macek T, Macková M, Pavlíková D et al (2002) Accumulation of cadmium by transgenic tobacco. Keynote Lect 1:42

Macnicol RD, Beckett PHT (1985) Critical tissue concentrations of potentially toxic elements. Plant Soil 85(1):107–129

Manara A (2012) Plant responses to heavy metal toxicity. In: Furini A (ed) Plants and heavy metals. Springer, Dordrecht, pp 27–53

Marschner H (1999) Mineral nutrition in higher plants. Academic, London

Marschner P (2012) Marschner's mineral nutrition of higher plants. Academic, London

Matsumoto H (1991) Biochemical mechanism of the toxicity of aluminum and the sequestration of aluminum in plant cells. In: Wright RJ, Baligar VC, Murrmann RP (eds) Plant-soil interactions at low pH. Kluwer Academic Publishers, Dordrecht, pp 825–838

Matsumoto H, Hirasawa E, Morimura S et al (1976) Localization of aluminium in tea leaves. Plant Cell Physiol 29:1133–1140

Matysik J, Alia B, Mohanty P (2002) Molecular mechanisms of quenching of reactive oxygen species by proline under stress in plants. Curr Sci 82:525–532

Maywald F, Weigel HJ (1997) Biochemistry and molecular biology of heavy metals accumulation in higher plants. Landbauforschung-Volkerode 47(3):103–126

McGrath SP (1982) The uptake and translocation of tri- and hexavalent chromium and effects on the growth of oat, in flowing nutrient solution and in soil. New Phytol 92(3):381–390

McGrath SP (1998) Phytoextraction for soil remediation. In: Brooks RR (ed) Plants that hyperaccumulate heavy metals. CAB Intern, Oxford, pp 261–287

McIntyre T (2003) Phytoremediation of heavy metals from soils. Adv Biochem Eng Biotechnol 8:97–123

McKenna IM, Chaney RL, Williams FM (1993) The effects of cadmium and zinc interactions on the accumulation and tissue distribution of zinc and cadmium in lettuce and spinach. Environ Pollut 79(2):113–120

McKenney DJ, Vriesacker JR (1985) Effect of cadmium contamination on denitrification processes in Brookston clay and Fox sandy loam. Environ Pollut Ser A 38(3):221–233

McLaughlin MJ, Andrews SJ, Smart MK et al (1998) Effects of sulfate on cadmium uptake by Swiss chard: I. Effects of complexation and calcium competition in nutrient solutions. Plant Soil 202:211–216

McNear DH, Afton SE, Caruso JA (2012) Exploring the structural basis for selenium/mercury antagonism in *Allium fistulosum*. Metallomics 4(3):267–276

Mhathre GN, Chaphekar SB (1984) Response of young plants to mercury. Water Air Soil Pollut 21(1–4):1–8

Migocka M, Papierniak A, Kosatka E et al (2011) Comparative study of the active cadmium efflux systems operating at the plasma membrane and tonoplast of cucumber root cells. J Exp Bot 62(14):4903–4916

Mildvan AS (1970) Metals in enzymes catalysis. In: Boyer DD (ed) The enzymes, vol II. Academic, London, pp 445–536

Monferrán MV, Wunderlin DA (2013) Biochemistry of metals/metalloids toward remediation process. In: Gupta DK, Corpas FJ, Palma JM (eds) Heavy metal stress in plants. Springer, Berlin/Heidelberg, pp 43–71

Montanini B, Blaudez D, Jeandroz S et al (2007) Phylogenetic and functional analysis of the cation diffusion facilitator (CDF) family: improved signature and prediction of substrate specificity. BMC Genomics 8(1):1–16

Moreira OC, Rios PF, Barrabin H (2005) Inhibition of plasma membrane Ca^{2+}-ATPase by CrATP. LaATP but not CrATP stabilizes the Ca^{2+}-occluded state. BBA-Bioenerg 3:411–419

Morel M, Crouzet J, Gravot A et al (2009) AtHMA3, a P1B-ATPase allowing Cd/Zn/Co/Pb vacuolar storage in *Arabidopsis*. Plant Physiol 149(2):894–904

Motesharezadeh B, Savaghebi GH (2012) Interaction between cadmium and lead and the effects of these on the concentration of zinc and manganese in sunflower. Int J Environ Res 6(3):793–800
Mukhopadhyah N, Aery NC (2000) Effect of Cr(III) and Cr(VI) on the growth and physiology of *Triticum aestivum* plants during early seedling growth. Biologia 55(4):403–408
Nagajyoti PC, Lee KD, Sreekanth TV (2010) Heavy metals, occurrence and toxicity for plants: a review. Environ Chem Lett 8(3):199–216
Naidu BP, Paleg LG, Aspinall D et al (1991) Amino acid and glycine betaine accumulation in cold-stressed wheat seedlings. Phytochemistry 30(2):407–409
Nakanishi H, Ogawa I, Ishimaru Y et al (2006) Iron deficiency enhances cadmium uptake and translocation mediated by the Fe^{2+} transporters OsIRT1 and OsIRT2 in rice. Soil Sci Plant Nutr 52(4):464–469
Nan Z, Zhao C, Li J et al (2002) Relations between soil properties and selected heavy metal concentrations in spring wheat (*Triticum aestivum* L.) grown in contaminated soils. Water Air Soil Pollut 133(1–4):205–213
Nedkovska M, Atanassov AI (1995) Metallothionein genes and expression for heavy metal resistance. Biotechnol Biotech Equip 11(2):11–16
Nieboer E, Richardson DH (1980) The replacement of the nondescript term 'heavy metals' by a biologically and chemically significant classification of metal ions. Environ Pollut B 1(1):3–26
Nies DH (1992) Resistance to cadmium, cobalt, zinc, and nickel in microbes. Plasmid 27:17–28
Nishida S, Mizuno T, Obata H (2008) Involvement of histidine-rich domain of ZIP family transporter TjZNT1 in metal ion specificity. Plant Physiol Biochem 46(5):601–606
Niu ZC, Zhang XS, Wang ZW et al (2011) Field controlled experiments of mercury accumulation in crops from air and soil. Environ Pollut 159:2684–2689
Nocito FF, Espen L, Crema B et al (2008) Cadmium induces acidosis in maize root cells. New Phytol 179(3):700–711
O'Halloran TV, Culotta VC (2000) Metallochaperones, an intracellular shuttle service for metal ions. J Biol Chem 275(33):25057–25060
Ozoliniya GR (1986) Influence of copper, cobalt and silver ions on total growth of grass roots. Mikroelem in USSR 27:72
Paleg LG, Aspinall D (1981) The physiology and biochemistry of drought resistance in plants. Academic, London
Palmiter RD, Findley SD (1995) Cloning and functional characterization of a mammalian zinc transporter that confers resistance to zinc. EMBO J 14(4):639–649
Panchal J, Aery NC (2008) Effects of manganese waste on growth, nodulation, proline levels, and enzymatic activities in *Vigna unguiculata* (L.) Wal. Toxicol Environ Chem 90(6):1063–1072
Papernik L, Bethea A, Singleton T et al (2001) Physiological basis of reduced Al tolerance in ditelosomic lines of Chinese Spring wheat. Planta 212:829–834
Papoyan A, Kochian LV (2004) Identification of *Thlaspi caerulescens* genes that may be involved in heavy metal hyperaccumulation and tolerance. Characterization of a novel heavy metal transporting ATPase. Plant Physiol 136(3):3814–3823
Patra M, Sharma A (2000) Mercury toxicity in plants. Bot Rev 66(3):379–422
Patra M, Bhowmik N, Bandopadhyay B et al (2004) Comparison of mercury, lead and arsenic with respect to genotoxic effects on plant systems and the development of genetic tolerance. Environ Exp Bot 52(3):199–223
Pavlíková D, Macek T, Macková M et al (2004) Cadmium tolerance and accumulation in transgenic tobacco plants with a yeast metallothionein combined with a polyhistidine tail. Int Biodeterior Biodegrad 54(2):233–237
Peralta-Videa JR, Gardea-Torresdey JL, Gomez E et al (2002) Effect of mixed cadmium, copper, nickel and zinc at different pHs upon alfalfa growth and heavy metal uptake. Environ Pollut 119(3):291–301
Pich A, Manteuffel R, Hillmer S et al (2001) Fe homeostasis in plant cells: does nicotianamine play multiple roles in the regulation of cytoplasmic Fe concentration? Planta 213(6):967–976

Pineros MA, Kochian LV (2001) A patch-clamp study on the physiology of aluminum toxicity and aluminum tolerance in maize. Identification and characterization of Al (3+)-induced anion channels. Plant Physiol 125:292–305

Pineros M, Tester M (1995) Characterization of a voltage dependent Ca^{2+}-selective channel from wheat roots. Planta 195:478–488

Poschenrieder C, Cabot C, Martos S et al (2013) Do toxic ions induce hormesis in plants? Plant Sci 212:15–25

Prasad MNV (1995) Cadmium toxicity and tolerance in vascular plants. Environ Exp Bot 35:525–545

Prasad MNV (2005) Heavy metal stress in plants: Metallomics approach for phytoremediation. In: Chakraborty U, Chakraborty B (eds) Stress biology. Narosa Publishing House, New Delhi, pp 21–26

Ramos I, Esteban E, Lucena JJ et al (2002) Cadmium uptake and subcellular distribution in plants of *Lactuca* sp. Cd-Mn interaction. Plant Sci 162:761–767

Ramos-Díaz A, Brito-Argáez L, Munnik T et al (2007) Aluminum inhibits phosphatidic acid formation by blocking the phospholipase C pathway. Planta 225(2):393–401

Rana D, Aery NC (1999) Effect of Al stress on the biochemical constituents during early seedling growth of mustard. Bionature 19(2):47–50

Rana DK, Aery NC (2000) Influence of aluminium on the growth performance and chlorophyll contents of wheat. Vasundhra 5:63–68

Raskin I, Ensley BD (2000) Phytoremediation of toxic metals-using plants to clean up the environment. Wiley, New York

Ratte HT (1999) Bioaccumulation and toxicity of silver compounds: a review. Environ Toxicol Chem 18(1):89–108

Rauser WE (1999) Structure and function of metal chelators produced by plants: the case for organic acids, amino acids, phytin and metallothioneins. Cell Biochem Biophys 31:19–48

Reisinger S, Schiavon M, Terry N et al (2008) Heavy metal tolerance and accumulation in Indian mustard (*Brassica juncea* L.) expressing bacterial gamma-glutamylcysteine synthetase or glutathione synthetase. Int J Phytorem 10(5):440–454

Rengel Z, Elliott DC (1992) Mechanism of aluminium inhibition of net $^{45}Ca^{2+}$ uptake by *Amaranthus* protoplasts. Plant Physiol 98:632–638

Revathi S, Venugopal S (2013) Physiological and biochemical mechanisms of heavy metal tolerance. Int J Environ Sci 3(5):1339–1352

Roderer G (1984) Toxic effects in plant organisms. In: Grandjean P (ed) Biological effects of organolead compounds. CRC Press, Boca Raton, pp 63–95

Ryan PR, Kochian LV (1993) Interaction between aluminum toxicity and calcium uptake at the root apex in near-isogenic lines of wheat (*Triticum aestivum* L.) differing in aluminum tolerance. Plant Physiol 102:975–982

Ryan P, DiTomaso J, Kochian L (1993) Aluminum toxicity in roots: An investigation of spatial sensitivity and the role of the root cap. J Exp Bot 44:437–446

Ryu SK, Park JS, Lee ISK (2003) Purification and characterization of a copper binding protein from Asian periwinkle *Littorina brevicula*. Comp Biochem Physiol 134(1):101–107

Salisbury FB, Ross C (1978) Plant physiology. Wadsworth Publishing Company, Belmont

Salt DE, Prince RC, Pickering IJ et al (1995) Mechanisms of cadmium mobility and accumulation in Indian mustard. Plant Physiol 109(4):1427–1433

Salt DE, Prince RC, Baker AJM et al (1999) Zinc ligands in the metal accumulator *Thlaspi caerulescens* as determined using X-ray absorption spectroscopy. Environ Sci Tech 33(5):713–717

Sappin-Didier V, Vansuyts G, Mench M et al (2005) Cadmium availability at different soil pH to transgenic tobacco overexpressing ferritin. Plant Soil 270(1):189–197

Sarkar S, Aery NC (1988) Studies on the deranged physiology of soybean under the influence of heavy metal stress. II Effects of cadmium on growth performance. Ind Bot Control 5:113–118

Sarret G, Saumitou-Laprade P, Bert V, Proux O et al (2002) Forms of zinc accumulated in the hyperaccumulator *Arabidopsis halleri*. Plant Physiol 130:1815–1826

Schiavon M, Pilon-Smits EAH, Wirtz M et al (2008) Interaction between chromium and sulfur metabolism in *Brassica juncea*. J Environ Qual 37:1536–1545
Schmohl N, Horst WJ (2000) Cell wall pectin content modulates aluminium sensitivity of *Zea mays* (L.) cells grown in suspension culture. Plant Cell Environ 23(7):735–742
Schützendübel A, Polle A (2002) Plant responses to abiotic stresses: heavy metal-induced oxidative stress and protection by mycorrhization. J Exp Bot 53(372):1351–1365
Schwarzerová K, Zelenková S, Nick P et al (2002) Aluminum-induced rapid changes in the microtubular cytoskeleton of tobacco cell lines. Plant Cell Physiol 43(2):207–216
Senden MN, Van der Meer AM, Verburg TG et al (1995) Citric acid in tomato plant roots and its effect on cadmium uptake and distribution. Plant Soil 171:333–339
Seregin IV, Ivanov VB (2001) Physiological aspects of cadmium and lead toxic effects on higher plants. Russ J Plant Physiol 48(4):523–544
Sharma SS, Dietz KJ (2006) The significance of amino acids and amino acid-derived molecules in plant responses and adaptation to heavy metal stress. J Exp Bot 57(4):711–726
Sharma SS, Dietz KJ (2009) The relationship between metal toxicity and cellular redox imbalance. Trends Plant Sci 14(1):43–50
Sharma P, Dubey RS (2005) Lead toxicity in plants. Braz J Plant Physiol 17(1):35–52
Sharma SS, Dietz KJ, Mimura T (2016) Vacuolar compartmentalization as indispensable component of heavy metal detoxification in plants. Plant Cell Environ 39(5):1112–1126
Siegel BZ, Lasconia M, Yaeger E et al (1984) The phytotoxicity of mercury vapor. Water Air Soil Pollut 23:15–24
Silva IR, Smyth TJ, Moxley DF et al (2000) Aluminum accumulation at nuclei of cells in the root tip. Fluorescence detection using lumogallion and confocal laser scanning microscopy. Plant Physiol 123(2):543–552
Silver S (1996) Bacterial resistances to toxic metal ions-a review. Gene 179(1):9–19
Singh AD, Aery NC (1993) Accumulation and seasonal variation in the zinc and lead content of certain plant species. Acta Ecol 15(1):51–57
Singh N, Ma LQ (2006) Arsenic speciation, and arsenic and phosphate distribution in arsenic hyperaccumulator *Pteris vittata* L. and non-hyperaccumulator *Pteris ensiformis* L. Environ Pollut 141(2):238–246
Sinha P, Dube BK, Chatterjee C (2006) Manganese stress alters phytotoxic effects of chromium in green gram physiology (*Vigna radiata* L.) cv. PU 19. Environ Exp Bot 57:131–138
Sivaguru M, Horst W (1998) The distal part of the transition zone is the most aluminum-sensitive apical root zone of maize. Plant Physiol 116:155–163
Sivaguru M, Baluska F, Volkmann D et al (1999) Impacts of aluminium on the cytoskeleton of the maize root apex, short-term effects on the distal part of the transition zone. Plant Physiol 119:1073–1082
Sivaguru M, Fujiwara T, Samaj J et al (2000) Aluminum-induced 1– > 3beta-D-glucan inhibits cell-to-cell trafficking of molecules through plasmodesmata. A new mechanism of aluminum toxicity in plants. Plant Physiol 124:991–1006
Sivaguru M, Pike S, Gassmann W et al (2003) Aluminum rapidly depolymerizes cortical microtubules and depolarizes the plasma membrane: evidence that these responses are mediated by a glutamate receptor. Plant Cell Physiol 44:667–675
Smeyers-Verbeke J, de Graeve M, Francois M et al (1978) Cd uptake by intact wheat plants. Plant Cell Environ 1(4):291–296
Sors TG, Martin CP, Salt DE (2009) Characterization of selenocysteine methyltransferases from Astragalus species with contrasting selenium accumulation capacity. Plant J 59(1):110–122
Sorterberg A (1980) The effect of some heavy metals on oats in pot experiments with three different soil types. J Sci Agric Soc Finl 46:277–288
Stohs SJ, Bagchi D (1995) Oxidative mechanisms in the toxicity of metal ions. Free Radic Biol Med 18(2):321–336
Stolt JP, Sneller FE, Bryngelsson T et al (2003) Phytochelatin and cadmium accumulation in wheat. Environ Exp Bot 49(1):21–28

Su YH, McGrath SP, Zhu YG et al (2008) Highly efficient xylem transport of arsenite in the arsenic hyperaccumulator *Pteris vittata*. New Phytol 180(2):434–441

Sun R, Zhou Q, Jin C (2006) Cadmium accumulation in relation to organic acids in leaves of *Solanum nigrum* L as a newly found cadmium hyperaccumulator. Plant Soil 285(1–2):125–134

Suzuki N (2005) Alleviation by calcium of cadmium-induced root growth inhibition in *Arabidopsis* seedlings. Plant Biotechnol 22:19–25

Tabuchi A, Matsumoto H (2001) Changes in cell wall properties of wheat (*Triticum aestivum*) roots during aluminum-induced growth inhibition. Physiol Plant 112:353–358

Taiz L, Zeiger E (2002) Plant Physiology. Sinauer Associates, Sunderland

Takahashi M, Terada Y, Nakai I et al (2003) Role of nicotianamine in the intracellular delivery of metals and plant reproductive development. Plant Cell 15(6):1263–1280

Talke IN, Hanikenne M, Krämer U (2006) Zinc-dependent global transcriptional control, transcriptional deregulation, and higher gene copy number for genes in metal homeostasis of the hyperaccumulator *Arabidopsis halleri*. Plant Physiol 142(1):148–167

Tangahu BV, Sheikh Abdullah SR, Basri H et al (2011) A review on heavy metals (As, Pb, and Hg) uptake by plants through phytoremediation. Int J Chem Eng. doi:10.1155/2011/939161

Thomine S, Wang R, Ward JM et al (2000) Cadmium and iron transport by members of a plant metal transporter family in *Arabidopsis* with homology to Nramp genes. Proc Natl Acad Sci U S A 97(9):4991–4996

Thomine S, Lelièvre F, Debarbieux E et al (2003) AtNRAMP3, a multispecific vacuolar metal transporter involved in plant responses to iron deficiency. Plant J 34(5):685–695

Tiagi YD, Aery NC (1986) Biogeochemical studies at the Khetri copper deposits of Rajasthan, India. J Geochem Explor 26(3):267–274

Tripathi RD, Srivastava S, Mishra S et al (2007) Arsenic hazards: strategies for tolerance and remediation by plants. TRENDS Biotechnol 25(4):158–165

Tukendorf A, Baszynki T (1991) The *in vivo* effect of cadmium on photochemical activities in chloroplasts of runner bean plants. Acta Physiol Plant 13:51

Turner MA (1973) Effect of cadmium treatment on cadmium and zinc uptake by selected vegetable species. J Environ Qual 2(1):118–119

Turner MA, Rust RH (1971) Effect of chromium on growth and mineral nutrition of soybeans. Soil Sci Soc Am Proc 35(5):755–758

van de Mortel JE, Villanueva LA, Schat H et al (2006) Large expression differences in genes for iron and zinc homeostasis, stress response, and lignin biosynthesis distinguish roots of *Arabidopsis thaliana* and the related metal hyperaccumulator *Thlaspi caerulescens*. Plant Physiol 142(3):1127–1147

Varga A, Záray G, Fodor F et al (1997) Study of interaction of iron and lead during their uptake process in wheat roots by total-reflection X-ray fluorescence spectrometry. Spectrochim Acta B At Spectrosc 52(7):1027–1032

Vazquez MD, Lopez J, Carballeira A (1999) Uptake of heavy metals to the extracellular and intracellular compartments in three species of aquatic bryophyte. Ecotoxicol Environ Safe 44(1):12–24

Verkleij JA, Sneller FE, Schat H (2003) Metallothioneins and phytochelatins: ecophysiological aspects. In: Abrol YP, Ahmad A (eds) Sulphur in plants. Springer, Dordrecht, pp 163–176

Vernay P, Gauthier-Moussard C, Hitmi A (2007) Interaction of bioaccumulation of heavy metal chromium with water relation, mineral nutrition and photosynthesis in developed leaves of *Lolium perenne* L. Chemosphere 68(8):1563–1575

Vert G, Grotz N, Dédaldéchamp F et al (2002) IRT1, an Arabidopsis transporter essential for iron uptake from the soil and for plant growth. Plant Cell 14:1223–1233

Wallace A, Romney EM, Alexander GV et al (1977) Some interactions in plants among cadmium, other heavy metals, and chelating agents. Agron J 69(1):18–20

Wang Y, Greger M (2004) Clonal differences in mercury tolerance, accumulation, and distribution in willow. J Environ Qual 33(5):1779–1785

Wang JL, Li T, Liu GY et al (2016) Unraveling the role of dark septate endophyte (DSE) colonizing maize (*Zea mays*) under cadmium stress: Physiological, cytological and genetic aspects. Sci Rep 6. doi:10.1038/srep22028

Wedepohl KH (1974) Handbook of geochemistry. Springer Verlag, Berlin

Williams DJ, Hall KB (2000) Experimental and computational studies of the G[UUCG]C RNA tetraloop. J Mol Biol 297(5):1045–1061

Williams LE, Pittman JK, Hall JL (2000) Emerging mechanisms for heavy metal transport in plants. Biochim et Biophys Acta- Biomembr 1465(1):104–126

Wilson DO, Cline JF (1966) Removal of plutonium-239, tungsten-185 and lead-210 from soils. Nature 209:941–942

Winge DR, Nielson KB, Gray WR et al (1985) Yeast metallothionein. Sequence and metal-binding properties. J Biol Chem 260(27):14464–14470

Wojas S, Ruszczyńska A, Bulska E et al (2007) Ca^{2+}-dependent plant response to Pb^{2+} is regulated by LCT1. Environ Pollut 147(3):584–592

Woźny A (1998) Lead in plants- penetration, distribution, reactions. In: Kabata-Pendias A, Szteke B (eds) Lead in the environment-ecological and analytical problems, PAN Kom. Nauk 21:171–180

Wu F, Zhang G, Yu J (2003) Interaction of cadmium and four microelements for uptake and translocation in different barley genotypes. Commun Soil Sci Plant Anal 34(13–14):2003–2020

Xu J, Yin H, Li X (2009) Protective effects of proline against cadmium toxicity in micropropagated hyperaccumulator, *Solanum nigrum* L. Plant Cell Rep 28(2):325–333

Yamamoto Y, Kobayashi Y, Matsumoto H (2001) Lipid peroxidation is an early symptom triggered by aluminum, but not the primary cause of elongation inhibition in pea roots. Plant Physiol 125(1):199–208

Yamamoto Y, Kobayashi Y, Devi SR et al (2002) Aluminum toxicity is associated with mitochondrial dysfunction and the production of reactive oxygen species in plant cells. Plant Physiol 128(1):63–72

Yang XE, Long XX, Ni WZ (2002) Physiological and molecular mechanisms of heavy metal uptake by hyperaccumulating plant species. J Plant Nutr Fertil 8(1):8–15

Yang X, Feng Y, He Z et al (2005) Molecular mechanisms of heavy metal hyperaccumulation and phytoremediation. J Trace Elem Med Bio 18(4):339–353

Yang J, Liu Z, Wan X et al (2016) Interaction between sulfur and lead in toxicity, iron plaque formation and lead accumulation in rice plant. Ecotoxicol Environ Saf 128:206–212

Yathavakilla SK, Caruso JA (2007) A study of Se-Hg antagonism in *Glycine max* (soybean) roots by size exclusion and reversed phase HPLC–ICPMS. Anal Bioanal Chem 389(3):715–723

Zayed AM, Terry N (2003) Chromium in the environment: factors affecting biological remediation. Plant Soil 249(1):139–156

Zayed A, Lytle CM, Qian JH et al (1998) Chromium accumulation, translocation and chemical speciation in vegetable crops. Planta 206(2):293–299

Zhou J, Ma D, Pan J et al (2008) Application of multivariate statistical approach to identify heavy metal sources in sediment and waters: a case study in Yangzhong, China. Environ Geol 54(2):373–380

Zimdahl RL (1976) Entry and movement in vegetation of lead derived from air and soil sources. J Air Pollut Control Assoc 26(7):655–660

Heavy Metal Accumulation Potential and Tolerance in Tree and Grass Species

8

Arideep Mukherjee, Shashi Bhushan Agrawal, and Madhoolika Agrawal

Abstract

Heavy metals (HM) are nondegradable, persistent, and toxic elements. Heavy metal contamination is a global concern and a major health hazard throughout the world. However, plants have evolved different mechanisms to combat these stresses and even have potential to accumulate them in large quantity. Natural flora has differential ability to metal tolerance with some plants showing induced tolerance at metal-enriched medium, whereas others find it difficult to sustain. Bioaccumulation potential of trees and grasses will be assessed based on peer-reviewed publications. Plants will be classified based on their accumulation potential of different heavy metals. Metal accumulation ability in plants and sources of heavy metals will be assessed for different regions of the world. Anthropogenic input of heavy metals from different sources significantly affects the biogeochemical cycling. Heavy metal depositions considerably affect the plant response. Heavy metal tolerance, uptake, and accumulation in different parts of the plant and possible metal-chelating compounds and their role in metal chelation in trees and grasses will be discussed. Morphological, physiological, biochemical, and molecular biomarkers of heavy metal tolerance or sensitivity among the trees and grasses will be correlated with bioaccumulation potential and heavy metal tolerance. This article will provide a broad overview of higher plant abilities and tolerance capabilities in heavy metal accumulation.

Keywords

Heavy metal • Bioaccumulation • Trees • Grasses • Tolerance • Chelation

A. Mukherjee • S.B. Agrawal • M. Agrawal (✉)
Laboratory of Air Pollution and Global Climate Change, Department of Botany, Banaras Hindu University, Varanasi 221005, Uttar Pradesh, India
e-mail: arideep@gmail.com; madhoo.agrawal@gmail.com

A. Singh et al. (eds.), *Plant Responses to Xenobiotics*,
DOI 10.1007/978-981-10-2860-1_8

8.1 Introduction

Heavy metals (HM) are ubiquitous in our environment and are persistent in nature. Plants have the potential to accumulate these metals and can even concentrate them in large quantities, several fold higher than in their natural environment (Verbruggen et al. 2009). Xenobiotic compounds in the environment are transformed and accumulated in plants based on their physiological status and surrounding environment. Plants that are capable of accumulating these compounds in large quantities are highly valuable for biomonitoring purpose as well as for removing these compounds from contaminated soil. Plants have unique ability to accumulate one or more metals, but no records are available for a plant which can accumulate all the metals. This necessitates the identification of plants which can accumulate multi-metal combinations.

Most of the HM contamination in the environment is from historical contaminations in soil naturally or from disturbances caused by mining activities (Madejón et al. 2002; Conesa et al. 2006; Jozefczak et al. 2012; Bech et al. 2012). Due to an increase in anthropogenic activity, normal metal cycling has been altered. Anthropogenic metal contaminations are mostly from industrial emissions, traffic emissions, waste burning, industrial dumping, agricultural practices, burning of fossil fuel, and sewage discharge (Table 8.1) (El-Hasan et al. 2002; Kummer et al. 2009). Apart from metal contamination in soil, deposition through air pollution in urban environment has become a huge problem in major cities of the world. With increase in number of automobiles throughout the world, urban centers are now facing excessive metal contaminations in ambient air (de Vives et al. 2006; Fujiwara et al. 2011; Tomaševića et al. 2011).

The use of plant species for biomonitoring of different types of pollutants is well known (Tomasevic et al. 2004; Cucu-Man and Steinnes 2013; Mukherjee and Agrawal 2016). Since industrial revolution, lichens and mosses have been extensively utilized for biomonitoring of HM. But in recent times due to their absence in harsh summer climate, lack of availability in large quantity, and higher sensitivity to air pollutants, trees and grasses are extensively used for assessing metal accumulation potential for phytoremediation and as a specific indicator of metal pollution (Krämer 2010; Aničič et al. 2011). Both evergreen and deciduous trees, perennial grasses, forbs, and legumes have been extensively studied for bioaccumulation mechanism to identify suitable indicator both in the natural environment as well as in artificial conditions (Hall 2002; Krämer 2010; Zeng et al. 2011; Bech et al. 2012; Ribeiro de Souza et al. 2012; Thapa et al. 2012).

Metal uptake in plant is regulated by several internal factors such as root and stem anatomy, xylem loading, cellular detoxification mechanism, physiological condition of plant, and external factors such as soil pH, soil texture, topography, nearby pollution sources, dry and wet deposition, and land use pattern (de Vives et al. 2006; Verbruggen et al. 2009; Ugolini et al. 2013).

Metals are present in earth crust and some of them are important for plants such as zinc (Zn), cobalt (Co), copper (Cu), iron (Fe), manganese (Mn), nickel (Ni), molybdenum (Mo), selenium (Se), and boron (B) for proper growth and development. But higher concentrations of any of the metals are toxic to plants, whereas

Table 8.1 Different sources of heavy metals and metalloids

Metal/ metalloids	Sources	References
Pb	Gasoline, automobile emissions, abrasion of tire treads, brake linings, mining and Pb ore smelting, fertilizers, pesticides, and pigments	Yu et al. (2007), Kummer et al. (2009) and de Souza et al. (2012)
Cu	Traffic exhaust, tires, and brake wear	Fujiwara et al. (2011)
Zn	Traffic exhaust, tire and brake wear, fossil fuel combustion, lubricating motor oil, and tires	Fujiwara et al. (2011)
Cd	Abrasion of tire treads and brake linings	Kummer et al. (2009)
Ni	Abrasion of tire treads and brake linings, fossil fuel combustion, power plant, industry dumping areas, and automobile part corrosion	Kummer et al. (2009) and El-Hasan et al. (2002)
Bi	Nonferrous metallurgy of copper	Cucu-Man and Steinnes (2013)
In	Copper metallurgy	Nriagu and Pacyna (1988)
As	Copper metallurgy, power plant	Nriagu and Pacyna (1988) and Cucu-Man and Steinnes (2013)
Sn	Combustion of liquid fuels, coal burning, nonferrous metallurgy of Cu–Ni, and waste combustion	Nriagu and Pacyna (1988)
V	Power plant	Cucu-Man and Steinnes (2013)
Cr	Power plant, industry dumping areas, and automobile part corrosion	Cucu-Man and Steinnes (2013)
Co	Traffic emissions, motor vehicle tire wheel	El-Hasan et al. (2002)
Mn	Industrial dumping areas and automobile part corrosion, auto workshops, electroplating industries, gasoline combustion	El-Hasan et al. (2002) and Shi et al. (2012)
Fe	Automobile	El-Hasan et al. (2002)
Ag	Photographic processing effluents, sewage sludge, biocide	Hajar et al. (2014)
Ba	Internal combustion engines	
Sb	Abrasion of vehicle brake linings, mining and smelting activities, electronic device	Klumpp et al. (2009) and Bech et al. (2012)

metals or metalloids like arsenic (As), mercury (Hg), lead (Pb), and cadmium (Cd) are toxic to plants even at low concentrations, and functions of these metals or metalloids are yet to be identified in plants (Buendía-González et al. 2010; Krämer 2010; Zeng et al. 2011). The metal accumulation potential under different environmental conditions may vary for different plant species, which is directly correlated with plant's ability to accumulate or exclude HM. Plants response to metals, and their accumulation depends upon metal concentrations in soil, location of metal accumulation in plant parts, changes in antioxidative defense response, presence of different chelator molecules, metal homeostasis, and change in gene expression (Hall 2002; Cobbett and Goldsbrough 2002; Sun et al. 2007; Chao et al. 2010; Zeng et al. 2011; Ribeiro de Souza et al. 2012; Manara 2012; Li et al. 2015).

In this article, we have tried to summarize potential accumulators of HM belonging to different plant groups. Trees and grasses have been the major focus of our study as they have wide evolutionary adaptation and are present in several different environments ranging from arid, coastal, forest, urban region and throughout the world. We have also tried to understand the metal accumulation in different plant parts as leaf, bark, and wood in the case of tree species, and for grasses metal accumulation in shoot and roots has been characterized. Metal tolerance in hyperaccumulator plants as well as in other metal-tolerant species was screened for identifying different metal-chelating compounds and their role in metal tolerance. Morphological, biochemical, anatomical, physiological, and molecular mechanisms in HM-tolerant plants were assessed and compared with sensitive plants to identify the specific responses of plants with HM stress tolerance and bioaccumulation.

8.2 Heavy Metal Bioaccumulation in Tree Species

Trees are perennial, long-living plants having a large surface area compared to other groups of plants. Because of their large size and longer persistence, they have unique adaptability and large accumulation potential to environmental xenobiotics. Much to their advantage, trees have large canopy area and extensive root and shoot system to accumulate and store a significant amount of xenobiotics (Fig. 8.1). Compared to other life forms, accumulation potential of tree species for heavy metals is less studied. It is mostly due to its long period of growth and slow growing rates compared to grasses and crop plants. Here we have tried to identify heavy metal accumulation potential in different parts of tree with major focus on leaf and bark in different regions of the world.

8.2.1 Leaf

Metal concentrations exhibit significant variations between different species in different regions of the world. Major factor controlling metal accumulation in foliage is direct atmospheric wet and dry depositions, atmospheric dust, local soil contamination, bioavailability of metal in soil, soil texture, soil pH, land use type, bioaccumulation potential, local flora, and transfer of metal within the plant (Cicek and Koparal 2004; Tomasevic et al. 2004; Tomaševića et al. 2011; Norouzi et al. 2015). For assessment of atmospheric heavy metal pollution at Belgrade, Serbia, two tree species, horse chestnut (*Aesculus hippocastanum*) and linden (*Tilia* sp.) leaves, were evaluated for their bioaccumulation potential to 9 HM in heavy traffic areas, park, and botanical garden. Authors found that levels of HM were above the reference standard for plants in most of the leaf samples with higher accumulation of Cu (110.2 mg/g DW), Pb (20.3 mg/g DW), and Cd (4.9 mg/g DW) in horse chestnut compared to linden (Table 8.2). Several-fold higher accumulations of HM were also noticed in this study at mature stage compared to the beginning of the vegetation period (Tomasevic et al. 2004). Significant degree of variations in HM content in

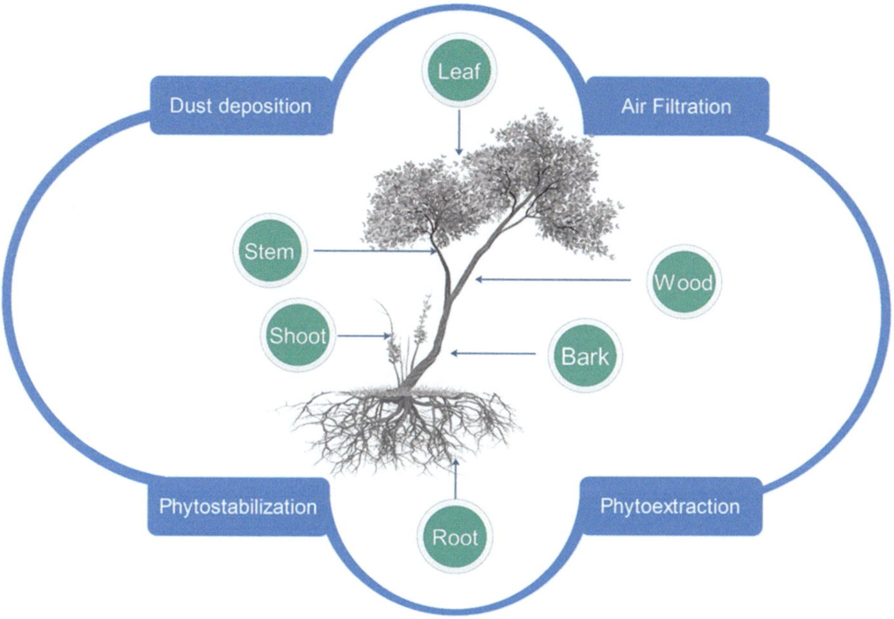

Fig. 8.1 Schematic diagram of possible HM bioaccumulation parts in plant and their role in phytoremediation

leaves were observed in four different tree species (*Acer platanoides*, *Aesculus hippocastanum*, *Betula pendula*, and *Tilia cordata*) when compared among each other in Belgrade, Serbia. Pb, Cr, Mn, Al, Cd, and Fe were higher in *A. platanoides*, whereas Ba, Ni, and Zn were higher in *B. pendula* and *T. cordata* showed higher accumulation of Cu and Sr (Tomaševića et al. 2011). When plants are exposed to similar sources of HM under similar environmental conditions, accumulation patterns are still governed by individual species characteristics. Higher metal accumulation potential of trees around point source was extensively studied to identify trees that can accumulate metals and also reduce air pollution (Table 8.2). In one such study around thermal power plant in Kütahya Province, Turkey, Cicek and Koparal (2004) estimated metal accumulation potential in leaves of *Salix alba* L., *Populus tremula* L., *Robinia pseudoacacia* L., *Quercus infectoria* L., and *Pinus nigra* around 10 km radius of the thermal power plant. They found significant accumulation of Cd, Cr, Cu, and Pb in all the selected plants with gradual decrease in accumulation, with increasing distance from power plant. Santos-Jallath et al. (2012) observed significant amount of As accumulation in leaves of *Nicotiana glauca* (91.94 mg/kg) and in *Tecoma stans* (9.22 mg/kg), whereas values were lower in *Prosopis* sp. (6.94 mg/kg) and *Casuarina* sp. (3.95 mg/kg) around tailings dams at Queretaro, Mexico.

Sawidis et al. (2011) compared bioaccumulation of Cr, Cu, Fe, and Pb in leaf samples of *Platanus orientalis* L. and *Pinus nigra* Arn. around three European cities (Salzburg, Belgrade, and Thessaloniki) and found an increase in bioaccumulation of all the studied metals in leaf of plants grown at polluted areas in comparison to

Table 8.2 Summary of HM bioaccumulation in leaves of different tree species

Location	Area	Plants	Result	Conclusion	References
Belgrade, Serbia and Montenegro	Urban	*Aesculus hippocastanum*	Fe being three-fold, Cu ten-fold, Pb 20-fold, and Cd several hundred times higher than the value at reference site	HM levels were above toxic levels in plants	Tomasevic et al. (2004)
Kütahya Province, Turkey	Tuncbilek Thermal Power Plant	*Salix alba* L., *Populus tremula* L., *Robinia pseudoacacia* L., *Quercus infectoria* L., *Pinus nigra* Arn. ssp. *pallasiana* (Lamb) Holmboe.	HM ranged from 0.1 to 2.73 for Cd, 0.6 to 110 for Cr, 2.1 to 59 for Cu, 0.1 to 55 for Pb, 1.9 to 81.7 for Ni, and 1.7 to 222.4 mg kg^{-1} for Zn	All the plants showed significant amount of metal bioaccumulation which was maximum for Zn and least for Cd	Cicek and Koparal (2004)
Belgrade, Serbia	Urban	*Acer platanoides*, *Aesculus hippocastanum*, *Betula pendula*, *Tilia cordata*	*Acer platanoides* showed highest bioaccumulation of Pb, Cr, and Cd, whereas *B. pendula* accumulated Ba, Ni, and Zn; *T. cordata* accumulated more Cu and Sr	*Acer platanoides* as a useful biomarker for urban HM pollution	Tomaševića et al. (2011)
Belgrade, Serbia	Urban	*A. hippocastanum* L. *Tilia* spp.	Significant seasonal variations in Pb accumulation	*A. hippocastanum* as a useful biomonitor for HM monitoring	Tomašević et al. (2013)
Isfahan City, Iran	Urban	*Platanus orientalis* L.	Higher Fe and Mn accumulation followed by Zn, Cu, and Ni	Useful indicator of air pollution for all the heavy metals except Pb	Norouzi et al. (2015)
Florence, Italy	Urban	*Quercus ilex* L.	High capture potential for Pb, Fe, Mn, Cr, and Ba in old leaves	Difference in metal accumulation with leaf age	Ugolini et al. (2013)

Belgrade (Serbia)	Urban	*Aesculus hippocastanum*, *Tilia* spp.	*Aesculus hippocastanum* showed higher bioaccumulation than *Tilia* spp.	*Aesculus hippocastanum* can be utilized in biomonitoring for Pb and Cu	Aničič et al. (2011)
Queretaro, Mexico	Tailings dams	*Nicotiana glauca*; *Prosopis* sp., *Casuarina* sp.	*Nicotiana glauca* showed higher accumulation for As, Cd, Pb, Cu, and Zn	*Nicotiana glauca* useful for bioaccumulation around HM-contaminated soil	Santos-Jallath et al. (2012)
Cartagena-La Union Mining District, Spain	Mine tailings	*Pinus halepensis*; *Tetraclinis articulate*	Both species showed higher Mn and Zn accumulation	Foliar concentrations indicate higher metal tolerance in these species	Parraga-Aguado et al. (2014)
Sarawak, Malaysia	Lambir Hills National Park	*Dryobalanops lanceolata*; *Macaranga* spp.	*Macaranga* spp. showed better bioaccumulation for Al, B, Ca, Cr, Fe, K, Mn, P, Pb, Ti, and V than *D. lanceolata*	Study showed direct relationship between HM accumulation and anthropogenic input in forest	Breaulmann et al. (2002)
Denizli City, Turkey	Urban and rural	*Robinia pseudoacacia*	Higher metal content at industrial and traffic site compared to rural site	*Robinia pseudoacacia* as a suitable bioindicator of HM pollution	Celik et al. (2005)
Antalya City, Turkey	Urban, suburban, and rural	*Nerium oleander* L.	Significant correlation between metal accumulation and surface soil metal content	Reliable biomonitor for HM pollution estimation	Aksoy and Ozturk (1997)
Naples, Italy	Urban	*Quercus ilex* L.	Significant correlation between leaf deposit and leaf tissue content of both Cu and Pb	Airborne pollutant and soil both contribute to leaf metal content	Alfani et al. (1996)
Salzburg, Austria; Belgrade, Serbia; and Thessaloniki, Greece	Polluted and nonpolluted	*Platanus orientalis* L. and *Pinus nigra* Arn	Both species showed significant accumulation of Pb, Fe, and Cu compared to nonpolluted area	Suitable for biomonitoring of airborne HM	Sawidis et al. (2011)

control. Leaf elemental composition data from eight urban sites in Naples, Italy, by Alfani et al. (1996) in *Quercus ilex* suggested accumulation of metals in leaf was both contributed by metal uptake from soil and from air pollution. Aksoy and Ozturk (1997) reported significant correlation between metal accumulation in leaf with soil surface metal in *Nerium oleander* L. (Oleander) at different urban, suburban, and rural areas of Antalya City, Turkey. The results also showed increase in metal accumulation with urbanization and found Pb to be a major source of HM pollution in the city.

Celik et al. (2005) assessed leaves of *Robinia pseudoacacia* L for heavy metal contamination in Denizli City, Turkey, and found higher accumulation of Pb, Cu, Fe, Zn, Mn, and Cd in samples collected from industrial sites followed by samples collected from road traffic. Breaulmann et al. (2002) assessed HM content in two tropical tree species *Dryobalanops lanceolata* and *Macaranga* spp. in Lambir Hills National Park and found *Macaranga* spp. have better bioaccumulation capacity for Al, B, Ca, Cr, Fe, K, Mn, P, Pb, Ti, and V than *D. lanceolata*. Parraga-Aguado et al. (2014) found higher accumulation of Mn and Zn in leaves of *Pinus halepensis* and *Tetraclinis articulata* around mine tailings in the Cartagena-La Union Mining District, Spain.

8.2.2 Bark and Wood

A critical factor which determines heavy metal uptake and bioaccumulation in bark is deposition in outer surface and translocation from root. Bark is a permanent feature mostly, whereas leaf sheds or falls after aging.

Fujiwara et al. (2011) in urban area of Buenos Aires, Argentina, found bark of green ash trees (*Fraxinus pennsylvanica*) as a suitable biomonitor for trace metal in ambient air compared to road dust. In this study, HM accumulation in tree bark helped in identifying different sources of contamination at monitoring sites. Enrichment factor for Cu, Pb, and Zn showed similar trend in bark and metal content in PM_{10} (particles < 10 μm in aerodynamic diameter), indicating the usefulness of this plant. *F. pennsylvanica* showed higher accumulation of Al and Fe and least for Cr and Ni in bark, indicating its differential potential for metal accumulation with different pollution sources. Similar experiment for air quality biomonitoring by using cypress tree (*Cupressus sempervirens* L.) bark samples was conducted in Amman City, Jordan, by El-Hasan et al. (2002). They found higher accumulation of metals in industrial and heavy traffic areas compared to residential and background sites. Cypress tree showed maximum accumulation of Zn (442–11 μg/kg) followed by Pb (445–22.6 μg/kg), Mn (56–4 μg/kg), and Cu (82.7–1.5 μg/kg) and least accumulation of Co (0.49–0.011 μg/kg) and Cd (0.83–0.069 μg/kg).

Epiphytic moss *H. cupressiforme* and oak tree bark (*Quercus* spp.) were compared for their metal accumulation by Cucu-Man and Steinnes (2013) at different sites in Eastern Romania and found similar magnitude of metal accumulation in both moss and tree bark and concluded that oak tree bark can be utilized as a suitable biomonitor in the absence of mosses. Oak tree bark showed higher accumulation of Zn (5–208 μg/kg), Pb (2.3–560 μg/kg), and Cr (1.1–31.3 μg/kg), whereas lower accumulation of Bi (0.007–0.19 μg/kg), Tl (0.013–0.32 μg/kg), and In

(0.0004–0.01 μg/kg). In different land use patterns as rural, industrial urban, and coastal in northwestern Portugal, Pacheco et al. (2002) compared olive tree (*Olea europaea* Linn.) bark metal accumulation with lichen (*Parmelia* spp.) and observed a strong correlation between metal accumulation in both species and suggested that bark samples can be utilized successfully in the absence of lichens as lichens are now mostly restricted or absent in urban environment with increase in air pollution.

Sawidis et al. (2011) compared bioaccumulation of Cr, Cu, Fe, and Pb in leaf and bark samples of *Platanus orientalis* L. and *Pinus nigra* Arn. around three European cities (Salzburg, Belgrade, and Thessaloniki) and found bark as a more suitable bioindicator compared to leaf. Between the trees, *P. nigra* showed better accumulation than *P. orientalis* (Table 8.3). In another study, Guéguen et al. (2011) estimated HM in tree bark in industrial zones of Strasbourg (France) and Kehl (Germany) and found higher accumulation around industrial and traffic sites. Compared to rural environment, maximum enrichment was observed around steel plant for Pb (148 ppm) followed by Cr (86.5 ppm), Ni (18.4 ppm), Sn (5.7 ppm), Cd (2 ppm), and Co (2 ppm) and least in As (1.7 ppm), whereas higher enrichment in Cd (2.86 ppm) and Mo (8.6 ppm) was found around waste incinerators.

Accumulation characteristics are also varied in different genotypes of the same plant as Mleczek et al. (2009) observed differences in HM accumulations of different metals on 12 different genotypes of willow (*Salix purpurea*) growing in the Potasze Forest Division Salicarium, Poland. Authors also found effect of height from where materials were collected on metal accumulation. Among the different genotypes, *S. purpurea* var. *angustifolia* Kerner showed maximum accumulation, whereas minimum accumulation was found for *S. purpurea* "Utilissima."

Time series analysis of HM in wood samples of *Caesalpinia peltophoroides* ("Sibipiruna") from 1971 to 2003 in Piracicaba City, Brazil, showed usefulness of this species as bioindicator of metal pollution (de Vives et al. 2006). The authors also found a significant decreasing pattern of HM content for Pb and Fe. Ranges of HM from 1971 to 2003 were < 0.39–10.39 μg/g for Ti, < 0.15–69.47 μg/g for Fe, 18.17–59.42 μg/g for Ba, and 9.89–46.54 μg/g for Pb. Pb specifically showed a negative trend toward the bark from the center (de Vives et al. 2006) (Table 8.3).

Beramendi-Orosco et al. (2013) analyzed HM in tree-ring sequences of *Prosopis juliflora* around different urban sites at different distances from a previously active Cu smelter facility at San Luis Potosi, Mexico, and found maximum accumulation of Zn (up to 120 mg/kg) followed by Cu (up to 9.6 mg/kg) and Pb (up to 1.4 mg/kg) with enrichment factor up to 26, 8.6, and 1.4, respectively, for Zn, Cu, and Pb compared to background site. They found increasing trend in metal concentration with time for Zn at residential site, whereas Cu accumulation response was similar to Cu production. The results clearly highlight the usefulness of tree-ring measurement for the assessment of long-term HM monitoring around different land use patterns.

Seedling response of the three tree species (*Mimosa caesalpiniaefolia*, *Erythrina speciosa*, and *Schizolobium parahyba*) to Pb treatment showed no changes in seed germination, but among the three species, *M. caesalpiniaefolia* was more resistant to Pb treatment. Both *E. speciosa* and *S. parahyba* showed linear increase in Pb bioaccumulation in shoot with Pb concentrations in soil, whereas increase in bioaccumulation in root was similar for all three species (Table 8.3) (Ribeiro de Souza et al. 2012).

Table 8.3 Summary of HM bioaccumulation in tree bark/wood in different tree species

Location	Area	Plants	Result	Conclusion	References
Northwestern Portugal	Rural, urban industrial, and coastal	*Olea europaea* Linn.	Bark of *Olea europaea* Linn. showed significant correlation with lichen in HM accumulation	Bark can be utilized for long-term biomonitoring in the absence of lichen	Pacheco et al. (2002)
Strasbourg (France) and Kehl (Germany)	Urban and suburban	*Tilia* sp.	Significant variation in metal accumulation with sources	Useful in monitoring different sources of HM pollution	Guéguen et al. (2011)
Salzburg, Austria; Belgrade, Serbia; and Thessaloniki, Greece	Polluted and nonpolluted	*Platanus orientalis* L. and *Pinus nigra* Arn.	*Pinus nigra* showed higher accumulation of Cr, Cu, Fe, and Pb compared to *P. orientalis*	Bark had higher metal accumulation compared to leaf in both species	Sawidis et al. (2011)
Salicarium, Poland	Forest	12 *Salix* genotype	HM bioaccumulation varied with species and different metals	*S. purpurea* var. *angustifolia* Kerner identified as most potent HM accumulator	Mleczek et al. (2009)
Piracicaba City, Brazil	Urban and country side	*Caesalpinia peltophoroides*	Significant accumulation of Pb 9.89 and 46.54 μg g − 1 in wood	Significant association between bioaccumulation and source	de Vives et al. (2006)
San Luis Potosi, Mexico	Around Cu smelter	*Prosopis juliflora*	High bioaccumulation potential for Zn, Cu, and Pb in wood	*P. juliflora* useful bioindicator for Cu pollution in urban and industrial areas	Beramendi-Orosco et al. (2013)

Greenhouse pot experiment	Seedling response to elevated increasing Pb concentrations (0250, 500 and 1000 mg kg1)	*Mimosa caesalpiniaefolia*, *Erythrina speciosa*, and *Schizolobium parahyba*	*M. caesalpiniaefolia* showed maximum tolerance to Pb, whereas *E. speciosa* and *S. parahyba* showed moderate tolerance	*M. caesalpiniaefolia* showed marked phytostabilization potential for Pb	Ribeiro de Souza et al. (2012)
Leizhou Peninsula, China	Mangrove	*Rhizophora apiculata*	Significant correlation between Pb accumulation in tree ring with Pb emission	Declining trend in metal accumulation with time due to changes in source emission	Yu et al. (2007)
Buenos Aires, Argentina	Urban	*Fraxinus pennsylvanica*	Metal bioaccumulation correlated with traffic sources	Useful air pollution indicator for different HM sources	Fujiwara et al. (2011)
Amman City, Jordan	Urban	*Cupressus sempervirens* L.	Metal accumulation varied in bark with pollution sources	*C. sempervirens* bark as a useful biomonitor for urban air pollution	El-Hasan et al. (2002)

8.3 Metal Accumulation in Grasses

Grasses though have lower biomass compared to tree and a lower life span and have tremendous potential to accumulate high concentration of metals in root, stem, and leaves. Most of the hyperaccumulators reported are from grass families (Table 8.4).

Madejón et al. (2002), in the spill-affected soils of the Guadiamar floodplain in SW Spain, recorded significant bioaccumulation of HM in Bermuda grass (*Cynodon dactylon*) and wild sorghum (*Sorghum halepense*). All metals were higher in the aboveground parts of the plants in sludge cover soil; average Fe concentrations were up to 4180 and 2446 mg/kg in *Cynodon* and *Sorghum*, respectively. Both grasses showed significant accumulation of As and Pb in sludge cover soil compared to control soil, which was 96 and 109 times higher for As and 95 and 76 times for Pb, respectively, in *Cynodon* and *Sorghum*. For Zn accumulation, there were no obvious differences between the two plants (Madejón et al. 2002). Onder et al. (2007) reported significant correlations between soil and grass for Pb (0.44), Cu (0.70), V (0.55), and Ni (0.68) at eight different sites in Konya City, Turkey, and found local traffic along with sugar factory and chrome-magnesite factory as the major sources of HM pollution in the city, with grasses showing higher accumulation capacity for the most of the metals. Similar observations were also made by Fakayode and Onianwa (2002) around Ikeja Industrial Estate, Lagos, Nigeria, for Guinea grass (*Panicum maximum*) where strong significant correlations were observed between soil and grass contents of Cd (0.83), Ni (0.90), Mn (0.94), and Pb (0.73). Around Ikeja Industrial Estate, Lagos, Nigeria, *Panicum maximum* showed higher accumulation factor (representing the ratio of average concentrations of a metal at the polluted site to that of control site) for Cd (34.1) followed by Ni (23.4), Cr (23), and Mn (12.3), whereas lower values were obtained for Zn (7.2), Cu (8.7), and Pb (9.8) (Table 8.5).

Mine soils are one of the major sources of metals around the world. Wei et al. (2015) characterized bioaccumulation pattern at the Xikuangshan Sb mine, China, where soil concentrations for Sb and As were ranged from 229 to 1472 mg/kg and 36.8 to 464 mg/kg, respectively. General trend of both Sb and As accumulation in grasses showed maximum concentrations in roots followed by stems and the least in leaves. As these metals have no known function in plants, their accumulation has been mostly restricted to root and stem, although leaves of *Fagopyrum dibotrys* showed maximum As accumulation (217 mg/kg) in this study. In leaves and stem, Sb was mostly accumulated in the form of Sb^{III} than Sb^{V}; the proportion of Sb^{III} was 31–85 % compared to 0.6–35 % of Sb^{V} in stem in the most cases. As also showed similar pattern with much higher levels of As^{III} in stem and leaves compared to As^{V}, but an opposite trend was observed in the case of root.

Bech et al. (2012) studied metal accumulation potential in eight herbaceous species (*Agrostis capillaris* L., *Echium vulgare* L., *Verbascum* sp., *Sonchus asper* L. Hill, *Veronica persica* Poiret, *Barbarea verna* (Mill) Asch., *Poa annua* L., *Erodium cicutarium* L.), growing around a former Sb mine in the Ribes Valley in NE Spain. Among plants, *A. capillaris* L. at low pH site showed higher As (888 μg/g), Sb (402 μg/g), and Pb (765 μg/g) accumulation in root, whereas *S. asper* at

Table 8.4 HM and their known or potential hyperaccumulators with their respective plant families

Metal/metalloids	Hyperaccumulator plant	Family	References
Cd	*Arabidopsis halleri*	Brassicaceae	Krämer (2010)
	Noccaea caerulescens	Brassicaceae	Krämer (2010)
	Arabis paniculata	Brassicaceae	Zeng et al. (2011)
	Sedum alfredii	Crassulaceae	Liu et al. (2008)
	Thlaspi caerulescens	Brassicaceae	Liu et al. (2008)
	Prosopis laevigata	Fabaceae	Buendía-González et al. (2010)
	Bidens pilosa L.	Asteraceae	Sun et al. (2009)
	Solanum nigrum L.	Solanaceae	Sun et al. (2007)
Zn	*Arabis paniculata*	Brassicaceae	Zeng et al. (2011)
	Noccaea caerulescens	Brassicaceae	Krämer (2010)
	Arabidopsis halleri	Brassicaceae	Krämer (2010)
	Sedum alfredii Hance	Crassulaceae	Krämer (2010)
	Potentilla griffithii	Rosaceae	Chao et al. (2010)
			Hu et al. (2009)
Ni	*Cardamine resedifolia*	Brassicaceae	Krämer (2010)
	Alyssum lesbiacum	Brassicaceae	Adamidis et al. (2014)
	Bornmuellera sp.	Brassicaceae	Krämer (2010)
	Cochlearia sp.	Brassicaceae	Krämer (2010)
	Peltaria emarginata	Brassicaceae	Krämer (2010)
	Streptanthus polygaloides	Brassicaceae	Krämer (2010)
	Berkheya coddii Rossler	Asteraceae	Moradi et al. (2009)
As	*Isatis capadocica*	Brassicaceae	Krämer (2010)
	Hesperis persica	Brassicaceae	Krämer (2010)
Se	*Stanleya pinnata*	Brassicaceae	Krämer (2010)
Cr	*Leersia hexandra*	Poaceae	Zhang et al. (2007)
	Spartina argentinensis	Poaceae	Redondo-Gomez et al. (2011)
	Prosopis laevigata	Fabaceae	Buendía-González et al. (2010)
Fe	*Imperata cylindrica*	Poaceae	Fuente et al. (2016)
Cr	*Leersia hexandra*	Poaceae	Zhang et al. (2007)
Se	*Astragalus selenium*	Fabaceae	DeTar et al. (2015)
Cu	*Commelina communis*	Commelinaceae	Wang et al. (2004)

high pH site showed higher Pb (997 μg/g) accumulation in root. Other species did not show significant metal accumulation for most of the metals (Table 8.5).

For biomonitoring of heavy metals in ambient air at 100 sites in 11 cities from 2000 to 2002, Klumpp et al. (2009) used Italian ryegrass (*Lolium multiflorum* Lam.) and found relationships between metal accumulation and pollution sources around

Table 8.5 Summary of studies related to HM bioaccumulation in grasses

Location	Area	Plants	Result	Conclusion	References
11 European cities	100 sites comprising of urban, suburban, industrial, and traffic-exposed and reference site	*Lolium multiflorum* Lam.	Site-wise variation in metal accumulation for most of the metals	Useful bioindicator for toxic compounds in particulate matter	Klumpp et al. (2009)
Faenza, Italy	Urban and rural	*Rosa rugosa*	Positive significant association between foliar accumulation and soil HM levels	Suitable biomonitor for to assess HM levels in soil	Calzoni et al. (2007)
Cartagena-La Union Mining District (SE Spain)	Mine tailings	Grass land community	Different kinds of plant communities with different seasonal strategies were colonizing high metal-polluted tailings	High bioaccumulation potential in grasses which was regulated by soil pH	Conesa et al. (2007)
Cartagena-La Unión mining district (SE Spain)	Mine tailings	*Helichrysum decumbens*; *Limonium* sp., *Zygophyllum fabago* L., *Piptatherum miliaceum* (L.) Cosson; *Z. fabago* L., *Hyparrhenia hirta* (L.) Stapt, *Tamarix* sp., and *Phragmites australis* (Cav.) Trin. ex Steud.	*Zygophyllum fabago* showed maximum accumulation of Zn (530 mg/kg); *Helichrysum decumbens* of Pb (390 mg/kg); *Tamarix* sp. of Cu (11 mg/kg)	Native plants have higher bioaccumulation capacity	Conesa et al. (2006)

Xikuangshan, China	Sb mine	*Imperata cylindrical*, *Rumex patientia*, *Oplismenus undulatifolius* (A.) Bea, *Erigeron annuus*(L.) Pers, *Fagopyrum dibotrys*	Maximum accumulation of As was in the leaves of *F. dibotrys* (217 mg/kg), whereas *D. erythrosora*, *F. dibotrys*, *Rumex patientia*, and *Oplismenus undulatifolius* (A.) Bea accumulated high concentrations of Sb in roots	Higher accumulation of Sb in all plants and organic Sb was only accumulated in leaf tissue	Wei et al. (2015)
Lagos, Nigeria	Industrial area	*Panicum maximum*	Positive correlation was observed for Cd, Ni, Mn, and Pb between soil HM and plant bioaccumulation	*P. maximum* is a suitable bioindicator for selected metals	Fakayode and Onianwa (2002)
Konya, Turkey	Urban	*Lolium perenne* L.	Significant positive correlation between Pb, Cu, V, and Ni in soil with *L. perenne*	Bioaccumulation of different heavy metals	Onder et al. (2007)
Aznalcollar mine spill (SW Spain)	Aznalcollar pyrite mine	*Cynodon dactylon* and *Sorghum halepense*	Both grasses showed higher accumulation of Cd in leaf	Suitable for stabilization of spill-affected soils	Madejón et al. (2002)
Wrocław Metropolitan Area, SW Poland	Urban	*Polygonum aviculare*	*P. aviculare* showed higher accumulation of most HM	Reliable indicator for urban biomonitoring	Polechonska et al. (2013)

(continued)

Table 8.5 (continued)

Location	Area	Plants	Result	Conclusion	References
Guilin (in the north of Guangxi Province), China	Electroplating factory	*Leersia hexandra* Swartz	High accumulation capacity of both Cr (III) and Cr (VI)	Phytoremediation of Cr-contaminated soil	Zhang et al. (2007)
Malacca, Malaysia.	________	*Stevia rebaudiana*	High tolerance to HM in leaves, stem, and flower	Useful bioindicator	Hajar et al. (2014)
Huelva, Spain	River bank	*Imperata cylindrica* (L.) P. Beauv	Fe accumulation in different parts of cells and tissue	Higher accumulation in epidermis, cortex, and stem	Fuente et al. (2016)

different sites and cities. Higher bioaccumulation was observed for Fe and Zn and the least for Cd, as authors concluded that Cd concentrations are on decline in most of the cities in Europe. For assessing the bioindicator of urban air pollution in Wrocław Metropolitan Area, Poland, Polechonska et al. (2013) utilized *Polygonum aviculare* on 79 different sites with varying pollution levels. They found higher accumulation of Fe, Cd, Cu, Pb, Ni, and Zn that was mainly attributed to higher vehicular pollution. Bioaccumulation factor (ratio of metal concentrations of root to shoot) was higher for Cd (0.6) and Zn (0.5) and the least for Mn (0.05) and Fe (0.04), whereas translocation factor was higher for Mn (2.5) followed by Cu (1.2), Fe (1.2), and Zn (1.2) and lesser for Pb (0.6), Cd (0.7), and Ni (0.8) indicating that nonessential metals were mostly remained in roots and their transfer to aboveground parts was restricted, which helped in proper growth and functioning of plant. It also implicates the usefulness of this plant as HM bioindicator or for remediation of contaminated soil.

Hyperaccumulator *Commelina communis* leaves accumulated Cu up to 1046 and 1119 μg/g DW with exposure of 8 days with 1 mM Cu treatment and after 6 days with 10 mM Cu treatment, respectively, whereas nonaccumulator species died after 6 days of treatment (Table 8.5) (Wang et al. 2004).

8.4 Heavy Metal Accumulation in Different Plant Parts

8.4.1 Root

Plants have evolve complex mechanisms to combat HM stress. The primary site for absorption of HM in plant is root. Different plant species have adapted differentially as some plants accumulate excessive amount of metals in root tissues which prevent transfer of HM to aboveground parts and result in proper growth and physiology of plants, whereas in other group of plants, HM after absorption through root are transferred to shoot and either remain in stem or transported and stored in leaf tissue. The difference in the pattern of accumulation is complex and may depend upon the different mechanisms that developed under different sets of conditions and plants' own evolutionary history. Plants can prevent absorption of metals by avoidance mechanism or can accumulate and store them in separate compartments to combat their negative effects in the cell (Cseh 2002).

It has been shown that nonessential or toxic metals are mostly accumulated in roots and plants prevent their translocation to aboveground parts as Polechonska et al. (2013) reported higher accumulation of Cd, Ni, and Pb in roots of *Polygonum aviculare* exposed to urban and highway environments in Poland. Different plants have different abilities to tolerate high metal concentrations even in hyperaccumulators as Zhang et al. (2007) observed in chromium hyperaccumulator *Leersia hexandra* Swartz that in comparison to control the root growth was normal up to Cr concentration below 20 mg/L, but at a higher concentration of 60 mg/L, a significant reduction of 58.5 % was observed in the root biomass. In roots of this Cr hyperaccumulator, Cr concentrations were up to 18,656 mg/kg in root. In Ni hyperaccumulator, *A. lesbiacum* showed interpopulation variation in root length with Ni treatment (Adamidis et al. 2014).

Among ten species studied growing in the tailings dams, only two plants, *Opuntia lasiacantha* (12.54 mg/kg) and *Cenchrus ciliaris* (4.63 mg/kg), showed higher accumulation of As in root than aboveground parts (Santos-Jallath et al. 2012). Other plants in this study like *Maurandya antirrhiniflora* and *Prosopis* sp. showed higher accumulation of Pb (103.41 mg/kg) and Cu (44.51 mg/kg), respectively, in root. Apart from Zn, all other metals showed higher accumulation in root compared to leaf in *Cenchrus ciliaris.*

Buendía-González et al. (2010) in Cr hyperaccumulator *Prosopis laevigata* also reported reduction in root elongation with increase in concentrations of Cd (II) and Cr (VI). The increase in concentration of Cd from 0.3 to 2.2 mM and Cr from 0.5 to 3.4 mM resulted in reduction of root elongation by 78.9–93.1 % and 46.8–61.5 %, respectively, in *P. laevigata.* Accumulation of Cd and Cr in roots of this plant was up to 21,437 and 8090 mg/kg of dry weight at 0.65 mM Cd treatment and 3.4 mM Cr treatment, respectively. Root biomass of in *S. nigrum* was reduced by up to 52 % with Cd treatments in soil (Sun et al. 2007). In *S. nigrum*, the Cd accumulation was more in leaf compared to root, whereas in *S. melongena* root tissue showed higher Cd levels than leaf. The behaviors of these two plants based on metal accumulation were due to differential antioxidative capacity of plants to cope up with enhanced metal tolerance (Sun et al. 2007). In zinc and cadmium hyperaccumulator plant, *Potentilla griffithii* roots showed Zn and Cd accumulation of 14,060 and 9098 mg/kg of DW, respectively, at 160 and 40 mg/L treatments of Zn and Cd, although maximum accumulation of Zn was found in petiole (19,600 mg/kg of DW). In roots, Zn was mostly accumulated in xylem parenchyma cells and epidermal cells, whereas, for Cd, maximum accumulation was observed around rhizodermal cell, cortex cell, and in central cylinder (Hu et al. 2009). Study also concluded that main storage of Zn and Cd in roots was around the cell wall.

Compared to control plant, mean root biomass of Se hyperaccumulator *A. racemosus* was decreased by 27 % with Mo treatment of 16 mg/L, whereas in *A. bisulcatus* root biomass was increased by 77 % at the same concentration but decreased at higher concentrations of Mo of 32 and 48 mg/L by 30 and 50 %, respectively (DeTar et al. 2015). In non-Se accumulator, *A. drummondii* and *A. convallarius* both showed reduction in mean root biomass by 35, 80, and 77 % and 30, 18, and 59 %, respectively, when treated with Mo concentration of 16, 32, and 48 mg/L (DeTar et al. 2015). In Fe hyperaccumulator *Imperata cylindrica*, Fuente et al. (2016) recorded higher Fe accumulation in root and rhizome compared to leaf tissue. They mostly accumulated in the apoplastic region in the form of iron oxide biominerals (Fuente et al. 2016).

Ribeiro de Souza et al. (2012) tested seedling response of three leguminous woody species, *Mimosa caesalpiniaefolia*, *Erythrina speciosa*, and *Schizolobium parahyba*, to Pb treatment in a greenhouse pot experiment and found higher accumulation of Pb in root (85–90 %) compared to shoot, but no changes in root DW were observed at higher Pb concentration in any of the three tested plants. Only *S. parahyba* showed an elevation of 37 % in root mycorrhizal colonization at highest

Pb concentration. Authors also found decrease in nitrogenase enzyme activity in root nodules of *E. speciosa* with an increase in Pb treatment. Results clearly indicated that higher accumulation in root tissue acts as a resistant mechanism to cope up with high Pb levels in soil.

Metals are mostly accumulated in different oxidation states, which is directly related to their transport mechanism and level of toxicity. In perennial herbs, As was mostly accumulated in the form of As^{V} in roots of the majority of herbs, whereas As^{III} was less in most cases around Xikuangshan Sb mine, China (Wei et al. 2015). High level of Sb was accumulated in roots of *Fagopyrum dibotrys* (D. Don), *Rumex patientia*, and *Oplismenus undulatifolius* (A) Bea around Xikuangshan Sb mine, China, indicating that selected plants were lacking the ability to transfer Sb to aboveground parts (Wei et al. 2015). Bech et al. (2012) reported higher Pb accumulation in *Sonchus asper* L. Hill (997 μg/g DW), *Agrostis capillaris* L. (765 μg/g DW), *Poa annua* L. (306 μg/g DW), *Veronica persica* Poiret (141 μg/g DW), and *Echium vulgare* L.(77 μg/g DW) growing around a former Sb mine in the Ribes Valley in NE Spain. Among other plants, *Agrostis capillaris* L. showed maximum accumulation of Sb (402 μg/g DW) and As (888 μg/g DW) in root. Cr hyperaccumulator, *Spartina argentinensis*, when treated with Cr^{6+} from 0 to 20 mmol/L, showed a linear increase in Cr concentration in root with treatment ($r = 0.94$, $P < 0.01$) (Redondo-Gomez et al. 2011).

8.4.2 Shoot

Metal uptake via root system from soil is either stored in roots or subsequently transported to the aboveground parts through the xylem. The ability of plants to transport and store excess metals in aboveground parts depends upon effective metal transport system, vacuolar compartmentalization, antioxidative defense response, and physiological status of the plant (Krämer 2010).

Leersia hexandra Swartz, a chromium hyperaccumulator plant, showed mean Cr concentration of 237 mg/kg in stem, which was very low compared to leaf, but mean ratio of Cr concentration in stems to that in pond water was very high and was up to 297.41 times with a maximum value of 517.86, although in nutrient solution culture, stem showed maximum Cr accumulation of 2976 mg/kg (Zhang et al. 2007). This plant clearly showed higher transfer rate of Cr from stem to leaf tissue. When this plant was artificially grown on liquid culture with treatment of Cr, the shoot biomass decreased linearly with increase in Cr (III) concentrations in the medium. Compared to Cr (III), Cr (VI) showed marked effect on stem DW with a reduction of 50.4 % with respect to the control. Sun et al. (2007) reported 32–49 % decrease in stem biomass in *Solanum nigrum* with increasing Cd treatment in soil, whereas in Cd hyperaccumulator plant *Bidens pilosa* L, at soil Cd level of 8 mg/kg, stems accumulated Cd up to 110.5 mg /kg of DW.

Ribeiro de Souza et al. (2012) studied seedling response of tree species on accumulation and translocation of Pb and found linear increase in Pb accumulation with Pb concentrations in soil in *E. speciosa* and *S. parahyba*. In *M. caesalpiniaefolia*,

Pb accumulation was maximum in stem, as this plant showed better tolerance to Pb and higher translocation of Pb from root to shoot. Results also revealed that under low Pb concentrations in soil, accumulation of Pb was higher in stem compared to root, but in soil with high Pb concentrations, reverse pattern was observed with higher Pb content in root than in stem. Indigenous plants growing around the Xikuangshan Sb mine, China, showed higher proportion of Sb^{III} in the stems (31–85 %), compared to Sb^{V} (0.6–35 %). Similar trend was also observed for As, which indicates that the same metals in different oxidation states have different transport mechanisms and accumulation (Wei et al. 2015). Polechonska et al. (2013) reported differential behavior of metal accumulation pattern in shoots, as Cu, Fe, Zn, and Mn concentrations were all higher in shoots, whereas Cd, Ni, and Pb concentrations were significantly higher in root of *Polygonum aviculare* growing under an urban environment.

In most of the heavy metal hyperaccumulators, shoot-root ratios of metal concentrations are generally above 1 as these plants mostly accumulate higher metal in aboveground parts due to high metal absorption rate by roots, limited storage in root, higher rates of metal translocation to stem, metal detoxification, and chelation in leaf and stem (Krämer 2010). In *Arabis paniculata* Franch, a Zn/Cd hyperaccumulator, levels of Zn and Cd were 14,716 and 2383 mg/kg, respectively, at treatment of 2000 μM Zn and 250 μM Cd (Zeng et al. 2011). In Ni hyperaccumulator, *Alyssum lesbiacum*, a clear significant relationship was observed between Ni accumulation and Ni tolerance (Adamidis et al. 2014). In Cr hyperaccumulator plant, *Prosopis laevigata* seedlings when grown on culture media having 0.65 mM Cd (II) and 3.4 mM Cr (VI) showed a marked accumulation of 8176 and of 5461 mg/kg DW of Cd and Cr, respectively, in shoots, although there was a strong reduction in shoot size with percent reduction of 28.37, 44.68, 56.03, and 75.46 exposed to Cd (II) concentrations of 0.3, 0.65, 1.3, and 2.2 mM (Buendía-González et al. 2010). But when exposed to 0.5, 1.0, 2.0, and 3.4 mM Cr (VI) concentrations, shoot size reduction was comparatively less with values of 3.55, 15.60, 22.70, and 34.04 %, respectively. The results highlighted that Cd (II) was more toxic compared to Cr (VI) in this experiment (Buendía-González et al. 2010).

An interesting result was observed in *Bidens pilosa* L. by Sun et al. (2009) that when Cd concentrations were lower (8 and 16 mg/kg) in soil, the plant tends to increase shoot dry mass by 3.7 and 9.8 %, respectively, compared with control, suggesting that lower Cd level can induce shoot growth, but at higher concentrations above 50 mg/kg in soil, a negative trend was observed.

In zinc hyperaccumulator, *Sedum alfredii* Hance, when grown in hydroponic growth in 500 M $Zn(NO_3)_2$ solution, Zn accumulation in shoot was approximately 14.5 times higher after 8 days treatment with maximum Zn accumulation of 14749.7 mg/kg compared to nonhyperaccumulator ecotype which showed only 4.2 times increase with maximum Zn accumulation of 454.6 mg/kg. Se hyperaccumulator *Astragalus racemosus* showed decrease in mean shoot biomass by 43 % ($p < 0.05$) with 16 mg/L Mo, but at higher Mo concentration of 32 or 48 mg/L of Mo, no significant changes were observed compared to the control, but in another hyperaccumulator, *Astragalus bisulcatus*, shoot biomass was increased by 16 % (NS), 17 %

(NS), and 38 % ($p < 0.05$) with increasing Mo concentration from 16, 32, and 48 mg/L, respectively. On the other hand, both nonaccumulator *Astragalus drummondii* and *Astragalus convallarius* showed 44 % ($p < 0.05$), 60 % ($p < 0.05$), and 56 % ($p < 0.05$) and no effect, 40 % (NS), and 55 % ($p < 0.05$) reduction in mean shoot biomass, respectively, with 16, 32, and 48 mg/L of Mo treatment (DeTar et al. 2015). Shoot Mo did not show any relationship with Se hyperaccumulation, with concentrations ranging from 1000 to 3000 mg/kg DW among four selected plants with the highest accumulation in nonaccumulator *A. drummondii* (DeTar et al. 2015). Hu et al. (2009) studied distribution of zinc and cadmium in hyperaccumulator *Potentilla griffithii* and found positive relationship between bioaccumulation in shoot with concentrations of Zn and Cu in nutrient solution. Maximum concentration of Cd and Zn in shoot was 1670 and 16,900 mg/kg DW at 20 and 160 mg/L Cd and Zn treatment, respectively. Further, higher concentrations of Cd (40 mg/L) did not result in significant accumulation of Cd in shoot of *P. griffithii*.

In seedling response of tree species to Pb treatment experiment, Ribeiro de Souza et al. (2012) found linear increase in Pb accumulation in shoots with an increase of Pb content in soil, but in *M. caesalpiniaefolia* shoots, Pb concentrations were only increased in stems. Different herbaceous plant species around a former Sb mine in NE Spain showed limited accumulation of most metal in shoot apart from *P. annua*, *E. vulgare*, *S. asper*, and *B. verna*, where average shoot concentrations for As and Sb ranged between 5.5 and 23 mg/kg and 1.21 and 4.9 mg/kg, respectively (Bech et al. 2012).

8.5 Metal-Chelating Compounds and Their Role in Metal Chelation

There are several metal-chelating or metal-stabilizing compounds, present in nature, having potential to bind with metals and immobilize them inside the cell shown in Fig. 8.2 (Cobbett and Goldsbrough 2002; Hall 2002; Li et al. 2015; Prasad 2004). They all have specific ability to bind with metal based on their special structure and symmetry and from metal-bound complexes (Prasad 2004). Their concentrations in the cell depend largely upon the amount or quantity of metals in surrounding environment. Not all plants are reported to have similar types of compound, and their expressions solely depend upon the availability of metal, specific cellular location, and specific plant tissue; they may act on same metal or different groups of metals. HM inside cells can react with different cellular component or produce oxidative stress inside the cell, to combat HM-generated stress as several different chelators have been identified in plants that can bind and neutralize them for proper cellular metabolism and plant growth.

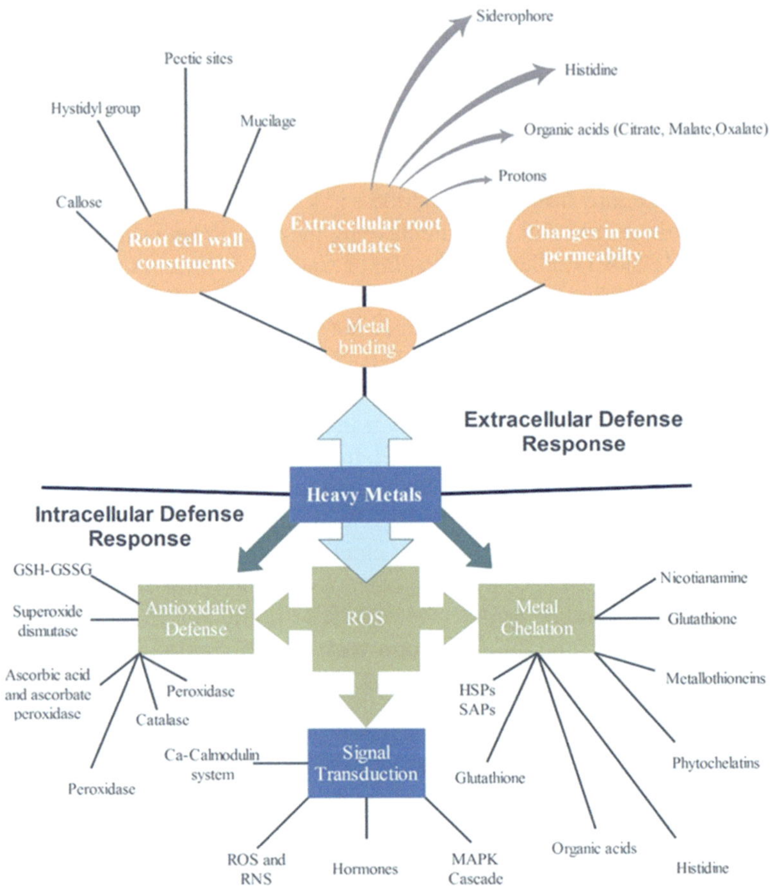

Fig. 8.2 Different pathways for intracellular and extracellular HM response and tolerance mechanisms in plant

8.5.1 Metallothioneins

(MTs): MTs are metal-binding, cysteine-rich polypeptide with several types based on structure and sequence of amino acids, spacers between two cysteines, and sequence of amino acids in C- and N-terminal unit of MTs (Cobbett and Goldsbrough 2002; Li et al. 2015). They have been reported from different groups of plants and animals, although less reports are available in other organisms. MTs are primarily known for their role in metal detoxification by chelation, homeostasis of essential metals, and scavenging of ROS as well as in other cellular metabolism as a part of the regulation of other enzymes, transcription factors, and stress-related proteins (Table 8.6) (Cobbett and Goldsbrough 2002; Li et al. 2015). The activity of MT is due to the presence of cysteine sulfhydryl groups in MT as they act as metal binders through mercaptide bonds (Cobbett and Goldsbrough 2002). Only few metals have been so far known to bind with MT like Cu, Cd, and Zn (Hall 2002). Different MTs as MT1-4 are reported in different groups of organisms and their induction depends upon specific HM (Cobbett and Goldsbrough 2002; Hall 2002; Verbruggen et al. 2009).

8.5.2 Phytochelatins

(PCs): Compared to MTs, PCs are more abundant in plants and made up of a 2–11 repeating units of the γ-glutamate–cysteine dipeptide with a terminal glycine, synthesized from precursor glutathione by an enzyme phytochelatin synthase (Cobbett and Goldsbrough 2002; Hall 2002; Prasad 2004). Different PCs are derived from common structure with modification in terminal glycine amino acid, where it is replaced by other amino acids such as alanine, serine, and glycine (Cobbett and Goldsbrough 2002; Prasad 2004). Apart from its role in HM chelation, it is also important in scavenging of ROS and transferring of essential metals in the cell and spatially in metal homeostasis. Prasad (2004) identified several PC activators such as Cd, Cu, Hg, Ni, Zn, Sn, Sb, Ag, and AsO_4^{3-} (Table 8.6). PCs levels are well regulated in the cell and their concentrations are regulated with an increase in metal uptake. After binding with metal, PC metal complexes are stored inside vacuoles and further immobilized. Interactions and fate of PCs with HM are less known and most of the research has focused on PCs mechanism with Cd tolerance. Apart from Cd, not much is known about interactions of PCs with other HM in higher plants.

8.5.3 Proline

Proline is a multifunctional amino acid which plays role as an osmolyte, a metal chelator, an antioxidant, maintenance of $NAD(P^+)/NAD(P)H$ ratio, and a signaling molecule (Hayat et al. 2012; Sharmila and Saradhi 2002). Most of the study revealed that proline helps in HM tolerance by chelation, by inducing formation of phytochelatins and scavenging of ROS generated due to metal accumulation (Hayat et al. 2012; Sun et al. 2007). Among metals, it has been shown that Co, Cd, Zn, and Pb can induce proline formation (Hayat et al. 2012; Sharmila and Saradhi 2002). Sun et al. (2007) observed an increase in proline accumulation at the rate of 131–184 % in leaf tissue of *S. nigrum* due to Cd treatment and concluded that the accumulation of free proline might be responsible for tolerance mechanism in this plant against Cd stress.

8.5.4 Histidine

Histidine is an amino acid which has also been reported to have high metal-binding potential specifically for Ni (Hall 2002; Krämer 2010). Krämer et al. (1996) in their Ni exposure experiment compared *A. lesbiacum* (Ni hyperaccumulator) with *A. montanum* (nonhyperaccumulator) with high concentration of Ni in hydroponic culture, found a 36-fold increase in histidine concentration in the xylem sap of Ni hyperaccumulator, and confirmed the role of histidine as a Ni chelator by transferring Ni accumulated in root to stem. Similar observations were also made in other Ni hyperaccumulators such as *A. murale* and *A. bertolonii*. Linear increase in histidine concentrations in the xylem sap was also found in the case of Co in the same plant; therefore, a similar type of metals can be chelated by histidine (Table 8.6) (Krämer et al. 1996). The chelation is the major mechanism by which histidine responds to Ni, as 97 % of histidine was found in the form of Ni–histidine complex in *A. lesbiacum* (Krämer et al. 1996).

Table 8.6 Common HM chelators with their specific activity and metal binder

Metal chelators	Structure	Activity	Binding metal	References
Metallothioneins (MTs)	Cysteine-rich proteins	Heavy metal detoxification and protecting against intracellular oxidative damages, ROS scavenging	Cd^{2+}, Zn^{2+}, and Cu^{2+} tolerance	Li et al. (2015) and Cobbett and Goldsbrough (2002)
Phytochelatins (PCs)	Cysteine-rich peptides	Metal detoxification, metal ion homeostasis	Cd, Pb, Ag, Hg, Cu, Zn	Cobbett and Goldsbrough (2002) and Manara (2012)
Proline	Imino acid	Metal chelator	Co, Cd, Zn, Pb	Hayat et al. (2012)
Histidine	Amino acid	Metal chelator	Ni, Zn, Cd	Krämer (2010)
Nicotianamine	S-adenosyl-L-methionine	Metal chelator, transport, metal homeostasis	Fe, Cu	Krämer (2010)
Malate	Organic acids	Metal chelator	Cd, Zn	Manara (2012)
Citrate	Organic acids	Metal chelator	Cd, Fe, Ni	Manara (2012) and Verbruggen et al. (2009)
Glutathione	γ-glutamyl cysteinylglycine tripeptide	Metal chelator, substrate for PCs, antioxidant	Zn, Co, Ni	Verbruggen et al. (2009), Yadav (2010) and Jozefczak et al. (2012)
Cysteine	Amino acid	Metal chelator, substrate for GSH	Cd, Zn, Cu	Jozefczak et al. (2012)
Ferritins	Multimeric proteins	Scavenge Fe	Fe	Manara (2012)
Phytate (myo-inositol hexakisphosphate)	Phosphate monomer	Metal chelator	Mn, Zn, Fe	Manara (2012)
HSPs	Protein	Molecular chaperones		Hall (2002)

8.5.5 Nicotianamine

(NA): NA is a nonprotein amino acid formed by the precursor S-adenosylmethionine by enzyme nicotianamine synthase (Kim et al. 2005). NA has been reported to be present in a wide variety of plants and known for its role in metal homeostasis and chelation, but its role as a Ni chelator is extensive (Kim et al. 2005; Krämer 2010). Pianelli et al. (2005) expressed nicotianamine synthase cDNA (TcNAS1) in *Arabidopsis thaliana* isolated from metal hyperaccumulator *Thlaspi caerulescens* and found 100 times more accumulation of NA compared to the wild-type plant, and as a result higher Ni accumulation and tolerance were observed in this plant when grown in high Ni-contaminated soil. The authors also found higher accumulation of Ni (2.5 mg/g of DW) in leaves of a NA over accumulating transgenic lines of *A. thaliana*. When barley nicotianamine synthase gene was constitutively overexpress in transgenic *Arabidopsis* and tobacco plants, an elevated metal tolerance response was observed with an increase in NA content in these plants (Kim et al. 2005). These results further confirm the role of NA in metal tolerance and homeostasis, although only limited information is available about its behavior with other metals and how NA binds with metals and regulates their transfer with in plant.

8.5.6 Glutathione

(GSH): GSH is a tripeptide made up of three amino acids glutamate, cysteine, and glycine and synthesized by enzymes γ-glutamylcysteine synthetase (GSH1) and glutathione synthetase (GSH2) (Jozefczak et al. 2012; Verbruggen et al. 2009). GSH is well known for its function in antioxidative defense, metal homeostasis, and signal transduction (Jozefczak et al. 2012; Yadav 2010). GSH directly binds to metal by formation of mercaptide bond and forms GSH–metal complex; at the same time, it also acts as a precursor of PCs (Jozefczak et al. 2012; Yadav 2010). Many studies have highlighted an increase in GSH concentrations with an increase in metal stress (Krämer 2010; Verbruggen et al. 2009). GSH role as an antioxidant and as a metal chelator makes it an important constituent of the cell against HM stress tolerance in different plant groups (Jozefczak et al. 2012).

8.5.7 Cysteine

(Cys): Cys is an amino acid with thiol side chain, which gives it the potential to bind metals which makes it as a potent HM chelator (Jozefczak et al. 2012; Zagorchev et al. 2013). Kumar et al. (2002) found an increase in free cysteine concentrations in *Cassia siamea* Lamk in fly ash containing soil rich in HM.

8.5.8 Organic Acids

Organic acids such as malate, oxalate, and citrate have the potential to chelate HM and transport HM organic acid-bound complexes to different parts of the plant tissue (Manara 2012). They have been known for their role in transfer of metal ions in plants. Citrate mostly binds with Fe for its transport (Pierre and Gautier-Luneau 2000), but most of the organic chelators have shown to bind with Ni, Cd, and Zn (Manara 2012; Verbruggen et al. 2009).

8.5.9 Phytate

Phosphate group of phytic acid has the ability to bind with metal cations and chelate them (Bohn et al. 2008). High concentrations of phytate is generally present in seed as a source of phosphate. Phytate is made up of M*yo*-inositol hexakisphosphate having central inositol ring attached to six phosphate group which binds metal cations (Bohn et al. 2008; Manara 2012). Phytate can bind with several metal cations such as Ca^{2+}, Cu^{2+}, Fe^{3+}, Zn^{2+}, Ni^{2+}, Co^{2+}, and Mn^{2+} (Bohn et al. 2008).

8.5.10 Other HM Chelators

Apart from chelators mentioned above, several other chelating compounds have been identified such as *O*-acetylserine, heat shock proteins (HSPs), and mugineic acid (Table 8.6). They have been reported as metal chelators in some plant species, and their mechanisms of action and roles in specific metal tolerance are yet to be identified.

8.6 Morphological, Physiological, Biochemical, and Molecular Biomarkers of Heavy Metal Tolerance

Plants are versatile organism which can adapt in versatile environmental conditions and maintain their proper growth and development. HM tolerance of higher plants is mostly regulated by maintaining normal cellular metabolism and physiology. Most of the studies related to HM stress have identified different biochemical responses for maintaining cellular redox potential, scavenging ROS, metal chelation, enhancement in stress-related protein, and induction in stress-related metabolic pathways (Krämer 2010). HM stress is counteracted by overexpression of numerous stress-related proteins, induction of antioxidative pathways, and activation of various signaling proteins related to stress regulations (Fig. 8.2) (Thapa et al. 2012). Most of the hyperaccumulator studies have shown higher metal accumulation by making several physiological adjustments such as restricting metal accumulation in root, changing uptake rate of metals from soil, increasing concentrations of antioxidant in leaf, enhancing metal uptake in the cellular level by changing

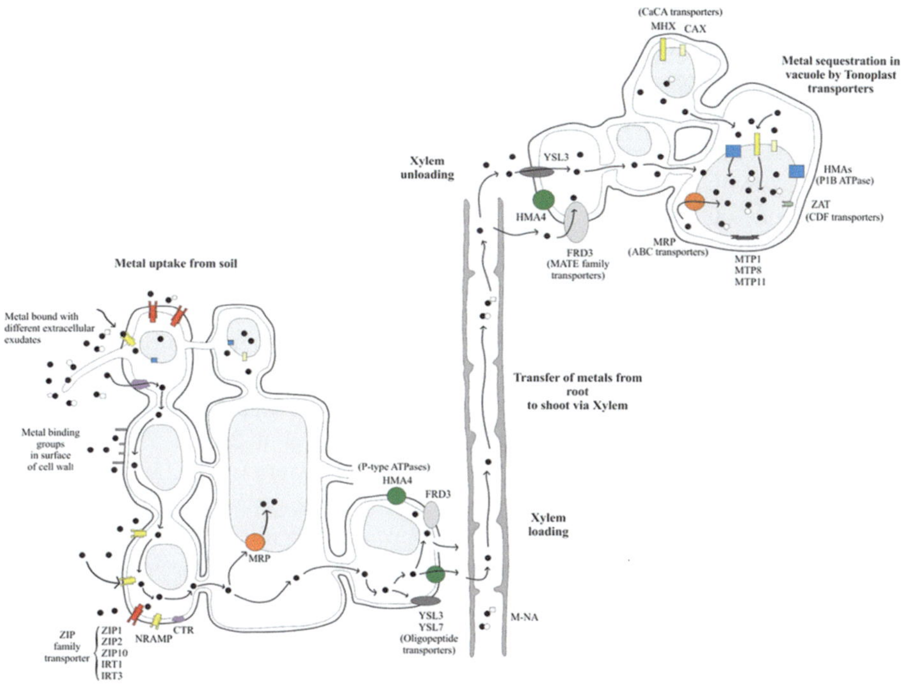

Fig. 8.3 Schematic diagram showing HM uptake by root, their transport to aboveground parts via xylem loading, and sequestration in vacuoles with different membrane transporters, transport proteins, and various intracellular and extracellular chelators of HM. (*CAX* cation exchangers; *CaCA* Ca2+/cation antiporter; *CDF* cation diffusion facilitators; *HM* heavy metals; *HMAs* heavy metal transporting ATPases; *M-NA* metal-nicotianamine; *YSL* yellow strip 1-like proteins; *ZIP* zinc-regulated transporter, iron-regulated transporter proteins; *MATE* multidrug and toxic compound extrusion; *CTR* copper transporter; *NRAMP* natural resistance-associated macrophage proteins; *MRP* multidrug resistance-associated proteins; *FRD* ferric reductase defective; *MTP* metal transporter proteins; *MHX* magnesium proton exchangers; *ZAT* zinc transporter of *Arabidopsis Thaliana*)

permeability of metal transporter, and enhancing the xylem loading (Fig. 8.3) (Krämer 2010). Exclusion and avoidance of HM in plants are the result of mechanisms such as reduction in metal uptake, sequestration of HM in higher concentrations at specific part of the plant, increase in concentrations of metal chelators, and activation of stress-related signaling pathways (Fig. 8.3) (Thapa et al. 2012). Several plant hormones like jasmonic acid, salicylic acid, abscisic acid, and ethylene have been known to be induced by HM stress response (Thapa et al. 2012).

Zeng et al. (2011) analyzed different expressions of proteins in Zn/Cd hyperaccumulator *A. paniculata* Franch by exposing plants to low and high doses of both Cd and Zn and identified 17 out of 19 Zn-responsive and 16 out of 18 Cd-responsive proteins. Proteins expressed due to Zn stress in the plant were mostly related to energy metabolism, photosynthesis, and auxin biosynthesis and those required for metabolism of normal growth and for protein misfolding correction.

Specific proteins in shoot associated with Zn stress were upregulated belonging to oxygen-evolving enhancer protein 1–2 (PSBO) involved in photosynthesis, ATB2 involved in energy metabolism, 3-isopropylmalate dehydrogenase (IMDH) involved in amino acid metabolism, meprin and TRAF homology domain-containing protein and luminal-binding protein 1 precursor both involved in protein metabolism, nitrilase 2 involved in auxin biosynthesis, ribulose bisphosphate carboxylase (RuBisCO) involved in carbon fixation, phosphoserine aminotransferase involved in amino acid metabolism, and glutamine synthetase involved in nitrogen metabolism, serine/threonine–protein phosphatase PP1 isozyme 2 involved in signaling (Zeng et al. 2011). Proteins which were specifically expressed due to Zn stress are cytochrome P450 76A1 involved in xenobiotic defense, protein disulfide isomerase 2 precursor and U-box domain-containing protein 16 involved in protein metabolism, mitogen-activated protein kinase 2 involved in cell signaling, and F-box protein PP2-A15 involved in cellular transportation (Zeng et al. 2011).

Specific proteins in shoot associated with Cd stress were upregulated belonging to ribulose-5-phosphate 3-epimerase involved in photosynthesis, superoxide dismutase, glutathione transferase and ascorbate peroxidase all involved in antioxidative defense, 25.3 kDa vesicle transport protein involved in protein transport, pentatricopeptide repeat-containing protein involved in RNA processing, and trehalose-6-phosphate synthase 3 involved in carbohydrate metabolism.

Specific proteins in shoot associated with Cd stress which was downregulated is RuBisCO, whereas specifically expressed proteins are cytochrome P450 76A1 involved in xenobiotic defense, mitochondrial ribosomal protein S3 involved in cell development, protein thylakoid formation 1 involved in cell signaling, G2/mitotic-specific cyclin S13-6 involved in cell cycle, and nucleoside diphosphate kinase 1 involved in energy metabolism (Zeng et al. 2011).

Level of malondialdehyde (MDA) is a good indicator of metal tolerance in plants; with an increase in stress, higher MDA levels are observed due to membrane damage by both generation of ROS and direct effect of metals on membrane integrity. Sun et al. (2007) found a 2.3–2.6-fold higher MDA content in leaves of non-metal-tolerant *S. melongena* compared to the metal-resistant *S. nigrum*, whereas in *Bidens pilosa*, MDA content showed marked elevation only after Cd concentrations in soil were 50 and 100 mg/kg, respectively, compared with the control (Sun et al. 2009). Treatments from 1 μM to 10 mM of Cu in hydroponic culture resulted in 77–126 % and 77–114 % increase in MDA concentrations in hyperaccumulator and nonaccumulator species of *Commelina communis*, respectively (Wang et al. 2004). Pb contents in soil negatively affect plants by increasing foliar MDA content in *S. parahyba* seedling, but in *M. caesalpiniaefolia* and *E. speciosa* seedling, no significant variations were observed in MDA content with Pb treatment (Ribeiro de Souza et al. 2012).

Photosynthetic pigments are also good markers of metal stress as Sun et al. (2007) reported a negative relationship between total chlorophyll content and Cd concentration in the soils. In *Bidens pilosa* L, Sun et al. (2009) also found a negative trend in total chlorophyll content and Cd concentrations in soil; decreases in chlorophyll content were 22.2 % and 23.3 % with respect to control at Cd level of 50 and

100 mg/kg, respectively. After 15 days of treatment with Cr (VI) in cordgrass, *Spartina argentinensis*, all pigments such as Chla *a* (r = −0.95), Chl *b* (r = −0.96), and carotenoids (r = −0.97) showed negative significant correlation with increasing Cr concentrations with percent reduction of 68, 58, and 47 %, respectively, with respect to the control (Redondo-Gómez et al. 2011). Pb content in soil negatively affects chl *a* and total carotenoid content in leaves of *S. parahyba* seedling and total chlorophyll and chl *b* content in *E. speciosa* (Ribeiro de Souza et al. 2012).

Several antioxidative enzymes are involved in metal stress tolerance. Cd treatment significantly changed enzymatic activities with increase in SOD in root, POD in leaf, and CAT in both leaves and root in *S. nigrum* compared to *S. melongena*. This higher induction of antioxidative enzymes led to more tolerance to Cd in *S. nigrum* compared to *S. melongena*. *S. nigrum* also showed two-fold higher PCs compared to control (Sun et al. 2007). Different plant parts also showed variations in the activities of antioxidative enzymes under Cd stress. SOD activity was lower in the roots of both *S. nigrum* and *S. melongena* compared to leaves (Sun et al. 2007).

POD activity increased by 1.33-, 3.33-, and 6.33-fold with Cd levels of 32, 50, and 100 mg/kg, respectively, compared to control in *Bidens pilosa*, but SOD activity was found to increase by 58.7 % only at 100 mg/kg soil Cd concentration (Sun et al. 2009). SOD activity increased linearly in *Commelina communis* plant with increasing Cu treatment from 1 μM to10 mM. There were increases of 160 % and 197 % in SOD activity in Cu hyperaccumulator and nonaccumulator *C. communis*, respectively. In the same experiment, Cu hyperaccumulator plant did not show any change in guaiacol peroxidase (GPX) activity, whereas a significant correlation was (with r = 0.74–0.96) observed between GPX activity and Cu treatment in the nonaccumulator plant. GPX activity was also higher in nonaccumulator compared to hyperaccumulator plant.

These results indicate that responses of plants to metal stress depend upon several antioxidants in cell, and the overall tolerance response is an outcome of combined interactions of different antioxidants (Wang et al. 2004).

Concentrations of soluble protein increase with treatment of Cu in both Cu hyperaccumulator and nonaccumulator of *Commelina communis*, although the treatment effect was more marked in nonaccumulator (Wang et al. 2004). Increase in foliar ammonia and nitrate content was shown to be influenced by soil Pb concentrations of 250 and 500 mg/kg in seedling of *E. speciosa*, whereas in *M. caesalpiniaefolia* and *S. parahyba*, foliar nitrate content decreased at 250 and 500 mg/kg of Pb concentrations (Ribeiro de Souza et al. 2012).

Gene expression study in both Zn hyperaccumulator and nonhyperaccumulator ecotype of *S. alfredii* revealed higher gene expression in hyperaccumulator compared to nonhyperaccumulator in ten tested genes (Chao et al. 2010). They also reported 22 gene fragments that were modified by Zn in hyperaccumulator, among them 73.6 % were known in database and matched with known proteins. The identified proteins in the study are RuBisCO subunit-binding protein beta subunit precursor, n-protein ligases, cystatin, cationic amino acid transporter, 60S ribosomal protein L21, ubiquitin protein ligase, and nitrate reductase 1 (Chao et al. 2010).

In *Potentilla griffithii* at lower concentration of metals, major storage of metals occurred in leaf epidermal and bundle sheath of vein in leaves, but at higher concentrations, metals were transferred to mesophyll cells. In Zn and Cd treatment of 160 mg/L and 20 mg/L, respectively, accumulation was predominantly in the vacuoles of epidermal and vascular bundle sheath cells, but at higher concentration of Cd (40 mg/L), accumulation was also observed in vacuoles of mesophyll cells (Fig. 8.3). In the root with similar treatment of Zn and Cd showed deposition of Zn on the walls of epidermis, endodermis, and xylem parenchyma, while Cd deposited on wall of epidermis, cortex, endodermis, and parenchyma of vascular cylinder (Hu et al. 2009). In Fe hyperaccumulator, *Imperata cylindrica*, Fe was mostly present in intercellular space, cells of the xylem and phloem, and in epidermal cell in roots, rhizomes, and leaves (Fuente et al. 2016). Moradi et al. (2009) compared root responses between Ni hyperaccumulator plant *Berkheya coddii* Rossler and nonaccumulator plant *Cicer arietinum* L. to soil Ni heterogeneity. Obvious changes in root morphology of *B. coddii* were observed as increasing in root thickness, lower biomass, reduction in number of fine roots, and reduction in branching were marked, whereas in *C. arietinum* reduction in root biomass was prominent.

Significant negative correlation was observed between values of F_v/F_m ($r = -0.99$), quantum efficiency of PSII ($r = -0.96$), net photosynthetic rate ($r = -0.93$), stomatal conductance ($r = -0.93$), and water use efficiency ($r = -0.90$) with increasing Cr^{6+} treatments after 9 days of treatment in cordgrass, *Spartina argentinensis* (Redondo-Gómez et al. 2011). High Pb concentrations in soil did not affect mycorrhizal association with native arbuscular mycorrhiza fungi in all studied tree seedling, but nitrogenase activity in root nodules of *E. speciosa* was negatively affected (Ribeiro de Souza et al. 2012).

8.7 Conclusions

Significant variations were found in HM bioaccumulation potential in different groups of plants. Compared to grasses, trees showed limited accumulation of HM in leaf, bark, and wood, although due to the larger surface area, higher biomass, and longer life span, they can accumulate very high concentrations of metals. Leaf, bark, and wood samples showed great variations in accumulation for overall metal or for individual metal; therefore, their role in biomonitoring is of great importance especially in urban environments, where other groups of plants are lacking. Most of the studies highlighted the usefulness of tree species in the identification of different metal sources. Grasses showed differential accumulation dependent upon species and growth conditions. Most of the hyperaccumulator plants belong to grass families and have huge capabilities to accumulate HM in high concentrations. Metal accumulation in both roots and shoots showed species-specific responses with some plants accumulating higher HM in aboveground parts, while others favoring more accumulation in belowground parts. Increase in concentrations of antioxidants and metal chelators, alteration in physiology, anatomical modifications, changes in gene expression, alteration in specific proteins, and several other mechanisms were identified for metal tolerance in higher plants. Identification of multi-metal-tolerant species and their mechanisms of tolerance under interactive action of different metals should be attempted in future studies. Such studies would help in identifying suitable species for growing in heavy metal-contaminated sites for phytoremediation as well as for environmental monitoring.

Acknowledgments The authors are thankful to the Head of the Department of Botany for providing library and laboratory facilities. Department of Science and Technology, New Delhi, is acknowledged for Inspire Fellowship to Arideep Mukherjee.

References

Adamidis GC, Aloupi M, Kazakou E, Dimitrakopoulos PG (2014) Intra-specific variation in Ni tolerance, accumulation and translocation in the Ni-hyperaccumulator *Alyssum lesbiacum*. Chemosphere 95:496–502

Aksoy A, Ozturk MA (1997) *Nerium oleander* L. as a biomonitor of lead and other heavy metal pollution in Mediterranean environments. Sci Total Environ 205:145–150

Alfani A, Maisto G, Iovieno P, Rutgliano FA, Bartoli G (1996) Leaf contamination by atmospheric pollutants as assessed by elemental analysis of leaf tissue, leaf surface deposit and soil. J Plant Physiol 148:243–248

Aničič M, Spasič T, Tomaševič M, Rajšič S, Tasič M (2011) Trace elements accumulation and temporal trends in leaves of urban deciduous trees (*Aesculus hippocastanum* and *Tilia* spp.). Ecol Indic 11:824–830

Bech J, Corrales I, Tume P, Barceló J, Duran P, Roca N, Poschenrieder C (2012) Accumulation of antimony and other potentially toxic elements in plants around a former antimony mine located in the Ribes Valley (Eastern Pyrenees). J Geochem Explor 113:100–105

Beramendi-Orosco LE, Rodriguez-Estrada ML, Morton-Bermea O, Romero FM, Gonzalez-Hernandez G, Hernandez-Alvarez E (2013) Correlations between metals in tree-rings of *Prosopis juliflora* as indicators of sources of heavy metal contamination. Appl Geochem 39:78–84

Bohn L, Meyer AS, Rasmussen SK (2008) Phytate: impact on environment and human nutrition. A challenge for molecular breeding. J Zhejiang Univ Sci B 9:165–191

Breaulmann G, Markert B, Weckert V, Herpin U, Yoneda R, Ogino K (2002) Heavy metals in emergent trees and pioneers from tropical forest with special reference to forest fires and local pollution sources in Sarawak, Malaysia. Sci Total Environ 285:107–115

Buendía-González L, Orozco-Villafuerte J, Cruz-Sosa F, Barrera-Díaz CE, Vernon-Carter EJ (2010) *Prosopis laevigata* a potential chromium (VI) and cadmium (II) hyperaccumulator desert plant. Bioresour Technol 101:5862–5867

Calzoni GL, Antognoni F, Pari E, Fonti P, Gnes A, Speranza A (2007) Active biomonitoring of heavy metal pollution using *Rosa rugosa* plants. Environ Pollut 149:239–245

Celik A, Kartal AA, Akdoğan A, Kaska Y (2005) Determining the heavy metal pollution in Denizli (Turkey) by using *Robinio pseudo-acacia* L. Environ Int 31:105–112

Chao Y, Zhang M, Feng Y, Yang X, Islam E (2010) cDNA-AFLP analysis of inducible gene expression in zinc hyperaccumulator *Sedum alfredii* Hance under zinc induction. Environ Exp Bot 68:107–112

Cicek A, Koparal AS (2004) Accumulation of sulfur and heavy metals in soil and tree leaves sampled from the surroundings of Tunc ¸bilek Thermal Power Plant. Chemosphere 57:1031–1036

Cobbett C, Goldsbrough P (2002) Phytochelatins and metallothioneins: roles in heavy metal detoxification and homeostasis. Annu Rev Plant Biol 53:159–182

Conesa HM, Faz A, Arnaldos R (2006) Heavy metal accumulation and tolerance in plants from mine tailings of the semiarid Cartagena-La Union Mining District (SE Spain). Sci Total Environ 366:1–11

Conesa HM, Garcia G, Faz A, Arnaldos R (2007) Dynamics of metal tolerant plant communities' development in mine tailings from the Cartagena-La Union Mining District (SE Spain) and their interest for further revegetation purposes. Chemosphere 68:1180–1185

Cseh E (2002) Metal permeability, transport and efflux in plants. In: Prasad MNV, Strazalka K (eds) Physiology and biochemistry of metal toxicity and tolerance in plants. Kluwer Academic Publishers, Dordrecht, pp 1–36

Cucu-Man S, Steinnes S (2013) Analysis of selected biomonitors to evaluate the suitability for their complementary use in monitoring trace element atmospheric deposition, Environ. Monit Assess 185:7775–7791

de Vives AES, Moreira S, Brienza SMB, Medeiros JGS, Filho MT, Zucchi OLAD, Filho VFN (2006) Monitoring of the environmental pollution by trace element analysis in tree-rings using synchrotron radiation total reflection X-ray fluorescence. Spectrochim Acta Part B: Atom Spectrosc 61:1170–1174

DeTar RA, Alford ÉR, Pilon-Smits EA (2015) Molybdenum accumulation, tolerance and molybdenum–selenium–sulfur interactions in *Astragalus* selenium hyperaccumulator and nonaccumulator species. J Plant Physiol 183:32–40

El-Hasan T, Al-Omari H, Jiries A, Al-Nasir F (2002) Cypress tree (*Cupressus Semervirens* L.) bark as an indicator for heavy metal pollution in the atmosphere of Amman City, Jordan. Environ Int 28:513–519

Fakayode SO, Onianwa PC (2002) Heavy metals contamination of soil, and bioaccumulation in Guinea grass (*Panicum maximum*) around Ikeja Industrial estate, Lagos, Nigeria. Environ Geol 43:145–150

Fuente V, Rufo L, Juárez BH, Menéndez N, García-Hernández M, Salas-Colera E, Espinosa A (2016) Formation of biomineral iron oxides compounds in a Fe hyperaccumulator plant: *Imperata cylindrica* (L.) P. Beauv. J Struct Biol 193:23–32

Fujiwara FG, Gómez DR, Dawidowskia L, Perelman P, Faggi A (2011) Metals associated with airborne particulate matter in road dust and tree bark collected in a megacity (Buenos Aires, Argentina). Ecol Indic 11:240–247

Guéguen F, Stille P, Millet M (2011) Air quality assessment by tree bark biomonitoring in urban, industrial and rural environments of the Rhine valley: PCDD/Fs, PCBs and trace metal evidence. Chemosphere 85:195–212

Hajar EWI, Sulaimanb AZB, Sakinahb AMM (2014) Assessment of heavy metals tolerance in leaves, stems and flowers of *Stevia rebaudiana* plant. Procedia Environ Sci 20:386–393

Hall JL (2002) Cellular mechanisms for heavy metal detoxification and tolerance. J Exp Bot 53:1–11

Hayat S, Hayat Q, Alyemeni MN, Wani AS, Pichtel J, Ahmad A (2012) Role of proline under changing environments: a review. Plant Signal Behav 7:1456–1466

Hu PJ, Qiu RL, Senthilkumar P, Jiang D, Chen ZW, Tang YT, Liu FJ (2009) Tolerance, accumulation and distribution of zinc and cadmium in hyperaccumulator *Potentilla griffithii*. Environ Exp Bot 66:317–325

Jozefczak M, Remans T, Vangronsveld J, Cuypers A (2012) Glutathione is a key player in metal-induced oxidative stress defenses. Int J Mol Sci 13:3145–3175

Kim S, Takahashi M, Higuchi K, Tsunoda K, Nakanishi H, Yoshimura E, Mori S, Nishizawa NK (2005) Increased nicotianamine biosynthesis confers enhanced tolerance of high levels of metals, in particular nickel, to plants. Plant Cell Physiol 46:1809–1818

Klumpp A, Ansel W, Klumpp G, Breuer J, Vergne P, Sanz MJ, Rasmussen S, Ro-Poulsen H, Ribas Artola A, Peñuelas J, He S, Garrec JP, Calatayud V (2009) Airborne trace element pollution in 11 European cities assessed by exposure of standardised ryegrass cultures. Atmos Environ 43:329–339

Krämer U (2010) Metal hyperaccumulation in plants. Annu Rev Plant Biol 61:517–534

Krämer U, Cotter-Howells JD, Charnock JM, Baker AJM, Smith JAC (1996) Free histidine as a metal chelator in plants that hyperaccumulate nickel. Nature 379:635–638
Kumar A, Vajpayee P, Ali MB, Tripathi RD, Singh N, Rai UN, Singh SN (2002) Biochemical responses of *Cassia siamea* Lamk. grown on coal combustion residue (Fly-ash). Bull Environ Contam Toxicol 68:675–683
Kummer U, Pacyna J, Pacyna E, Friedrich R (2009) Assessment of heavy metal releases from the use phase of road transport in Europe. Atmos Environ 43:640–647
Li Y, Chen YY, Yang SG, Tian WM (2015) Cloning and characterization of HbMT2a, a metallothionein gene from *Hevea brasiliensis* Muell. Arg differently responds to abiotic stress and heavy metals. Biochem Biophys Res Commun 461:95–101
Liu MQ, Yanai J, Jiang RF, Zhang F, McGrath SP, Zhao FJ (2008) Does cadmium play a physiological role in the hyperaccumulator *Thlaspi caerulescens*? Chemosphere 71:1276–1283
Madejón P, Murillo JM, Marañón T, Cabrera F, López R (2002) Bioaccumulation of As, Cd, Cu, Fe and Pb in wild grasses affected by the Aznalcóllar mine spill (SW Spain). Sci Total Environ 290:105–120
Manara A (2012) Plant responses to heavy metal toxicity in plants and heavy metals. In: Furini A (ed) SpringerBriefs molecular science, Springer, Dordrecht, Netherlands, 2012 pp. 27–53
Mleczek M, Rissmann I, Rutkowski P, Kaczmarek Z, Golinski P (2009) Accumulation of selected heavy metals by different genotypes of *Salix*. Environ Exp Bot 66:289–296
Moradi AB, Conesa HM, Robinson BH, Lehmann E, Kaestner A, Schulin R (2009) Root responses to soil Ni heterogeneity in a hyperaccumulator and a non-accumulator species. Environ Pollut 157:2189–2196
Mukherjee A, Agrawal M (2016) Pollution response score of tree species in relation to ambient air quality in an urban area. Bull Environ Contam Toxicol 96:197–202
Norouzi S, Khademi H, Cano AF, Acosta JA (2015) Using plane tree leaves for biomonitoring of dust borne heavy metals: a case study from Isfahan, central Iran. Ecol Indic 57:64–73
Nriagu J, Pacyna J (1988) Quantitative assessment of worldwide contamination of air, water and soils by trace metals. Nature 333:134–139
Onder S, Dursun S, Gezgin S, Demirbas A (2007) Determination of heavy metal pollution in grass and soil of city centre green areas (Konya, Turkey). Pol J Environ Stud 16:145–154
Pacheco AM, Barros LIC, Freitas MC, Reis MA, Hipólito C, Oliveira OR (2002) An evaluation of olive-tree barks for the biological monitoring of airborne trace-elements at ground level. Environ Pollut 120:79–86
Parraga-Aguado I, Querejeta J-I, González-Alcaraz M-N et al (2014) Usefulness of pioneer vegetation for the phytomanagement of metal(loid)s enriched tailings: grasses vs. shrubs vs. trees. J Environ Manage 133:51–58
Pianelli K, Mari S, Marques L, Lebrun M, Czernic P (2005) Nicotianamine over-accumulation confers resistance to nickel in *Arabidopsis thaliana*. Transgen Res 14:739–748
Pierre JL, Gautier-Luneau I (2000) Iron and citric acid: a fuzzy chemistry of ubiquitous biological relevance. Biometals 13:91–96
Polechonska M, Zawadzki K, Samecka-Cymerman A, Kolon K, Klink A, Krawczyk J, Kempers AJ (2013) Evaluation of the bioindicator suitability of *Polygonum aviculare* in urban area. Ecol Ind 24:552–556
Prasad MNV (2004) Metallothioneins, metal binding complexes and metal sequestration in plants. In: Prasad MNV (ed) Heavy metal stress in plants: from biomolecules to ecosystems, 2nd edn. Springer, Berlin, pp 47–83
Redondo-Gomez S, Mateos-Naranjo E, Vecino-Bueno I, Feldman SR (2011) Accumulation and tolerance characteristics of chromium in a cordgrass Cr-hyperaccumulator, *Spartina argentinensis*. J Haz Mater 185:862–869
Ribeiro de Souza SC, Adrián López de Andrade S, Anjos de Souza L, Schiavinato MA (2012) Lead tolerance and phytoremediation potential of Brazilian leguminous tree species at the seedling stage. J Environ Manage 15:299–307

Santos-Jallath J, Castro-Rodr ı A, Huezo-Casillas J, TorresBustillos L (2012) Arsenic and heavy metals in native plants at tailings impoundments in Queretaro, Mexico. Phys Chem Earth 37–39:10–17

Sawidis T, Breuste J, Mitrovic M, Pavlovic P, Tsigaridas K (2011) Trees as bioindicator of heavy metal pollution in three European cities. Environ Pollut 159(12):3560–3570

Sharmila P, Saradhi PP (2002) Proline accumulation in heavy metal stressed plants: an adaptive strategy. In: Prasad MNV, Strzałka K (eds) Physiology and biochemistry of metal toxicity and tolerance in plants. Kulwer Academic, Netherlands, pp 179–199

Shi G, Chen Z, Teng J, Bi C, Zhou D, Sun C, Li Y, Xu S (2012) Fluxes, variability and sources of cadmium, lead, arsenic and mercury in dry atmospheric depositions in urban, suburban and rural areas. Environ Res 113:28–32

Sun RL, Zhou QX, Sun FH, Jin CX (2007) Antioxidative defense and proline/phytochelatin accumulation in a newly discovered Cd-hyperaccumulator, *Solanum nigrum* L. Environ Exp Bot 60:468–476

Sun YB, Zhou QX, Wang L, Liu WT (2009) Cadmium tolerance and accumulation characteristics of *Bidens pilosa* L. as a potential Cd-hyperaccumulator. J Hazard Mater 161:808–814

Thapa G, Sadhukhan A, Panda SK, Sahoo L (2012) Molecular mechanistic model of plant heavy metal tolerance. Biometals 25:489–505

Tomasevic M, Rajsic S, Dordevic D, Tasic M, Krstic J, Novakovic V (2004) Heavy metals accumulation in tree leaves from urban areas. Environ Chem Lett 2:151–154

Tomašević M, Antanasijević D, Aničić M, Deljanin I, Perić-Grujić A et al (2013) Lead concentrations and isotope ratios in urban tree leaves. Ecol Indic 24:504–509

Tomaševića M, Aničić M, Jovanović L, Perić-Grujić A, Ristić M (2011) Deciduous tree leaves in trace elements biomonitoring: a contribution to methodology. Ecol Indic 11:1689–1695

Ugolini F, Tognetti R, Raschi A, Bacci L (2013) *Quercus ilex* L. as bioaccumulator for heavy metals in urban areas: effectiveness of leaf washing with distilled water and considerations on the trees distance from traffic. Urban For Urban Green 12:576–558

Verbruggen N, Hermans C, Schat H (2009) Molecular mechanisms of metal hyperaccumulation in plants. New Phytol 181:759–776

Wang H, Shan X-Q, Wen B, Zhang S, Wang Z-J (2004) Responses of antioxidative enzymes to accumulation of copper in a copper hyperaccumulator of *Commelina communis*. Arch Environ Con Tox 47:185–192

Wei C, Ge Z, Chu W, Feng R (2015) Speciation of antimony and arsenic in the soils and plants in an old antimony mine. Environ Exp Bot 109:31–39

Yadav S (2010) Heavy metals toxicity in plants: an overview on the role of glutathione and phytochelatins in heavy metal stress tolerance of plants. S Afr J Bot 76:167–179

Yu KF, Kamber BS, Lawrence MG, Greig A, Zhao JX (2007) High-precision analysis on annual variations of heavy metals, lead isotopes and rare earth elements in mangrove tree rings by inductively coupled plasma mass spectrometry. Nucl Instrum Methods Phys Res B Beam Interact Mater Atoms 255:399–408

Zagorchev L, Seal CE, Kranner I, Odjakova M (2013) A central role for thiols in plant tolerance to abiotic stress. Int J Mol Sci 14:7405–7432

Zeng XW, Qiu RL, Ying RR, Tang YT, Tang L, Fang XH (2011) The differentially-expressed proteome in Zn/Cd hyperaccumulator *Arabis paniculata* Franch. in response to Zn and Cd. Chemosphere 82:321–328

Zhang XH, Liu J, Huang HT, Chen J, Zhu YN, Wang DQ (2007) Chromium accumulation by the hyperaccumulator plant *Leersia hexandra* Swartz. Chemosphere 67:1138–1143

Microbial-Mediated Management of Organic Xenobiotic Pollutants in Agricultural Lands

9

Anjali Singh, Shivani Chaudhary, Bhawna Dubey, and Vishal Prasad

Abstract

Contamination of agricultural soil by organic xenobiotic compounds is becoming a serious problem in most of the developed and developing countries. Chemicals foreign to an organism or chemicals not natural to an ecosystem are considered as xenobiotic for an organism, and upon exposure they impose toxicity threats to the organism. The term is mainly used in the context of pollutants such as chemical fertilizers, pesticides, dyes, dioxins, polychlorinated biphenyls (PCBs), and polycyclic aromatic hydrocarbons (PAHs) and their side effect on the biota. High concentrations of these xenobiotics create a biological imbalance in soil leading to surface and groundwater pollution. Inside plants they block the functional groups of biologically important molecules like enzymes, transport system of nutrient ions, polynucleotides, etc. The degradation of these organic xenobiotic pollutants in nature is a serious challenge and microorganisms have been observed to play a vital role in their degradation. They transform hazardous organic xenobiotic compound into harmless or less hazardous form, generally carbon dioxide, water, methane, and nitrogen. The different groups of microbes produce different types of enzymes and organic acids that act on recalcitrant compounds and degrade them to simpler forms. As a consequence of biodegradation of xenobiotic compounds, microorganisms are helpful to overcome environmental pollution and considered as eco-friendly. This chapter tries to elaborate some of the mechanisms employed by the microorganisms to carry out the xenobiotic degradation and remediation process along with different genera of microbes involved in the process.

A. Singh • S. Chaudhary • V. Prasad (✉)
Institute of Environment and Sustainable Development, Banaras Hindu University, Varanasi 221005, Uttar Pradesh, India
e-mail: ways2good@gmail.com

B. Dubey
Amity Institute of Environmental Sciences, Amity University, Gautam Buddha Nagar, Noida 201303, Uttar Pradesh, India

A. Singh et al. (eds.), *Plant Responses to Xenobiotics*,
DOI 10.1007/978-981-10-2860-1_9

Keywords
Biodegradation • Xenobiotic • Microorganisms • Polychlorinated biphenyls • Polycyclic aromatic hydrocarbons

9.1 Introduction

Agriculture plays a transcendent role in the world's economy both in developed and developing countries. In developing countries like India, agriculture is a primary contributor to gross domestic product (GDP). In most of the Asian, African, and South American countries, the populations are heavily dependent on agriculture. Besides this, over two-thirds of the world populations living below poverty line are directly or indirectly dependent on agriculture (Baloch and Haseeb 1996). With an accelerating growth in population, industrialization, and urbanization, the agricultural lands are decreasing accordingly. This unavoided condition compels human to produce more and more food grains. Therefore humans have developed several improved agronomic practices and agronomic chemicals such as fertilizers and pesticides of chemical origin (Iovdijova and Bencko 2010; Azaizeh et al. 2011; Agrawal and Shahi 2015). The use of these xenobiotic chemicals in the form of fertilizers and pesticides has beneficial effects, on the one hand, and on the other hand, they impose serious environmental problems.

A xenobiotic is a broad term alluding to any chemical which is foreign to an organism, or a chemical which is not a natural component of the ecosystem or organisms exposed to it. Xenobiotic broadly refers to the synthetic chemical compounds which are unnatural and come through foreign sources in our environment such as pesticides, chemical fertilizers, refrigerants, solvents, and other organic and inorganic compounds (Baloch and Haseeb 1996; Iovdijova and Bencko 2010; Park et al. 2014; Agrawal and Shahi 2015). Xenobiotic compounds are chemicals made by humans that are present in the environment at unnaturally high concentrations. An organic chemical compound that is normal to one organism may act as a xenobiotic to another. The substances which are present in very high concentration than usual also come under xenobiotic category. However, the term is mainly used in the context of pollutants such as chemical fertilizers, pesticides, dyes, dioxins, polychlorinated biphenyls (PCBs), and polyaromatic hydrocarbons and their side effects on the biota (Park et al. 2014; Agrawal and Shahi 2015; Chandra and Singh 2015).

Although both organic and inorganic chemical compounds come under xenobiotics, it has been supposed that organic xenobiotics are mainly responsible for chemical-induced toxicity in agricultural land as well as in human population. According to Park et al. (2014), approximately 90 % of group I agents (classified by International Agency for Research on Cancer) are derived from organic xenobiotics, which are carcinogenic in nature. Thus xenobiotic compound can also be referred to organic chemical compounds, which are strange and unnatural and mimic natural biochemicals which are essential for life. Xenobiotic compounds may not be recognized by biochemical processes in plants and microorganisms and are thus resistant to degradation (Iovdijova and Bencko 2010).

9.2 Xenobiotics: The Stubborn Compounds

There are two types of xenobiotic compounds; they are either biodegradable or nondegradable that is recalcitrant in nature. Biodegradable xenobiotic compounds are those that are broken down to harmless forms by the activity of microbes or other reactions, while recalcitrant compounds are resistant to any degradation process. The recalcitrant xenobiotic compounds can be classified into various groups like alkyl benzyl sulfonates, halocarbons, oil mixtures, polychlorinated biphenyls, synthetic polymers, polychlorinated hydrocarbons, pesticides, etc., on the basis of their physicochemical characteristics which decide its stability in soil environment, its biological availability, and how it will be revamped from the point of application to the environment of soil and the organisms it contains (Chandra and Singh 2015). The probable health hazards of a xenobiotic compound are attributed to its tenacity in the environment as well as its lethality. They tend to concentrate in the environment which leads their bioaccumulation and biomagnification. Agricultural soil contamination by xenobiotic organic compounds is a severe complication in most industrialized countries, leading to an intense contamination of soil worldwide. The contamination of soil by xenobiotic organic compounds such as chlordane, dioxins, aldrin, lindane, dichlorodiphenyltrichloroethane (DDT), parathion, disulfoton, paraquat, polychlorinated biphenyls (PCBs), polycyclic aromatic hydrocarbons (PAHs), etc., creates a serious problem in most of the developed countries (Solouki et al. 2011). For example, PAHs released throughout the world as industrial by-products of coal processing, asphalt production, fossil fuel combustion, and wood preservation into the environment can efficiently accumulate in rice straw used as cattle feed, which can finally be conveyed to human beings through the food chain. The anaerobic and reducing soil environment at the time of rice growth season in paddy soils is similar to that of a wetland system, thus admitting PAHs to be efficiently dispersed in soil-plant-water systems (Su and Zhu 2007; Azaizeh et al. 2011). The same happens to various other persistent xenobiotic pollutants. There are several human activities which add xenobiotic pollutant deposition in soils including industrial activities and combustion of fuel, and the most important contribution is the use of pesticides and chemical fertilizers for high-production agricultural lands (Alkorta and Garbisu 2001; Azaizeh et al. 2011). Pesticides including insecticides, herbicides, fungicides, bactericides, algaecides, etc., are chemical compounds used for protection of crops and to control pest, but they are apparently the most extensively dispersed contaminants in our environment after the green revolution, because millions of tons of pesticides are generated and spread out each year globally (Schwitzguébel et al. 2011). The physicochemical characteristics are one of the main reasons of the persistence of these organic xenobiotic pollutants in the soil environment as well as in the biosphere. The xenobiotics possess several physical and chemical properties like absorption, adsorption, reactivity, volatility, water solubility, degradation by microbes, etc., which decide their persistence in the environment (Varsha et al. 2012; Ferradji et al. 2014). Copious persistent organic pollutants (POPs) such as aldrin, DDT, dieldrin, dinoseb salts, HCH, heptachlor, chlordane, mercury, arsenic compounds, etc., are perceived as a possible health hazard due to their inborn chemical stability, potential and intense toxicity, recalcitrance, and carcinogenicity. The transformed products also pose growing concerns because they may be present in soil at higher levels than the parent pesticide itself (Baloch and Haseeb 1996; Patel et al. 2012; Agrawal and Shahi 2015). In some cases these products are more toxic and more mobile, representing a greater risk to the environment than parent molecules (Schwitzguébel et al. 2011). Microorganisms are generally deliberated to be the perfect signal of soil contamination due to their affectionate association with the environment of soil and by occupying large surface area. Commonly, microorganisms are very susceptible to lower concentrations of pollutants and expeditiously react to soil disruption. The soil-disturbing condition may result in the altered activity and diversity of microorganisms which lead to a reduction in soil quality (Schröter-Kermani et al. 2006; Chandra and Singh 2015).

Microorganism mediates conversion of pollutants into nonhazardous or less hazardous forms through a process termed as biodegradation (Agrawal and Shahi 2015). Microorganisms are considered as nature's recyclers, transforming hazardous organic xenobiotic compounds to harmless compounds, generally carbon dioxide, water, methane, and nitrogen compounds (Jain et al. 2005; Azaizeh et al. 2011). Numerous works done by researchers reported the suitable use of various organisms like bacteria, mycorrhizal fungi, algae, and actinomycetes for effective biodegradation of pollutants (Vidali 2001; Leung 2004; Agarwal et al. 2009; Agrawal and Shahi 2015). Most organisms, particularly bacteria, are known for their detoxifying capabilities. They mineralize, convert, or debilitate the contaminants (Tropel and Van Der Meer 2004).

9.2.1 Major Xenobiotics and Their Sources in Agricultural Land

Major organic xenobiotic compounds causing environmental pollution and persistent in the environment include PCBs (polychlorinated biphenyls), PAHs (polycyclic aromatic hydrocarbons), trinitrotoluene, trichloroethylene, nitroglycerine, etc. (Eapen et al. 2007; Agarwal et al. 2009; Fatta-Kassinos et al. 2011). Some of the representative xenobiotic compounds found commonly in soil are depicted in Fig. 9.1.

Aromatic compounds enter into the soil environment through numerous sources which include the use of detergents, pesticides, drugs, dyes, etc., posing dangerous threats to living organisms. Release of several PAHs through industrial processes seems to be one of the major carcinogenic agents persisting in the soil (Visioli 2015; Hernández-Castellanos et al. 2013). Different species of microbes including bacteria, fungi, yeast, and algae can act on these xenobiotic compounds and degrade them. Compounds which are degraded by bacteria include benzene, toluene, xylene, ethylbenzene, etc. Benzene is one of the persistent organic xenobiotics with the presence of a thermodynamically stable ring in its structure (Dıaz and Prieto 2000; Chandra and Singh 2015). Groups like methyl, chloro, nitro, amino, and sulfonyl in benzene ring cause recalcitrancy of these compounds. Phenol is one of the widespread chemical pollutants which are persistent in the environment due to its

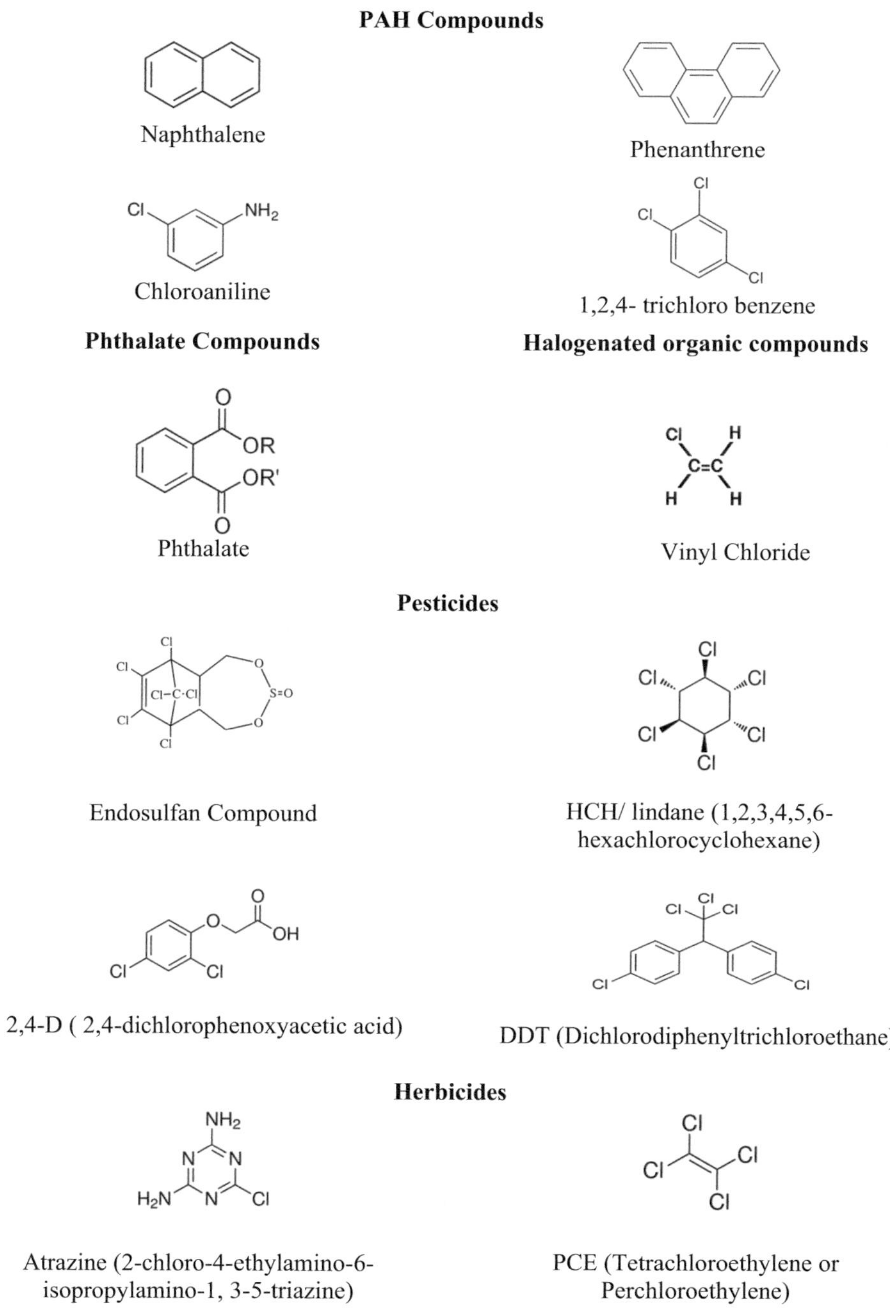

Fig. 9.1 Some of the common xenobiotic compounds found in the soil of agricultural lands

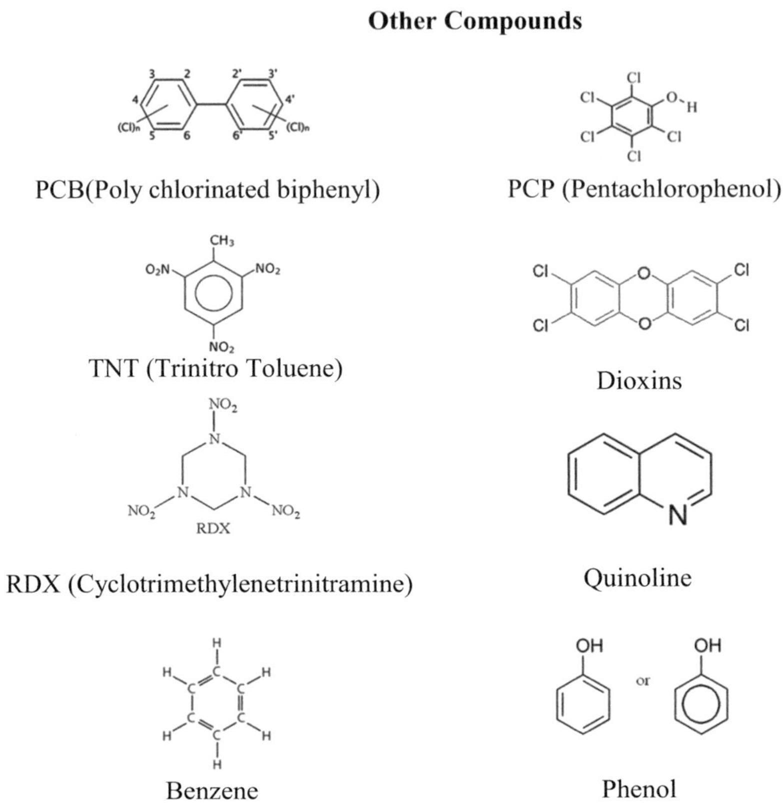

Fig. 9.1 (continued)

toxicity even at low concentration. It enters the environment from the chemical and pharma industrial effluents. It affects the environment in several ways including ozone depletion, reducing visibility and adding acidic air to the atmosphere. Hence phenol degradation is necessary to reduce all these effects. The contamination of the agricultural soil with PAHs creates a serious threat of introduction of these xenobiotics into the human food chain. These PAHs get entry into the humans mainly by inhalation of particulates carrying PAHs, alimentary consumption of contaminated food products, and direct association with polluted soils (Fatta-Kassinos et al. 2011; Hernández-Castellanos et al. 2013; Visioli 2015; Agarwal et al. 2009).

Other persistent organic xenobiotics such as PCBs are the compounds which are chiefly used in transformer oil, dielectric fluid in condensers, heat exchange medium, hydraulic fluid, plasticizers, etc. PCBs have been revealed in most ecosystems and particularly in soils and sediments of industrial areas. Such pervasive presence is elucidated by their high chemical stability and low water solubility, which contribute to their persistence in the environment. They enter the food chain and cause several harmful effects on different organisms. A number of plants have been shown to acquire PCBs, such as *Cucurbita pepo* grown in PCB-contaminated soil which has been shown to bioaccumulate Aroclors 1254 and 1260 through root uptake and translocation (Iovdijova and Bencko 2010; Åslund et al. 2008). In spite of being banned, today also the PCBs are released into the environment. These PCBs are added to the environment from various waste sites and from offensive dumping of PCB wastes, such as old transformer fluids, and disposal of PCB-containing products into landfills, which are not designed to hold harmful wastes (Fatta-Kassinos et al. 2011; Agrawal and Shahi 2015).

9.2.2 Consequences of Xenobiotics on Agricultural Land

Among all of the organic xenobiotic compounds, pesticides had a unique position and have been purposely supplemented to the environment for controlling the pest and weeds to improve agricultural production (Baloch and Haseeb 1996). Among the chemicals used for the above purpose, most of them are not selective but are commonly harmful to many nontarget organisms habitating the environment including plants, animals, and microorganisms, in addition to human. Xenobiotic pesticides such as chlordane, DDT, aldrin, etc., when sprayed or applied to the crop fields get mixed with the soil components whether living or nonliving (Fig. 9.2). The pesticides which are not bound to the soil components get run off over the soil or leached through the soil, resulting in contamination of groundwater, rivers, and reservoirs, or they will be vaporized causing the contamination of the atmosphere (Fig. 9.2). The phenomenon of progressive increase in the concentration of a xenobiotic compound, when the substance is passed through the food chain from one trophic level to another, is called as biomagnification or bioaccumulation. For example, the insecticide DDT is absorbed constantly by plants and microorganism. When these plants and microbes are eaten by birds and fishes, this pesticide enters in the food chain and, because of its recalcitrant nature, gets accumulated and transferred from one trophic level to another. Thus, DDT may discover its access into assorted animals, in addition to human (Solouki et al. 2011; Agrawal and Shahi 2015). Instead of organochlorine pesticides, some of the other groups of pesticides such as organophosphorous, carbamates, etc., are also reported for their harmful effects on biota. Some herbicides, used for controlling weed, are becoming more common than the other pesticides. They are extremely toxic for plant and animals, although most of the herbicides are highly selective and show moderate toxicity to mammals. One of the most common herbicide paraquats (bipyridium compound) is also creating a major threat to the environment. Paraquat compound is water soluble but nonselective herbicide. After application, it takes electrons from the photosynthetic process of plants and also produces hydrogen peroxide, superoxide radical, and other free radicals which are highly toxic to energy relocation and other necessary cellular processes and incursion to cell membrane (Baloch and Haseeb 1996; Iovdijova and Bencko 2010; Chandra and Singh 2015).

Pesticide application is one of the most important sources of pollution in agricultural land. The soil contaminated with pesticides and other POP applications results into their direct uptake by plants and movement from roots to shoots. The pesticides

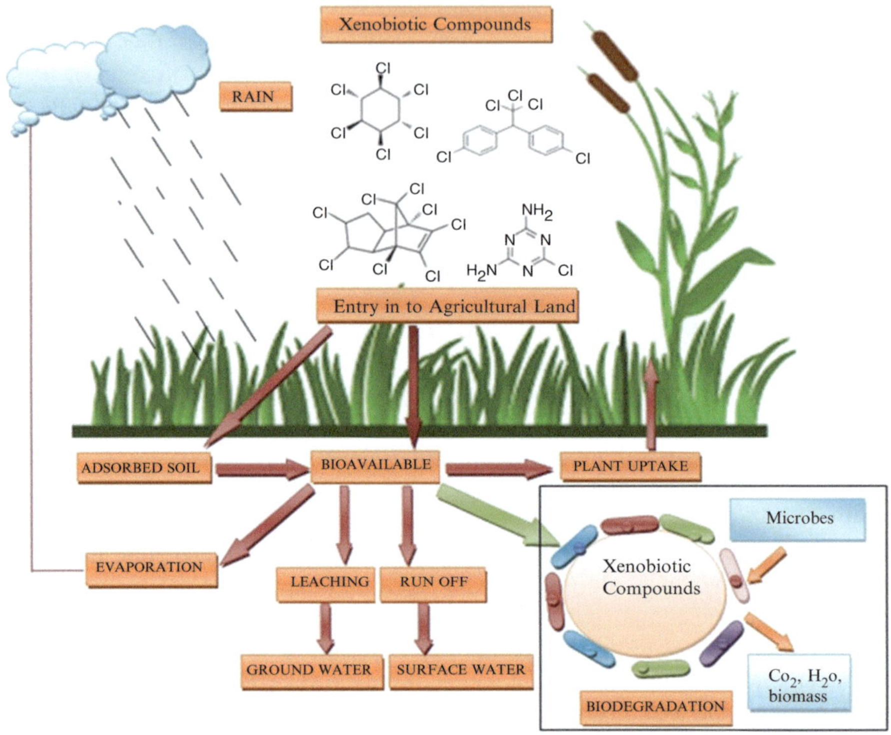

Fig. 9.2 Schematic diagram showing different pathways of xenobiotic compounds spread over an agricultural land

move from the roots by translocation and get accumulated in the leaves and fruits. These crops when consumed by cattle or humans or other wildlife cause a massive increase in the concentrations of these chemicals through biomagnifications (Langenbach 2013; Chandra and Singh 2015). Large amounts of fertilizers are continuously added to soils in intensive farming systems to fulfill satisfactory nitrogen, phosphorus, and potassium for the growth of plants. The compounds applied to provide these elements contain trace amounts of heavy metals as contaminants which may considerably increase their concentration in the soil after continued fertilizer application. The concentration of the metals in agricultural soils may be significantly increased also by the use of pesticides, use of sewage sludge, addition of micronutrient, aerial emission, and farm slurries (Azaizeh et al. 2011).

The high concentration of trace heavy metals affects soil microorganism and microbial arbitrated soil processes such as litter decomposition, enzyme activities, nitrogen mineralization, and biogeochemical cycle of carbon, nitrogen, etc. Soil respiration, a common signal of soil microbial activity, is also very susceptible to high concentration of heavy metals like cadmium, arsenic, and mercury (Arora et al. 2009; Chien et al. 2014). On the basis of fertilization effect, the changes in a complex of soil microorganisms were classified into four types. The first one is zone of homeostasis under which the fertilizer concentration range is in between 0 and 100 Kg ha^{-1} which does not change the complex of soil microorganisms. The second type is zone of stress and the concentration range is 100–10,000 Kg ha^{-1} which provides considerable changes in a complex of soil microbes. These changes occur on those microbes which are involved in some soil processes. The third one is zone of resistance and its concentration range is above 10,000 Kg ha^{-1} which contributes to considerable qualitative changes in the composition of active soil microbes. The last one is zone of repression and the concentration range is much more than 10,000 Kg ha^{-1} in which microorganism's growth and development are strongly affected (Zvyagintsev 1990; Agrawal and Shahi 2015).

High doses of nitrogenous fertilizers create biological imbalance in soil which expressed a sharp increase in the amount of denitrifying and humus-decomposing bacteria by 10–11 times and cellulose-decomposing bacteria by 1.5 times and decrease in free-living nitrogen-fixing bacteria by more than 8 times (Jain et al. 2005). It was also found that influence of fertilization up to 960 Kg ha^{-1} of N, K_2O, and P_2O_5, disturbed urease, protease, and catalase activity of soil (Ferradji et al. 2014).

In plants through root uptake, these xenobiotics go to cellular level. The toxic metals block the functional groups of biologically important molecules, like enzymes; transport system of nutrient ions, polynucleotides; etc. Displacing the essential metal ions inactivates the essential enzymes, disrupting the integrity of cell and organelle membrane (Ross 1994; Baloch and Haseeb 1996; Agrawal and Shahi 2015).

9.3 Role of Microbes in Biodegradation of Xenobiotics

Half of the biomass of our planet is represented by microorganisms especially bacteria, and these bacteria are very helpful in degradation of xenobiotic organic pollutants (Curtis and Reinhard 1994). The environment is continuously disturbed by human activity through introduction of several xenobiotic chemicals in the biosphere. Increased industrialization in association with expeditious urbanization and modern agricultural practices has accelerated the levels of pollutants in the agricultural environment, with an ensuing harmful impact on human health. These xenobiotic pollutants not only harm crop plants and degrade fertility of agricultural soil, but they also harm soil microorganisms which reside in this soil. Cleaning up of the toxic organic xenobiotic compounds from agricultural environment is a compelling problem, which warrants versatile accesses for attaining suitable solutions. Microorganisms have potential to degrade xenobiotic pollutants by their unique metabolic pathways because they exploit toxic compounds as their new carbon source, and by this way, they detoxify toxic xenobiotic compounds (Copley 2000). Due to the biodegradation of xenobiotic compounds, microorganisms are helpful to overcome environmental pollution and considered as eco-friendly (Agrawal and Shahi 2015). Due to its low cost, in comparison to physical and chemical methods, biodegradation is considered as a viable solution for the degradation of organic xenobiotic pollutants (Baloch and Haseeb 1996; Eapen et al. 2007; Park et al. 2014; Agrawal and Shahi 2015). Biodegradation or biological degradation is defined as a phenomenon of biological transformation of organic compounds by living organisms, chiefly the microorganisms. It mainly involves the conversion of complex organic molecules to simpler and mostly nontoxic forms. The use of microorganisms including bacteria, fungi, algae, etc., has been given much attention for cleaning up of xenobiotic pollutant in the past few years (Eapen et al. 2007; Chandra and Singh 2015). The microorganisms act like scavengers during the bioremediation process. The whole bioremediation process can be divided into three categories, i.e., mineralization, biotransformation, and co-metabolism. By passing through these three processes, the complex organic xenobiotic compounds are broken down into simpler inorganic compounds. The typical products of biodegradation are carbon dioxide, water, and ammonia. Sometimes a group of microorganisms known as consortium can together also be used to degrade organic xenobiotic compounds wholly or partially. The different groups of microbes produce different types of enzymes and organic acids that act on recalcitrant compounds and degrade them to simpler forms. These smaller-size compounds are again taken up by other microbes and degraded completely (Agrawal and Shahi 2015; Monica et al. 2012). Microbes have also been found useful for the biodegradation of some of the toxic organic pollutants which are highly recalcitrant in nature (Tripathi 2012).

There are chiefly two methods by which microbes degrade xenobiotic compounds: one is aerobic biodegradation and the other is anaerobic biodegradation. The bacteria carrying out aerobic biodegradation of xenobiotic compounds belong to the genera *Pseudomonas*, *Bacillus*, *Micrococcus*, *Gordonia*, *Pandoraea*, *Moraxella*, *Escherichia*, *Rhodococcus*, and *Sphingobium*, while the bacteria which bring about anaerobic biodegradation are from genera *Desulfovibrio*, *Desulfotomaculum*, *Methanospirillum*, *Pelotomaculum*, *Methanosaeta*, *Syntrophobacter*, etc. (Chowdhury et al. 2008; Varsha et al. 2012). *Pseudomonas* species have been most widely studied among all the above microbes because of their dominant nature in debasing divergent types of xenobiotic compounds including polycyclic aromatic compounds to a wide range of fertilizer and pesticides (Cao et al. 2009). *Pseudomonas fluorescence* SM1 strain showed a very good potential of remediation of some heavy metals and phenolics from heavily polluted sites (Wasi et al. 2013).

9.3.1 Microbial-Associated Degradation Pathway(s) for Xenobiotic Compounds

Remediation with the help of microbes is one of the effective methods for removal of organic xenobiotic pollutants, as it is less expensive and can selectively achieve complete destruction of organic pollutants (Alexander 1977; Cookson Jr. 1995). Generally, the biodegradation process depends upon the pollutant oxidation state. The xenobiotic compound can be either acceptor of electron or its donor. In case of bacterial respiration, the very common electron acceptor is oxygen (Cao et al.

2009). Microbes degrade xenobiotic compound by two processes, i.e., aerobic biodegradation and anaerobic biodegradation. Carbon dioxide is the final product after biodegradation of xenobiotic compound in aerobic condition, while in the absence of oxygen, anaerobic condition occurs and methane is produced as the degradation product.

9.3.1.1 Aerobic Biodegradation Pathway

By aerobic biodegradation some of the xenobiotics like chlorinated aliphatic compounds, toluene, petroleum, benzene, phenol, pyrene, naphthalene, chloroanilines, fluorine, dichlorobenzene, pentachlorophenol, etc., are degraded. Bacteria have the potential to grow on these chemicals and by producing enzymes degrade toxic xenobiotic compounds into nontoxic compounds (Cao et al. 2009; Agrawal and Shahi 2015; Shimao 2001).

$$\text{Xenobiotic compound} + O_2 \xrightarrow{\text{Microbes and their enzymes}} CO_2 + H_2O + \text{biomass} + \text{residue}(s)$$

The aerobic biodegradation of organic xenobiotic is completed in three stages. Mineralization is the first stage of aerobic biodegradation in which biodegradable compounds are broken into inorganic compounds such as carbon dioxide, water, methane, and nitrogen. The mineralization process gets completed when all the biodegradable biomass is consumed and all the carbon is converted into carbon dioxide. Biotransformation is the next stage in which the organic chemicals only go through small structural alterations. Co-metabolism is the third stage in which the primary substrate induces the production of an enzyme which alters the molecular structure of another compound. During co-metabolism the organisms do not get benefitted; they only help in minor modification of compound. There are several microbial enzymes which play a significant role in biodegradation of organic xenobiotic pollutants under aerobic condition. The oxidoreductase is one of them. These enzymes cleave chemical bonds and shift the electrons from donor to acceptor. During this chemical reaction, harmful pollutants are converted to harmless compound. These enzymes mainly detoxify toxic xenobiotics such as anilinic and phenolic compounds into humic substances by polymerization or copolymerization (Park et al. 2014; Agrawal and Shahi 2015). The oxidation reaction has an enormous significance in enzymatic reaction of aerobic biodegradation and is governed by enzyme oxygenases. Oxygenase enzyme helps in the metabolization of organic pollutants by increasing their reactivity and water solubility and breaking the ring (Arora et al. 2009; Agrawal and Shahi 2015).

9.3.1.2 Anaerobic Biodegradation Pathway

Some organic xenobiotic pollutants are not deteriorated by the aerobic biodegradation process. The reason behind this is the recalcitrant nature of these compounds. Due to increasing halogenations as well as substitution of sulfur and nitro groups on the aromatic ring, the electrophilicity of the molecule is increased, which makes these compounds withstand the electrophilic intrusion by oxygenases under aerobic condition. Due to this reason, these compounds are highly recalcitrant and persistent in nature. Chlorinated dioxins, polychlorinated biphenyls (PCBs), and some pesticides like DDT, aldrin, chlordane, etc., are that kind of compounds which are not easily degraded and persist in the environment.

$$\text{Xenobiotic compounds} \xrightarrow[\text{enzymes}]{\text{Microbes and their}} CO_2 + CH_4 + H_2O + \text{biomass} + \text{residue(s)}$$

For achieving reduction of highly persistent halogenated xenobiotics from the environment, anaerobic bacteria play a very significant role. They perform reductive dehalogenation by anaerobic respiration and reduce the degree of chlorination and make the product more accessible for mineralization by bacteria under aerobic condition. Reductive dehalogenation is the initial process of degradation of organic pollutants during anaerobic degradation. Under this process organic substrate acts as electron donors. By accepting electrons organic xenobiotics allow the anaerobic bacteria to transfer electrons to these compounds (Liu et al. 2014a; Jayasekara et al. 2005). Some organic xenobiotic compounds are listed in Table 9.1 with the respective bacterial species responsible for their biodegradation and the sites of isolation of these bacterial isolates.

Bioremediation of chlorinated aromatic compounds is also reported by several researchers under anaerobic condition (Kazumi et al. 1995; Vargas et al. 2000; Ferradji et al. 2014). Microbes adopt various paths simultaneously for the degradation of chlorine atoms leading to the formation of phenol and finally convert it into carbon dioxide and methane (Agrawal and Shahi 2015).

During anaerobic condition a microbial enzyme dehalogenase plays a very significant role in the remediation of chlorinated pollutant (Copley 2000; Agrawal and Shahi 2015). A very good example for this process has been reported by Magnuson et al. (1998). They reported the partial purification of two reductive dehalogenase enzymes from the bacterium *Dehalococcoides ethenogenes* and both are membrane proteins. The first enzyme perchloroethylene-reductive dehalogenase shows the ability to reduce perchloroethylene (PCE) and convert PCE to trichloroethene (TCE), while the other enzyme TCE-reductive dehalogenase has the potential to reduce TCE and convert TCE to trans-dichlorothene, cis-dichloroethene, 1,1-dichloroethane, and vinyl chloride.

In case of organophosphate pesticide bioremediation, microbial enzyme phosphotriesterases (PTEs) have an important role. They hydrolyze and detoxify the pesticides which are organophosphoric in nature. These degrade the toxicity of organic phosphate by decreasing its ability to inactivate acetylcholine esterase (Singh and Walker 2006; Theriot and Grunden 2011).

A study carried out by Zeinali et al. (2007) reported about a thermophilic bacterium, *Nocardia otitidiscaviarum* TSH1, which has the capability of hydrocarbon degradation. They found this bacterial strain was able to grow on PAHs, phenol, and straight-chain aliphatic hydrocarbons and used these compounds as their sole source of carbon and energy. Due to having extremely lipophilic cell surfaces, this bacterial strain was better suited for direct uptake of highly hydrophobic hydrocarbons. This strain was also investigated for the production of biosurfactants. Another study by Zaidi et al. (2006) reported that the plant *Brassica juncea* growing in Ni-contaminated

Table 9.1 List of some organic xenobiotic compounds with their degrading bacterial species and their sites of isolation

Organic xenobiotic compounds	Bacteria carrying out its biodegradation	Site of isolation	References
Pesticides			
Endosulfan compounds	*Achromobacter xylosoxidans* strain C8B	Bacteria were isolated from soil through selective enrichment technique	Singh and Singh (2011)
	Stenotrophomonas maltophilia, *Rhodococcus erythropolis*	Mixed culture isolated from a pesticide-contaminated soil	Kumar et al. (2007)
	Klebsiella oxytoca KE-8	Isolated from endosulfan-polluted soil	Kwon et al. (2005)
HCH/lindane (1,2,3,4,5,6-hexachlorocyclohexane)	*Streptomyces* sp. M7	Bacterial strain was isolated from wastewater sediment samples from a copper filter plant	Cuozzo et al. (2009)
	Pseudomonas strains	Bacterial strain was isolated from agricultural soil, possessed hexachlorocyclohexane-degrading ability	Nawab et al. (2003)
DDT (dichlorodiphenyltrichloroethane)	*Pseudoxanthobacter liyangensis*	Strain isolated from DDT-contaminated soil	Liu et al. (2014b)
	Serratia marcescens DT-1P	Bacterial isolate was collected from DDT-contaminated soil	Bidlan and Manonmani (2002)
Diuron DCMU (3-(3,4-dichlorophenyl)-1,1-dimethylurea	*Streptomyces* sp.	Strain isolated from agricultural and nonagricultural soils	Castillo et al. (2006)
	Arthrobacter sp.	Bacterial strain was isolated from a soil	Widehem et al. (2002)

(continued)

Table 9.1 (continued)

Organic xenobiotic compounds	Bacteria carrying out its biodegradation	Site of isolation	References
Herbicides			
Atrazine (2-chloro-4-ethylamino-6-isopropylamino-1,3-5-triazine)	*Rhodococcus* sp.	Strain was collected from agricultural soil	Umar et al. (2012)
	Nocardioides sp. SP12	Strain was isolated from atrazine-treated maize rhizospheric soil	Piutti et al. (2003)
Other compounds			
PCB (polychlorinated biphenyl)	*Rhodococcus erythropolis* U23A	Bacterial strain was isolated from PCB-contaminated rhizospheric soil	Toussaint et al. (2012)
	Achromobacter sp.	Strain isolated from petroleum-contaminated soil, which is capable of degrading biphenyl	Hong et al. (2009)
	Enterobacter sp. LY402	Isolated from contaminated soil	Jia et al. (2008)
Dioxins	*Paenibacillus* sp. KBC101	Strain was isolated from soil	Sakai et al. (2005)
Quinoline	*Pseudomonas mendocina* strain NSYSU	Strain was isolated from dioxin-contaminated soil	Lin et al. (2014)
	Bacillus sp. Q2	Bacteria were isolated from petroleum-contaminated soil	Tuo et al. (2012)
TNT (Trinitrotoluene)	*Pseudomonas* sp.	Strain was isolated from TNT-contaminated soil	Chien et al. (2014)

Phenol	*Bacillus cereus*	Bacterial species was isolated from TNT-contaminated soil	Mercimek et al. (2013)
	Acinetobacter calcoaceticus P23	Bacteria were isolated from the rhizosphere of duckweed (*Lemna aoukikusa*)	Yamaga et al. (2010)
	Pseudomonas sp.	Bacteria were isolated from bioreactor sludge of petroleum-refining wastewater treatment	Dong et al. (2008)
	Alcaligenes faecalis	Strain was isolated from forest soil in the presence of phenol and a high salt concentration	Bastos et al. (2000)
PAH compounds			
Naphthalene	*Streptomyces* sp.	Isolated from soil surface	Ferradji et al. (2014)
Chloroaniline	*Novosphingobium naphhalenivorans* sp.	Bacteria isolated from PAH-contaminated soil	Suzuki and Hiraishi (2007)
1,2,4-trichlorobenzene (1,2,4-TCB)	*Acinetobacter baumannii* CA2, *Pseudomonas putida* CA16, and *Klebsiella* sp. CA17	Bacterial strains were isolated from agricultural soil	Vangnai and Petchkroh (2007)
	Bordetella sp.	Bacteria were isolated from soil which was polluted with chlorinated benzenes	Wang et al. (2007)

soil showed low toxicity indication when inoculated with bacterium *Bacillus subtilis* SJ-101, because this rhizobacterium has the capability to share the load of Ni metal by accumulation and also enhance the biomass formation of the plant.

9.4 Conclusions

The most pressing need of countries with their rapidly ever-increasing population is to produce more food, fiber, and industrial raw material, simultaneously maintaining productivity of natural resources. The present trend of increasing food productions is due to tremendous use of agrochemicals including different types of chemical fertilizers and variety of pesticides. In return, by using these agrochemicals, the farmers get one-time increased production and depletion of soil fertility for future generation. Xenobiotics reach the soil surface as a result of intended or unintended use; its higher concentration affects soil microbial number and microbial-mediated soil processes such as biogeochemical cycles, nitrogen mineralization, enzyme activities, litter decomposition, etc. The decomposition rate of organic matter is also reduced by high level of trace metals in soils (Baloch and Haseeb 1996; Jain et al. 2005; Azaizeh et al. 2011; Ferradji et al. 2014; Agrawal and Shahi 2015). Numerous sites with hazardous waste have been created worldwide as a result of accumulation of xenobiotics in cultivable soils over the years. The remedial strategies used to clean up sites contaminated by these xenobiotic compounds using various physical and chemical measures are not cost effective or adequate enough. This has speeded up the search for newer methods involving biological organisms or methods for degradation and removal of such pollutants. The pressing need for cleanup of sites contaminated with the xenobiotics has led to the discovery of a diverse range of microorganisms having capabilities to utilize these xenobiotics as substrates. These microbes have the property to either mineralize these xenobiotics or convert them into harmless products, and by doing so, they clean up the contaminated or polluted environment. Investigation and elucidation of the range of microbial diversity are vital for developing effective and environment friendly "green" technologies (Lin et al. 2014; Chandra and Singh 2015). To intersect and understand the interactions between a microbe and a xenobiotic, various cross-disciplinary studies will be required to benefit the environment. Such interdisciplinary approaches will be effective enough in unmasking the environmental processes and will lead to finding of diverse microbial populations for efficient and effective removal of xenobiotics from contaminated environment.

References

Agrawal N, Shahi SK (2015) An environmental cleanup strategy-microbial transformation of xenobiotic compounds. Int J Curr Microbiol Appl Sci 4(4):429–461

Agarwal T, Khillare P, Shridhar V, Ray S (2009) Pattern, sources and toxic potential of PAHs in the agricultural soils of Delhi, India. J Hazard Mater 163(2):1033–1039

Alexander M (1977) Introduction to soil microbiology, vol Ed. 2. John Wiley & Sons, New York

Alkorta I, Garbisu C (2001) Phytoremediation of organic contaminants in soils. Bioresour Technol 79(3):273–276

Arora PK, Kumar M, Chauhan A, Raghava GP, Jain RK (2009) OxDBase: a database of oxygenases involved in biodegradation. BMC Res Notes 2(1):67

Åslund MLW, Rutter A, Reimer KJ, Zeeb BA (2008) The effects of repeated planting, planting density, and specific transfer pathways on PCB uptake by *Cucurbita pepo* grown in field conditions. Sci Total Environ 405(1):14–25

Azaizeh H, Castro PM, Kidd P (2011) Biodegradation of organic Xenobiotic pollutants in the Rhizosphere. In: Org Xenob plants. Springer, Dordrecht, pp 191–215

Baloch UK, Haseeb M (1996) Xenobiotics in the third world agricultural environment. Environmental Xenobiotics. TJ Press, London, pp 64–81

Bastos AER, Moon DH, Rossi A, Trevors JT, Tsai SM (2000) Salt-tolerant phenol-degrading microorganisms isolated from Amazonian soil samples. Arch Microbiol 174(5):346–352

Bidlan R, Manonmani H (2002) Aerobic degradation of dichlorodiphenyltrichloroethane (DDT) by *Serratia marcescens* DT-1P. Process Biochem 38(1):49–56

Cao B, Nagarajan K, Loh K-C (2009) Biodegradation of aromatic compounds: current status and opportunities for biomolecular approaches. Appl Microbiol Biotechnol 85(2):207–228

Castillo M, Felis N, Aragon P, Cuesta G, Sabater C (2006) Biodegradation of the herbicide diuron by streptomycetes isolated from soil. Int Biodeterior Biodegrad 58(3):196–202

Chandra S, Singh N (2015) Enhanced bioremediation techniques for agricultural soils. Int J Curr Res Acad Rev 3(7):166–173

Chien CC, Kao CM, Chen DY, Chen SC, Chen CC (2014) Biotransformation of trinitrotoluene (TNT) by *Pseudomonas* spp. isolated from a TNT-contaminated environment. Environ Toxicol Chem 33(5):1059–1063

Chowdhury S, Mishra M, Adarsh V, Mukherjee A, Thakur AR, Chaudhuri SR (2008) Novel metal accumulator and protease secretor microbes from east Calcutta Wetland. Am J Biochem Biotechnol 4(3):255–264

Cookson JT Jr (1995) Bioremediation engineering: design and application. McGraw-Hill, Inc, New York

Copley SD (2000) Evolution of a metabolic pathway for degradation of a toxic xenobiotic: the patchwork approach. Trends Biochem Sci 25(6):261–265

Cuozzo SA, Rollán GG, Abate CM, Amoroso MJ (2009) Specific dechlorinase activity in lindane degradation by *Streptomyces* sp. M7. World J Microbiol Biotechnol 25(9):1539–1546

Curtis GP, Reinhard M (1994) Reductive dehalogenation of hexachloroethane, carbon tetrachloride, and bromoform by anthrahydroquinone disulfonate and humic acid. Environ Sci Technol 28(13):2393–2401

Dıaz E, Prieto MA (2000) Bacterial promoters triggering biodegradation of aromatic pollutants. Curr Opin Biotechnol 11(5):467–475

Dong X, Hong Q, He L, Jiang X, Li S (2008) Characterization of phenol-degrading bacterial strains isolated from natural soil. Int Biodeter Biodegr 62(3):257–262

Eapen S, Singh S, D'souza S (2007) Advances in development of transgenic plants for remediation of xenobiotic pollutants. Biotechnol Adv 25(5):442–451

Fatta-Kassinos D, Kalavrouziotis I, Koukoulakis P, Vasquez M (2011) The risks associated with wastewater reuse and xenobiotics in the agroecological environment. Sci Total Environ 409(19):3555–3563

Ferradji FZ, Mnif S, Badis A, Rebbani S, Fodil D, Eddouaouda K, Sayadi S (2014) Naphthalene and crude oil degradation by biosurfactant producing *Streptomyces* spp. isolated from Mitidja plain soil (North of Algeria). Int Biodeterior Biodegrad 86:300–308

Hernández-Castellanos B, Ortíz-Ceballos A, Martínez-Hernández S, Noa-Carrazana J, Luna-Guido M, Dendooven L, Contreras-Ramos S (2013) Removal of benzo (a) pyrene from soil using an endogeic earthworm *Pontoscolex corethrurus*. Appl Soil Ecol 70:62–69

Hong Q, Dong X, He L, Jiang X, Li S (2009) Isolation of a biphenyl-degrading bacterium, *Achromobacter* sp. BP3, and cloning of the bph gene cluster. Int Biodeter Biodegrad 63(4):365–370

Iovdijova A, Bencko V (2010) Potential risk of exposure to selected xenobiotic residues and their fate in the food chain – Part. I: Classification of xenobiotics. Ann Agric Environ Med 17(2):183–192

Jain RK, Kapur M, Labana S, Sarma P, Lal B, Bhattacharya D, Thakur IS (2005) Microbial diversity: application on micro-organisms for the biodegradation of xenobiotics. Curr Sci 89:101–112

Jayasekara R, Harding I, Bowater I, Lonergan G (2005) Biodegradability of a selected range of polymers and polymer blends and standard methods for assessment of biodegradation. J Polym Environ 13(3):231–251

Jia L-Y, Zheng A-P, Xu L, Huang X-D, Zhang Q, Yang F-L (2008) Isolation and characterization of comprehensive polychlorinated biphenyl degrading bacterium, *Enterobacter* sp. LY402. J Microbiol Biotechnol 18(5):952–957

Kazumi J, Häggblom M, Young L (1995) Diversity of anaerobic microbial processes in chlorobenzoate degradation: nitrate, iron, sulfate and carbonate as electron acceptors. Appl Microbiol Biotechnol 43(5):929–936

Kumar K, Devi SS, Krishnamurthi K, Kanade GS, Chakrabarti T (2007) Enrichment and isolation of endosulfan degrading and detoxifying bacteria. Chemosphere 68(2):317–322

Kwon G-S, Sohn H-Y, Shin K-S, Kim E, Seo B-I (2005) Biodegradation of the organochlorine insecticide, endosulfan, and the toxic metabolite, endosulfan sulfate, by *Klebsiella oxytoca* KE-8. Appl Microbiol Biotechnol 67(6):845–850

Langenbach T (2013) Persistence and bioaccumulation of persistent organic pollutants (POPs). Bioremediation-Active and passive approaches. InTech, pp 307–331

Leung M (2004) Bioremediation: techniques for cleaning up a mess. Bio Tech J 2:18–22

Lin W, Chang-Chien G, Kao C, Newman L, Wong T, Liu J (2014) Biodegradation of Polychlorinated Dibenzo–Dioxins by Strain NSYSU. J Environ Qual 43(1):349–357

Liu H, Yao J, Yuan Z, Shang Y, Chen H, Wang F, Masakorala K, Yu C, Cai M, Blake RE (2014a) Isolation and characterization of crude-oil-degrading bacteria from oil-water mixture in Dagang oilfield, China. Int Biodeterior Biodegrad 87:52–59

Liu X-M, Chen K, Meng C, Zhang L, Zhu J-C, Huang X, Li S-P, Jiang J-D (2014b) *Pseudoxanthobacter liyangensis* sp. nov., isolated from dichlorodiphenyltrichloroethane-contaminated soil. Int J Syst Evol Microbiol 64(10):3390–3394

Magnuson JK, Stern RV, Gossett JM, Zinder SH, Burris DR (1998) Reductive dechlorination of tetrachloroethene to ethene by a two-component enzyme pathway. Appl Environ Microbiol 64(4):1270–1275

Mercimek HA, Dincer S, Guzeldag G, Ozsavli A, Matyar F (2013) Aerobic biodegradation of 2, 4, 6-trinitrotoluene (TNT) by *Bacillus cereus* isolated from contaminated soil. Microb Ecol 66(3):512–521

Monica S, Karthik L, Mythili S, Sathiavelu A (2012) Formulation of effective microbial consortia and its application for sewage treatment. J Microb Biochem Technol 3(3):051–055

Nawab A, Aleem A, Malik A (2003) Determination of organochlorine pesticides in agricultural soil with special reference to γ-HCH degradation by *Pseudomonas* strains. Bioresour Technol 88(1):41–46

Park Y-C, Lee S, Cho M-H (2014) The simplest flowchart stating the mechanisms for organic xenobiotics-induced toxicity: can it possibly be accepted as a "central dogma" for toxic mechanisms? Toxicol Res 30(3):179

Patel V, Jain S, Madamwar D (2012) Naphthalene degradation by bacterial consortium (DV-AL) developed from Alang-Sosiya ship breaking yard, Gujarat, India. Bioresour Technol 107:122–130

Piutti S, Semon E, Landry D, Hartmann A, Dousset S, Lichtfouse E, Topp E, Soulas G, Martin Laurent F (2003) Isolation and characterisation of Nocardioides sp. SP12, an atrazine-degrading bacterial strain possessing the gene trzN from bulk- and maize rhizosphere soil. FEMS Microbiol Lett 221(1):111–117

Ross S (1994) Sources and forms of potentially toxic metals in soil-plant systems. Toxic metals in soil-plant systems. Wiley, Chichester, pp 3–26

Sakai M, Ezaki S, Suzuki N, Kurane R (2005) Isolation and characterization of a novel polychlorinated biphenyl-degrading bacterium, *Paenibacillus* sp. KBC101. Appl Microbiol Biotechnol 68(1):111–116

Schröter-Kermani C, Kreft D, Schilling B, Herrchen M, Wagner G (2006) Polycyclic aromatic hydrocarbons in pine and spruce shoots – temporal trends and spatial distribution. J Environ Monit 8(8):806–811

Schwitzguébel J-P, Page V, Martins-Dias S, Davies LC, Vasilyeva G, Strijakova E (2011) Using plants to remove foreign compounds from contaminated water and soil. In: Schroder P, Collins CD (eds) Organic Xenobiotics and plants. Springer, Dordrecht, pp 149–189

Shimao M (2001) Biodegradation of plastics. Curr Opin Biotechnol 12(3):242–247

Singh NS, Singh DK (2011) Biodegradation of endosulfan and endosulfan sulfate by *Achromobacter xylosoxidans* strain C8B in broth medium. Biodegradation 22(5):845–857

Singh BK, Walker A (2006) Microbial degradation of organophosphorus compounds. FEMS Microbiol Rev 30(3):428–471

Solouki T, Khalvati MA, Miladi M, Zekavat B (2011) State-of-the-art chemical analyses: Xenobiotics, plant proteomics, and residues in plant based products. In: Schroder P, Collins CD (eds) Organic Xenobiotics and plants. Springer, Dordrecht, pp 261–306

Su Y-H, Zhu Y-G (2007) Transport mechanisms for the uptake of organic compounds by rice (Oryza sativa) roots. Environ Pollut 148(1):94–100

Suzuki S, Hiraishi A (2007) *Novosphingobium naphthalenivorans* sp. nov., a naphthalene-degrading bacterium isolated from polychlorinated-dioxin-contaminated environments. J Gen Appl Microbiol 53(4):221–228

Theriot CM, Grunden AM (2011) Hydrolysis of organophosphorus compounds by microbial enzymes. Appl Microbiol Biotechnol 89(1):35–43

Toussaint J-P, Pham TTM, Barriault D, Sylvestre M (2012) Plant exudates promote PCB degradation by a rhodococcal rhizobacteria. Appl Microbiol Biotechnol 95(6):1589–1603

Tripathi BD (2012) A short term study on toxic effects of distillery sludge amendment on microbiological and enzymatic properties of agricultural soil in a tropical city. J Earth Sci Clim Change 2:1–8

Tropel D, Van Der Meer JR (2004) Bacterial transcriptional regulators for degradation pathways of aromatic compounds. Microbiol Mol Biol Rev 68(3):474–500

Tuo B-H, Yan J-B, Fan B-A, Z-H Y, Liu J-Z (2012) Biodegradation characteristics and bioaugmentation potential of a novel quinoline-degrading strain of *Bacillus* sp. isolated from petroleum-contaminated soil. Bioresour Technol 107:55–60

Umar AF, Tahir F, Larkin MJ, Oyawoye OM, Musa BL, Yerima MB, Agbo EB (2012) AtzABC catabolic gene probe from novel atrazine-degrading *Rhodococcus* strain isolated from a Nigerian agricultural soil. Adv Microbiol 2:593–597

Vangnai AS, Petchkroh W (2007) Biodegradation of 4-chloroaniline by bacteria enriched from soil. FEMS Microbiol Lett 268(2):209–216

Vargas C, Song B, Camps M, Häggblom M (2000) Anaerobic degradation of fluorinated aromatic compounds. Appl Microbiol Biotechnol 53(3):342–347

Varsha Y, Naga Deepthi CH, Chenna S (2012) An emphasis on xenobiotic degradation in environmental clean up. J Biorem Biodegrad 2(4):1–10

Vidali M (2001) Bioremediation. an overview. Pure Appl Chem 73(7):1163–1172

Visioli F (2015) Xenobiotics and human health: a new view of their pharma-nutritional role. PharmaNutrition 3(2):60–64

Wang F, Grundmann S, Schmid M, Dörfler U, Roherer S, Munch JC, Hartmann A, Jiang X, Schroll R (2007) Isolation and characterization of 1, 2, 4-trichlorobenzene mineralizing *Bordetella* sp. and its bioremediation potential in soil. Chemosphere 67(5):896–902

Wasi S, Tabrez S, Ahmad M (2013) Toxicological effects of major environmental pollutants: an overview. Environ Monit Assess 185(3):2585–2593

Widehem P, Aıt-Aıssa S, Tixier C, Sancelme M, Veschambre H, Truffaut N (2002) Isolation, characterization and diuron transformation capacities of a bacterial strain *Arthrobacter* sp. N2. Chemosphere 46(4):527–534

Yamaga F, Washio K, Morikawa M (2010) Sustainable biodegradation of phenol by Acinetobacter calcoaceticus P23 isolated from the rhizosphere of duckweed *Lemna aoukikusa*. Environ Sci & Technol 44(16):6470–6474

Zaidi S, Usmani S, Singh BR, Musarrat J (2006) Significance of Bacillus subtilis strain SJ-101 as a bioinoculant for concurrent plant growth promotion and nickel accumulation in *Brassica juncea*. Chemosphere 64(6):991–997

Zeinali M, Vossoughi M, Ardestani SK, Babanezhad E, Masoumian M (2007) Hydrocarbon degradation by thermophilic *Nocardia otitidiscaviarum* strain TSH1: physiological aspects. J Basic Microbiol 47(6):534–539

Zvyagintsev D (1990) Soil microorganisms and environment protection. Agrochem Soil Sci 39(3–4):283–285

Metals from Mining and Metallurgical Industries and Their Toxicological Impacts on Plants

10

Meenu Gautam, Divya Pandey, S.B. Agrawal, and Madhoolika Agrawal

Abstract

The growth of population, increasing urbanization and rising standards of human have contributed to increase in both quantity and variety of solid wastes generated by agricultural, domestic and industrial activities. Industrial wastes contributed more than 85 % of solid waste generation globally. Metals are the major component of almost all the industrial activities but their mining, extraction, purification and various manufacturing processes generate mining and metallurgical wastes having enormous environmental and health impacts. This chapter aims to describe the metals in solid wastes from mining and metallurgical industries and their toxicological impacts on plant community. Industrial wastes are composed of a wide range of essential macro- and micronutrients such as Na, K, Ca, Mg, Mn, Fe, Cu, Zn, Ni, Co, and Mo which are required by plants for their growth and development. But the concentrations of micronutrients in plants when they exceed certain thresholds may interfere with plant metabolic activities leading to the reduction in their productivity. Similarly, non-essential metals and metalloids such as Cd, Pb, As, Al, Bi, Cr, Hg, Ti and Si at elevated concentrations in plants cause phytotoxic effects and lead to food chain contamination. These wastes are generated in huge quantities and discarded without any proper pretreatment; therefore, chances of contamination of environmental components are obvious. This chapter also suggests the possible and better management opportunities including site restoration by rehabilitation and phytoremediation of

M. Gautam • S.B. Agrawal • M. Agrawal (✉)
Laboratory of Air Pollution and Global Climate Change, Department of Botany, Institute of Science, Banaras Hindu University, Varanasi 221005, Uttar Pradesh, India
e-mail: madhoo.agrawal@gmail.com

D. Pandey
Stockholm Environment Institute at York, University of York, Heslington, York YO10 5DD, United Kingdom

A. Singh et al. (eds.), *Plant Responses to Xenobiotics*,
DOI 10.1007/978-981-10-2860-1_10

metal-contaminated sites using native and medicinal plant species to reduce food chain contamination and an ultimate risk to human health.

Keywords
Mining • Metallurgy • Xenobiotics • Metals • Plants

10.1 Introduction

Xenobiotics in residues from agricultural, domestic and industrial sectors are significant environmental pollutants and are the major concerns for safeguarding the human and ecological health of various ecosystems. Common examples of organic xenobiotics are soap, detergents, disinfectants, herbicides, insecticides, vinegar, spices, fats, oils, etc., whereas inorganic xenobiotics include inorganic fertilizers, acidic and basic compounds and metals (Tyus 2012). Metals such as zinc, copper, iron, magnesium, nickel, manganese, molybdenum, etc. are essential for living organism but when present in higher concentration than usual, cause toxic effects (Hodson 2012). Toxic metals include cadmium, lead, chromium and mercury, which are foreign to biological systems are referred as xenobiotic metals (Solenkova et al. 2014). Metals are of significant importance because of their widespread application in manufacturing and infrastructure developments going on throughout the world leading to increased waste generation, and, hence, the metallic wastes from mining and metallurgical industries may pose significant threats to plant species and ecosystems.

Rapid industrialization and urbanization have resulted in an enormous increase in solid wastes due to a variety of activities. Out of ≈ 12 billion tonnes of solid wastes generated during 2002, 11 billion tonnes were contributed by industrial wastes (Yoshizawa et al. 2004). As per the statistics on waste generation in India given by Pappu et al. (2007), the highest proportion of annual solid waste was contributed by the agricultural sector (147.5 MT) followed by thermal power plants (in terms of coal combustion residues) (112 MT), mining and metallurgical industries (99 MT) and municipal solid wastes (48 MT) (Fig. 10.1). Mining and metallurgical industries are of considerable importance in providing great diversity of minerals for industrial and household activities, thus contributing the major proportion of the world's economy.

10.2 Mining Industries

Mining is a process where extraction of materials from the ground takes place in order to recover metalliferous (bauxite, lead, zinc, copper, etc.) and non-metalliferous (sand, stone, kaolinite, phosphate, limestone, rock salt, slate and sulphur) ores and fuel (coal and oil). Mining process can be categorized into surface, underground and in situ mining.

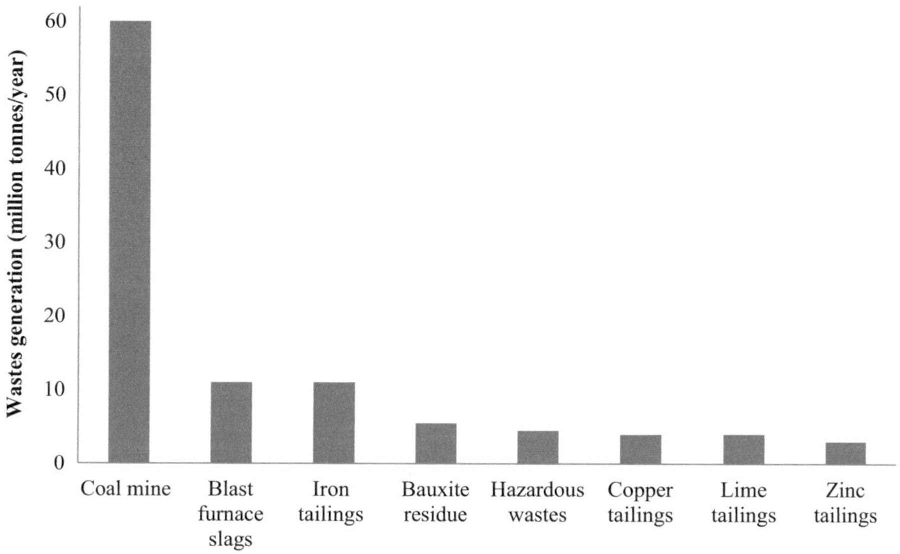

Fig. 10.1 Solid waste generation in India from mining and metallurgical industries (Modified from Pappu et al. 2007)

10.2.1 Surface Mining

Mining of mineral deposits from either at or close to the earth's surface involves removing surface vegetation, topsoil and layers of bedrock in order to reach buried mineral deposits.

10.2.2 Underground Mining

Underground mining consists of digging tunnels or shafts into the earth to reach buried ore deposits. Ores for processing and waste rock for disposal are brought to the surface through the tunnels and shafts.

10.2.3 In Situ Mining

In situ mining is a method of extracting minerals from an orebody that is left in place rather than being broken up and removed. The process involves a series of wells that are drilled into the orebody, and solvents are injected through certain wells and withdrawn through others. In situ mining is an advanced technique, providing an alternative with less environmental impact than conventional surface and underground mining.

During the process of mining, large quantities of solid wastes are generated and are categorized as rock wastes, overburden, sludge, tailings and spoils (Fig. 10.2).

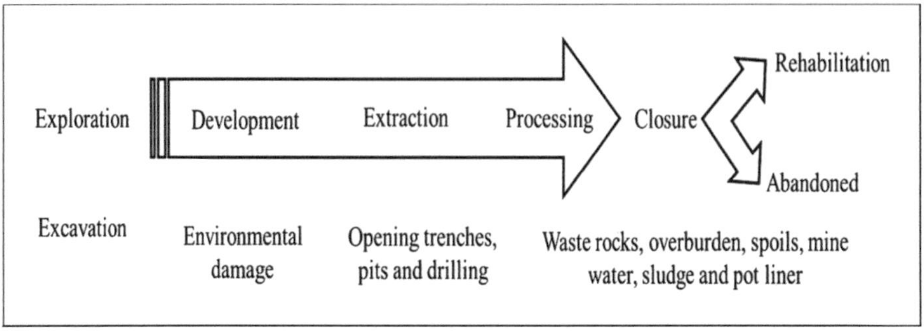

Fig. 10.2 Schematic representation of the main steps and waste generation during mining

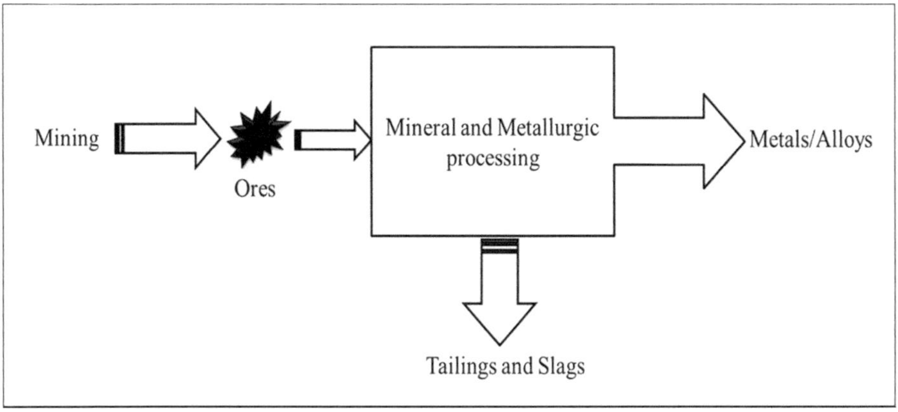

Fig. 10.3 Schematic representation of metallurgical operations and waste generation

Mining industries that contribute a major proportion of the gross domestic product (GDP) of the world are bauxite, coal, copper, diamond, gold, iron ore, natural gas, nickel, oil shale, opal, petroleum, rare earth elements, silver, uranium, zinc and lead.

10.3 Metallurgical Industries

A metallurgical industry involves mechanical, physical and chemical methods of producing a pure form of metals or alloys from ores. There are mainly three types of metallurgical operations, namely, pyrometallurgical process, where smelting, refining and roasting of extracted ores are performed; hydrometallurgical operation, where production of phosphoric acid by phosphate digestion takes place and electrometallurgical process which is an electrolytic process of metal refining. Waste materials generated from metallurgical industries are slags, tailings, red mud, sludge and filter residues (Fig. 10.3).

This chapter particularly aims to describe the metals and metalloids in solid wastes from mining and metallurgical industries and their toxic impacts on plant community structure. Information on better management options including phytoremediation, reclaimation of polluted sites and potential reuse of these wastes are also discussed.

10.4 Characterization and Estimation of Metal Contents in Solid Wastes Generated from Mining and Metallurgical Industries

The wastes are characterized, in terms of chemical and mineralogical compositions. Chemical composition is determined by digesting the material in appropriate acids and analysing by atomic absorption spectroscopy (AAS) and inductively coupled plasma and mass spectroscopy (ICP-MS). Other methods of analysis include potentiometric titration, conductometric titration and colorimetric methods using a spectrophotometer (Willard et al. 1988). Mineralogical compositions of the wastes can be determined by X-ray diffraction (XRD), scanning electron microscopy (SEM), microprobe, image analyser (IA), proton-induced X-ray analyser (PIXE), energy-dispersive X-ray analyser (EDX), secondary ion mass spectrometer (SIMS), laser ionization mass spectrometer (LIMS), infrared analysis (IRA) and cathode luminescence (Rao 2011).

Waste materials generated from mining and metallurgical processings (slags, tailings, overburden, rocks, filter residues, sludge and red mud) are heterogeneous geologic materials, which have been deposited in surrounding areas without any proper pretreatments. Physico-chemical properties of these wastes depend upon the mineralogy, geochemistry, particle size of mine materials, moisture content, type of processes used in extraction, purification and refining of materials (Hassinger 1997). Generated wastes are composed of a wide range of particle size fractions varying from coarse mine wastes to slimes (Ritcey 1989). These wastes generally have extreme pH values (acidic to alkaline), reduced concentrations of essential plant nutrients (nitrogen (N), phosphorous (P), potassium (K) and other micronutrients), low organic matter, extremely low microbial activities, high cation exchange capacity, poor water holding capacity, high bulk density and elevated levels of heavy metals (Conesa and Faz 2011).

Metals are an important component in industrial wastes that are dispersed in soil, surface and groundwater leading to environmental risks to adjoining areas (Santos-Jallath et al. 2012; Wójcik et al. 2014). Based on previous studies, concentrations of selected metals such as sodium (Na), potassium (K), calcium (Ca), manganese (Mn), magnesium (Mg), iron (Fe), cobalt (Co), cadmium (Cd), nickel (Ni), copper (Cu), zinc (Zn), arsenic (As), lead (Pb), chromium (Cr), mercury (Hg), aluminium (Al), molybdenum (Mo), bismuth (Bi), titanium (Ti) and silica (Si) in mining (metalliferous, non-metalliferous and fuel) and metallurgical wastes are presented in Table 10.1, which shows a wide range of their concentrations in discharged wastes resulting in contamination of not only soil at dumping site but also affecting nearby areas. Essential macro- and micronutrients such as Na, K, Ca, Mg, Fe, Zn, Mn, Co,

Table 10.1 Range of selected metal concentrations present in solid waste dump from mining and metallurgical industries

Metals	Metalliferous mining (ppm)	Non-metalliferous mining (ppm)	Fuel mining (ppm)	Metallurgical industries (ppm)
Na	<0.50–3480.0[1,9]	0.50–30.00[10]	1.40–2.50[17]	0.3–121.80[19,24]
K	21.90–57.40[1]	7.90–36.50[11,10]	3.30–12.70[17]	0.5–91.0[19,18]
Ca	0.60–4548.0[1,3]	31.70–552.60[12,10]	2.0–36.90[17]	11.00–541.0[38,21]
Mg	2.79–6030.0[2,9]	0.70–29.40[10,11]	2.60–9.40[17]	0–380.0[24,22]
Mn	0.20–7111.0[1,3]	0.70[10]	0–993.0[17,16]	0.30–4000.0[20,28]
Fe	38.40–37,7671.0[1,4]	1.10–77.30[10,12]	26.50–50,691.0[17,16]	0.4–474.80[36,24]
Co	13.05–371.0[5,4]	≈79.0[10]	15.0–75.37[17]	2.30–6150.0[29,35]
Ni	73.50–1548.0[6,4]	16.0–243.0[13]	8.05–107.0[15,17]	0.70–2150.0[21,34]
Cu	0.30–2595.0[1,5]	28.10–99,999.0[11,12]	5.50–101.0[14,17]	0.50–41,900.0[19,33]
Zn	0.10–21,007.50[1,7]	46.0–14,1000.0[10,12]	6.60–213.11[14]	0.04–95,940.0[22,31]
Ti	3.90–12.40[1]	≈0.50[10]	4.10–19.60[17]	1.30–171.30[19,25]
Pb	0.20–5220.20[1,4]	111.0–1533.0[12]	0.90–38.40[14,16]	0.17–14,7700.0[23,30]
Cd	0.05–1811.66[6,3]	48.20–823.50[11,12]	0–11.50[14,15]	0.01–2402.0[28,32]
Bi	<0.05–0.40[1]	–	–	≈21.40[37]
As	0.20–80,000.0[2,9]	3.70–290.0[12]	0–81.50[14,17]	0.02–86,000.0[27,29]
Al	91.40–12,594.0[1,5]	1.60–122.10[10]	62.30–278.70[17]	10.00–810.0[26,36]
Si	23.00–646.70[8,1]	5.70–698.80[10]	83.30–444.0[17]	26.00–3016.0[36,18]
Cr	200.0–3073.0[4]	0.70–352.0[10,11]	9.67–164.0[15,17]	1.0–3000.0[38]
Hg	0.01–18.30[2,3]	–	0.10–0.22[17]	2.63–9300.0[29]

Superscript numbers are citation of references where first and second number represents minimum and maximum concentrations of metals, respectively

[1]Filippi et al. (2015), [2]Rola et al. (2015), [3]Mathiyazhagan et al. (2015), [4]Nawab et al. 2015a, [5]Gutiérrez-Gutiérrez et al. (2015), [6]Cele and Maboeta (2016), [7]Bacchetta et al. (2015), [8]Mohanty et al. (2010), [9]Palumbo-Roe et al. (2007), [10]Hamzah et al. (2011), [11]Jellali et al. (2010), [12]Boulet and Larocque (1998), [13]Singh and Hendry (2013),[14]Gholizadeh et al. (2015), [15]Juwarkar and Jambhulkar (2008), [16]Pandey et al. (2016), [17]Qureshi et al. (2016), [18]Remon et al. (2005), [19]Lopez et al. (1997), [20]Shen and Forssberg (2003), [21]Huaiwei and Xin (2011), [22]Costa et al. 2016, [23]Liu et al. (2007), [24]Liu et al. (2009), [25]Samal et al. (2015), [26]Sarkar and Mazumder (2015), [27]Ene and Pantelică (2011), [28]Leonard (1978), [29]Guo et al. (2014), [30]Cabala and Teper (2007), [31]Niemeyer et al. (2010), [32]Douay et al. (2009), [33]Křĭbek et al. (2010), [34]Adamo et al. (2002), [35]Narendrula et al. (2012), [36]Galindo et al. (2015), [37]Li et al. (2014) and [38]Jacobs and Testa 2005

Cu, Ni and Mo are important for plant growth and development, but their concentrations in either agricultural soil or plants beyond certain limit may cause toxic effects on physico-chemical and biological properties of soil and plant's performance growing on heavily contaminated soil.

10.5 Toxic Impacts of Metals on Crop and Medicinal Plants

Contamination of agricultural lands with heavy metals in the vicinity of industries has become a major environmental concern. Such toxic elements are considered as soil pollutants due to their acute and (or) chronic toxicity to plants. Metal pollution

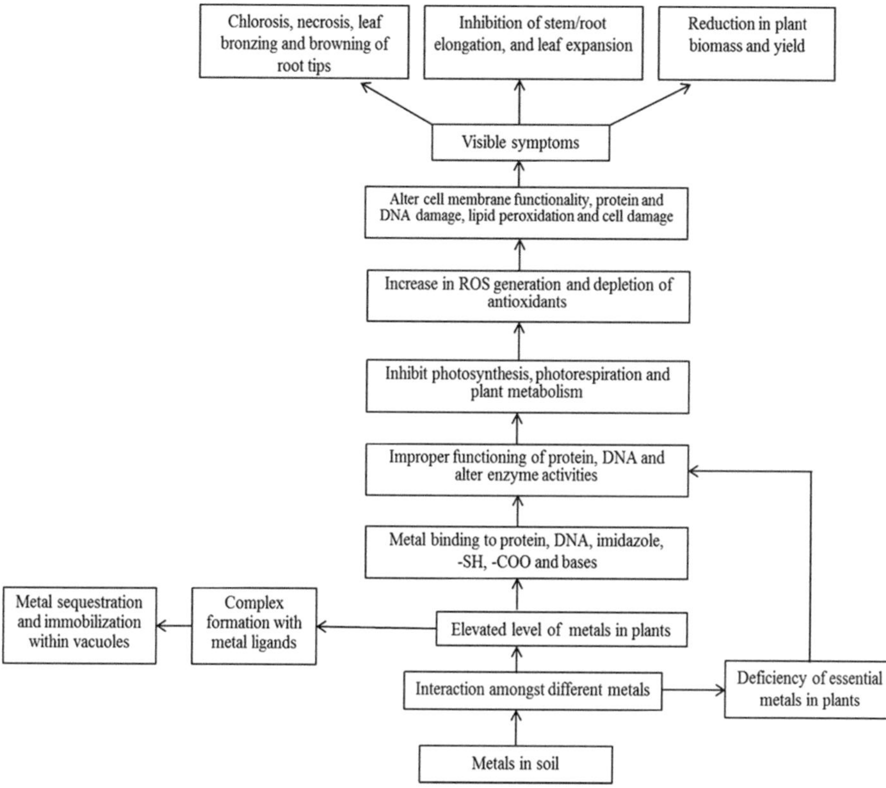

Fig. 10.4 Toxicological impacts of elevated concentrations of metals on plants

becomes a persistent problem, as, once released into the environment where organisms may get affected because metals are not destroyed, but they only transform from one oxidation state or organic complex to another or are gradually move into different components of the biosphere (Marques et al. 2009). Surface runoff and leaching from waste dumps pollute the groundwater (Baba and Tayfur 2011), while the dust spread by wind settles on agricultural crops from where they enter the food chain when consumed (Salomons 1995). Like other living organisms, plants are often sensitive both to deficiency and excess availability of some essential micronutrients. Higher concentrations of essential micronutrients are strongly toxic to the metabolic activities of plants (Fig. 10.4). Several researches have been conducted to assess the toxic effects of elevated metal concentrations on plants (Reeves and Baker 2000). Here, the potential implications of some metals on plants are discussed in detail:

10.5.1 Sodium

Sodium is an essential nutrient for the growth of plants, and it plays an important role in maintaining the turgor pressure inside plant cells (Jennings 1976), but its excessive concentrations produce toxic effects on older leaves, cause premature leaf senescence and reduce total photosynthetic leaf area (Munns 2002). Na salts are the major cause of salinity in soil that is a major challenge in many agricultural regions in the world (Pitman and Läuchli 2002). In heavily polluted sites, Na salts precipitate on leaves as water evaporates which in turn may result its higher concentration in plants (Bailey et al. 1999). Phytotoxic symptoms due to higher Na accumulation in plants are leaf burn, scorch and dead tissues which first appear on the outer edges of leaves and then move progressively inward between the veins towards the leaf centre with increase in severity (Table 10.2).

High Na concentration may alter normal growth and physiology of plants. Significant reductions in shoot length, leaf area and dry weight of brinjal (*Solanum melongena*) were recorded beyond 10 mmol L^{-1} NaCl concentration, whereas maximum yield reduction (88.0 %) was observed above 150 mmol L^{-1} NaCl concentration (Chartzoulakis and Loupassaki 1997). Similarly, leaf area development of cotton (*Gossypium hirsutum*) and bean (*Phaseolus vulgaris*) was strongly inhibited under Na stress, and such reduction in leaf area altered the photosynthesis and growth of these plants (Brugnoli and Lauteri 1991). Plant height and leaf elongation in tomato (*Lycopersicon esculentum*) also showed reductions with increase in NaCl in the nutrient solution (Montesano and Van Iersel 2007). In *Capsicum annuum*, plant height and total leaf area were reduced by 49.0 and 82.0 %, respectively, under NaCl concentrations above 50 mmol L^{-1} (Chartzoulakis and Klapaki 2000). High levels of NaCl in soil are reported to decrease the number of flowers and stem quality of *Gerbera jamesonii* (De Kreij and Van Os 1989) and *Rosa hybrida* (De Kreij and Van Den Berg 1990). Lee and Van Iersel (2008) found that *Chrysanthemum* sp. receiving 9 gL^{-1} NaCl showed reduction in shoot dry weight (76.0 %), stomatal conductance (90.0 %) and chlorophyll content (from 42.3 to 29.2 SPAD units), and a 4-day delayed flowering was also observed compared to the control plants. Montesano and Van Iersel (2007) also found significant reductions in leaf photosynthesis and chlorophyll content at NaCl concentrations above 4.1 gL^{-1}. Increased Na^+ uptake also interferes with uptake of K^+ in plants, thus causing K^+ deficiency (Montesano and Van Iersel 2007).

10.5.2 Potassium

Potassium is also an important macronutrient required for plant physiological and metabolic processes. The most important role of K in plants is to activate several enzymes participating in plant metabolism (Evans and Sorger 1966). Most plants require K in the range of 50–100 mM for their normal functioning (Epstein 1980). Due to either K deficiency or excessive uptake, plants show necrosis, chlorosis and leaf curling as visible symptoms (Table 10.2).

Critical concentrations of K for *L. esculentum*, *Helianthus annuus* and *Zea mays* have been found to be 1.0, 2.2 and 1.3 ppm, respectively, which means that K helps to increase their biomass that may reach up to 90 %, but above this concentration, no further increase in biomass takes place (Besford and Maw 1975; Spear et al. 1978; Tyner and Webb 1946). Excessive K uptake may reduce plant's ability to

Table 10.2 Phytotoxic threshold levels and visible symptoms of injuries in crop and medicinal plants under elevated levels of selected metals

Metals	Phytotoxic threshold[a] (ppm)	Phytotoxic symptoms	References
Na	NT	Interveinal chlorosis on young leaf, leaves scorching and flecking and reduced plant growth	Arizona cooperative extension (1998)
K	NT	Necrosis and chlorosis along leaf margins, curling of leaf	Arizona cooperative extension (1998)
Ca	NT	Necrosis between veins and chlorosis along leaf margins extending between veins in Christmas pattern with curling of leaf margins and puckering effects	Chang et al. (2004), Wissemeier 1993 and Arizona Cooperative Extension (ACE) (1998)
Mg	NT	Necrosis at the tip and margins of leaves forming a hooklike structure at leaf tip and reduction in plant growth	Brooks (1987) and Arizona cooperative extension (1998)
Mn	NT	Marginal chlorosis and necrosis on leaves, petioles and stems, leaf bronzing, shortening of internodes, crinkling in youngest leaf; browning of root tips and root cracking under severe Mn exposure	Kitano et al. (1997), Wu (1994), Horiguchi (1988) and Foy et al. (1978)
Fe	10–20	Dark-green, brown or purple foliage, brown spots on leaves, necrotic spots and chlorotic stippling, cupping of leaves, bronze speckle, stunted plant growth, weak stem and delayed flowering	Albano et al. (1996) and Broschat and Moore (2004)
Cu	20–100	Chlorotic and necrotic spots, yellow and purple coloration on the lower side of mid rib, plant growth retardation and inhibition of root elongation	Neelima and Reddy (2002) and Mahmood and Islam (2006)
Zn	100–400	Chlorosis in the younger leaves, which can extend to older leaves after prolonged exposure, purplish-red coloration on leaves and growth stunting	Prasad et al. 1999 and Romero-Puertas et al. (2004)
Ni	10–100	Necrosis, uniform interveinal chlorosis and yellowish-white discoloration on older leaves; reduction in root growth and leaf area	Gajewska et al. (2006) and Ishtiaq and Mahmood (2012)
Co	NT	Red-brown discoloration first in the veins and later in petioles and stems, premature leaf closure; reduction in plant growth	Li et al. (2009) and Chatterjee and Chatterjee (2000)
Mo	NT	Chlorosis, marginal leaf scorch, abscission, yellowing and browning of leaves; reduced tillering	Osman (2012)

(continued)

Table 10.2 (continued)

Metals	Phytotoxic threshold[a] (ppm)	Phytotoxic symptoms	References
Al	NT	Chlorosis, dark-green leaves, purpling of abaxial leaf surface, petiole and stem, yellowing and death of leaf tips, curling or rolling of young leaves; stunted growth and delayed plant maturity	Foy et al. (1978) and Singh (2005)
Cd	5–30	Chlorosis, necrosis, brown stunted roots and leaf margin, reddish veins and petioles, curled leaves, leaf rolls and purple coloration of whole plant; growth inhibition	Nawab et al. (2015b) and Mohanpuria et al. (2007)
Pb	30–300	Chlorosis, dark-green leaves, adverse effect on growth, inhibits seed germination, leaf expansion, root and stem elongation	Sharma and Dubey (2005) and Reddy et al. (2005)
As	5–20	Red-brown necrotic spots on older leaves, yellow browning of roots; growth reduction	Meharg and Macnair (1992)
Cr	5–30	Chlorosis in young leaves resulting in wilting	Panda and Choudhury (2005) and Dube et al. (2004)
Ti	NT	Stunted growth of plants	Singh (2005)
Bi	NT	Necrosis on foliage; inhibition of radicle growth.	Dayan et al. (1999) and Galindo et al. (1999)
Hg	1–3	Chlorosis; reduced seedling, root and shoot growth	Patra and Sharma (2000) and Gao et al. (2010)
Si	NT	Yellow streaked and brittle leaves; stunted plant growth	Côté-Beaulieu et al. (2009)

[a]Alloway and Ayres (1997); [NT] no phytotoxic threshold levels have been established

uptake P from soils as observed by Karlen et al. (1987) in the case of maize/corn (*Z. mays*). Rodrigues et al. (2016) showed that accumulation of K in *Jatropha curcas* shoot showed a strong correlation with increase in stomatal conductance and transpiration that ultimately cause reductions in water use efficiency and Na^+ content under high relative humidity (80.0 %). The study also showed that the supply of K^+ in growing medium strongly restricted Na^+ uptake and transport to the shoot because of a strong competitive interaction between K^+ and Na^+ ions in the growth medium.

10.5.3 Calcium

Calcium is an essential macronutrient which plays a significant role in preserving the structural and functional integrity of cell membranes, stabilizes cell wall structures, regulates ion transport as well as controls cell wall enzymatic activities (Marschner 1995). But excessive Ca uptake by plants may produce phytotoxic effects such as necrosis, chlorosis, leaf curling and puckering (Table 10.2). Studies indicate that higher Ca accumulation in tomato leads to the development of yellowish flecks or gold spots around the calyx and on fruit due to formation of calcium oxalate crystals (De Kreij et al. 1992). Through a hydroponic experiment, Nichols and Beardsell (1981) also showed that increase in levels of Ca induced necrotic spots on leaves and caused reduction in the growth of *Grevillea* sp.

Addition of $CaSO_4$ (gypsum) to the soil plays a significant role in reducing heavy metal toxicity (Illera et al. 2004), but more than 25.0 % gypsum may cause significant reduction in yield of crop plants due to imbalanced K/Ca and Mg/Ca ratios (Van Alphen and de los Ríos Romero F 1971). Hernando et al. (1965) also reported that higher Ca accumulation caused poor growth of corn at 80.0 % soil moisture in the field, whereas wheat (*Triticum aestivum*) growth is reduced when gypsum content in soil was 25.0 %. Bureau and Roederer (1960) also suggested that crop cultivation in the soil of Tunisia with 30.0 % gypsum content may cause toxic effects on plant growth and development. Smith and Robertson (1962) observed that wheat grown in soil with higher gypsum input showed wilting during spring time due to reduced uptake of soil moisture by plants. Explants of *Chrysanthemum morifolium*, a medicinal plant treated with different levels of Ca that showed variable callogenesis and callus growth, were negatively affected by high levels of Ca due to inhibition of enzyme activities, magnesium uptake and protein synthesis (Borgatto et al. 2002).

10.5.4 Magnesium

Magnesium is another nutrient essential for plant growth and development. It is a component of chlorophyll and also plays an important role in plant respiration and energy metabolism, but becomes toxic when available in excess. In serpentine soil (soil which is derived from ultramafic rock, have high pH and are rich in Cr, Co, Ni, Fe and Mg, but deficient in macronutrients such as N, P, K and Ca), Mg phytotoxicity is the most common cause of "serpentine syndrome", resulting in reduction of plant's growth and development due to high Mg/Ca ratio (Brooks 1987). Certain visible toxic symptoms caused by Mg phytotoxicity are presented in Table 10.2. Oat (*Avena sativa*) plant in serpentine soil is more susceptible to Mg toxicity, which is caused by lowering of Ca uptake in plants due to their antagonistic behaviour.

Even though it is an essential component of chlorophyll, elevated levels of Mg may impair photosynthesis by inhibiting K^+ transport from cytosol to stroma and possibly interfere with Mg homeostasis within the chloroplast (Shaul 2002). Wu et al. (1991) showed that although Mg is very important for tea plants, but its higher concentration may adversely affect the growth and development of plants by altering their metabolic processes. Tea (UPASI-9) plant receiving above 5000 ppm of Mg supplement in soil showed coppery colour development on the leaf surface, and plant death occurred under long-term exposure (Venkatesan and Jayaganesh 2010). Increased concentration of Mg^{2+} in cytosol blocked K^+ channel across the membrane of chloroplasts, thereby inhibiting H^+ ion removal from chloroplasts stroma, resulting in its acidification which cause oxidative damage to plant cells (Wu et al. 1991). Wilkinson and Ohki (1988) reported that accumulation of Mg in plants reduced total chlorophyll and carotenoids contents by altering pigment synthetic pathway. Amino acid contents in plant showed a significant decline under its elevated doses above 1000 ppm due to hindered amino acid transport pathway (Ma et al. 2005).

10.5.5 Manganese

Manganese is an essential micronutrient for plant's growth and development but can be detrimental if available in excessive amounts in soil. There is no regulatory limit for Mn in agricultural soil, whereas for crop and medicinal plants, permissible limits are 500 ppm (FAO/WHO 2001) and 200 ppm (WHO 1998), respectively. Clark (1982) reported that excessive Mn in growth medium may interfere with the absorption, translocation and utilization of other minerals (Ca, Mg, Fe and P) by a plant which may lead to Mn toxicity (Table 10.2). Common phytotoxic symptoms are chlorosis (marginal and interveinal), necrotic brown spots as observed on leaves of *Brassica* sp., *Lactuca sativa, H. vulgäre, G. hirsutum* and *Tagetes erecta* (Bachman and Miller 1995; Albano et al. 1996; Führs et al. 2008). Kang and Fox (1980) reported loss of apical dominance and enhanced formation of auxiliary shoots (witches' broom) in *Vigna unguiculata* as symptoms of Mn toxicity.

Maksimović et al. (2012) reported that Mn at 100 μM concentration caused significant reduction in root and shoot biomass of cucumber (*Cucumis sativus*) as compared to 0.5 μM dose of Mn in the growth medium. Similarly, marked reductions in the dry weight of plant and leaf area were observed in *Oryza sativa*, *Lolium perenne* and *Populus* sp. at 583, 150 and 1000 μM of Mn concentrations, respectively (Lidon and Teixeira 2000; Lei et al. 2007; Mora et al. 2009). Excess Mn in plant tissues may produce errors during the mitochondrial replication by inducing mitochondrial mutations and inhibiting total protein synthesis (Foy et al. 1978). In cotton, Mn toxicity has been associated with increase in the activities of indoleacetic acid oxidase, peroxidase and polyphenol oxidase and reduced catalase, ascorbic acid oxidase and glutathione oxidase activities with lower ATP content and respiration rate (Morgan et al. 1976). Furthermore, in rice (*O. sativa*) seedlings under Mn stress, superoxide radical was increased preferentially in roots, while H_2O_2 content was found to be increased in shoots.

10.5.6 Iron

Iron is an essential nutrient for all plants with significant biological role in chlorophyll biosynthesis and photosynthesis; also it is the most limiting nutrient for plant growth primarily due to low solubility of oxidized ferric form in aerobic environment (Guerinot and Yi 1994). No standard maximum allowable limit for Fe in soil has been recommended because of its abundance in mineral soil; however, the expression of Fe toxicity symptoms in leaf tissues may occur under flooded conditions due to reduction of the Fe^{3+} to Fe^{2+} (Becker and Asch 2005). Fe is an integral component of many enzymes and proteins including heme and iron sulphur proteins (Marschner 1995). For crop and medicinal plants, permissible limits of Fe are 425.5 ppm (FAO/WHO 2001) and 20 ppm (WHO 1998), respectively. Iron phytotoxicity occurs only when it reaches beyond a threshold level which is characterized by preliminary symptoms such as necrosis, chlorotic stippling, cupping of leaves, bronze speckle, stunted growth, weak stem and delayed flowering (Table 10.2).

Iron content in medicinal plants consumed in UAE ranged between 26.96 and 1046.25 mg kg^{-1} (Abou-Arab and Abou Donia 2000). High iron levels often cause Mn deficiency in plants because of their competitive behaviour. In wheat, root- and shoot dry weights were found to decrease at 100 ppm Fe concentration in soil (Fageria and Rabelo 1987). Inhibitory effects of elevated Fe concentration on root elongation and photosynthetic pigments in *Sinapis alba* were reported (Fargašová 2001). Kampfenkel et al. (1995) observed brown spots on the leaf surface of *Nicotiana plumbaginifolia* and 40.0 % reduction in photosynthetic rate due to foliar accumulation of Fe. Iron toxicity in soybean (*Glycine max*) caused reduction in photosynthesis rate and yield due to increase in oxidative stress and ascorbate peroxidase activity (Sinha et al. 1997). Excess Fe leads to free radical production which alters the cellular structure irreversibly and damages membranes, DNA and protein structures (de Dorlodot et al. 2005).

10.5.7 Cobalt

Cobalt is a transition element, essential for several enzymes and coenzymes participating in plant metabolism. The maximum allowable range for Co in agricultural soil is 20–50 ppm (Kabata-Pendias and Sadurski 2004), whereas, for crop plants, permissible limit is 50 ppm (FAO/WHO 2001). For medicinal plants, no permissible limit has been specified. Cobalt affects the growth and metabolism of plants by different degrees depending upon its concentration and form in the soil. Toxic effects of Co on morphology include leaf fall, inhibition of greening, discoloured veins, premature leaf closure and reduced shoot weight. The supranormal doses of Co in plants have relatively high toxic effects which are mostly reflected in growth inhibition of plants accompanied by chlorosis of young leaves and other disorders (Table 10.2).

Phytotoxicity study of Co on crop plants such as barley (*Hordeum vulgare*), oilseed rape (*B. napus*) and tomato has shown reductions in shoot growth and biomass (Li et al. 2009). The higher foliar concentration of Co leads to lowering of essential mineral nutrients and photosynthesis rate and disturbance in the structural integrity of chloroplasts. High Co concentration (500 ppm) was found to reduce germination percentage and seedling growth of *T. aestivum* with 97.0 and 83.0 % reductions in root and shoot length, respectively (Gang et al. 2013). Chatterjee and Chatterjee (2000) reported that excess Co in cauliflower (*Brassica oleracea*) restricted the foliar uptake of Fe, P, S, Mn, Zn and Cu, altered the biosynthesis of chlorophyll and protein and reduced the catalase activity. Water potential and transpiration rate in cauliflower were increased significantly, while diffusive resistance and relative water content increased upon exposure to excess Co (Chatterjee and Chatterjee 2000). Palit et al. (1994) observed that Co affects photosystem (PS-II) by inhibiting either the reaction centre or components of PS-II. Moreover, in C_4 and CAM plants, Co hindered the fixation of CO_2 by inhibiting the activities of photosynthetic enzymes. Cobalt acts as a preprophase poison and thus retards the process of karyokinesis and cytokinesis, and higher concentrations of Co may hamper RNA synthesis and decrease DNA and RNA contents probably by modifying the activities of endo- and exonucleases (Palit et al. 1994).

10.5.8 Nickel

Nickel is an essential micronutrient required for plant normal functioning, but due to industrial activities, extent of soil contamination with Ni is so high that in some areas it is causing serious damages to agricultural crops (Frank et al. 1982). For agricultural soil, maximum allowable range of Ni is 20–60 ppm (Kabata-Pendias and Sadurski 2004), whereas for crop and medicinal plants, permissible limits are 67 ppm (FAO/WHO 2001) and 1.5 ppm (WHO 2005), respectively. Necrosis, chlorosis, inhibition of seed germination, reduced root and shoot growth, poorly developed branches, deformed plant parts and abnormal flowering are the common symptoms when foliar Ni concentration exceeds its phytotoxic threshold level (Table 10.2).

Ahmad et al. (2011) reported significant reductions in plant biomass, achene yield and foliar concentrations of essential nutrients (Mn, Zn, Cu and Fe) in sunflower (*H. annuus*), under high Ni concentration. Excessive Ni accumulation in crop and medicinal plants are reported to inhibit photosynthesis and transpiration rates (Sheoran et al. 1990). Progressive impairment of photosynthetic machinery coupled with oxidative damages in *Amaranthus paniculatus* was observed with increasing Ni treatment in the solution (Pietrini et al. 2015). Nickel concentration ranging from 0.01 to 10 ppm dry weight is considered essential for plant metabolism, regulation of lipid content and as an important constituent of enzymes such as urease, hydrogenase, superoxide dismutase (SOD) and glyoxalases (Küpper and Kroneck 2007). Gajewska et al. (2006) found significant reductions in wheat growth and proline accumulation along with significant decline in SOD and CAT activities at 200 μM Ni concentration. Molas (1998) reported significant decreases in number and size of chloroplasts, grana, thylakoids and plasto globuli in leaves of *B. oleracea* grown in soil containing $NiSO_4.7H_2O$ (10–20 g m^{-3}).

10.5.9 Copper

It is an essential trace element for all lower as well as higher plants with several roles in metabolic processes (Narula et al. 2005), but its increased concentration may produce toxic effects on plants (Mittler et al. 2004). The maximum allowable range of Cu for agricultural soil is 60–150 ppm (Kabata-Pendias and Sadurski 2004), whereas permissible limits for crop and medicinal plants are 73.30 ppm (FAO/WHO 2001) and 10 ppm (WHO 2005), respectively. Chlorosis, necrosis, purple coloration of midrib and reduction in plant growth are common symptoms observed due to Cu phytotoxicity (Table 10.2).

Excessive accumulation of Cu in roots caused root system damage (Atanassova and Zapryanova 2009), photosynthetic inhibition and plasma membrane permeability damage (Narula et al. 2005). Khatun et al. (2008) reported reductions in plant growth parameters, biomass and pigment contents in *Withania somnifera* above 10 μM of $CuSO_4.5H_20$ solution, whereas significant decreases in root and shoot biomass were observed in *Z. mays* treated with 10 μM of Cu. In *Solanum nigrum*, relative fresh weight, number of leaves, root and shoot lengths were reduced with increase in $CuSO_4$ level from 50 to 200 μM (Al-Khateeb and Al-Qwasemeh 2014). In barley leaves, Cu inhibited pigment synthesis and retarded chlorophyll integration into photosystems (Caspi et al. 1999). The reduction in pigment contents was attributed to the interaction of Cu to –SH group of enzymes during chlorophyll biosynthesis (Nyitrai et al. 2003). High level of Cu interferes with protein formation, photosynthetic processes and enzyme activities and alters plasma membrane permeability (Al-Khateeb and Al-Qwasemeh 2014). Ouzounidou et al. (1997) reported that Cu affects the ultrastructure of meristematic cells, altering the ribosomal RNA precursor biosynthesis and thus reducing the wheat growth.

10.5.10 Zinc

It is an important constituent of metalloenzyme and acts as a cofactor for several enzymes including anhydrases, dehydrogenases, oxidases and peroxidases (Hewitt 1983). It also plays an important role in regulating the nitrogen metabolism, cell multiplication, auxin synthesis and photosynthesis (Doncheva et al. 2001). The maximum allowable range of Zn for agricultural soil is 100–300 ppm (Kabata-Pendias and Sadurski 2004), whereas for crop and medicinal plants, permissible limits are 99.40 ppm (FAO/WHO 2001) and 50 ppm (WHO 1998), respectively. General symptoms of Zn phytotoxicity when its concentration exceeds the threshold level are chlorosis, necrosis, wilting, purplish colour patches, stunting of shoot, curling and rolling of young leaves and death of leaf tips (Table 10.2).

Zn toxicity is reported to cause nutrient (Fe, Mn and Cu) deficiencies in shoot due to hindered transference of these nutrients from root to shoot of geranium (Lee et al. 1996). Zinc accumulates to a greater extent in roots than in shoot and hence interferes with root growth and elongation and thereby limits plant's uptake of water and nutrients (Disante et al. 2010). White et al. (1979) reported a 50.0 % reduction in soybean biomass at 450 ppm Zn concentration, whereas 620–860 ppm of Zn caused ≈ 20 % reduction in foliar biomass of trifoliate leaves. Disintegration of cell organelles, disruption of membranes, condensation of chromatin material and increase in number of nucleoli were major events observed in pigeon pea (*Cajanus cajan*) during Zn toxicity (Sresty and Rao 1999).

10.5.11 Titanium

Although titanium is present in the soil in relatively higher concentrations, majority of Ti is poorly available for plants, due to the insoluble nature of the form of minerals (TiO_2 or $FeTiO_3$) in water (Dumon and Ernst 1988). Lower concentration of Ti might participate in plant metabolism as a redox catalyst and has a significant biological role in plant functioning, but at higher concentrations, deleterious effects on plant performances such as reduction in growth, yield and nutrient uptake of Zn, Mg and Fe, alteration in normal physiological functioning and chromosomal aberration with observed phytotoxic symptoms are reported (Geilmann 1920) (Table 10.2).

Burke et al. (2015) reported no negative effects of TiO_2 at low concentrations, but had strong negative effects on plant growth such as reduced root growth and elongation under elevated Ti concentration in crop and medicinal plants (Boonyanitipong et al. 2011). Phytotoxic effects on oat biomass at the concentration of 18 ppm Ti in nutrient solution were observed by Kužel et al. (2003). Yield in terms of grain weight in barley was reduced under foliar spray of 18 ppm Ti solution (Tlustoš et al 2005). The length of the petiole of strawberries was reduced, and hardness of fruits was increased under increasing applications of Ti (50, 100 or 150 mg kg^{-1}) (Choi et al. 2015). Feizi et al (2012) found reductions in plant growth and leaf carbon content in soybean plants under TiO_2 treatments. Also, the hydraulic conductivity of cell wall and diameter of root cell wall pores were reduced from 6.6 to 3.0 nm. Ghosh et al. (2010) found negative effects of Ti on plant growth, cell elongation and transpiration. Pakrashi et al. (2014) showed that TiO_2 nanoparticle is capable of inducing genotoxicity in plants even at a low concentration (12.5 μg mL^{-1}) due to internalization of particles, resulting in oxidative stress due to ROS generation which can damage cell structures and DNA. A dose-dependent decrease in the mitotic index (69 to 21) and increase in chromosomal aberrations, DNA damage and ROS generation were observed in onion (*Allium cepa*) root tips treated with Ti nanoparticles at four different concentrations (12.5, 25, 50, 100 μg mL^{-1}) (Pakrashi et al. 2014).

10.5.12 Lead

Lead is one of the most abundant toxic elements in soil with the half-life of 740–5900 years (Iimura et al. 1977). Its allowable range in agricultural soil is 20–300 ppm (Kabata-Pendias and Sadurski 2004), whereas permissible limits for crop and medicinal plants are 0.3 ppm (FAO/WHO 2001) and 10 ppm (WHO 1998), respectively. General symptoms due to Pb toxicity in plants are chlorosis, necrosis, inhibition of root growth, blackening of root system and underdeveloped shoot (Table 10.2).

Increase in treatments of Pb from 0 to 3 mM in aqueous solution is known to reduced seed germination and biomass in *T. aestivum*, with 50.0 % reduction in their values at 3 mM of Pb (Lamhamdi et al. 2011). Inhibition of seedling growth under high Pb levels was reported in rice (Mishra and Choudhuri 1999), maize (Małkowski et al. 2002) and medicinal plants (Street et al. 2007). In *Sinapis arvensis*, Pb at 1500 μM reduced seed germination and plant biomass by 10.23 and 23.0 %, respectively, whereas Pb treatments beyond 400 μM showed more than 50.0 % reduction in biomass as well as water content of *B. juncea* (Zaier et al. 2010). Lead accumulation beyond its permissible level caused inhibition of leaf expansion, root and stem elongation in *A. cepa* (Gruenhage and Jager 1985) and *H. vulgare* (Juwarkar and Sinde 1986). Low amount of Pb (0.005 ppm) has also been reported to cause significant reduction in the growth of lettuce (*L. sativa*) and carrot (*Daucus carota*) roots, primarily due to Pb-induced simulation of indol-3 acetic acid (IAA) oxidation (Barker 1972). Toxicity due to Pb alters photosynthetic and enzymatic activities, water balance and mineral uptake by the plant, which affects their normal physiological functioning (Sharma and Dubey 2005). A high Pb level in soil induced abnormal plant morphology such as irregular radial thickening in pea (*Pisum sativum*) roots, cell walls of the endodermis and lignification of the cortical parenchyma (Paivoke 1983). High Pb concentrations inhibited the activities of enzymes at cellular level by reacting with their sulfhydryl groups and induced oxidative stress by increasing the production of ROS in plants (Reddy et al. 2005).

10.5.13 Cadmium

Ranked seventh amongst top toxins affecting environment and living beings, Cd is of major environmental concern to agriculture system in the vicinity of industries because of its longer residence period (>1000 years) in soil (Nazar et al. 2012). Although Cd is a non-essential metal for medicinal and crop plants, it is an extremely significant pollutant due to its high toxicity and large solubility in water, resulting in an easy uptake by plants when grown in soil either supplemented or contaminated with Cd. For uncontaminated or agricultural soil, allowable range of Cd is 1–5 ppm (Kabata-Pendias and Sadurski 2004), whereas for crop and medicinal plants, permissible limits are 0.2 ppm (FAO/WHO 2001) and 0.3 ppm (WHO 2005), respectively. Cadmium, beyond its phytotoxic thresholds for crop and medicinal plants cause chlorosis, growth inhibition, leaf rolls and stunting, browning of root tips, biomass reduction and finally death which are the main and easily visible symptoms of Cd toxicity when grown in soil containing high levels of Cd (Table 10.2).

Accumulation of Cd in edible plants may cause several physiological, biochemical and structural changes (Feng et al. 2010). In *Rhazya stricta*, a traditional medicine used in treatment of diabetes mellitus, skin infections and stomach disorders, total concentration of Cd was found 9.63 ppm which caused chlorosis and growth reduction of the plant (Nawab et al. 2015b). In *C. sativus*, Cd at 5 M concentration or higher induced Fe(II) deficiency by inhibiting root Fe(III) reductase, which affects photosynthesis (Alcantara et al. 1994). Cd caused alteration in uptake and transport of essential nutrients by plants either by affecting the availability of minerals from the soil or by reducing the soil microbial population (Moreno et al. 1999). Cd toxicity can affect the plasma membrane permeability thereby altering water balance and stomatal opening in mung bean (Hossain et al. 2010). Fodor et al. (1995) observed reduced ATPase activity in roots of wheat and sunflower and altered membrane functionality inducing lipid peroxidation at higher Cd levels. Cd accumulation at 200 μM reduced nitrogen fixation and primary ammonia assimilation in nodules of soybean plants (Chikile et al. 2013), while 100 mM Cd uptake by mustard and soybean plants inhibited nitrate reductase activity (Balestrasse et al. 2003).

10.5.14 Bismuth

Bismuth exists in number of oxidation states but its trivalent forms (bismuth, bismuthinite and bismite) are most stable, abundant and rarely occur alone (Das et al. 2006). Very little is known about the phytotoxic effects of Bi on crop and medicinal plants due to less information on biocoordination chemistry of Bi(III) with proteins, enzymes and cell membranes. Galindo et al. (1999) observed an inhibition of radical growth due to heterocyclic Bi compounds with necrotic effects on foliage of lettuce and cucumber. Similar effects were observed on *Sorghum bicolor* by Rimando et al. (1998). Seed germination was inhibited by 50 % in lettuce at 16.1, 34.0 and 49.7 μM of different Bi compounds, whereas it was 4.6, 7.5 and 12.4 μM in *Trifolium pratense*. Cespedes et al. (2003) reported an inhibition of hypocotyl growth of *L. sativa* and *T. pratense* beyond 15 μM of Bi treatment. Nagata (2015) showed higher germination rate, total dry weight and root length of *Arabidopsis thaliana* under low concentration of Bi; however, these parameters reduced significantly above 1.0 μM Bi concentrations.

10.5.15 Arsenic

Arsenic is the most toxic metalloid widely distributed in environment as As(III) and As(V) which are ubiquitous and toxic to many life forms (Tripathi et al. 2007). The range of maximum allowable concentration of As for agricultural soil is 15–20 ppm (Kabata-Pendias and Sadurski 2004), whereas for crop and medicinal plants, permissible limits are 1.0 ppm (WHO 1992) and 5.0 ppm (WHO 1998), respectively. Growth reduction, interveinal necrosis and chlorosis have been reported as easily visible symptoms due to As phytotoxicity (Table 10.2).

Plants exposed to As undergo severe stress such as growth inhibition, improper physiological functioning and finally leading to death (Stoeva et al. 2005). It is an analogue of phosphate (P) and transported across the plasmalemma through phosphate transport systems (Smith et al. 2010). Liu et al. (2005) reported a significant decline in seed germination, biomass production and grain yield with an increase in As concentrations (0–16 mgL^{-1}) in growing medium for six varieties of *T. aestivum*. The straight head disease is a physiological disorder of *O. sativa* due to As toxicity characterized by sterile florets/spikelets leading to reduced grain yield (Smith et al. 2010). Mazumdar et al. (2015) showed significant reductions in shoot length (34.6 %), shoot biomass (27.0 %) and essential oil yield (0.08 %) in *Ocimum basilicum* under As stress. Arsenic accumulation in plants caused chloroplast membrane damage, reduced stomatal conductance and interfered with water and essential nutrient uptake by plant leading to improper functioning of photosynthetic process (Stoeva and Bineva 2003) and altered plant metabolic activities (Mokgalaka-Matlala et al. 2008). Elevated levels of As in plants thus cause considerable cellular damage through lipid peroxidation and protein and deoxyribonucleic acid damage (Pitzschke et al. 2006). In thylakoids, As may create a condition where energy level exceeds the amounts that can be dissipated by metabolic pathways of chloroplasts; as a result the electron transport system in the thylakoid membranes is impeded, and toxic symptoms develop (Stoeva et al. 2003).

10.5.16 Aluminium

Aluminium is a metalloid contributing about 7.0 % of the earth's crust and exists in non-reactive state and produces no toxic effects on plants. However, in acidic environment, Al turns to soluble forms and readily uptake by plants thus producing phytotoxic effects with certain visible symptoms such as chlorosis, yellowing, curling or rolling of young leaves and stunted growth (Table 10.2). Aluminium toxicity has been considered as a main limiting factor in crop production due to inhibition of root growth and metabolic alteration in plant cells (Inostroza-Blancheteau et al. 2012). Significant reductions in fresh weights of cotyledons, hypocotyl and radicles were observed in *J. curcas* at 2 and 3 mM of Al treatments (Ou-yang et al. 2014). It alters morphology of root cells, resulting in thick, stunted, brittle and poorly developed root system thereby affecting nutrient and water uptake (Matsumoto 2000). Net photosynthetic rate, transpiration rate and stomatal conductance reduced significantly under Al stress which could be attributed to significant reductions in length, width and area of stomata in leaves of *Scutellaria baicalensis* (YaMin et al. 2011). Greatest cell damage and ROS generation were found in the distal transition zone in roots of *Z. mays* and *S. bicolor* (Sivaguru et al. 2013). Based on previous report, cell wall is considered a major site of Al accumulation with 85–90 % of total Al accumulation roots of *H. vulgare* (Zhu et al. 2013). Higher hemicellulose content was found in wheat on exposure to 10 μM Al for 6 h (Tabuchi and Matsumoto 2001). Aluminium ions possess higher affinity (560-folds) for phosphatidylcholine by replacing Ca^{2+}resulting in inhibition of H^{+}-ATPase activity, alteration of membrane fluidity and phospholipid packing (Ahn et al. 2001). Binding of Al to nuclear materials results in their condensation and inhibition of cell division, nuclear aberration and micronuclear and binuclear cells in *H. vulgare* (Zhang 1995).

10.5.17 Molybdenum

Molybdenum is a transition element, which exists in several oxidation states ranging from zero to VI. Mo(VI) form is most commonly found in agricultural soils and is essential for growth of plants (Bergeaux 1976). The maximum allowable range of Mo for agricultural soil is 4–10 ppm (Kabata-Pendias and Sadurski 2004), whereas no permissible limits are specified for crop and medicinal plants. Importance of Mo in growth and development of higher plants was first shown by Arnon and Stout (1939). Though required only in small amounts, it has a large role within the plant system. Molybdenum itself is not biologically active but is rather predominantly found to be an integral part of an organic pterin complex called the molybdenum cofactor (MoCo) (Mendel and Hansch 2002). Brenchley (1948) reported that heavier dressing of molybdate in soil suppressed the growth of plants with an appearance of golden colour toxic symptoms due to molybdenum poisoning. Mo toxicity leads to marginal leaf scorch, abscission, yellowing and browning of leaves and depressed tillering (Table 10.2).

For barley and oats, toxic effects of Mo were observed when it reached 135 and 200 ppm, respectively (Davis et al. 1978; Hunter and Vergnano 1953). Significant reductions in maize seedling growth (Kovács et al. 2015) and grain yield of wheat and barley (Gupta 1971) were observed in soil containing excessive levels of Mo. Foliar application of 40 g Mo ha^{-1} at 25 days after plant emergence resulted in higher reduction in acetylene and nitrate reductase activities in bean (Vieira et al. 1998).

10.5.18 Chromium

This is one of the most common contaminant in soil, water and sediments mainly due to industrial activities. Amongst its different valance states, Cr (III) and Cr (VI) are most stable and common in terrestrial environment (Kimbrough et al. 1999). In agricultural soil, 20–500 ppm is a maximum allowable range of Cr (Kabata-Pendias and Sadurski 2004), whereas permissible limits set by WHO are 2.3 and 1.5 ppm for crop (FAO/WHO 2001) and medicinal plants (WHO 1998), respectively. It is a non-essential element which produces toxic effects on plant's growth and development as its concentrations reaches beyond the phytotoxic threshold. Phytotoxic effects of Cr are characterized by reduced plant growth and chlorosis in young leaves followed by wilting.

Seed germination is a first physiological process affected by Cr (Peralta et al. 2001). Reductions of 51.1 and 57 % in seed germination of *Hibiscus esculentus* (Amin et al. 2013) and sugarcane (*Saccharum officinarum*) bud germination (Jain et al. 2000) were observed under 100 and 80 ppm Cr treatments, respectively. Similarly, seed germination and total plant biomass in different cultivars of *T. aestivum* showed significant reductions with increase in concentrations of Cr from 0 to 125 ppm (Datta et al. 2011). Reduced seed germination under Cr stress could be ascribed to inhibition of amylases and enhancement in protease activities (Zeid

2001). Elevated Cr accumulation by *Vetiveria zizanoides* caused reductions in root length, biomass and essential oil yield (Prasad et al. 2014). Chromium stress affects photosynthesis in terms of CO_2 fixation, electron transport, photophosphorylation and enzyme activities (Clijsters and Van Assche 1985). Higher Cr exposure to plants may disrupt the defence mechanism by inactivating enzymatic and non-enzymatic antioxidants (Gwóźdź et al. 1997). Chromium stress alters normal plant metabolism by altering pigment's production such as chlorophyll and anthocyanin (Boonyapookana et al. 2002), elevating production of glutathione and ascorbic acid (Shanker et al. 2005) and by altering metabolic pool to channelize the production of new metabolites, which either exhibit resistance or tolerance to Cr stress (Schmfger 2001).

10.5.19 Mercury

Mercury is a rare element, ubiquitously distributed in the environment in trace amounts in two ionic forms Hg^{2+} and Hg^{+} amongst which Hg^{2+} is highly water-soluble and strongly phytotoxic (Goldwater 1971). In agricultural soil, maximum allowable range of Hg is 0.5–5 ppm (Kabata-Pendias and Sadurski 2004), whereas for crop and medicinal plants, its permissible limits are 0.3 (FAO/WHO 2001) and 0.2 ppm (WHO 1998), respectively. Above permissible limit, it produces toxic effects on human health through crop and medicinal plants leading to contamination of food chain and causing disease such as "Minamata". Concentrations of Hg beyond its phytotoxic levels can induce visible injuries such as chlorosis and reduction in growth and yield (Table 10.2).

Seed germination of bean showed a significant reduction when exposed to 2-methoxy ethyl mercuric chloride and mercurous chloride (Semu et al 1985). Use of mercury-based pesticide treatments in agricultural fields caused damaging effects on wheat crops characterized by hypertrophy of roots and coleoptile of seedlings, inhibition of cell division in the apical meristem of plumule and extreme cell enlargement of existing cells (Purdy 1956). These adverse effects of Hg on seed germination and growth could be attributed to Hg interference with –SH system in living cells (Sass 1937). Exposure of green gram (*Vigna radiata*) to Hg caused reductions in its biomass by 95.0 % with significant inhibition of α- amylase activity (Varshney 1990). Godbold (1991) reported significant reductions in K, Mg and Mn contents in root due to Hg induced Fe and Ca accumulation. Substitution of Hg in central atom of chlorophyll is a damaging mechanism thus affecting light and dark photosynthesis (Krupa and Baszymski 1995). An elevated level of Hg uptake by plants can bind to water channel proteins, thus inducing leaf stomatal closure and physical obstruction in water flow. Also, anatomical distortion in root and stem structures was observed at 2.0 mM $HgCl_2$ treatment to *T. aestivum* (Zhang and Tyerman 1999). Setia et al. (1994) reported reductions in cell sizes, cell wall thickness and number of vascular bundles in plants. Vijay et al. (1988) showed marked inhibitions in amino-transferase and β-amylase activities in *H. vulgare* at 50 ppm Hg treatment. Similarly, Mahajan and Dua (1993) showed a significant inhibition in endo- β-1-3-glucanase activity in *Brassica campestris* due to Hg toxicity. Somatic mutation, inhibition in spindle formation during cell division and chromosomal aberration were observed in monocot as well as dicot plants under Hg stress (De Flora et al. 1994). A high level of Hg interferes with mitochondrial activity and induced oxidative stress by triggering ROS generation which led to disruption of biomembrane lipids and cellular metabolism in cucumber seedlings (Cargnelutti et al. 2006).

10.5.20 Silicon

Silicon is the second most abundant element after oxygen both on the surface of the earth's crust and in soils in the form of silicic acid at concentrations normally ranging from 0.1 to 0.6 mM (Epstein 1999). Although, Si is not recognized as an essential element for plant but its beneficial effects on the growth, development, yield and disease resistance in many crop plants such as maize, rice and some cyperaceous plants have been observed within certain limits (Liang et al. 2005; Ma and Yamaji 2006).

Like all the metals we discussed above, silica compounds when present in excess start interfering with the plant's normal development and defence mechanisms. Côté-Beaulieu et al. (2009) observed yellow streaks on foliage and brittle leaves of wheat treated with monomethyl and dimethyl silicic acid followed by a stunted growth after 10 days of treatments (Table 10.2). Dimethyl silicic acid at 0.1 mM concentration was found sufficient to reduce growth and development of wheat plant (Côté-Beaulieu et al. 2009). Reduction in length of third and second leaves of rice and wheat plants, respectively, was observed at 20 mM of silicic acid which could be due to the formation of polymer and then changes to gel at higher concentration of Si resulting in later period of Si deficiency (Hossain et al. 2002). It has been reported that Si promotes cell wall extensibility in the growing zone and decreased the cell wall extensibility in the basal zone of isolated stellar tissues in the roots of *S. bicolor*, implying that Si plays a significant role in enhancing root elongation and protecting the stele by hardening the cell wall of the stele and endodermal tissues (Hattori et al. 2003). Similar observation was made in roots of rice plant by Hossain et al. (2002).

10.6 Alteration in Plant Community Structure in Response to Metals from Mining and Metallurgical Industries

During mining and metallurgical activities, significant land areas are degraded, and existing habitats are replaced by solid wastes such as tailings, slags, red mud and sludge that contain several metals. Soil is the main terrestrial sink for such toxic and persistent industrial pollutants, and it cause alteration in the vegetation structure and physico-chemical properties of the soil (Adriano 2001). These activities are directly responsible for destruction of vegetation cover and deterioration of soil quality (Conesa et al. 2011).

Major visible symptoms of environmental stress due to industrial activities are changes in vegetation structure which also variably or invariably alters animal communities and threatens the natural biodiversity in the area. Hence, research on changes in vegetation pattern, structure of plant communities and their dynamics are useful in assessing the degree of environmental contamination and degradation. Dutta and Agrawal (2003) observed the plant growth performance, biomass accumulation and net primary productivity (NPP) of some exotic species on wasteland of coal mining area and observed significant biochemical responses in *Eucalyptus hybrid* and *Acacia auriculiformis*, whereas toxic components of mine spoils absorbed through the roots of *Cassia siamea* resulted in its reduced above-ground biomass. Impact of mining on plant communities in the district of Villa de la Paz was studied by Espinosa-Reyes et al. (2014) where they observed that the plants in the proximity of 0.3 km from mining industry were characterized by lower diversity with species richness of 13 compared to the reference site (10 km from mining industry). The most polluted sites were dominated by plant species such as *Parthenium incanum*, *Larrea tridentata*, *Zaluzania triloba*, *Jatropha dioica*, *Dyssodia acerosa*, *Zinnia acerosa* and *Bahia absinthifolia* (Espinosa-Reyes et al. 2014). Pandey et al. (2014) showed the effects of coal mining activities on plant community structure where minimum numbers of herbaceous species (19) were found in both Raniganj and Jharia coalfields compared to the reference site (Central Institute of Mining and Fuel Research (CIMFR)). Both the coalfields were dominated by *Alternanthera paronychioides* and *Cynodon dactylon*, whereas *Achyranthes aspera*, *Convolvulus* sp., *Dichanthium* sp., *Eclipta alba* and *Solanum* sp. were the most sensitive species present only at reference site and completely vanished from coal mining areas. *Eragrostis cynosuroides* and *Setaria glauca* were identified as polluphilic species only found at coal fields. Pandey et al. (2014) also observed a significant reduction in numbers of woody species in coal fields, while *Butea monosperma*, *Ficus benghalensis*, *Ficus religiosa* and *Psidium guajava* were dominant species. Moreover, canonical correspondence analysis of the study revealed that main drivers of herbaceous community structure in mining affected areas were soil total organic carbon and nitrogen, whereas woody layer community was influenced mainly by soil sulphate and phosphorus contents. The changes in species diversity indicated an increase in proportion of resistant herbs and grasses in response to altered soil characteristics due to mining activities (Pandey et al. 2014). Morrey et al. (1988) performed an analysis of vegetation composition in relation to physico-chemical variation in soil near metalliferous mining industry and found that soil pH was a main driving component in determining the species distribution affecting 51.0 % of floristic variation, whereas 43.8, 19.7 and 44.6 % of floristic variations were influenced by soil concentrations of phytoavailable Zn, Ca and Pb, respectively.

Koptsik et al. (2003) reported reductions in number, height, diameter at breast height and crown density of living trees, whereas number of standing dead increased with decrease in distance from Zn-Cu smelter in the Kola Peninsula, Russia. Moreover, reductions in the number of plant species from 13 to 5 per 100 m^2, plant cover from 100 to 20 % and total plant biomass from 1.0 to 0.15 kg m^{-2} were found at highly contaminated sites near to Zn-Cu smelter compared to reference site. Narayan et al. (1994) assessed the vegetation characteristics at different distances from HINDALCO Industries Ltd., Renukoot, an important aluminium smelter in India. It was found that important value index (IVI) of sensitive species such as *Achyranthes aspera*, *Cassia tora* and *Eclipta alba* decreased and those of tolerant species such as *Alternanthera* sp., *Cynodon* sp., *Cyprus* sp. and *Sida* sp. increased with decrease in distance from the industry. Species richness and Shannon-Wiener index though increased, while concentration of dominance reduced on moving from 1 to 11 km from the industrial premise. A quadrat study of vegetation cover was carried out by Remon et al. (2005) at solid waste dumps from the iron and steel industry at Firminy, Loire, France. It was found that 30 plant species belonging to 11 families were present, and most of these species were perennial forbs and grasses from family Asteraceae and Poaceae. Despite of the taxonomic diversity at that site, the vegetation cover was not uniform, and, inside each quadrant, the covered ground surface, the number of taxons and the type of dominant species were highly variable (Remon et al. 2005).

10.7 Risk of Food Chain Contamination

Soil contamination by anthropogenic activities results in multiparametric consequences on the quality of living beings. Disposal of residues from these industries has resulted in contamination of surrounding areas thus converting them into land not suitable for agricultural practice. The productivity of crop and medicinal plants growing in contaminated agricultural soil or cultivation of these plants in soil amended with industrial wastes can be reduced due to elevated metal uptake (Alloway et al. 1990; Pruvot et al. 2006). Solid waste dumps which are naturally invaded by endemic species pose a potential threat of transfer of toxic metals into food chain through the accumulation of metals in above-ground plant parts. Amongst plants with special reference to high added values are medicinal plants which are commonly consumed in countries like Greece and some Mediterranean regions which are collected from contaminated sites (Pullaiah 2006). This raises a question of how safe it is for consumption of crop and medicinal plants collected from such contaminated areas (Fig. 10.5). Cultivation of edible plants mainly crop and vegetables for human or livestock consumption on contaminated sites can potentially lead to uptake, accumulation and biomagnification of toxic metals such as Cd, Pb, Hg, As, Cr, etc. with a resultant risk to human and animal health leading to serious systemic health issues (Gautam et al. 2016; Sharma and Agrawal 2005).

Zhuang et al. (2009) reported that in Asia's staple crop "rice", cultivated in the vicinity of Dabaoshan Mine, it contained many folds higher Pb (8-folds) and Cd (6.5-folds) concentrations in comparison to their maximum permissible limit (MPL) as per national safety standard for milled rice (NPSF 2002). Similarly, in corn grains grown in the vicinity of Pb-Zn mine, Liaoning, concentrations of Cd and Pb were found 1.5 and 2 times higher than their MPL, respectively (Gu et al. 2005).

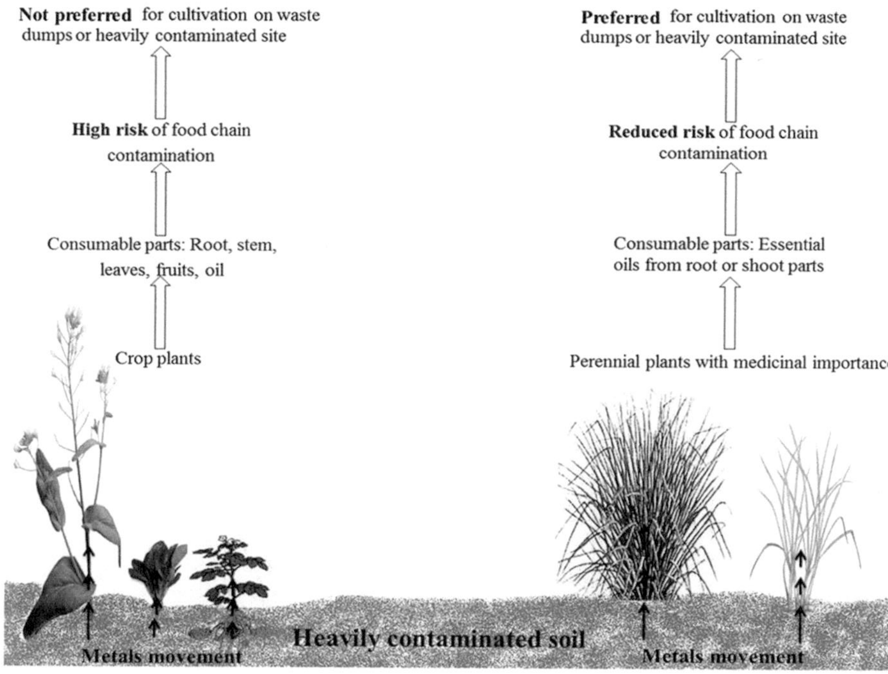

Fig. 10.5 Risk of food chain contamination through plants grown either on industrial waste dumps or meta contaminated soil

Concentration of Cd was assessed in leafy vegetables grown in a village near Dabaoshan Mine where it was found 2.5 times higher than its MPL in spinach and 3.8 times of MPL in carrot (Zhuang et al. 2009; Hernandez et al. 2003). Ok et al. (2011) reported 0.90 ppm of Cd accumulation in rice grain, grown in metal-contaminated paddy field in the vicinity of an abandoned metal mine in South Korea. Bose and Bhattacharyya (2012) found significantly higher concentration of Zn, an essential micronutrient in wheat grain grown in JNU and Chattarpur soil amended with industrial waste. In Bulgaria, large areas of agricultural land in the vicinity of Zn-Cu smelter (0.8–3 km) are contaminated with heavy metals resulting in contamination of medicinal plants with several times higher concentrations of Cd, Pb and Zn than their allowable limits. Also, essential oil yields from sage, basil, dill, cham, coriander, lemon balm and hyssop plants were increased with increase in distance from the source (Zheljazkov et al. 2008). Similarly, Angelova et al. (2006) showed that concentrations of Pb, Cu, Zn and Cd in root, stem and leaves were manyfolds higher in *Mentha piperita*, *Salvia officinalis* and *Salvia sclera* grown at 0.1 km from non-ferrous metal industry near Plovdiv, Bulgaria, compared to reference site (15 km from the industrial premise). Linalool content in volatile oil of sweet basil was found to reduce under concentrations of Cr (10 and 20 ppm), Cd (25 and 50 ppm), Pb (25 and 50 ppm) and Ni (25 and 50 ppm) compared to uncontaminated soil (Prasad et al. 2011). Affholder et al. (2013) also assessed the effects of heavy metals on *Rosmarinus officinalis* growing in the vicinity of the former metallurgical industry (1851–1952) compared to the reference site (a suburban area) and found that of the total volatile compounds, more than 50.0 % of compounds in essential oil were reduced significantly in plants grown at contaminated site. On contrary to this, 15 compounds including alpha pinene, camphene, myrcene, limonene, etc. were increased in plants at contaminated soil.

It is clear from the studies that metal accumulation not only cause quantitative and qualitative changes in yield of economically important plants but also affect their nutritional qualities. Edible parts from crop plants grown in the vicinity of industry often contain toxic levels of potentially toxic metals, whereas in case of medicinal plants, essential oil yield and composition were found less or unaffected by metal contaminants (Zheljazkov et al. 2008).

10.8 Technological Innovations in Management of Mining and Metallurgical Solid Wastes

In recent years, almost every country is facing the challenge of managing the huge quantity of wastes generated from mining and metallurgical industries because of their accumulation and suitable storage space constraints. Therefore, the wastes are usually dumped on land, either in wet or dry forms without proper pretreatments which occupy larger land areas leading to various environmental problems within and surrounding areas. Disposal of such a huge quantity of waste poses a big challenge because of lack of their cost-effective management practices. Several technological innovations on the applications of mining and metallurgical wastes have been suggested such as manufacture of ceramics, building materials, pigments, paints, adsorbents and catalysts (Pappu et al. 2007; Wang et al. 2008).

Mining wastes such as Fe, Cu, Zn and Al tailings, coal washeries and overburden wastes are used as raw materials in the recovery of expensive minerals and manufacture of construction materials for embankments of roadways, railways, rivers, dams, bricks, tiles, lightweight aggregates and fuel (Skarżyńska1995). The Pb-Zn tailings from upper Mississippi Valley mining district were used to prepare foamed building blocks, concrete beams and tiles, dense silicate bricks and aerated concrete (Hansen et al. 1968). Dean et al. (1986) also reported the utilization of Cu mill tailings for making building bricks. In the USA, Canada and Britain, mine wastes are used in manufacturing glasses and ceramics (Jacobi 1975). In India also, mine wastes are utilized in manufacture of glasses (IBM 2002; Kumar 2000). Metallurgical wastes such as slags, red mud and galvanizing residues are used in making cement, bricks, tiles, ceramics, blocks, paints and boards (Skarżyńska 1995). Slags from non-ferrous metal industries are used in improving the strength, morphology and abrasion resistance of cement, whereas ferroalloy industrial wastes are used in making high-strength and lightweight concrete (Bhattacharya et al. 2004). Gorai and Jana (2003) reported the use of Cu slags in preparing tiles, mine backfill and granular materials. Similarly, red mud from aluminium industries has been utilized as a substitute for ordinary clay for producing bricks, polymer, composites, wood substitute products, ceramic glazes and in metal recovery (Sglavo et al. 2000; Saxena and Mishra 2004).

The engineered techniques in management of industrial wastes are although advanced and highly efficient but are still at their initial stages of development. Particularly in developing countries like India, these technologies do not offer a cost-effective option at the moment. Conventional mechanical or physico-chemical treatments such as excavation, soil washing, solidification/stabilization, electrokinetic remediation and soil incineration also suffer from limitations like cost ineffective, require intensive labour, cause irreversible soil disturbances, etc. Therefore, rehabilitation of industrial waste dumps by revegetation is an environmentally benign process to safeguard the environment.

The approach of ecological restoration is the most accepted and cost-effective way to restore the ecological integrity of disturbed land due to mining and metallurgical activities. The goal of restoration is usually to develop a long-term sustainable management of residual dumps in industrial areas. It includes the management of all types of physical, chemical and biological disturbances in soils such as soil pH, fertility, microbial community and various soil nutrient cycles. Revegetation and reclamation of waste dumps are extremely difficult, due to physical or chemical limitations to plant growth and presence of potentially toxic concentrations of heavy metals in the spoil (Conesa et al. 2011). Such constraints can be resolved by adding suitable soil amenders such as sawdust, wood residues, sewage sludge and animal manures, as these amendments stimulate the microbial activities and add up nutrients (N, P, K) and organic carbon to the soil thereby reducing the phytotoxic effects of metals (Juwarkar and Jambhulkar 2008). Suitable soil amenders or microbial assisted revegetation minimizes the damages and helps in recovering the waste dumps by stabilization through development of extensive root systems. Once vegetation gets established on waste dumps, it improves soil organic matter and nutrient status, lower soil bulk density, moderate soil pH and enhances nutrient bioavailability in soil (Conesa et al. 2007; Mendez and Maier 2008). For revegetation, it is necessary to choose drought-resistant, metal-tolerant, fast-growing plants with dense canopies and root systems (Mendez and Maier 2008). Dutta and Agrawal (2002) suggested that indigenous plants should be preferred over exotic species for reclamation of the coal mine spoil dumps because indigenous plants easily fit into a fully functional ecosystem and adopt climatically also. Annual grasses are considered as a nurse crop for an early vegetation purpose offering superior tolerance to drought, low soil nutrients and other climatic stresses. Roots of grasses are fibrous that can slow erosion, and their soil-forming tendencies eventually produce a layer of organic soil, stabilize soil, conserve soil moisture and enable them to compete with weedy species. The initial vegetation cover must be allowed for the development of diverse self-sustaining plant communities (Singh et al. 2002; Xiuzhen et al. 2004).

Revegetation is a widely used technique for stabilization of dumps (Singh 1996) and maintaining ecological equilibrium of mining and metallurgical areas (Jørgensen 1994). Restoration or reclamation of industrial dumps coupled with phytoremediation triggers the stabilization of dumps with further reduced risk of environmental contamination (Salt et al. 1998). Juwarkar and Jambhulkar (2008) attempted a phytoremediation of coal mine spoil dumps by using effluent treated plant sludge, organic amendments and biofertilizer (*Rhizobium* sp., *Azotobacter* sp. and VAM spores) inoculation along with suitable plant species, to improve the physico-chemical properties of coal mine spoil and to reduce the metal toxicity in spoil. Microbial inoculation with organic amendments helped in reducing the concentrations of heavy metals such as chromium, zinc, copper, iron, manganese, lead, nickel and cadmium by 41.0, 43.0, 37.0, 37.0, 34.0, 39.0, 37.0 and 40 %, respectively. For the process of phytoremediation, it is preferable to use the plants which are metallophytes, pseudometallophytes and hyperaccumulators (Meagher 2000) as these plants have evolved biological mechanism to resist, tolerate and thrive in metalliferous soils. They are an optimal choice for restoration of mining and metallurgical closure for rehabilitation of metal-contaminated sites and underpinning for the development of environmental technologies such as phytoremediation of metals and making the substrate favourable for the flourishment of sensitive plant species (Adams and Lamoureux 2005).

Despite of the fact of human health risk, several researchers recommended many edible crops for phytoremediation purposes (Gupta et al. 2013). But utilization of crop plants for phytoremediation does not seem to be an intelligent option because heavy metals may enter into food chain through consumption by humans and animals (Vamerali et al. 2010). Contrary to this, many aromatic plants are not being consumed directly by humans and animals and hence can be grown for the production of essential oil, in case oils can strictly qualify the safety limits for toxic contaminants as have been shown in some studies (Lal et al. 2013; Zheljazkov et al. 2008). Gupta et al. (2013) also suggested the use of aromatic plants rather than non-aromatic edible crops for cultivation in metal-polluted land as a sustainable and environmental-friendly technique.

10.9 Conclusion

Mining and metallurgical industries are crucial for development of any country; however, wastes generated from these industries pose a threat to human and biological welfare. It is evident from several reports that extreme physico-chemical and biological properties and excessive amounts of metals affect adversely to native as well as exotic plant species including crop and medicinal plants. Both plant essential and non-essential metals when exceeding their phytotoxic threshold interfere with several metabolic processes, causing toxicity to the plants as exhibited by reduced growth, chlorosis, impaired photosynthesis and finally plant death. As industrial development continues, sustainable and environmental-friendly ways to manage these wastes remain a big challenge, but restoration of such polluted sites by rehabilitation and phytoremediation using native and medicinal plants could be a sustainable and environmental viable option for better management of these wastes. Such studies, however, need to be designed cautiously with utmost care to prevent any further contamination of food chain or environment.

Acknowledgements Divya Pandey and Meenu Gautam are thankful to the AXA Junior Research (Post-doctoral) Fellowship, UK, and Council of Scientific and Industrial Research (CSIR), India, respectively, in the form of Research Associateship and Senior Research Fellowship, respectively.

References

Abou-Arab AAK, Abou Donia MA (2000) Heavy metals in Egyptian spices and medicinal plants and the effect of processing on their levels. J Agric Food Chem 48(6):2300–2304

Adamo P, Dudka S, Wilson MJ, McHardy WJ (2002) Distribution of trace elements in soils from the Sudbury smelting area (Ontario, Canada). Water Air Soil Pollut 137(1–4):95–116

Adams PW, Lamoureux S (2005) A literature review of the use of native northern plants for the re-vegetation of arctic mine tailings and mine waste. Environ Nat Resour, Northwest Territories, pp 21–22. http://www.enr.gov.nt.ca/sites/default/files/wkssnorthern plants_re-vegetation-2005.pdf. Accessed on 12 Mar 2016

Adriano DC (ed) (2001) Trace elements in the terrestrial environments: biogeochemistry, bioavailability and risks of heavy metals, 2nd edn. Springer, New York

Affholder MC, Prudent P, Masotti V, Coulomb B, Rabier J, Nguyen-The B, Laffont-Schwob I (2013) Transfer of metals and metalloids from soil to shoots in wild rosemary (*Rosmarinus officinalis* L.) growing on a former lead smelter site: human exposure risk. Sci Total Environ 454:219–229

Ahmad MSA, Ashraf M, Hussain M (2011) Phytotoxic effects of nickel on yield and concentration of macro-and micro-nutrients in sunflower (*Helianthus annuus* L.) achenes. J Hazard Mater 185(2):1295–1303

Ahn SJ, Sivaguru M, Osawa H, Chung GC, Matsumoto H (2001) Aluminum inhibits the H^+-ATPase activity by permanently altering the plasma membrane surface potentials in squash roots. J Plant Physiol 126(4):1381–1390

Albano JP, Miller WB, Halbrooks MC (1996) Iron toxicity stress causes bronze speckle, a specific physiological disorder of marigold (Tagetes erecta L.). J Am Soc Hortic Sci 121:430–437

Alcántara E, Romera FJ, Cañete M, Manuel D (1994) Effects of heavy metals on both induction and function of root Fe (lll) reductase in Fe-deficient cucumber (*Cucumis sativus* L.) plants. J Exp Bot 45(12):1893–1898

Al-Khateeb W, Al-Qwasemeh H (2014) Cadmium, copper and zinc toxicity effects on growth, proline content and genetic stability of *Solanum nigrum* L., a crop wild relative for tomato; comparative study. J Physiol Mol Biol Plants 20(1):31–39

Alloway B, Ayres DC (eds) (1997) Chemical principles of environmental pollution, 2nd edn. CRC Press/University of London, London

Alloway BJ, Jackson AP, Morgan H (1990) The accumulation of cadmium by vegetables grown on soils contaminated from a variety of sources. Sci Total Environ 91:223–236

Amin H, Arain BA, Amin F, Surhio MA (2013) Phytotoxicity of chromium on germination, growth and biochemical attributes of *Hibiscus esculentus* L. Am J Plant Sci 4(12):2431–2439

Angelova V, Ivanov K, Ivanova R (2006) Heavy metal content in plants from family Lamiaceae cultivated in an industrially polluted region. J Herbs Spices Med Plants 11(4):37–46

Arizona Cooperative Extension (ACE) (1998) Diagnosing plant damage: key to symptoms of chemical disorders on individual plants. College of Agriculture, University of Arizona https://ag.arizona.edu/pubs/garden/mg/damage/key.html. Accessed on 12 Mar 2016

Arnon DI, Stout PR (1939) Molybdenum as an essential element for higher plants. J Plant Physiol 14(3):599

Atanassova B, Zapryanova N (2009) Influence of heavy metal stress on growth and flowering of Salvia Splendens Ker.-Gawl. Biotechnol Biotechnol Equip 23(1):173–176

Baba A, Tayfur G (2011) Groundwater contamination and its effect on health in Turkey. Environ Monit Assess 183:77–94

Bacchetta G, Cappai G, Carucci A, Tamburini E (2015) Use of native plants for the remediation of abandoned mine sites in Mediterranean semiarid environments. B Environ Con Tox 94(3):326–333

Bachman GR, Miller WB (1995) Iron chelate inducible iron/manganese toxicity in zonal geranium. J Plant Nutr 18(9):1917–1929

Bailey D, Bilderback T, Bir D (1999) Water considerations for container production of plants. In: Horticulture information leaflet. Raleigh: North Carolina State University, NC Cooperative Extension Service. http://faculty.caes.uga.edu/pthomas/hort4050.web/Hort4050web/HIL557.pdf.Accessed 12 Mar 2016

Balestrasse KB, Benavides MP, Gallego SM, Tomaro ML (2003) Effect of cadmium stress on nitrogen metabolism in nodules and roots of soybean plants. Funct Plant Biol 30(1):57–64

Barker WG (1972) Toxicity levels of mercury, lead, copper, and zinc in tissue culture systems of cauliflower, lettuce, potato, and carrot. Can J Bot 50(5):973–976

Becker M, Asch F (2005) Iron toxicity in rice – conditions and management concepts. J Plant Nutr Soil Sci 168(4):558–573

Bergeaux PJ (1976) The need for micronutrients in Georgia agriculture. Bull No 784, pp 1–18

Besford RT, Maw GA (1975) Effect of potassium nutrition on tomato plant growth and fruit development. Plant Soil 42(2):395–412

Bhattacharyya JK, Shekdar AV, Gaikwad SA (2004) Recyclability of some major industrial solid wastes. J Ind Assoc Environ Mange 31:71–75

Boonyanitipong PKB, Kumar P, Baruah S, Dutta J (2011) Toxicity of ZnO and TiO_2 nanoparticles on germinating rice seed *Oryza sativa* L. Int J Biosci Biochem Bioinform 1:282–285

Boonyapookana B, Upatham ES, Kruatrachue M, Pokethitiyook P, Singhakaew S (2002) Phytoaccumulation and phytotoxicity of cadmium and chromium in duckweed *Wolffia globosa*. Int J Phytorem 4(2):87–100

Borgatto F, Dias CTDS, Amaral AFC, Melo M (2002) Calcium, potassium and magnesium treatment of *Chrysanthemum morifolium* cv. "Bi Time" and callogenesis in vitro. Sci Agricola 59(4):689–693

Bose S, Bhattacharyya AK (2012) Effect of industrial sludge application on soil nitrogen and wheat plant response. Open J Soil Sci 2(02):133

Boulet MP, Larocque AC (1998) A comparative mineralogical and geochemical study of sulfide mine tailings at two sites in New Mexico, USA. Environ Geol 33(2–3):130–142

Brenchley WE (1948) The toxic action of molybdenum in relation to soils and crops. Ann App Biol 35(2):139–160

Brooks RR (1987) Serpentine and its vegetation: a multidisciplinary approach. Dioscorides Press. http://agris.fao.org/agris-search/search.do?recordID=US8918938. Accessed on 12 Mar 2016

Broschat TK, Moore KK (2004) Phytotoxicity of several iron fertilizers and their effects on Fe, Mn, Zn, Cu, and P content of African marigolds and zonal geraniums. Hortic Sci 39(3):595–598

Brugnoli E, Lauteri M (1991) Effects of salinity on stomatal conductance, photosynthetic capacity, and carbon isotope discrimination of salt-tolerant (*Gossypium hirsutum* L.) and salt-sensitive (*Phaseolus vulgaris* L.) C3 non-halophytes. J Plant Physiol 95:628–635

Bureau P, Roederer O (1960) Contribution a Fetude des sols gypseux du Sud Tunisien, Croutes et encroutements gypseux de la partie Sud de Golfe de Gabes. Bull Assoc Franc Etude Sol 8:145–149

Burke DJ, Pietrasiak N, Situ SF, Abenojar EC, Porche M, Kraj P, …, Samia ACS (2015) Iron oxide and titanium dioxide nanoparticle effects on plant performance and root associated microbes. Int J Mol Sci 16(10):23630–23650

Cabala J, Teper L (2007) Metalliferous constituents of rhizosphere soils contaminated by Zn–Pb mining in southern Poland. Water Air Soil Pollut 178(1–4):351–362
Cargnelutti D, Tabaldi LA, Spanevello RM, de Oliveira Jucoski G, Battisti V, Redin M, ..., Morsch VM (2006) Mercury toxicity induces oxidative stress in growing cucumber seedlings. Chemosphere 65(6):999–1006
Caspi V, Droppa M, Horváth G, Malkin S, Marder JB, Raskin VI (1999) The effect of copper on chlorophyll organization during greening of barley leaves. Photosynth Res 62(2):165–174
Cele EN, Maboeta M (2016) A greenhouse trial to investigate the ameliorative properties of biosolids and plants on physicochemical conditions of iron ore tailings: implications for an iron ore mine site remediation. J Environ Manag 165:167–174
Céspedes CL, Lemus A, Salazar JR, Cabrera A, Sharma P (2003) Herbicidal, plant growth inhibitory, and cytotoxic activities of bismuthines containing aromatic heterocycles. J Agric Food Chem 51(10):2923–2929
Chang YC, Grace-Martin K, Miller WB (2004) Efficacy of exogenous calcium applications for reducing upper leaf necrosis in Lilium Star Gazer. Hortic Sci 39(2):272–275
Chartzoulakis K, Klapaki G (2000) Response of two greenhouse pepper hybrids to NaCl salinity during different growth stages. Sci Hortic 86:247–260
Chartzoulakis KS, Loupassaki MH (1997) Effects of NaCl salinity on germination, growth, gas exchange and yield of greenhouse eggplant. Agric Water Manag 32(3):215–225
Chatterjee J, Chatterjee C (2000) Phytotoxicity of cobalt, chromium and copper in cauliflower. Environ Pollut 109(1):69–74
Chikile SJ, Sharma R, Mogallapu NR (2013) Cadmium toxicity on Soybean (*Glycine max* L.). Ind J App Pure Biol 28(2):257–264
Choi HG, Moon BY, Bekhzod K, Park KS, Kwon JK, Lee JH, ..., Kang NJ (2015) Effects of foliar fertilization containing titanium dioxide on growth, yield and quality of strawberries during cultivation. Hortic Environ Biotechnol 56(5):575–581
Clark RB (1982) Plant response to mineral element toxicity and deficiency [Potentially arable soils in the world]. In: Breeding plants for less favorable environments. International Council on Lethal Yellowing, USA. http://agris.fao.org/agris-search/search.do?recordID=US19830863455. Accessed on 12 Mar 2016
Clijsters H, Van Assche F (1985) Inhibition of photosynthesis by heavy metals. Photosynth Res 7(1):31–40
Conesa HM, Faz Á (2011) Metal uptake by spontaneous vegetation in acidic mine tailings from a semiarid area in south Spain: implications for revegetation and land management. Water Air Soil Pollut 215(1-4):221–227
Conesa HM, Schulin R, Nowack B (2007) A laboratory study on revegetation and metal uptake in native plant species from neutral mine tailings. Water Air Soil Pollut 183(1-4):201–212
Conesa HM, María-Cervantes A, Álvarez-Rogel J, González-Alcaraz MN (2011) Influence of soil properties on trace element availability and plant accumulation in a Mediterranean salt marsh polluted by mining wastes: implications for phytomanagement. Sci Total Environ 409(20):4470–4479
Costa G, Polettini A, Pomi R, Stramazzo A (2016) Leaching modelling of slurry-phase carbonated steel slag. J Hazard Mater 302:415–425
Côté-Beaulieu C, Chain F, Menzies JG, Kinrade SD, Bélanger RR (2009) Absorption of aqueous inorganic and organic silicon compounds by wheat and their effect on growth and powdery mildew control. Environ Exp Bot 65(2):155–161
Das AK, Chakraborty R, Cervera ML, de la Guardia M (2006) Analytical techniques for the determination of bismuth in solid environmental samples. Trends Anal Chem 25:599–608
Datta JK, Bandhyopadhyay A, Banerjee A, Mondal NK (2011) Phytotoxic effect of chromium on the germination, seedling growth of some wheat (*Triticum aestivum* L.) cultivars under laboratory condition. J Agric Technol 7(2):395–402
Davis RD, Beckett PHT, Wollan E (1978) Critical levels of twenty potentially toxic elements in young spring barley. J Plant Soil 49(2):395–408

Dayan FE, Hernandez A, Allen SN, Moraes R, Vroman JA, Avery MA, Duke SO (1999) Comparative phytotoxicity of artemisinin and several sesquiterpene analogues. Phytochemistry 50:607–614
De Dorlodot S, Lutts S, Bertin P (2005) Effects of ferrous iron toxicity on the growth and mineral composition of an interspecific rice. J Plant Nutr 28(1):1–20
De Flora S, Bennicelli C, Bagnasco M (1994) Genotoxicity of mercury compounds: a review. Mutat Res/Rev Gen Tox 317(1):57–79
De Kreij C, Van Den Berg THJM (1990) Nutrient uptake, production and quality of *Rosa hybrida* in rockwool as affected by Electrical Conductivity of the Nutrient Solution. In: Van Beusichem ML (ed) Plant nutrition: physiology and applications. Springer, Dordrecht, pp 519–523
De Kreij C, Van Os PC (1989) Production and quality of Gerbera in rockwool as affected by electrical conductivity of the nutrient solution. In: Proceedings of the seventh international congress on soilless culture, Flevohof, Netherlands, 13–21 May 1988
De Kreij C, Janse J, Van Goor BJ, Van Doesburg JDJ (1992) The incidence of calcium oxalate crystals in fruit walls of tomato (*Lycopersicon esculentum* Mill.) as affected by humidity, phosphate and calcium supply. J Hortic Sci 67(1):45–50
Dean KC, Froisland LJ, Shirts MB (1986) Utilization and stabilization of mineral waste. Bulletin 688. United States: Department of Interior Bureau of Mines, p 46
Disante KB, Fuentes D, Cortina J (2010) Sensitivity to zinc of Mediterranean woody species important for restoration. Sci Total Environ 408(10):2216–2225
Doncheva S, Stoynova Z, Velikova V (2001) Influence of succinate on zinc toxicity of pea plants. J Plant Nutr 24(6):789–804
Douay F, Pruvot C, Waterlot C, Fritsch C, Fourrier H, Loriette A, …, Scheifler R (2009) Contamination of woody habitat soils around a former lead smelter in the North of France. Sci Total Environ 407(21):5564–5577
Dube BK, Sinha P, Gopal R, Chatterjee C (2004) Chromium phytotoxicity alters metabolism in radish. J Veg Crop Prod 10(2):61–71
Dumon JC, Ernst WHO (1988) Titanium in plants. J Plant Physiol 133(2):203–209
Dutta RK, Agrawal M (2002) Effect of tree plantations on the soil characteristics and microbial activity of coal mine spoil land. J Trop Ecol 43(2):315–324
Dutta RK, Agrawal M (2003) Restoration of opencast coal mine spoil by planting exotic tree species: a case study in dry tropical region. Ecol Eng 21(2):143–151
Ene A, Pantelică A (2011) Characterization of metallurgical slags using low-level gamma-ray spectrometry and neutron activation analysis. Rom J Phys 56:7–8
Epstein E (1980) In: Spanswik RM, Lucas WJ, Dainty J (eds) Plant membrane transport: current conceptual Carlen issues. Elsevier/North Holland Biomedical Press, Amsterdam, p 32
Epstein E (1999) Silicon. Annu Rev Plant Biol 50(1):641–664
Espinosa-Reyes G, González-Mille DJ, Ilizaliturri-Hernández CA, Mejía-Saavedra J, Cilia-López VG, Costilla-Salazar R, et al (2014) Effect of mining activities in biotic communities of Villa de la Paz, San Luis Potosi, Mexico BioMed Res Int, pp 1–13
Evans HJ, Sorger GJ (1966) Role of mineral elements with emphasis on the univalent cations. Annu Rev Plant Physiol 17(1):47–76
Fageria NK, Rabelo NA (1987) Tolerance of rice cultivars to iron toxicity. J Plant Nutr 10(6):653–661
FAO/WHO (Codex Alimentarius Commission) (2001) Food additives and contaminants. Joint FAO/WHO Food Standards Program, ALINORM 01/12A:1-289
Fargašová A (2001) Phytotoxic effects of Cd, Zn, Pb, Cu and Fe on Sinapis alba L. seedlings and their accumulation in roots and shoots. Biol Planta 44(3):471–473
Feizi H, Moghaddam PR, Shahtahmassebi N, Fotovat A (2012) Impact of bulk and nanosized titanium dioxide (TiO2) on wheat seed germination and seedling growth. Biol Trace Elem Res 146(1):101–106

Feng J, Shi Q, Wang X, Wei M, Yang F, Xu H (2010) Silicon supplementation ameliorated the inhibition of photosynthesis and nitrate metabolism by cadmium (Cd) toxicity in *Cucumis sativus* L. Sci Hortic 123(4):521–530

Filippi M, Drahota P, Machovič V, Böhmová V, Mihaljevič M (2015) Arsenic mineralogy and mobility in the arsenic-rich historical mine waste dump. Sci Total Environ 536:713–728

Fodor E, Szabó-Nagy A, Erdei L (1995) The effects of cadmium on the fluidity and H^+-ATPase activity of plasma membrane from sunflower and wheat roots. J Plant Physiol 147(1):87–92

Foy CD, Chaney RLT, White MC (1978) The physiology of metal toxicity in plants. Annu Rev Plant Physiol 29(1):511–566

Frank R, Stonefield KI, Suda P, Potter JW (1982) Impact of nickel contamination on the production of vegetables on an organic soil, Ontario, Canada, 1980–1981. Sci Total Environ 26(1):41–65

Führs H, Hartwig M, Buitrago L, Heintz D, Van Dorsselaer A, Braun H, Horst W (2008) Early manganese-toxicity response in *Vigna unguiculata* L. – a proteomic and transcriptomic study. Proteomics 8:149–159

Gajewska E, Skłodowska M, Słaba M, Mazur J (2006) Effect of nickel on antioxidative enzyme activities, proline and chlorophyll contents in wheat shoots. Biol Planta 50(4):653–659

Galindo JC, Hernandez A, Dayan FE, Tellez M, Macias FA, Paul RN, Duke SO (1999) Dehrydozaluzanin C, a natural sesquiterpenolide, causes rapid plasma membrane leakage. Phytochemistry 52:805–813

Galindo R, Padilla I, Rodríguez O, Sánchez-Hernández R, López-Andrés S, López-Delgado A (2015) Characterization of solid wastes from aluminum tertiary sector: the current state of Spanish industry. J Miner Mater Charact Eng 3(02):55

Gang A, Vyas A, Vyas H (2013) Toxic effect of heavy metals on germination and seedling growth of wheat. J Environ Res Dev 8(2):206

Gao S, Ou-yang C, Tang L, Zhu JQ, Xu Y, Wang SH, Chen F (2010) Growth and antioxidant responses in *Jatropha curcas* seedling exposed to mercury toxicity. J Hazard Mater 182(1):591–597

Gautam M, Pandey D, Agrawal M (2016) Bioprocessing of metals from packaging wastes. In: Muthu SS (ed) Environmental footprints of packaging. Springer, Singapore, pp 139–164

Geilmann (1920) The occurrence of titanium in soil and plants. Zeitschrift für Landwirtschaft 68:107–124

Gholizadeh A, Borůvka L, Vašát R, Saberioon M, Klement A, Kratina J, …, Drábek O (2015) Estimation of potentially toxic elements contamination in anthropogenic soils on a brown coal mining dumpsite by reflectance spectroscopy: a case study. PloS one 10(2):117–457

Ghosh M, Bandyopadhyay M, Mukherjee A (2010) Genotoxicity of titanium dioxide (TiO2) nanoparticles at two trophic levels: Plant and human lymphocytes. Chemosphere 81:1253–1262

Godbold DL (1991) Mercury-induced root damage in spruce seedlings. Water Air Soil Pollut 56:823–831

Goldwater LJ (1971) Mercury in the environment. Sci Am, United States 224(5)

Gorai B, Jana RK (2003) Characteristics and utilisation of copper slag – a review. Resour Conserv Recycl 39(4):299–313

Gruenhage L, Jager HJ (1985) Effect of heavy metals on growth and heavy metal content of *Allium porrum* and *Pisum sativum*. Angew Bot 59:11–28

Gu JG, Lin QQ, Hu R, Zhuge YP, Zhou QX (2005) Translocation behavior of heavy metals in soil–plant system: a case study from Qingchengzi lead–zinc mine in Liaoning Province. J Agro-Environ Sci 24:634–637

Guerinot ML, Yi Y (1994) Iron: nutritious, noxious, and not readily available. J Plant Physiol 104(3):815

Guo X, Wang K, He M, Liu Z, Yang H, Li S (2014) Antimony smelting process generating solid wastes and dust: characterization and leaching behaviors. J Environ Sci 26(7):1549–1556

Gupta UC (1971) Boron and molybdenum nutrition of wheat, barley and oats grown in Prince Edward Island soils. Can J Soil Sci 51(3):415–422

Gupta AK, Verma SK, Khan K, Verma RK (2013) Phytoremediation using aromatic plants: a sustainable approach for remediation of heavy metals polluted sites. Int J Environ Sci Technol 47(18):10115–10116

Gutiérrez-Gutiérrez SC, Coulon F, Jiang Y, Wagland S (2015) Rare earth elements and critical metal content of extracted landfilled material and potential recovery opportunities. Waste Manag 42:128–136

Gwóźdź EA, Przymusiński R, Rucińska R, Deckert J (1997) Plant cell responses to heavy metals: molecular and physiological aspects. Acta Physiol Plant 19(4):459–465

Hamzah Z, Saat A, Bakar ZA, Wood AK (2011) Anthropogenic heavy metals, U-238 and Th-232 profiles in sediments from an abandoned tin mining lake in Malaysia. In: Proceedings of 3rd International Conference on Chemical, Biological and Environmental Engineering, Singapore, 2011

Hansen TC, Richards CW, Mindess S (1968) Production of high pressure steam-cured calcium silicate building materials from mining industry waste products. Part I: Program for Investigation and Report of Field Survey of Siliceous Mine Waste in California, Department of Civil Eng, Stanford University, p 83

Hassinger BW (1997) Erosion. In: Marcus JJ (ed) Mining environmental handbook: effects of mining on the environment and American environmental controls on mining. Imperial College Press, London, pp 136–140

Hattori T, Inanaga S, Tanimoto E, Lux A, Luxová M, Sugimoto Y (2003) Silicon-induced changes in viscoelastic properties of *Sorghum* root cell walls. Plant Cell Physiol 44(7):743–749

Hernandez L, Probst A, Probst JL, Ulrich E (2003) Heavy metal distribution in some French forest soils: evidence form atmospheric contamination. Sci Total Environ 312:195–219

Hernando V, Sanchez Conde MP, Contreras JG (1965) Study of the mineral nutrition of maize on soils rich in gypsum. Zolfo in Agricoltura, Palermo 1964:398–411

Hewitt EJ (1983). In: Robb DA, Pierpoint WS (eds) Metals and micronutrients: uptake and utilization by plants. Academic, London, pp 277–300

Hodson MJ (2012) Metal toxicity and tolerance in plants. Biochemical Society, Oxford Brooks University, United Kingdom. http://www.hodsons.org/MartinHodson/Biochemist12.pdf. Accessed on 22 May 2016

Horiguchi T (1988) Mechanism of manganese toxicity and tolerance of plants: IV. Effects of silicon on alleviation of manganese toxicity of rice plants. Soil Sci Plant Nutr 34(1):65–73

Hossain MT, Mori R, Soga K, Wakabayashi K, Kamisaka S, Fujii S, …, Hoson T (2002) Growth promotion and an increase in cell wall extensibility by silicon in rice and some other Poaceae seedlings. J Plant Res 115(1): 0023–0027

Hossain MA, Hasanuzzaman M, Fujita M (2010) Up-regulation of antioxidant and glyoxalase systems by exogenous glycinebetaine and proline in mung bean confer tolerance to cadmium stress. Physiol Mol Biol Plant 16(3):259–272

Huaiwei Z, Xin H (2011) An overview for the utilization of wastes from stainless steel industries. Resour Conserv Recycl 55(8):745–754

Hunter JG, Vergnano O (1953) Trace-element toxicities in oat plants. Ann App Biol 40(4):761–777

Iimura K, Ito H, Chino M, Morishita T, Hirata H (1977) Behavior of contaminant heavy metals in soil plant system. In: Proceedings of the international seminar on soil environment and fertility management in intensive agriculture, AGRIS, FAO of United Nations

Illera V, Garrido F, Serrano S, García-González MT (2004) Immobilization of the heavy metals Cd, Cu and Pb in an acid soil amended with gypsum-and lime-rich industrial by-products. Eur J Soil Sci 55(1):135–145

Indian Bureau of Mines. (2002) Indian minerals year book, Ministry of Mines, Government of India, India. http://ibm.nic.in/. Accessed on 12 Mar 2016

Inostroza-Blancheteau C, Rengel Z, Alberdi M, de la Luz MM, Aquea F, Arce-Johnson P, Reyes-Díaz M (2012) Molecular and physiological strategies to increase aluminum resistance in plants. Mol Biol Repor 39(3):2069–2079

Ishtiaq S, Mahmood S (2012) Phytotoxicity of nickel and its accumulation in tissues of three Vigna species at their early growth stages. J App Bot Food Qual 84(2):223
Jacobi JS (1975) Recovery and reuse of metals. In: Proceedings of Minerals and the environment. The institute of mining and metallurgy proceedings of international symposium, pp 291–301
Jacobs JA, Testa SM (2005) Overview of chromium (VI) in the environment: background and history. In: Guertin J, Jacobs JA, Avakian CP (eds) Chromium (VI) handbook. Independent environmental technical group, pp 1–21
Jain R, Srivastava S, Madan VK (2000) Influence of chromium on growth and cell division of sugarcane. Indian J Plant Physiol 5(3):228–231
Jellali S, Wahab MA, Anane M, Riahi K, Bousselmi L (2010) Phosphate mine wastes reuse for phosphorus removal from aqueous solutions under dynamic conditions. J Hazard Mater 184(1):226–233
Jennings BD (1976) The effects of sodium chloride on higher plants. Biol Rev 51(4):453–486
Jørgensen SE (1994) Models as instruments for combination of ecological theory and environmental practice. Ecol Model 75:5–20
Juwarkar AA, Jambhulkar HP (2008) Phytoremediation of coal mine spoil dump through integrated biotechnological approach. Biores Technol 99(11):4732–4741
Juwarkar AS, Sinde GB (1986) Interaction of Cd, Pb: effect on growth, yield and content of Cd, Pb in barley, *Hordeum vulgare*. Ind J Environ Health 28:218–234
Kabata-Pendias A, Sadurski W (2004) Trace elements and compounds in soil. In: Elements and their compounds in the environment: occurrence, analysis and biological relevance, 2nd edn, pp 79–99
Kampfenkel K, Van Montagu M, Inzé D (1995) Effects of iron excess on *Nicotiana plumbaginifolia* plants (implications to oxidative stress). J Plant Physiol 107(3):725–735
Kang BT, Fox RL (1980) A methodology for evaluating the manganese tolerance of cowpea (*Vigna unguiculata*) and some preliminary results of field trials. Field Crops Res 3:199–210
Karlen DL, Sadler EJ, Camp CR (1987) Dry matter, nitrogen, phosphorus, and potassium accumulation rates by corn on Norfolk loamy sand. J Agro 79(4):649–656
Khatun S, Ali MB, Hahn EJ, Paek KY (2008) Copper toxicity in *Withania somnifera*: growth and antioxidant enzymes responses of in vitro grown plants. Environ Exp Bot 64(3):279–285
Kimbrough DE, Cohen Y, Winer AM, Creelman L, Mabuni C (1999) A critical assessment of chromium in the environment. Crit Rev Int J Environ Sci Technol 29(1):1–46
Kitano T, Sasaki K, Amako K, Takahashi M (1997) PCR-assisted differential display of the expression of genes that are induced by manganese-deficiency in tomato roots. In: Plant nutrition for sustainable food production and environment. Springer, Dordrecht, pp 225–226
Koptsik S, Koptsik G, Livantsova S, Eruslankina L, Zhmelkova T, Vologdina Z (2003) Heavy metals in soils near the nickel smelter: chemistry, spatial variation, and impacts on plant diversity. J Environ Monitor 5(3):441–450
Kovács B, Puskás-Preszner A, Huzsvai L, Lévai L, Bódi É (2015) Effect of molybdenum treatment on molybdenum concentration and nitrate reduction in maize seedlings. J Plant Physiol Biochem 96:38–44
Kříbek B, Majer V, Veselovský F, Nyambe I (2010) Discrimination of lithogenic and anthropogenic sources of metals and sulphur in soils of the central-northern part of the Zambian Copperbelt Mining District: a topsoil vs. subsurface soil concept. J Geochem Explor 104(3):69–86
Krupa Z, Baszynski T (1995) Some aspects of heavy metals toxicity towards photosynthetic apparatus: direct and indirect effects on light and dark reactions. Acta Physiol 17:177–190
Kumar A (2000) Conveners report. In: The souvenir of mines environment and mineral conservation week (Goa and parts of Karnataka), p 8
Küpper H, Kroneck PM (2007) Nickel in the environment and its role in the metabolism of plants and cyanobacteria. Metal Ions Life Sci 2:31–62
KuželS, Hruby M, Cígler P, Tlustoš P, Van Nguyen P (2003) Mechanism of physiological effects of titanium leaf sprays on plants grown on soil. Biol Trace Elem Res 91(2):179–189

Lal K, Yadav RK, Kaur R, Bundela DS, Khan MI, Chaudhary M, …, Singh G (2013) Productivity, essential oil yield, and heavy metal accumulation in lemon grass (*Cymbopogon flexuosus*) under varied wastewater–groundwater irrigation regimes. Indus Crops Prod 45:270–278

Lamhamdi M, Bakrim A, Aarab A, Lafont R, Sayah F (2011) Lead phytotoxicity on wheat (*Triticum aestivum* L.) seed germination and seedlings growth. Comptes Rendus Biol 334(2):118–126

Lee MK, Van Iersel MW (2008) Sodium chloride effects on growth, morphology, and physiology of *Chrysanthemum* (*Chrysanthemum* × morifolium). Hortic Sci 43(6):1888–1891

Lee CW, Choi JM, Pak CH (1996) Micronutrient toxicity in seed geranium (*Pelargonium* × hortorum Bailey). J Am Soc Hortic Sci 121(1):77–82

Lei Y, Korpelainen H, Li C (2007) Physiological and biochemical responses to high Mn concentrations in two contrasting *Populus cathayana* populations. Chemosphere 68:686–694

Leonard RP (1978) Hazardous solid waste from metallurgical industries. Environ Health Persp 27:251

Li HF, Gray C, Mico C, Zhao FJ, McGrath SP (2009) Phytotoxicity and bioavailability of cobalt to plants in a range of soils. Chemosphere 75(7):979–986

Li Z, Feng X, Bi X, Li G, Lin Y, Sun G (2014) Probing the distribution and contamination levels of 10 trace metal/metalloids in soils near a Pb/Zn smelter in middle China. Environ Sci Pollut Res 21(6):4149–4162

Liang Y, Wong JWC, Wei L (2005) Silicon-mediated enhancement of cadmium tolerance in maize (*Zea mays* L.) grown in cadmium contaminated soil. Chemosphere 58(4):475–483

Lidon FC, Teixeira MG (2000) Rice tolerance to excess Mn: implications in the chloroplast lamellae and synthesis of a novel Mn protein. Plant Physiol Biochem 38:969–978

Liu X, Zhang S, Shan X, Zhu YG (2005) Toxicity of arsenate and arsenite on germination seedling growth and amylolytic activity of wheat. Chemosphere 61:293–301

Liu Y, Lin C, Wu Y (2007) Characterization of red mud derived from a combined Bayer Process and bauxite calcination method. J Hazard Mater 146(1):255–261

Liu W, Yang J, Xiao B (2009) Application of Bayer red mud for iron recovery and building material production from alumosilicate residues. J Hazard Mater 161(1):474–478

López I, Balocchi O, Lailhacar P (1997) Caracterización de sitios de crecimiento de seis especies pratenses nativas y naturalizadas del dominio húmedo de Chile. Agro Sur 25(1):62–80

Ma JF, Yamaji N (2006) Silicon uptake and accumulation in higher plants. Trends Plant Sci 11(8):392–397

Ma L, Ruan JY, Yang Y, Han W, Shi Y (2005) Effect of magnesium nutrition of the formation and transport of free amino acids in tea plants. In: Proceedings of international symposium on innovation in tea science and sustainable development in tea industry, pp 10–15

Mahajan A, Dua S (1993) Some characteristics of Indian rapeseed acid phosphatase. Physiol Biochem 20:86–89

Mahmood T, Islam KR (2006) Response of rice seedlings to copper toxicity and acidity. J Plant Nutrition 29(5):943–957

Maksimović JD, Mojović M, Maksimović V, Römheld V, Nikolic M (2012) Silicon ameliorates manganese toxicity in cucumber by decreasing hydroxyl radical accumulation in the leaf apoplast. J Exp Bot 63(7):2411–2420

Małkowski E, Kita A, Galas W, Karcz W, Kuperberg JM (2002) Lead distribution in corn seedlings (*Zea mays* L.) and its effect on growth and the concentrations of potassium and calcium. Plant Growth Regul 37(1):69–76

Marques AP, Rangel AO, Castro PM (2009) Remediation of heavy metal contaminated soils: phytoremediation as a potentially promising clean-up technology. Crit Rev Int J Environ Sci Technol 39(8):622–654

Marschner H (1995) Functions of mineral nutrients: macronutrients. In: Mineral nutrition of higher plants, 2nd edn. Academic, New York, pp 299–312

Mathiyazhagan N, Natarajan D, Suresh K (2015) Impact of waste dumps of mines on plants in genomic & bio molecules level. Mesop Environ J 2(1):33–45

Matsumoto H (2000) Cell biology of aluminum toxicity and tolerance in higher plants. Int Rev Cytol 200:1–46

Mazumdar M, Christiani DC, Biswas SK, Ibne-Hasan OS, Kapur K, Hug C (2015) Elevated sweat chloride levels due to arsenic toxicity. New Engl J Med 372(6):582–584

Meagher RB (2000) Phytoremediation of toxic elemental and organic pollutants. Curr Opin Plant Biol 3(2):153–162

Meharg AA, Macnair MR (1992) Suppression of the high affinity phosphate uptake system: a mechanism of arsenate tolerance in *Holcus lanatus* L. J Exp Bot 43(4):519–524

Mendel RR, Hänsch R (2002) Molybdoenzymes and molybdenum cofactor in plants. J Exp Bot 53(375):1689–1698

Mendez MO, Maier RM (2008) Phytostabilization of mine tailings in arid and semiarid environments-an emerging remediation technology. Environ Health Persp 116(3):278

Mishra A, Choudhuri MA (1999) Monitoring of phytotoxicity of lead and mercury from germination and early seedling growth indices in two rice cultivars. Water Air Soil Pollut 114(3-4):339–346

Mittler R, Vanderauwera S, Gollery M, Van Breusegem F (2004) Reactive oxygen gene network of plants. Trends Plant Sci 9(10):490–498

Mohanpuria P, Rana NK, Yadav SK (2007) Cadmium induced oxidative stress influence on glutathione metabolic genes of *Camellia sinensis* (L.) O. Kuntze. Environ Tox 22(4):368–374

Mohanty M, Dhal NK, Patra P, Das B, Reddy PSR (2010) Phytoremediation: a novel approach for utilization of iron-ore wastes. In: Whitacre DM (ed) Reviews of environmental contamination and toxicology, vol 206. Springer, New York, pp 29–47

Mokgalaka-Matlala NS, Flores-Tavizón E, Castillo-Michel H, Peralta-Videa JR, Gardea-Torresdey JL (2008) Toxicity of arsenic (III) and (V) on plant growth, element uptake, and total amylolytic activity of mesquite (*Prosopis juliflora* x *P. velutina*). Int J Phytorem 10(1):47–60

Molas J (1998) Changes in morphological and anatomical structure of cabbage (*Brassica oleracea* L.) outer leaves and in ultrastructure of their chloroplasts caused by an in vitro excess of nickel. Photosynth 34(4):513–522

Montesano F, Van Iersel MW (2007) Calcium can prevent toxic effects of Na^+ on tomato leaf photosynthesis but does not restore growth. J Am Soc Hortic Sci 132:310–318

Mora M, Rosas A, Ribera A, Rengel R (2009) Differential tolerance to Mn toxicity in perennial ryegrass genotypes: involvement of antioxidative enzymes and root exudation of carboxylates. J Plant Soil 320:79–89

Moreno JL, Hernández T, Garcia C (1999) Effects of cadmium-contaminated sewage sludge compost on dynamics of organic matter and microbial activity in an arid soil. Biol Fert Soils 28(3):230–237

Morgan PW, Taylor DM, Joham HE (1976) Manipulation of IAA-oxidase activity and auxin-deficiency symptoms in intact cotton plants with manganese nutrition. Physiol Planta 37(2):149–156

Morrey DR, Baker AJM, Cooke JA (1988) Floristic variation in plant communities on metalliferous mining residues in the northern and southern Pennines, England. Environ Geochem Health 10(1):11–20

Munns R (2002) Comparative physiology of salt and water stress. Plant Cell Environ 25(2):239–250

Nagata T (2015) Growth inhibition and IRT1 induction of Arabidopsis thaliana in response to bismuth. J Plant Biol 58(5):311–317

Narayan D, Agrawal M, Pandey J, Singh J (1994) Changes in vegetation characteristics downwind of an aluminium factory in India. Ann Bot 73(5):557–565

Narendrula R, Nkongolo KK, Beckett P (2012) Comparative soil metal analyses in sudbury (Ontario, Canada) and lubumbashi (Katanga, DR-Congo). Bull Environ Con Tox 88(2):187–192

Narula A, Kumar S, Srivastava PS (2005) Abiotic metal stress enhances diosgenin yield in *Dioscorea bulbifera* L. cultures. Plant Cell Rep 24(4):250–254

National Programme on Safety Food (NPSF) (2002) Safety milled rice criteria of PR China (NY5115-2002). China Ministry of Agriculture, Beijing [In Chinese]
Nawab J, Khan S, Shah MT, Khan K, Huang Q, Ali R (2015a) Quantification of heavy metals in mining affected soil and their bioaccumulation in native plant species. Int J Phytorem 17(9):801–813
Nawab J, Khan S, Shah MT, Qamar Z, Din I, Mahmood Q, …, Huang Q (2015b) Contamination of soil, medicinal, and fodder plants with lead and cadmium present in mine-affected areas, Northern Pakistan. Environ Monit Asses 187(9):1–14
Nazar R, Iqbal N, Masood A, Khan MIR, Syeed S, Khan NA (2012) Cadmium toxicity in plants and role of mineral nutrients in its alleviation. Am J Plant Sci 3(10):1476–1489
Neelima P, Reddy KJ (2002) Interaction of copper and cadmium with seedling growth and biochemical responses in *Solanum melongena*. Nat Environ Pollut Technol 1(3):285–290
Nichols DG, Beardsell DV (1981) Interactions of calcium, nitrogen and potassium with phosphorus on the symptoms of toxicity in Grevillea cv. 'poorinda firebird'. J Plant Soil 61(3):437–445
Niemeyer JC, Moreira-Santos M, Nogueira MA, Carvalho GM, Ribeiro R, Da Silva EM, Sousa JP (2010) Environmental risk assessment of a metal-contaminated area in the Tropics. Tier I: screening phase. J Soils Sediment 10(8):1557–1571
Nyitrai P, Bóka K, Gáspár L, Sárvári É, Lenti K, Keresztes Á (2003) Characterization of the stimulating effect of low-dose stressors in maize and bean seedlings. J Plant Physiol 160(10):1175–1183
Ok YS, Kim SC, Kim DK, Skousen JG, Lee JS, Cheong YW, …, Yang JE (2011) Ameliorants to immobilize Cd in rice paddy soils contaminated by abandoned metal mines in Korea. Environ Geochem Health 33(1):23–30
Osman KT (ed) (2012) Soils: principles, properties and management. Springer, Chittagong
Ou-yang C, Gao S, Mei LJ, Chung TW, Tang L, Wang SH, Chen F (2014) Effects of aluminum toxicity on the growth and antioxidant status in *Jatropha curcas* seedlings. J Med Plants Res 8(3):178–185
Ouzounidou G, Moustakas M, Eleftheriou EP (1997) Physiological and ultrastructural effects of cadmium on wheat (*Triticum aestivum* L.) leaves. Arch Environ Con Tox 32(2):154–160
Päivöke A (1983) The long-term effects of lead and arsenate on the growth and development, chlorophyll content and nitrogen fixation of the garden pea. Ann Bot Fenn 20(3):297–306
Pakrashi S, Jain N, Dalai S, Jayakumar J, Chandrasekaran PT, Raichur AM, …, Mukherjee A (2014) In vivo genotoxicity assessment of titanium dioxide nanoparticles by *Allium cepa* root tip assay at high exposure concentrations. PloS one 9(2):87789
Palit S, Sharma A, Talukder G (1994) Effects of cobalt on plants. The Bot Rev 60(2):149–181
Palumbo-Roe B, Klinck B, Cave M (2007) Arsenic speciation and mobility in mine wastes from a copper–arsenic mine in Devon, UK: a SEM, XAS, sequential chemical extraction study. Trace Metals Other Con Environ 9:441–471
Panda SK, Choudhury S (2005) Chromium stress in plants. Braz J Plant Physiol 17(1):95–102
Pandey B, Agrawal M, Singh S (2014) Assessment of air pollution around coal mining area: emphasizing on spatial distributions, seasonal variations and heavy metals, using cluster and principal component analysis. Atmos Pollut Res 5(1):79–86
Pandey B, Agrawal M, Singh S (2016) Ecological risk assessment of soil contamination by trace elements around coal mining area. J Soils Sediment 16(1):159–168
Pappu A, Saxena M, Asolekar SR (2007) Solid wastes generation in India and their recycling potential in building materials. Build Environ 42(6):2311–2320
Patra M, Sharma A (2000) Mercury toxicity in plants. Bot Rev 66:379–422
Peralta JR, Gardea-Torresdey JL, Tiemann KJ, Gomez E, Arteaga S, Rascon E, Parsons JG (2001) Uptake and effects of five heavy metals on seed germination and plant growth in alfalfa (*Medicago sativa* L.). Bul Environ Con Tox 66(6):727–734
Pietrini F, Iori V, Cheremisina A, Shevyakova NI, Radyukina N, Kuznetsov VV, Zacchini M (2015) Evaluation of nickel tolerance in *Amaranthus paniculatus* L. plants by measuring pho-

tosynthesis, oxidative status, antioxidative response and metal-binding molecule content. Environ Sci Pollut Res 22(1):482–494

Pitman MG, Läuchli A (2002) Global impact of salinity and agricultural ecosystems. In: Salinity: environment-plants-molecules. Springer, Dordrecht, pp 3–20

Pitzschke A, Forzani C, Hirt H (2006) Reactive oxygen species signaling in plants. Antioxid Redox Signal 6:1757–1764

Prasad KVSK, Saradhi PP, Sharmila P (1999) Concerted action of antioxidant enzymes and curtailed growth under zinc toxicity in *Brassica juncea*. Environ Exp Bot 42(1):1–10

Prasad A, Kumar S, Khaliq A, Pandey A (2011) Heavy metals and arbuscular mycorrhizal (AM) fungi can alter the yield and chemical composition of volatile oil of sweet basil (*Ocimum basilicum* L.). Biol Fert Soils 47(8):853–861

Prasad MSR, Babu BR, Ramesh KV, Trinath K (2014) Structural and magnetic studies on chromium substituted Ni-Zn nano ferrite synthesized by citrate Gel auto combustion method. J Supercond Nov Magn 27(12):2735–2745

Pruvot C, Douay F, Herve F, Waterlot C (2006) Heavy metals in soil, crops and grass as a source of human exposure in the former mining areas. J Soils Sediments 6:215–220

Pullaiah T (2006) Encyclopaedia of world medicinal plants. In: Daya books, Vol-III, Department of Botany, Shri Krishnadevaraya University, Anantpur, A.P. 515003. Regency Publication, New Delhi

Purdy LH (1956) The nature and extent of mercury injury and associated anatomical responses in wheat seedlings. Rev App Mycol 36:391

Qureshi A, Maurice C, Öhlander B (2016) Potential of coal mine waste rock for generating acid mine drainage. J Geochem Explor 160:44–54

Rao SR (2011) Resource recovery and recycling from metallurgical wastes. In: Waste manag series, vol 7. Department of mining, metals and materials engineering, McGill University, Montreal, Quebec, Canada. Elsevier, pp 13–34

Reddy AM, Kumar SG, Jyothsnakumari G, Thimmanaik S, Sudhakar C (2005) Lead induced changes in antioxidant metabolism of horsegram (*Macrotyloma uniflorum* (Lam.) Verdc.) and bengalgram (*Cicer arietinum* L.). Chemosphere 60(1):97–104

Reeves RD, Baker AJ (2000) Metal-accumulating plants. Phytoremediation of toxic metals: using plants to clean up the environment. Wiley, New York, pp 193–229

Remon E, Bouchardon JL, Cornier B, Guy B, Leclerc JC, Faure O (2005) Soil characteristics, heavy metal availability and vegetation recovery at a former metallurgical landfill: implications in risk assessment and site restoration. Environ Pollut 137(2):316–323

Rimando AM, Dayan F, Czarnota MA, Weston LA, Duke SO (1998) A new photosystem II electron-transfer inhibitor from *Sorghum bicolor*. J Nat Prod 61:927–930

Ritcey GM (1989) Tailings management: problems and solutions in the mining industry. Elsevier, Amsterdam/New York. ISBN 0-444-87374-0

Rodrigues CR, Silveira JA, Viégas RA, Moura RM, Aragão RM, Silva EN (2016) Combined effects of high relative humidity and K^+ supply mitigates damage caused by salt stress on growth, photosynthesis and ion homeostasis in *J. curcas* plants. Agric Water Manag 163:255–262

Rola K, Osyczka P, Nobis M, Drozd P (2015) How do soil factors determine vegetation structure and species richness in post-smelting dumps? Ecol Eng 75:332–342

Romero-Puertas MC, Rodríguez-Serrano M, Corpas FJ, Gomez MD, Del Rio LA, Sandalio LM (2004) Cadmium-induced subcellular accumulation of O^{2-} and H_2O_2 in pea leaves. Plant Cell Environ 27(9):1122–1134

Salomons W (1995) Environmental impact of metals derived from mining activities: processes, predictions, prevention. J Geochem Explor 52(1):5–23

Salt DE, Smith RD, Raskin I (1998) Phytoremediation. Annu Rev Plant Biol 49(1):643–668

Samal S, Ray AK, Bandopadhyay A (2015) Characterization and microstructure observation of sintered red mud–fly ash mixtures at various elevated temperature. J Clean Prod 101:368–376

Santos-Jallath J, Castro-Rodríguez A, Huezo-Casillas J, Torres-Bustillos L (2012) Arsenic and heavy metals in native plants at tailings impoundments in Queretaro, Mexico. Phys Chem Earth Parts A/B/C 37:10–17

Sarkar S, Mazumder D (2015) Solid waste management in steel industry-challenges and opportunities, world academy of science, engineering and technology. Int J Soc Behav Edu Econ Busi Indus Eng 9(3):984–987

Sass JE (1937) Histological and cytological studies of ethyl mercury phosphate poisoning in corn seedlings. J Phytopat 27:95–99

Saxena M, Mishra CR (2004) Processing of red mud for development of wood substitutes. In: Recycling, waste treatment and clean technology, vol 1. TMS Min Metals Mater, Spain, pp 371–380

Schmfger MEV (2001) Phytochelatins: complexation of metals and metalloids, studies on the phytochelatin synthase. Doctoral dissertation, PhD thesis, Munich University of Technology, Munich, Germany

Semu E, Singh BR, Selmer Olsen AR, Steenberg K (1985) Uptake of mercury from mercury-203-labeled mercury compounds by wheat and beans grown on an oxisol. Plant Soil 87:347–355

Setia RC, Rajna B, Bala R (1994) Anatomical changes in root and stem of wheat (*Triticum aestivum* L.) in response to different heavy metals. J Phytomor 44:95–104

Sglavo VM, Campostrini R, Maurina S, Carturan G, Monagheddu M, Budroni G, Cocco G (2000) Bauxite 'red mud'in the ceramic industry. Part 1: thermal behaviour. J Eur Ceram Soc 20(3):235–244

Shanker AK, Cervantes C, Loza-Tavera H, Avudainayagam S (2005) Chromium toxicity in plants. Environ Int 31(5):739–753

Sharma RK, Agrawal M (2005) Biological effects of heavy metals: an overview. J Environ Biol 26(2):301–313

Sharma P, Dubey RS (2005) Lead toxicity in plants. Braz J Plant Physiol 17(1):35–52

Shaul O (2002) Magnesium transport and function in plants: the tip of the iceberg. Biometals 15(3):307–321

Shen H, Forssberg E (2003) An overview of recovery of metals from slags. Waste Manag 23(10):933–949

Sheoran IS, Singal HR, Singh R (1990) Effect of cadmium and nickel on photosynthesis and the enzymes of the photosynthetic carbon reduction cycle in pigeonpea (*Cajanus cajan* L.). Photosynth Res 23(3):345–351

Singh JS (1996). In: An integrated ecological study on revegetation of mine spoil, Banaras Hindu University, Department of Botany

Singh VP (2005) Trace metals in plants. In: Metal toxicity and tolerance in plants and animals, Edition 1. Swaroop and Sons, New Delhi, p 5

Singh SP, Hendry MJ (2013) Solid-phase distribution and leaching behaviour of nickel and uranium in a uranium waste-rock piles. Water Air Soil Pollut 224(1):1–11

Singh AN, Raghubansh AS, Singh JS (2002) Plantations as a tool for mine spoil restoration. Curr Sci 82(12):1436–1441

Sinha S, Gupta M, Chandra P (1997) Oxidative stress induced by iron in *Hydrilla verticillata* (lf) royle: response of antioxidants. Ecotoxicol Environ Saf 38(3):286–291

Sivaguru M, Liu J, Kochian LV (2013) Targeted expression of Sb Mate in the root distal transition zone is responsible for *Sorghum* aluminum resistance. J Plant 76(2):297–307

Skarżyńska KM (1995) Reuse of coal mining wastes in civil engineering – Part 2: Utilization of minestone. Waste Manag 15(2):83–126

Smith R, Robertson VS (1962) Soil irrigation classification of shallow soils overlying gypsum beds, Northern Iraq. J Soil Sci 13:106–115

Smith SE, Christophersen HM, Pope S, Smith FA (2010) Arsenic uptake and toxicity in plants: integrating mycorrhizal influences. J Plant Soil 327:1–21

Solenkova NV, Newman JD, Berger JS, Thurston G, Hochman JS, Lamas GA (2014) Metal pollutants and cardiovascular disease: mechanisms and consequences of exposure. Am Heart J 168:812–822

Spear SN, Asher CJ, Edwards DG (1978) Response of cassava, sunflower, and maize to potassium concentration in solution I. Growth and plant potassium concentration. Field Crops Res 1:347–361

Sresty TVS, Rao KM (1999) Ultrastructural alterations in response to zinc and nickel stress in the root cells of pigeonpea. Environ Exp Bot 41(1):3–13

Stoeva N, Bineva T (2003) Oxidative changes and photosynthesis in Oat plants grown in As-contaminated soil. Bulg J Plant Physiol 29(1–2):87–95

Stoeva N, Berova M, Zlatev Z (2003) Physiological response of maize to arsenic contamination. Biol Planta 47(3):449–452

Stoeva N, Berova M, Vassilev A, Zlatev Z (2005) Effect of arsenic on some physiological parameters in bean plants. Biol Planta 49(2):293–296

Street RA, Kulkarni MG, Stirk WA, Southway C, Van Staden J (2007) Toxicity of metal elements on germination and seedling growth of widely used medicinal plants belonging to Hyacinthaceae. Bull Environ Con Tox 79(4):371–376

Tabuchi A, Matsumoto H (2001) Changes in cell-wall properties of wheat (*Triticum aestivum*) roots during aluminum-induced growth inhibition. Physiol Planta 112(3):353–358

Tlustoš P, Cígler P, Hrubý M, Kužel S, Száková J, Pavlíková D, Balík J (2005) The role of titanium in biomass production and its influence on essential elements contents in field growing crops. Plant Soil Environ 51:19–25

Tripathi RD, Srivastava S, Mishra S, Singh N, Tuli R, Gupta DK, Maathuis FJ (2007) Arsenic hazards: strategies for tolerance and remediation by plants. J Trends Biotechnol 25(4):158–165

Tyner EH, Webb JR (1946) Relation of corn yields to nutrient balance as revealed by leaf analysis. J Am Soc Agron, pp 173–185

Tyus HM (2012) Ecology and conservation of fishes. CRC Press/Taylor and Francis Group, Boca Raton/London/New York, pp 439. https://books.google.co.in/books?isbn=143989759X. Accessed on 10 Mar 2016

Vamerali T, Bandiera M, Mosca G (2010) Field crops for phytoremediation of metal-contaminated land: a review. Environ Chem Lett 8(1):1–17

Van Alphen JG, de los Ríos Romero F (1971) Gypsiferous soils: notes on their characteristics and management ILRI (Bulletin/International Institute for Land Reclamation and Improvement 12), p 44. http://library.wur.nl/WebQuery/wurpubs/personal/453924. Accessed on 12 Mar 2016

Varshney AK (1990) Differential response of *Phaseolus aureus* cultivars to mercury pretreatment on mobilization of N and protease activity during germination and seedling growth. Geobios 17:82–87

Venkatesan S, Jayaganesh S (2010) Characterisation of magnesium toxicity, its influence on amino acid synthesis pathway and biochemical parameters of Tea. Res J Phytochem 4:67–77

Vieira RF, Vieira C, Cardoso EJBN, Mosquim PR (1998) Foliar application of molybdenum in common bean. II Nitrogenase and nitrate reductase activities in a soil of low fertility. J Plant Nutr 21(10):2141–2151

Vijay S, Kant S, Bohra SP (1988) Ecophysiological studies on Indian desert plants, II: Physiological response to heavy metals in *Sorghum vulgare*. Pers Trans Indian Soc Desert Technol 59–67

Wang S, Ang HM, Tade MO (2008) Novel applications of red mud as coagulant, adsorbent and catalyst for environmentally benign processes. Chemosphere 72(11):1621–1635

White MC, Chaney RL, Decker AM (1979) Differential cultivar tolerance in soybean to phytotoxic levels of soil Zn. II. Range of Zn additions and the uptake and translocation of Zn, Mn, Fe, and P. J Agro 71(1):126–131

Wilkinson RE, Ohki K (1988) Influence of manganese deficiency and toxicity on isoprenoid syntheses. J Plant Physiol 87(4):841–846

Willard HH, Merritt LL Jr, Dean JA, Settle FA Jr (1988) Instrumental methods of analysis, 7th edition. Wadsworth Publishing Company, United States. http://www.osti.gov/scitech/biblio/5209599. Accessed in 12 Mar 2016

Wissemeier AH (1993) Marginal bract necrosis in poinsettia cultivars and the relationship to bract calcium nutrition. Gartenbauwissenschaft 58:158–163

Wójcik M, Sugier P, Siebielec G (2014) Metal accumulation strategies in plants spontaneously inhabiting Zn-Pb waste deposits. Sci Total Environ 487:313–322

World Health Organization (WHO) (1992) Guideline for drinking water quality, recommendation, vol 1, 2nd edn. World Health Organization, Geneva, p 41

World Health Organization (WHO) (1998) Quality control methods for medicinal plant materials. World Health Organization, Geneva

World Health Organization (WHO) (2005) Quality control methods for medicinal plant materials. World Health Organization, Geneva

Wu SH (1994) Effect of manganese excess on the soybean plant cultivated under various growth conditions. J Plant Nutr 17(6):991–1003

Wu W, Peters J, Berkowitz GA (1991) Surface charge-mediated effects of Mg^{2+} on K^+ flux across the chloroplast envelope are associated with regulation of stromal pH and photosynthesis. J Plant Physiol 97(2):580–587

Xiuzhen H, Dongmei ZHOU, Yujun WANG (2004) Study of ryegrass growth in copper mine tailing treated with peat and chemical fertilizer. Acta Pedol Sinica 41(4):645–648

YaMin L, JianZhong L, YongBo S, Feng Z (2011) Effects of aluminum stress on the stomata characteristics and photosynthesis of *Scutellaria baicalensis* Georgi seedlings. Med Plant 2(1):5–8

Yoshizawa S, Tanaka M, Shekdar AV (2004) Global trends in waste generation. Recyc Waste Treat Clean Technol. TMS Mineral, Metals and Materials Publishers, Spain, pp 1541–1552

Zaier H, Mudarra A, Kutscher D, de la Campa MF, Abdelly C, Sanz-Medel A (2010) Induced lead binding phytochelatins in *Brassica juncea* and *Sesuvium portulacastrum* investigated by orthogonal chromatography inductively coupled plasma-mass spectrometry and matrix assisted laser desorption ionization-time of flight-mass spectrometry. Analy Chimica Acta 671(1):48–54

Zeid IM (2001) Responses of *Phaseolus vulgaris* chromium and cobalt treatments. Biol Planta 44(1):111–115

Zhang Y (1995) Effects of aluminum chloride on the nucleus and nucleolus in root tip cells of *Hordeum vulgare*. Mut Res/Environ Mutag Rel Sub 335(2):137–142

Zhang WH, Tyerman SD (1999) Inhibition of water channels by $HgCl_2$ in intact wheat root cells. J Plant Physiol 120(3):849–858

Zheljazkov VD, Jeliazkova EA, Kovacheva N, Dzhurmanski A (2008) Metal uptake by medicinal plant species grown in soils contaminated by a smelter. Environ Exp Bot 64(3):207–216

Zhu XF, Lei GJ, Wang ZW, Shi YZ, Braam J, Li GX, Zheng SJ (2013) Coordination between apoplastic and symplastic detoxification confers plant aluminum resistance. J Plant Physiol 162(4):1947–1955

Zhuang P, McBride MB, Xia H, Li N, Li Z (2009) Health risk from heavy metals via consumption of food crops in the vicinity of Dabaoshan mine, south China. Sci Total Environ 407(5):1551–1561

The Risk Associated with the Xenobiotics Released Through Wastewater Reuse

11

Prabhat Kumar Srivastava, Parul Parihar, Rachana Singh, and Sheo Mohan Prasad

Abstract

Irrigation of our crop fields is indispensable to get sufficient crop yield, and increasing trend of reusing wastewater for agricultural practices has provided an answer of water scarcity. The wastewater is recycled and then used for drinking and agricultural practices, but concerns have been raised regarding the quality of water being used. The major concern is the presence of xenobiotics in recycled water that affects the crops' yield qualitatively as well as quantitatively and in turn affects human beings. Risk factors due to the presence of xenobiotics can have long-term effects on soil, thereby affecting plants and human health. Therefore, assessing the risk and thereafter managing the risk are very necessary.

Keywords

Xenobiotics • Wastewater • Pharmaceutical • Pesticide • Risk

P.K. Srivastava (✉)
Department of Botany, Ranjan Plant Physiology and Biochemistry Laboratory, University of Allahabad, Allahabad 211002, Uttar Pradesh, India

Pt. Ravi Shankar Tripathi Government Degree College, Bhaiyathan, Surajpur 497231, C.G., India
e-mail: prabhatsrivastava.au@gmail.com

P. Parihar • R. Singh • S.M. Prasad (✉)
Ranjan Plant Physiology and Biochemistry Laboratory, Department of Botany, University of Allahabad, Allahabad 211002, Uttar Pradesh, India
e-mail: profsmprasad@gmail.com

A. Singh et al. (eds.), *Plant Responses to Xenobiotics*,
DOI 10.1007/978-981-10-2860-1_11

11.1 Introduction

Today's world is facing serious challenge of water scarcity due to substantial shrinkage in rainwater, overexploitation of water resources for domestic and industrial purposes and human population burst. The component of drinking water used both for human beings and livestock is much less than those used in domestic and industrial purposes. Newer lifestyle has also threatened and destroyed our traditional water resources. Heavy emission of CO_2 throughout the globe probably has changed the climate, now being experienced in change in weather pattern in almost every part of the world. Wet months are becoming dry year after year. The groundwater level has fallen up to its lowest level as in some parts of Peninsular India, like Vidarbha, Marathwada and Telangana and its plains, like Bundelkhand (SNDRP 2016). Persisting drought and hence the decrease in groundwater level has raised concerns about the very existence of mankind. In these circumstances too, agriculture and livestock rearing cannot be given up as these have been the part and parcel of human civilization since time immemorial. Agricultural practices and livestock rearing do need water. Therefore, it becomes imperative to reuse and recycle the water once used in households and industries.

After heavy urbanization and industrialization in cities, plenty of water is discharged in our water bodies from our households after domestic and industrial usage. These raise two important issues: first, huge amount of water is flown uselessly when the whole world is facing one of the most severe water crises of the history, so this water must be reused after adequate treatment processes and, second constraint is the presence of pollutants in industrial and domestic wastewater, like heavy metals, pesticides and toxic substances raising the concerns if reused without adequate treatment processes. Ingestion of this wastewater via one or other methods may cause severe diseases in human beings. These pollutants are often referred as xenobiotics (Greek word, xenobiotic = stranger to life) when they get entry into the bodies of living beings. Therefore, adequate treatment of industrial and domestic wastewater is indispensable to secure the existence of mankind on the earth. Now, the governments have seriously started to combat the situations of water scarcity throughout the world. Further, lax in treatment procedure again raises concerns among environmentalists and biologists to think again. So, with prevailing water purifying technologies, only secondary, tertiary or even more times treated effluent water can be utilized for irrigation and drinking purposes to combat the water scarcity situations.

A xenobiotic substance is defined as an unnatural foreign chemical found within an organism, generally not expected to be present within that organism at significant concentrations for longer periods. Since the basic unit of life is the cell and the basic life processes of all the cells, their physiology and biochemistry are the same, xenobiotics are understood as substances foreign to the whole biological system, i.e. artificial substances which did not exist in nature before their artificial synthesis. They may be synthetic organochlorines such as pesticides, organic chemicals such as polyaromatic hydrocarbons (PAHs), coal and crude oil. Antibiotics, anti-inflammatory drugs, beta-blockers, cancer therapeutics, contraceptives, lipid regulator agents and supplementary hormones fall under pharmaceutical classes, most commonly detected in the treated effluents worldwide (Gebhardt and Schröder 2007). There are some narcotic substances, viz. amphetamines, cannabinoids, cocaine, opiates etc. that have recently been identified in wastewater. They are hardly removed or transformed in wastewater effluents (Petrovic et al. 2009). Different inorganic species, viz., chlorides and bromides of sodium, potassium and calcium as well as trace heavy metals have their own benefits and threats when they come into the contact of plants with the irrigation water. Some other chemicals show similar effects but at relatively higher concentrations. They are present in domestic wastewaters and include some plasticizers, pesticides and degradation products of some detergents. These groups of synthetic organic compounds are often called as xenobiotics and are produced from our households, pharmaceutical industries and hospitals and other industries. Another recent and important class of trace organic contaminants in wastewater is the endocrine-disrupting compounds (EDCs) (Liu et al. 2009). Natural and synthetic steroidal hormones have also been shown to induce biological effects even at parts per trillion concentrations. These EDCs and steroidal hormones are not adequately removed by conventional water treatment processes.

In addition to man-made chemicals, natural compounds could also become xenobiotics if they are taken up by another organism and have potent biological properties and special medicinal properties or a given organism is at risk of exposure to that natural compounds; for instance, the uptake of fish hormones by humans.

Any xenobiotic substance may enter into the environment in two main ways: direct and indirect. Manufacturers of the xenobiotics directly discharge concerned xenobiotic substances. Indirect discharge takes place after the biologically intended effects and is passed to the environment either in their complete or a modified state. However, with the increasing regulations by regulating agencies, direct discharge has become the practice of the past. There are also several other direct and indirect sources of xenobiotic substance into the environment that affect the living system (Fig. 11.1). One common indirect source of xenobiotic substance is their discharge in large pits before being pumped and applied to fields as fertilizers where many of them are washed away by rainfall to aquatic environments. Often the xenobiotic substances that are present in landfills are found in their original, most chemically active states.

Xenobiotics produce many biological effects so they are analyzed for risk factors, such as ecotoxicity, persistence in the environment or toxicity to humans. Xenobiotics may be grouped under heavy metals, antibiotics, carcinogens, drugs, environmental pollutants, food additives, hydrocarbons and pesticides and are systematically analyzed for associated risk factors.

11.2 Heavy Metals as Xenobiotics

Metal contaminants include Ag, As, Cd, Cr, Cu, Hg, Ni, Pb, Se and Zn. Other less common metallic contaminants include Al, Co, Cs, Mn, Mo, Sr and U. Living organisms do require varying amounts of some heavy metals like Co, Cu, Fe, Mn,

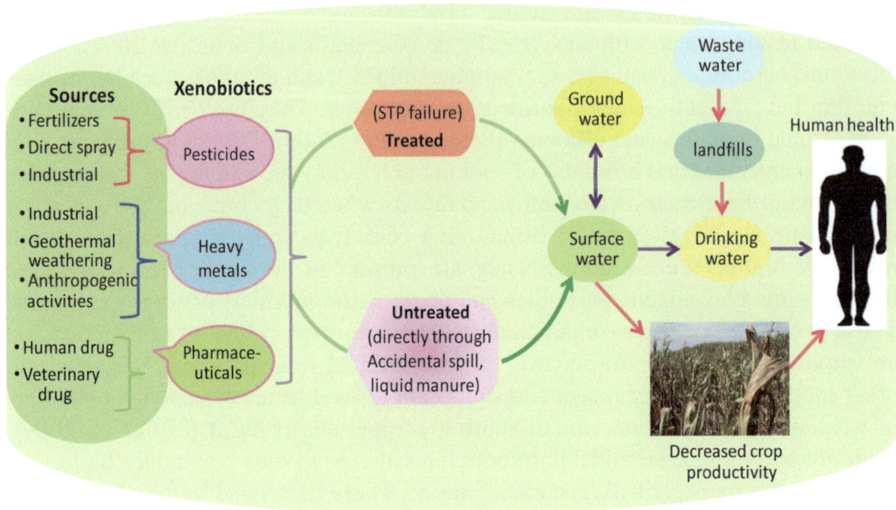

Fig. 11.1 Different ways of xenobiotic entry into the living system

Mo and Zn. But these metals too become toxic at their higher concentrations (Singh et al. 2016). Though heavy metals are essential constituents of the earth's crust, heavy metal contamination of soil and water resources has become an environmental concern. This takes place due to indiscriminate human activities and industrial wastes altering geochemical cycles and biochemical balances. Heavy metals are nonbiodegradable and persist in the environment for indefinite period (Yang et al. 2005; Singh et al. 2016). Untreated wastewater is commonly used for irrigation purposes in developing countries. Many times, heavy metal polluted waters are inadvertently mixed with ordinary irrigation water due to failure of treatment plants or excessive load of wastewater in them (Singh et al. 2009). Long-term use of such water for irrigation results in the accumulation of heavy metals in soils (Marshall et al. 2007) and thus changes the quality of the soil (Singh et al. 2009).

Heavy metals are known to cause harmful effects on many biochemical and physiological processes of plants such as photosynthesis, mineral nutrition and water uptake (Singh et al. 2013; Singh et al. 2016). The uptake of heavy metals by plants not only reduces their growth and biomass accumulation but also poses risk to human health and animals as heavy metals find entry into the food chain via primary producers (Bharwana et al. 2013). Besides these direct deleterious effects, heavy metals may also adversely affect plants indirectly by the production of excessive reactive oxygen species (ROS) (Singh et al. 2016). Excess ROS may damage biomolecules of cells like proteins, lipids and nucleic acids. Thus, heavy metals cause adverse impacts on flora from the molecular to the whole-plant level. Plants have evolved several detoxification mechanisms to minimize the damaging effects of heavy metals based on chelation, subcellular compartmentalization and a well-off antioxidative system (Salt et al. 1998).

Prolonged exposure to heavy metals such as As, Cd, Cu, Pb, Ni and Zn can cause adverse health effects in humans (McLaughlin et al. 1999; Singh et al. 2015). Pb and Hg also fall under toxic heavy metals. Their accumulation in the animal tissues can cause serious problems (Tchounwou et al. 2012). These metals primarily come from the industrial effluents and runoff from the soil to rivers and other water bodies and then to human food sources (Tchounwou et al. 2012). These toxins can be introduced into the human and animal bodies by consumption of foods and beverages and by skin exposure. Heavy metals disturb metabolic functions of animals in two ways: they may accumulate in vital organs, such as heart, liver, brain, kidney, bone etc. and upset their functions or they may displace various nutritional minerals from their original sites and hinder their biological functions.

The effects of various heavy metals on plants and human health have been enumerated in Table 11.1.

Table 11.1 Some xenobiotics and their effects on human and plants

Xenobiotics	Sources	Human ailments	Effects on plants
Heavy metals			
As	Rock weathering, smelting, thermal power plants, fuel burning	Skin, lung, bladder and kidney cancer, thickening and pigmentation of the skin, blackfoot disease in Taiwan	Inhibits plant growth by slowing or arresting expansion and biomass accumulation as well as affects plant reproductive capacity through losses in fertility, yield and fruit production (Garg and Singla 2011)
Cd	Zinc production, waste batteries, e-wastes, paint sludge, incinerations and fuel combustion, phosphate -ore implication and bio-industrial manure, artificial phosphate fertilizers	Short-term exposure causes diarrhea, stomach pains and vomiting; long-term exposure causes bone fracture (itai-itai or ouch-ouch disease), infertility, damage to the central nervous and immune system, psychological disorders, DNA damage and cancer development, accumulates in the kidneys and damages filtering mechanisms	Inhibits the biosynthesis of chlorophyll (Qian et al. 2009) and imparts deleterious effects on stomatal opening, transpiration and respiration processes (Sandalio et al. 2001) and decreases the photosynthetic rate (Durand et al. 2010) and hence declines the crop productivity

(continued)

Table 11.1 (continued)

Xenobiotics	Sources	Human ailments	Effects on plants
Cr	Mining, industrial coolants, chromium salt manufacturing, leather tanning	Respiratory tract irritant results in airway irritation, airway obstruction and lung, nasal or sinus cancer	Inhibits seed germination, disturbs pigment concentration and water uptake, causes nutrient imbalance (Panda and Choudhury 2005), changes ultrastructure, particularly vacuolation of root cell (Panda 2007) and induces oxidative stress
Cu	Mining, electroplating, smelting operations	Kidney failure and death occur with as little as 1 g of copper sulfate. Symptoms of copper overdose include nausea, vomiting, bloody diarrhea, fever, stomach pain, low blood pressure, anemia and heart problems	Interferes with the biosynthesis of photosynthetic machinery, modifies the pigments and protein composition and interacts with ions like Mn, Ca and Cl that are necessary for proper functioning of oxygen-evolving complex (Yruela 2005)
Hg	Batteries, chemical industries, chlor-alkali plants, thermal power plants, electrical appliances, fluorescent lamps, hospital wastes, damaged thermometers, barometers, sphygmomanometers, organic mercury (primarily methyl mercury) produced by specific bacterial organisms etc.	Inorganic form is not very harmful and may damage the kidney in certain cases; organic mercury (primarily methyl mercury) ingested by fish bioaccumulates there and later in human tissues	Changes the permeability of cell membrane, reacts with the sulfhydryl (SH) groups and phosphate groups, also replaces essential ions and disrupts functions of proteins (Patra and Sharma 2000 and Patra et al. 2004)

(continued)

Table 11.1 (continued)

Xenobiotics	Sources	Human ailments	Effects on plants
Ni	Smelting operations, thermal power plants, battery industry	Higher chances of development of lung cancer, nose cancer, laryngeal cancer and prostate cancer, sickness and dizziness after exposure to nickel gas, lung embolism, respiratory failure, birth defects, asthma and chronic bronchitis, allergic reactions such as skin rashes, mainly from jewelry, heart disorders	Inhibits mitotic activities (Madhava Rao and Sresty 2000), reductions in plant growth (Molas 2002) and adverse effects on fruit yield and quality (Gajewska et al. 2006)
Pb	Lead-acid batteries, paints, e-wastes, smelting, coal-based thermal power plants, ceramics, bangle industry	Pb(II) salts and organic lead are most harmful, toxic to a wide variety of organs in human to low-level exposure of lead, nervous system (changes I neurotransmitter levels), metabolism (impairment of vitamin D metabolism), reproductive system (irregular estrus (heat cycle in female dogs) and decreased sexual hormone level), immune system (impaired lymphocyte function and impaired antibody formation)	Distortion of chloroplast ultrastructure, obstructed electron transport, inhibition of Calvin cycle enzymes, impaired uptake of essential elements, such as Mg and Fe and induced deficiency of CO2 resulting from stomatal closure (Pourrut et al. 2011)
Zn	Smelting, electroplating	Excessive supplemental zinc increases the risk prostate cancer. Large doses of zinc can lower blood sugar in people with diabetes. Zinc use has been linked to shorter survival time in people with HIV/AIDs	Zn reduces yield and causes stunted growth which is due to the fall in photosynthetic performance, thereby affecting the photochemical reactions and also carbonic anhydrase activity, biosynthesis of chlorophyll and cell membrane integrity (Tsonev and Lidon 2012)

(continued)

Table 11.1 (continued)

Xenobiotics	Sources	Human ailments	Effects on plants
Pesticides			
Herbicides	Commonly used in agricultural fields to control the broadleaf weeds and annual grasses. However, indiscriminate use and continuous irrigation practices with non-recycled water have been reported to be the major source	Causes cancer, reproductive or developmental effects, or endocrine system effects	Negative impact on photosynthesis as revealed by a reduction in foliar chlorophyll and carotenoid contents, gas exchanges, and alteration in plastid structure Accumulation of soluble sugar and starch was observed in all organs (Saladin and Clement 2003) Affects the activities of nitrate reductase (NR) and nitrite reductase (NiR) and greatly inhibits glutamine synthetase (GS) and glutamate synthase (GOGAT). Accumulation of total N, protein, and amino acid (Nemat Alla et al. 2007)
		Agent Orange, a mixture of the herbicides 2,4-D and 2,4,5-T that was contaminated with 2,3,7,8-tetrachlorodibenzo-p-dioxin (TCDD), causes birth defects, cancers, liver disease and other illness	Inhibits the C4 photosynthetic enzymes like PEPC, MDH, PPDK, and Rubisco ((Nemat Alla et al. 2007) Inhibits important physiological (photosynthesis and nitrogen metabolism) processes (Prasad et al. 2015)

(continued)

Table 11.1 (continued)

Xenobiotics	Sources	Human ailments	Effects on plants
Insecticide		Organochlorine insecticides can cause a loss of sensation around the mouth, hypersensitivity to light, sound and touch, dizziness, tremors, nausea, vomiting, nervousness and confusion	Significant decrease under higher concentration in Chl *a* and *b*, total Chl, carotenoids and Chl/Car ratio, photosynthetic oxygen yield, photofixation of carbon ($^{14}CO_2$), photosynthetic electron transport activities and photorespiration (Mishra et al. 2008)
		Organochlorine insecticide, in Hopewell, Virginia, developed a variety of neurological symptoms, the most prominent of which became known as the 'Kepone shakes'; Organophosphate and carbamate exposure causes signs and symptoms of excess acetylcholine, such as increased salivation and perspiration, narrowing of the pupils, nausea, diarrhea, decrease in blood pressure, muscle weakness and fatigue	
		During prohibition, people consumed a homemade alcoholic drink made of Jamaican ginger that was contaminated with the organophosphate triorthocresyl phosphate (TOCP). More than 20,000 people were affected by a condition called 'Ginger Jake paralysis'. Pyrethroids can cause hyperexcitation, aggressiveness, uncoordination, whole-body tremors and seizures (Lah 2011)	
Human and veterinary drugs	Drugs are persistent or are metabolized only partially in the body. The unchanged substances as well as their metabolites are then excreted with urine and feces and enter wastewater treatment plants (WWTPs) by the way of sewage	Prescribed doses do not pose negative effect on human, but doses higher than those might cause severity	Blocks the biosynthesis of carotenoids, reduces plant height, shoot and root dry weight as well as affects Ca, Mg, K and N contents (Batchelder 1982) and inhibits glutathione S-transferase activity (Farkas et al. 2007) as well as phosphatase activity (Liu et al. 2009)

11.3 Sources of Heavy Metal Pollution

Sources of heavy metals in the environment are natural and anthropogenic, such as weathering of rocks, industrial effluents, organic wastes, sludge or municipal compost, pesticides, fertilizer transport, emissions from municipal waste incinerate and power generation and exudates, residues from metalliferous mines and smelting industries. Industrial effluents and sewerage are the sources of metallic pollution in the hydrosphere. Disposal of municipal wastes are considered as one of the major sources of heavy metal pollution to the soil. Disposal of metal-rich materials results in the formation of metalliferous mine spoils and metalliferous tailings that cause heavy metal contamination. Heavy metals can be carried to distant places if they are in gaseous or particulate form. Further, some agricultural practices, such as excess use of pesticides and fertilizers have contributed to increased concentrations of heavy metal in the soil.

11.3.1 Pesticides as Xenobiotics

Organisms like insects, weeds, microorganisms (microfungi, bacteria), rodents, nematodes, unwanted plants, and others that cause economic loss or damage to the physical well-being of humans as well as other organisms are known as pests.

Natural formulations or artificially synthesized chemicals used to control, eradicate or kill these pests are known as pesticides. Pesticides are often used in agrarian practices to protect crop plants from weeds, diseases and depredations from insects, fungi, mites and rodents. Pesticides are the only toxic chemicals deliberately released into the environment in large amounts. The use of pesticides became indispensable and an integral part of modern agriculture as food security for increasing population is also necessary. The use of pesticides cannot be ruled out in our agricultural fields because of steady but continuous rise in population and lesser availability of agricultural fields. Pesticides that are used in agriculture are transported to water bodies through runoff, drift and leaching, pollute them and increase the risk of exposure to nontarget organisms (Chen et al. 2007). Thus, soil and water pollutions due to pesticides have become a common concern among environmentalists. Some of the pesticides (organochlorines, e.g. DDT) are biomagnified in the terrestrial ecosystems, so they were banned worldwide according to the international treaties and conventions (Kutz et al. 1991). Some prominent pesticides which are frequently used in these days by the farmers include 2,4-D, aldrin, atrazine, butachlor, carbaryl, carbofuran, chlordane, chlorpyrifos, cypermethrin, DDT, dieldrin, dimethoate, endosulfan, glyphosate, heptachlor, lindane, malathion, monocrotophos, parathion, permethrin, phorate, triazophos, trifluralin etc. Most of the pesticides are nondegradable and accumulate in the environment.

The excessive use of pesticides has a serious impact on many beneficial microorganisms resulting in greater loss in crop productivity. Since most of the pesticides are nonbiodegradable, they have long residence time in water and soil and thus may enter and magnify at various trophic levels (Yadav 2010). Their prospects to cause adverse impacts on human and cattle health have been the subject of risk assessment studies and have led to the development of increasingly stringent regulations for the risk assessment of novel formulations and to control the use of existing compounds (Galloway and Handy 2003).

The organophosphorus pesticides were introduced in the 1970s as replacements of the persistent organochlorines after the tendency of DDT and its metabolites to accumulate in ecosystems and to cause health hazards, particularly in top predators (Murphy 1986). The increased use of organophosphorus pesticides is originally seen as a lesser threat to the environment, but over the time organophosphorus pesticides have become a serious environmental concern due to their high acute toxicity despite their low persistence. There are some 200 organophosphorus pesticides used in agriculture, forests, gardens, home and industrial sites to control the pests.

Organophosphorus pesticides are ubiquitous in the environment and are highly toxic to animals like fish, amphibians, rats etc. They inhibit acetylcholinesterase (AChE) enzyme of animals by binding it, resulting in neurological dysfunction and normal nerve impulse is checked. When AChE is inhibited, then neurons overexcite and respiratory control is lost due to asphyxiation. Oxidative desulfuration is necessary to achieve the greatest cholinesterase-inhibiting potencies of organophosphates. Oxidative desulfuration is mediated by mixed-function oxidases (MFO) which increase the toxicity of the pesticide concerned. Organophosphates are specific to a target animal; similarly animals are also specific to organophosphates, as many organophosphates do not kill a particular pest even if their highest concentrations are used while few of them can kill those particular pests at their recommended doses.

Many studies prove that oxon derivatives of organophosphates are significantly more toxic, sometimes up to 3000 times than their respective parental forms. In addition of being metabolized internally in liver cells, bacteria and other microorganisms can convert pesticides into sulfons or oxons (Hill 2003), thereby making them available in the environment (Schomburg et al. 1991; Domagalski 1996; Whitehead et al. 2005). A wide range of harmful environmental effects are linked to organophosphorus pesticides, including toxicity to domestic animals, freshwater fish, other aquatic organisms, birds, beneficial insects, plants and soil organisms etc. It has been shown to bioaccumulate in fish and synergistically reacts with other chemicals.

Pesticides may also generate ROS via their toxic effects on photosynthesis. Any adverse effect of pesticide may severely affect metabolic functions and overall growth performance of plants and other microflora, including cyanobacteria. Some of the chlorophyll biosynthetic enzymes are hindered and thereby ROS are generated which in turn damage macromolecules such as lipids, proteins and nucleic acids (Sheeba et al. 2011; Srivastava et al. 2012). There are several studies which show the worst effect of pesticides on growth and photosynthesis, nitrogen metabolism and amino acid metabolism in higher plants (Srivastava et al. 2012). Hormesis is a dose-dependent phenomenon of a toxicant, characterized by low-dose stimulation and high-dose inhibition (Calabrese and Baldwin 2003). Hormesis is seen up to 40% of toxicological experiments (Calabrese and Baldwin 2001).Calabrese and Baldwin

(2003) have carried out the dose range finding studies under the National Toxicology Program and suggested five doses of such types for each toxicant. However, the study was done on animals. They argued that the enhanced efficiency in the utilization of consumed nutrients was the primary cause of low-dose stimulation. Hammouda (1999) observed in his experiment with *Anabaena doliolum* that the sheathless heterocystous cyanobacterium was initially able to utilize low concentrations of carbofuran (a carbamate pesticide), whereas higher concentrations and the subsequent formation of hydrolytic breakdown products were toxic. Growth inhibition reached more than 50% when treated with 80 and 100 μg ml^{-1} of the insecticide. Lower concentrations of Bavistin (a fungicide) were observed to support the growth of *Tolypothrix scytonemoides* having maximum protein and pigment contents (Rajendran et al. 2007). The effects of pesticides on plants and human health have been enumerated in Table 11.1.

11.3.2 Pharmaceuticals

Pharmaceuticals are very frequently used for the treatment and prevention of diseases and epidemics in human beings and livestock. Pharmaceuticals have beneficial biological effects, but many pharmaceuticals are often discharged as such in water bodies due to lax in regulatory frameworks. They may exert unwanted biological effects there. Sometimes, they are not considered to be potentially harmful due to inadequate experience with environmental issues. Pharmaceuticals include antibiotics, analgesics and anti-inflammatory drugs, anticonvulsant drugs, beta-blockers, blood lipid regulators, hormones, chemicals used for disinfection and endocrine-disrupting compounds, X-ray contrast media, cytostatic drugs (chemotherapy), oral contraceptives and veterinary pharmaceuticals (Kot-Wasik et al. 2007). Specifically, drugs such as antibiotics are considered as xenobiotics in humans. As the name suggests, 'antibiotics' are chemicals that are produced naturally or semisynthetically in response to the organisms and possess antibacterial activity. Antibiotics are designed to affect mainly microorganisms and bacteria. Therefore, this makes them potentially hazardous to other such organisms found in the environment. Since the launch of penicillin in 1928, hundreds of other antibiotics have come for the treatments and other purposes in human beings and livestock. Antibiotics themselves fall into several categories: anthelmintic, anti-infective, antibacterial, antimicrobial, antiparasitic, antiseptic etc. One estimate says that a total of about 1–2 lakh tons $year^{-1}$ of approximately 3000 pharmacologically active compounds are used worldwide (Kumar et al. 2012). Generally, microorganisms like bacteria and microalgae are at more risks for antibiotics as the lethal doses are 2–3 orders lesser than those for higher trophic levels (Wollenberger et al. 2000).

After the antibiotics are administered, they may undergo conjugation with sugars present in the liver. After excretion, microbes can rapidly degrade the conjugated sugars and compounds become bioactive again (Renner 2002). Additionally, antibiotics are often poorly adsorbed in the bodies of humans and animals and most of them are excreted in unchanged bioactive forms through feces and urine into the drainage system (Kim and Aga 2007; Du and Liu 2012). These compounds reach aquatic and agricultural fields without adequate treatment of sewage system by the existing wastewater treatment methods. Thus, irrigation water may contain number of pharmaceuticals with different concentrations (Du and Liu 2012; Grassi et al. 2013). Physical and chemical properties of pharmaceuticals like water solubility, dissociation constant sorption-desorption process, stability and binding with the soil and the partitioning coefficients at various pH values can affect the mobility of antibiotics in the soil environment. Further, mobility of pharmaceuticals along the soil column is negatively proportional to the availability of organic matter in the soil (Chefetz et al. 2008). Therefore, pharmaceuticals are transported to groundwater after intensive irrigation. Thus, most rivers, lakes, and coastal waters along populated sites are often polluted with pharmaceuticals and their degradation products. Their traces can be detected in groundwater in populated areas (Daneshvar et al. 2012; Matamoros et al. 2012). When such water is used for crop irrigation, along with land application of manure, these compounds are introduced into arable lands and crops (Fatta-Kassinos et al. 2011). In addition to these, land application of composted animal dung without any treatment as a supplement to fertilizer is often a common practice in many countries, like India (Kumar et al. 2012), which is the source of pharmaceuticals in the environment.

There are apprehensions to the rise of new bacterial strains resistant to antibiotics after frequent and indiscriminate usage of antibiotics. Low concentrations (ng L^{-1} to μg L^{-1} order) but continuous presence of pharmacologically active compounds and their degradation products in ground and surface waters make bacterial populations resistant against multiple antibiotics resulting in potential risk to human (Harris et al. 2012; Grassi et al. 2013). Resistance can also be acquired through horizontal gene transfer via conjugation, transformation and transduction; and resistance can be attained in other bacteria as well (Fatta-Kassinos et al. 2011). Bacterial populations exposed to antibiotics in the gut of animals were found to be five times more likely to be resistant to any given antibiotic-resistant microbial populations. Recently, Goss et al. (2013) reviewed the use of organic amendments and concluded that humans are at the greatest risk because of their exposure to pathogens having multiple resistances. As a result, human beings and animals are presently facing 'post-antibiotic era'. Organizations like the World Health Organization (WHO) are of the view to use antibiotics for limited purposes and banning them as growth-promoting additives (Kumar et al. 2012).

Pharmaceuticals affect plant growth and yields (Jjemba 2002; Fatta-Kassinos et al. 2011; Goss et al. 2013). Accumulation of pharmacologically active compounds and their metabolites in agroecosystems can result in crop toxicity and soil microorganisms, particularly bacteria may get infected (Du and Liu 2012). Bacterial exposure to antibiotics may decrease their biodegradation capacity (Richardson 2012) affecting the plant development indirectly due to nutrient scarcity (Cooper et al. 2008). Some studies have also shown that the presence of low level of pharmaceuticals in arable soils can induce toxic or other kinds of effects on terrestrial and aquatic organisms (Jjemba 2002; Fatta-Kassinos et al. 2011; Kumar et al. 2012). Phytotoxicity of pharmaceuticals depends on their physicochemical properties, soil types, soil organic matter content, soil pH and prevailing climatic conditions. Studies have also proved that pharmaceutical phytotoxicity may vary with the plant species.

Antibiotics may be accumulated by roots and stems of certain plant species (Migliore et al. 1995). The accumulation of antibiotics in crops and vegetables poses great risks to human beings and livestock (Du and Liu 2012). When pharmaceutical residues enter into primary producers of the ecosystem, they travel throughout the food chain and reach animals and humans (Fatta-Kassinos et al. 2011).

Several workers like Batchelder (1981), Kong et al. (2007), Farkas et al. (2009), Migliore et al. (1995, 1996, 1998, 2003), Shenker et al. (2011), Liu et al. (2009) and Xie et al. (2010) worked with multiple antibiotics, e.g. tetracyclines, sulfonamides, oxytetracycline, sulfadimethoxine, enrofloxacin and carbamazepine on several plants like *Phaseolus vulgaris*, *Medicago sativa*, *Panicum miliaceum*, *Pisum sativum*, *Zea mays*, *Hordeum distichum*, *Amaranthus retroflexus*, *Plantago major*, *Rumex acetosella*, *Cucumis sativus*, *Lactuca sativa*, *Raphanus sativus*, *Oryza sativa*, *Cichorium endivia and Triticum aestivum* and concluded that these antibiotics posed inhibitory effects on these plants in one or other ways on their different morphological and cytological parameters – growth and development of roots and shoot, biomass production of the plants, the number/length of leaves, seed germination, cell mitotic division, the frequency of micronucleus, chromosomal aberration and sister chromatid exchange.

Folic acid is involved in the synthesis of purines, precursors of cytokinins and abscisic acid. Sulfadimethoxine competes with folic acid and gets accumulated (Migliore et al. 1998). Many times antibiotics like chlortetracycline and enrofloxacin showed hormesis effects, i.e. low-dose stimulation. The effects of human and veterinary drugs on plants and human health have been enumerated in Table 11.1.

11.4 Treatment of Wastewater Is Necessary

Wastewater reuse has become today's global demand because human population has reached over 7 billion. Conventional markers of pollution in drinking water have been the biological oxygen demand (BOD), chemical oxygen demand (COD), pH, total suspended solids, heavy metals and microbiological load (i.e. viruses, bacteria and protozoa). Inorganic elements are often found in salt form and often difficult to be removed during most processes of wastewater treatment. They are often highly water-soluble and are not susceptible to biological degradation. But, as the human life became easier for living after several inventions in the field of science and technology, the level and variety of pollutants of water bodies increased substantially, providing possibilities to reassess water qualities on several newer parameters for safer reuse of wastewater. Therefore, proper quantification of heavy metals, pesticides, pharmaceutical metabolites and other substances like potential endocrine-disrupting compounds (EDCs), illicit drugs and personal care products must be done. Without quantification of pollutants and their proper removal, reuse of wastewater for agricultural purposes poses serious threat to plants and hence human health and total environment (Novoa-Muñoz et al. 2008). Pathogens and microorganisms in water sources have been the major concerns of medical professionals, environmentalists and policy makers; and there is little or no worry about other deadly inorganic compound chemicals, heavy metals, xenobiotics, pesticides, inorganic compounds, persistent organic substances like EDCs, pharmaceuticals, drug metabolites, transformation products etc., which do persist even after conventional treatment of municipal wastewater. Several studies have warned that recently included nanoparticles of several metals are hazardous enough to plants directly and indirectly by affecting beneficial microbes that function in element cycling, pollutant degradation and plant growth (Gajjar et al. 2009).

Adequate treatment of wastewater is necessary, and only then the treated wastewater can be reused for irrigation and drinking purposes. When the pollutants are insoluble and do not make homogeneous phase with water then after covering some distances, impurities automatically silt down. Further, tributaries of rivers discharge huge amount of water in their main rivers; therefore, the level of pollution becomes substantially down. This is a natural method of water treatment/purification. But, in the modern world, the level of pollution became much high, the variety of pollutants became much diverse, the way and free flow of river water have not remained unchecked, and this natural method cannot be relied upon. Almost every city in India has its own water treatment plants, where domestic and industrial wastewater is treated before their discharge into water bodies. But there are so many industrial units in India which discharge their water without treatment process. There is no stringent regulatory framework in India that could check them. In this situation the National Green Tribunal and Supreme Court like apex institutions have intervened and ordered to close down such units which openly discharge heavy polluted water in rivers like Yamuna and Ganga.

Presently, treatment processes for urban wastewater are niggling and often fail to remove contaminants completely, and thus our terrestrial and aquatic environment are getting polluted through disposal and petty reuse applications. Inorganic elements are very difficult to be removed by advanced oxidation processes since they are highly water-soluble and cannot be effectively treated by membrane processes. In this way, the knowledge gaps in order to assess the impact of wastewater reuse for irrigation and drinking purposes must be filled and must be shared with our policy makers who can formulate adequate legislations for the regulatory purposes. Technological sophistication is needed vis-à-vis analytical chemistry, chromatographic methods etc., for adequate treatment processes enabling identification and quantification along with separation of a number of organic xenobiotic compounds in treated wastewater. Technological improvements made us capable to reuse the severely polluted industrial/sewage wastewater even for drinking purposes, as recently seen in New Delhi, India's national capital and one of the most populated cities and worst polluted cities in the world. The treated wastewater may be an attractive and promising source of irrigation water in many states of India conflicting over river water distribution issues.

11.5 Regulations to Improve the Water Quality

There is an urgent need of newer and flexible regulatory mechanism in the age of industrial and technological improvements which produce and discharge newer chemicals day by day. Substances fall under xenobiotic category must undergo extensive risk evaluation, such as toxicity to humans, ecotoxicity or persistence in the environment, before they are registered for sale. There are several institutional bodies in India, like the Ministry of Water Resources, Water Quality Assessment Authority, Central Ground Water Board, Central Pollution Control Board, Ministry of Environment and Forests, Centre for Science and Environment, State Pollution Control Boards, Central Water Commission etc. for risk evaluation that assess and quantify the level of pollutants in water with the help of other scientific bodies. These bodies have laboratories that are purely dedicated to testing of water quality, for instance, Central Water Commission (CWC) has been assigned for analyzing water samples that are collected from different sites under observation and thereafter, for analyzing in a three-tier laboratory system. The I-level labs are situated on major rivers and analyze seven water quality parameters, while II-level labs have been established in selected divisional offices of CWC and analyze 25 parameters related to water quality and III-level labs have been provided with the capability of analyzing about 41 parameters of water quality with special emphasis on analyses of heavy metals, toxic compounds and pesticides. Similar to CWC, Central Ground Water Board (CGWB) also monitors water quality, but it is specialized for testing groundwater and generating documentation work for chemical constituents of water on regional scale. Apart from this, CGWB also operates observations for wells and water constituents and analyzes water quality of deeper aquifer by tentative drilling operations, and these collected samples are then tested in laboratories for the quality. After the preparation of documents, these bodies share their data with the government and the Central Parliament formulates laws and legislations like 'The Water (Prevention and Control of Pollution) Act, 1974'; 'The Water (Prevention and Control of Pollution) Rules, 1975'; 'The Water (Prevention and Control of Pollution) Cess Act, 1977'; 'The Water (Prevention and Control of Pollution) Cess Rules, 1978'; and 'The Environment (Protection) Act, 1986'. However, many times these institutions work in uncoordinated manner resulting in doubtful data. Furthermore, pharmaceuticals and personal care products (PPCPs) and other recently detected compounds have not yet been the area of interest, both for treatment plants and workers studying in these areas.

The Ministry of Forest and Environment in India exercises its ecological impact assessment to test the sustainability and viability from ecological point of view. The Ministry of Forest and Environment in India monitors the formulation and implementation of programs and policies for preventing and abating entry of different kinds of pollutants in rivers and water bodies. It also monitors risks to the terrestrial ecosystem as untreated, partially treated or adequately treated wastewater ultimately goes to soil. However, in general, in context to the soil, environmental risks due to heavy metals have been sufficiently studied, but studies are very scarce for other substances.

11.6 Conclusion

Xenobiotics have to be removed for the wastewater reclamation to assure existing per capita use of water and its reuse for the environment and plant, animal, and human health protection. Newer contaminants are introduced day by day in our ecosystem with the technical innovations, but at the same time, technical, regulatory and economic constraints related to water treatment process do not remove xenobiotics adequately leading to various human and animal health hazards. Governments have to step in to synchronize the risk evaluation data provided by different monitoring agencies and only then to formulate legislation regarding adequate removal of xenobiotics from wastewater and to reuse it, not only for irrigation purpose but also for drinking purpose too.

References

Batchelder AR (1981) Chlortetracycline and oxytetracycline effects on plant growth and development in liquid cultures. J Environ Qual 10:515–518

Batchelder AR (1982) Chlortetracycline and oxytetracycline effects on plant growth and development in soil systems. J Environ Qual 11:675–676

Bharwana SA, Ali S, Farooq MA et al. (2013) Alleviation of lead toxicity by silicon is related to elevated photosynthesis, antioxidant enzymes suppressed lead uptake and oxidative stress in cotton. J Biorem Biodegrad 4:187. doi:10.4172/2155-6199.1000187

Calabrese EJ, Baldwin LA (2001) Hormesis: U-shaped dose responses and their centrality in toxicology. Trends Pharmacol Sci 22:285–291

Calabrese EJ, Baldwin LA (2003) Hormetic dose-response model is more common than the threshold model in toxicology. Toxicol Sci 71:246–250

Chefetz B, Mualem T, Ben-Ari J (2008) Sorption and mobility of pharmaceutical compounds in soil irrigated with reclaimed wastewater. Chemosphere 73:1335–1343

Chen Z, Cuin TA, Zhou M et al. (2007) Compatible solute accumulation and stress-mitigating effects in barley genotypes contrasting in their salt tolerance. J Exp Bot 58:4245–4255

Cooper ER, Siewicki TC, Phillips K (2008) Preliminary risk assessment database and risk ranking of pharmaceuticals in the environment. Sci Total Environ 398:26–33

Daneshvar A, Svanfelt J, Kronberg L et al. (2012) Neglected sources of pharmaceuticals in river water-footprints of a Reggae festival. J Environ Monit 14:596–603

Domagalski J (1996) Pesticides and pesticide degradation products in storm-water runoff: Sacramento River Basin, California. Water Resour Bull 32:953–964

Du L, Liu W (2012) Occurrence, fate and ecotoxicity of antibiotics in agro-ecosystems. A review. Agron Sustain Dev 32:309–327

Durand TC, Sergeant K, Planchon S et al. (2010) Acute metal stress in *Populus tremula* x *P. alba* (717-1B4 genotype): leaf and cambial proteome changes induced by Cd^{2+}. Proteomics 10:349–368

Farkas MH, Berry JO, Aga DS (2007) Chlortetracycline detoxification in maize via induction of glutathione *S*-transferases after antibiotic exposure. Environ Sci Technol 41:1450–1456

Farkas MH, Mojica E-RE, Patel M et al. (2009) Development of a rapid biolistic assay to determine changes in relative levels of intracellular calcium in leaves following tetracycline uptake by pinto bean plants. Analyst 134:1594–1600

Fatta-Kassinos D, Kalavrouziotis IK, Koukoulakis PH et al. (2011) The risks associated with wastewater reuse and xenobiotics in the agroecological environment. Sci Total Environ 409:3555–3563

Gajewska E, Sklodowska M, Slaba M et al. (2006) Effect of nickel on antioxidative enzyme activities, proline and chlorophyll contents in wheat shoots. Biol Plant 50:653–659

Gajjar P, Pettee B, Britt DW et al. (2009) Antimicrobial activities of commercial nanoparticles against an environmental soil microbe, *Pseudomonas putida* KT2440. J Biol Eng 3:9–13

Galloway T, Handy R (2003) Immunotoxicity of organophosphorous pesticides. Ecotoxicology 12:345–363

Garg N, Singla P (2011) Arsenic toxicity in crop plants: physiological effects and tolerance mechanisms. Environ Chem Lett 9:303–321

Gebhardt W, Schröder H Fr (2007) Liquid chromatography-(tandem) mass spectrometry for the follow-up of the elimination of persistent pharmaceuticals during wastewater treatment applying biological wastewater treatment and advanced oxidation. J Chromatogr 1160:34–43

Goss MJ, Tubeileh A, Goorahoo D (2013) A review of the use of organic amendments and the risk to human health. In: Donald LS (ed) Advances in agronomy, Academic, San Diego, 12, pp 275–379

Grassi M, Rizzo L, Farina A (2013) Endocrine disruptors compounds, pharmaceuticals and personal care products in urban wastewater: implications for agricultural reuse and their removal by adsorption process. Environ Sci Pollut Res 20:3616–3628

Hammouda O (1999) Response of the paddy field cyanobacterium *Anabaena doliolum* to carbofuran. Ecotoxicol Environ Saf 44:215–219

Harris SJ, Cormican M, Cummins E (2012) Antimicrobial residues and antimicrobial-resistant bacteria: impact on the microbial environment and risk to human health – a review. Hum Ecol Risk Assess 18:767–809

Hill EF (2003) Wildlife toxicology of organophosphorus and carbamate pesticides. In: Hoffman DJ, Rattner BA, Burton GA Jr, Cairns J Jr (eds) Handbook of Ecotoxicology. Lewis Publishers, Boca Raton, pp 281–312

Jjemba PK (2002) The potential impact of veterinary and human therapeutic agents in manure and biosolids on plants grown on arable land: a review. Agric Ecosyst Environ 93:267–278

Kim S, Aga DS (2007) Potential ecological and human health impacts of antibiotics and antibiotic-resistant bacteria from wastewater treatment plants. J Toxicol Environ Health B 10:559–573

Kong WD, Zhu YG, Liang YC et al. (2007) Uptake of oxytetracycline and its phytotoxicity to alfalfa (*Medicago sativa* L.). Environ Pollut 147:187–193

Kot-Wasik A, Debska J, Namiesnik J (2007) Analytical techniques in studies of the environmental fate of pharmaceuticals and personal care products. TrAC Trends Anal Chem 26:557–568

Kumar R, Lee J, Cho J (2012) Fate, occurrence, and toxicity of veterinary antibiotics in environment. J Korean Soc Appl Biol Chem 55(6):701–709

Kutz FW, Wood PH, Bottimore DP (1991) Organochlorine pesticides and polychlorinated biphenyls in human adipose tissue. Rev Environ Contam Toxicol 120:1–82

Liu F, Ying GG, Tao R et al. (2009) Effects of six selected antibiotics on plant growth and soil microbial and enzymatic activities. Environ Pollut 157:1636–1642

Madhava Rao KV, Sresty TV (2000) Antioxidative parameters in the seedlings of pigeon pea (*Cajanus cajan* (L.) Millspaugh) in response to Zn and Ni stresses. Plant Sci 157:113–128

Marshall FM, Holden J, Ghose, C, et al. (2007) Contaminated irrigation water and food safety for the urban and peri-urban poor: appropriate measures for monitoring and control from field research in India and Zambia. Research project Department for International Development SPRU, University of Sussex. www.pollutionandfood.net

Matamoros V, Arias CA, Nguyen LX et al. (2012) Occurrence and behavior of emerging contaminants in surface water and a restored wetland. Chemosphere 88:1083–1089

McLaughlin MJ, Parker DR, Clark JM (1999) Metal and micronutrients-food safety issues. Field Crop Res 60:143–163

Migliore L, Brambilla G, Cozzolino S et al. (1995) Effect on plants of sulphadimethoxine used in intensive farming (*Panicum miliaceum*, *Pisum sativum* and *Zea mays*). Agric Ecosyst Environ 52:103–110

Migliore L, Brambilla G, Casoria P et al. (1996) Effect of sulphadimethoxine contamination on barley (*Hordeum distichum* L., Poaceae, Liliopsida). Agric Ecosyst Environ 60:121–128

Migliore L, Civitareale C, Cozzolino S et al. (1998) Laboratory models to evaluate phytotoxicity of sulphadimethoxine on terrestrial plants. Chemosphere 37:2957–2961

Migliore L, Cozzolino S, Fiori M (2003) Phytotoxicity to and uptake of enrofloxacin in crop plants. Chemosphere 52:1233–1244

Mishra V, Srivastava G, Prasad SM et al. (2008) Growth, photosynthetic pigments and photosynthetic activity on seedling stage of *Vigna unguiculata* in response to UV-B and dimethoate. Pestic Biochem Physiol 92:30–37

Molas J (2002) Changes of chloroplast ultrastructure and total chlorophyll concentration in cabbage leaves caused by excess of organic Ni(II) complexes. Environ Exp Bot 47:115–126

Murphy PM (1986) Effect of light and atmospheric carbon dioxide concentration on nitrogen fixation by herbage legumes. Plant Soil 95:399–409

Nemat Alla MM, Hassan NM, El-Bastawisy ZM (2007) Differential influence of herbicide treatments on activity and kinetic parameters of C_4 photosynthetic enzymes from *Zea mays*. Pesticide Biochem Physiol 89:198–205

Novoa-Munoz JC, Pontevedra-Pombal X, Martinez-Cortizas A et al. (2008) Mercury accumulation in upland acid forest ecosystems nearby a coal fired power-plant in Southwest Europe (Galicia NW Spain). Sci Total Environ 394:303–312

Panda SK (2007) Chromium-mediated oxidative stress and ultrastructural changes in root cells of developing rice seedlings. J Plant Physiol 164:1419–1428

Panda SK, Choudhury S (2005) Chromium stress in plants. Braz J Plant Physiol 17:95–102

Patra M, Sharma A (2000) Mercury toxicity in plants. Bot Rev 66:379–422

Patra M, Bhowmik N, Bandopadhyay B et al. (2004) Comparison of mercury, lead and arsenic with respect to genotoxic effects on plant systems and the development of genetic tolerance. Environ Exp Bot 52:199–223

Petrovic M, Alda MJLD, Diaz-Cruz S et al. (2009) Fate and removal of pharmaceuticals and illicit drugs in conventional and membrane bioreactor wastewater treatment plants and by riverbank filtration. Trans R Soc A 367:3979–4003

Pourrut B, Shahid M, Dumat C et al. (2011) Lead uptake, toxicity, and detoxification in plants. Rev Environ Contam Toxicol 213:113–136

Prasad SM, Kumar S, Parihar P et al. (2015) Evaluating the combined effects of pretilachlor and UV-B on two *Azolla* species. Pestic Biochem Physiol 128:45–56

Qian H, Li J, Sun L et al. (2009) Combined effect of copper and cadmium on *Chlorella vulgaris* growth and photosynthesis-related gene transcription. Aquat Toxicol 94:56–66

Rajendran UM, Kathirvel E, Anand N (2007) Desiccation-induced changes in antioxidant enzymes, fatty acids and amino acids in the cyanobacterium *Tolypothix scytonemoides*. World J Microbiol Biotechnol 23:251–257

Renner R (2002) Do cattle growth hormones pose an environmental risk? Environ Sci Technol 36:194A–197A

Richardson SD (2012) Environmental mass spectrometry: emerging contaminants and current issues. Anal Chem 84:747–778

Saladin GC, Clement MC (2003) Stress effects of flumioxazin herbicide on grapevine (*Vitis vinifera* L.) grown *in vitro*. Plant Cell Rep 21:1221–1227

Salt DE, Smith RD, Raskin I (1998) Phytoremediation. Annu Rev Plant Biol 49:643–668

Sandalio LM, Dalurzo HC, Gomez M et al. (2001) Cadmium-induced changes in the growth and oxidative metabolism of pea plants. J Exp Bot 52:2115–2126

Schomburg CJ, Glotfelty DE, Seiber JN (1991) Pesticide occurrence and distribution in fog collected near Monterey, California. Environ Sci Technol 25:155–160

Sheeba, Singh VP, Srivastava PK et al. (2011) Differential physiological and biochemical responses of two cyanobacteria *Nostoc muscorum* and *Phormidium foveolarum* against oxyfluorfen and UV-B radiation. Ecotoxicol Environ Saf 74:1981–1993

Shenker M, Harush D, Ben-Ari J et al. (2011) Uptake of carbamazepine by cucumber plants – a case study related to irrigation with reclaimed wastewater. Chemosphere 82:905–910
Singh N, Ma LQ, Vu JC et al. (2009) Effects of arsenic on nitrate metabolism in arsenic hyperaccumulating and non-hyperaccumulating ferns. Environ Pollut 157:2300–2305
Singh VP, Srivatava PK, Prasad SM (2013) Nitric oxide alleviates arsenic-induced toxic effects in ridged *Luffa* seedlings. Plant Physiol Biochem 71:155–163
Singh R, Singh S, Parihar P et al. (2015) Arsenic contamination, consequences and remediation techniques: a review. Ecotoxicol Environ Saf 112:247–270
Singh S, Parihar P, Singh R et al. (2016) Heavy metal tolerance in plants: role of transcriptomics, proteomics, metabolomics, and ionomics. Front Plant Sci 6. doi: 10.3389/fpls.2015.01143
South Asia Network on Dams, Rivers and People. (2016) DRP News Bulletin. Is India facing its worst-ever water crisis?
Srivastava PK, Singh VP, Prasad SM (2012) Compatibility of ascorbate-glutathione cycle enzymes in cyanobacteria against low and high UV-B exposures, simultaneously exposed to low and high doses of chlorpyrifos. Ecotoxicol Environ Saf 74:1981–1993
Tchounwou PB, Yedjou CG, Patlolla AK (2012) Heavy metals toxicity and the environment. NIH Public Access 101:133–164
Tsonev T, Lidon FJC (2012) Zinc in plants – an overview. Emir J Food Agric 24:322–333
Whitehead A, Anderson SL, Ramirez A et al. (2005) Cholinesterases in aquatic biomonitoring: assay optimization and species specific characterization for a California native fish. Ecotoxicology 14:597–606
Wollenberger L, Halling-Sørensen B, Kusk KO (2000) Acute and chronic toxicity of veterinary antibiotics to *Daphnia magna*. Chemosphere 40:723–730
Xie X, Zhou Q, He Z et al. (2010) Physiological and potential genetic toxicity of chlortetracycline as an emerging pollutant in wheat (*Triticum aestivum* L.). Environ Toxicol Chem 29:922–928
Yadav SK (2010) Heavy metals toxicity in plants: an overview on the role of glutathione and phytochelatins in heavy metal stress tolerance of plants. S Afr J Bot 76:167–179
Yang X, Feng Y, He Z, Stoffell PJ (2005) Molecular mechanisms of heavy metal hyperaccumulation and phytoremediation. J Trace Elem Med Biol 18:339–353
Yruela I (2005) Copper in plants. Braz J Plant Physiol 17:145–156

Silver Nanoparticle in Agroecosystem: Applicability on Plant and Risk-Benefit Assessment

12

Rima Kumari and D.P. Singh

Abstract

Engineered nanomaterials are the major components among the broad range of xenobiotic particles. Nowadays scientists have gained higher attention on environmental nanomaterial exposures to elucidate its effects on natural ecosystem. Most of the studies on nanoparticles are concerned with silver nanoparticle, exhibited wide applications in various fields, i.e., in agricultural field, in biotechnology and bioengineering, in textile industries, in wastewater treatment plants, as well as in cosmetic products. Silver nanoparticle plays a significant role in smart and modern agriculture due to its antimicrobial and pesticidal activity. Along with positive aspects, the possible toxic effects of silver nanoparticle on human and other living organism as well as on environment must not be overlooked. Exposure to silver nanoparticle could exhibit an adverse effect on human cells, causing argyria, liver and kidney damage, respiratory problems, eye irritation, heart problem, etc. The interaction of nanosilver particles (AgNPs) to the plant – soil system – may influence the toxicity in ecologically important bacteria soil biota and other living organisms. So detailed risk-benefit assessment is required to predict the environmental effect of nanosized silver particles (AgNPs) in the foreseeable future.

This particular research review highlighted the insight of nanosilver to assess its applicability on agricultural practices and to understand its possible risk impacts. In this chapter, both two issues, i.e., applicability and risk assessment related to the use of nanosilver in modern agriculture, are studied: (i) silver nanoparticle as antimicrobial, antifungal, pesticidal, and nanofertilizer effect on plants and (ii) in water treatment plant and (iii) risk assessment of the vast use of silver nanoparticle and their entry into the environment on biological life.

R. Kumari • D.P. Singh (✉)
Department of Environmental Science, Babasaheb Bhimrao Ambedkar University, Lucknow 226025, Uttar Pradesh, India
e-mail: dpsingh_lko@yahoo.com; dpsinghbbaulko@gmail.com

A. Singh et al. (eds.), *Plant Responses to Xenobiotics*,
DOI 10.1007/978-981-10-2860-1_12

Keywords
Silver nanoparticle • Agriculture • Antimicrobial agent • Water treatment • Biological life

12.1 Introduction

Nanotechnology is the promising field of interdisciplinary sciences in the past few decades. The products of nanotechnology, nanoparticles (particles in size range 1–100 nm) due to its small size, have very high bioactivity. Nanoparticles are rapidly gaining interest by researchers because of its varied technological applications in industrial, medicinal, and agricultural fields. From the past decades, among the different kinds of nanomaterial, the nanosilver is extensively applied in the field of bioscience and biomedicine as biological tagging, biomarker, biomedical diagnostics purpose, antimicrobial application, and pharmaceuticals, in minimizing environmental pollution-related problems, targeted release of therapeutic drugs, and biosafety control, in treatment of cancer cells, as biosensor, and in material science (Salata 2004). It is estimated that of all the nanoparticles, silver nanoparticles (AgNPs) have approached the high demand of marketing and commercialization, ranging from disinfecting medical devices and pest resistance purpose to wastewater treatment plants. Its diverse properties like catalysis, good conductivity, chemical stability, and magnetic and optical polarizability electrical conductivity make these materials superior and indispensible (Abbasi et al. 2016).

Agriculture is the major resource for livelihood in most of the developing countries. The explosive growth of world population demands on agricultural productivity. Nanotechnology in the field of agriculture focuses currently on involvement of silver nanoparticles with unique properties to boost crop productivity. Silver nanoparticles are the most studied and utilized nanoparticles in the field of agricultural research to improve the efficiency, yield, and sustainability of agricultural crops. It has long been known to have strong pesticidal, antifungal, antiviral, and bactericidal effects (Chen and Schluesener 2008). Due to its broad spectrum of antimicrobial activities, silver nanoparticles have the prospect to increase food quality, global food production, plant protection, detection and regulation of plant diseases, monitoring of plant growth, and pest control for "sustainable agricultural development" (Kim et al. 2012; Khan and Rizvi 2014). Silver nanoparticles are highly stable and very well dispersive in aqueous solution. It is being used as foliar spray to stop fungi, molds, rot, and several other plant diseases (Singh et al. 2015). Moreover, silver nanoparticles are also used as an excellent plant-growth stimulator. It provides novel tool for the management of diseases, rapid disease detection, and minimizing nutrient losses in fertilization through an optimized nutrient management (Pérez-de-Luque and Rubiales 2009). Apart from multiple beneficiary aspects of silver nanoparticle for sustainable growth of plant, there are few possible risks associated with these silver nanoparticles. Silver nanoparticles in agricultural soil affect many bacterial communities which are beneficial/harmful for plant and

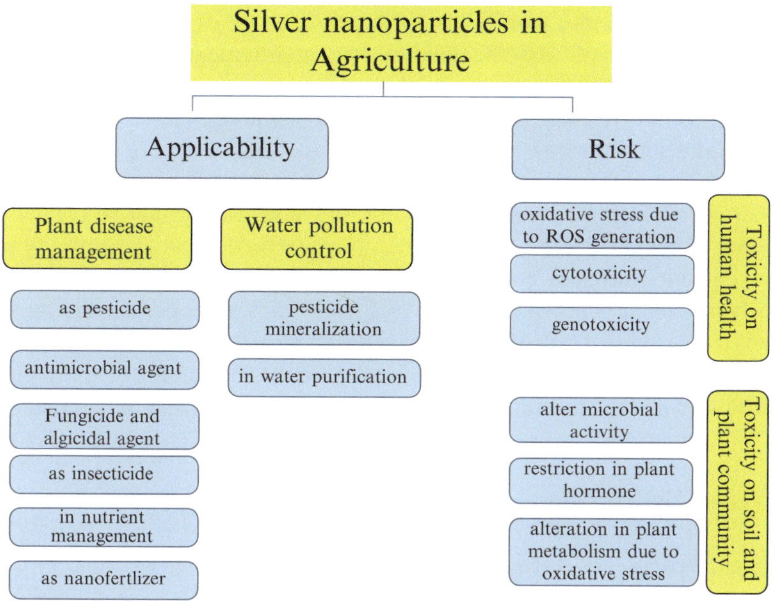

Fig. 12.1 Applicability of silver nanoparticles in agriculture: risk-benefit assessment on biological system

environment (Panyala et al. 2008). The extensive use of nanosilver can lead to development of resistance among pathogens and induced a serious threat to microbial diversity (Faunce and Watal 2010). Silver nanoparticle induced dose-dependent toxicity response that will be differing from each microorganism. Yang et al. (2014) have reported the toxic response of silver nanoparticles on some nitrifying bacterial species which play an important role in nitrification, and this sensitivity depends on the size of nanoparticles. Gavanji et al. (2012) reported the destructive effect of silver nanoparticles on beneficial Fungi *Trichoderma viride and T. harzianum.* These fungi act as biocontrol and growth-promoting agents for many crop plants due to their antagonistic properties against plant pathogens. Based on these reports, it is essential that the usage of silver nanoparticles should be in a good management in environment in optimized and regulated way because microorganisms play a very significant role in environment and any changes/damage in their growth can cause many problems for environment and biological life. This review article provides an overview of the risk-benefit assessments of most demanded and highly used silver nanoparticle in agricultural application (diagrammatically represented in Fig. 12.1).

12.2 Applicability of Silver Nanoparticles as Different Tools

Agricultural production is reduced worldwide every year due to various plant diseases which passes major challenges for agronomists to control these plant diseases. Conventionally, pesticide applications release approx. of ~2×10^6 metric tons of chemical-based pesticides worldwide, 90 % of which are dispersed to the air during application and as runoff and affected both the environmental hazards and application costs to the farmer (Stephenson 2003). Consistent long-term use of synthetic pesticides also has risk of developing resistance in pathogen, reduces soil biodiversity, and diminishes nitrogen fixation, increase bioaccumulation of pesticides. In the past decades, several studies have been carried out to develop harmless pest management that pose fewer environmental hazards and have focused to overcome the lack of ongoing demands of synthetic pesticide. Nanotechnological applicability in crop disease protection offers a great promise in the management of insects and pathogens. Silver nanoparticle is very effective against phytopathogens with low toxicity and leads to broad range of applicability in pesticidal activity. It is efficiently used for site-targeted delivery of important agrochemical products and for diagnosis purpose tools in case of prior detection of plant diseases (Chowdappa and Shivakumar 2013).

12.2.1 As Microbicides

Silver nanoparticles act as strong antimicrobial agent due to strong inhibitory effects against various bacterial organisms (Clement and Jarrett 1994). Nanosilver exhibits high level of toxicity to the microorganism and lower toxicity to the mammalian cells. It was observed that the microbe-killing effects of silver nanoparticles were size dependent (Raza et al. 2017). Silver nanoparticles, mainly in the size, which ranged from 1 to 10 nm, attach to the surface of cell membrane and drastically disturb its proper function like respiration and permeability (Morones et al. 2005). Applicability of silver nanoparticles against various microorganisms, i.e., *Staphylococcus aureus*, *Pseudomonas aeruginosa*, *Bacillus subtilis*, *Salmonella typhi*, *Syphilis typhus*, *Escherichia coli*, and *Vibrio cholera*, has been well investigated by several researchers (Rai et al. 2009; Zhou et al. 2012; Rajeshkumar and Malarkodi 2014; Franci et al. 2015).

12.2.2 As Fungicides

Silver nanoparticles are also responsible to develop eco-friendly fungicides because of its strong fungal-killing properties, used at targeting area of pathogen and inhibiting their metabolic processes (Park et al. 2006). The negative effect of nanosilver against 18 different commercially harmful fungi was reported by Kim et al. (2012), and they found dose-specific inhibition effect of nanosilver on these fungi. The applicability of nanosilver for its antifungal activity was also reported against *Sclerotinia minor*, *Rhizoctonia solani*, and *Sclerotinia sclerotiorum* by Min et al. (2009). Jo et al. (2009) also investigated antifungal potency of silver nanoparticles against two plant-pathogenic fungi, *Magnaporthe grisea* and *Bipolaris sorokiniana.* The in vitro and in vivo efficacy of silver nanoparticles against powdery mildew was reported by Lamsal et al. (2011). They showed strong inhibitory effect of nanosilver on fungal hyphae and conidial germination with minimal effect on cucumbers and pumpkins.

12.2.3 As Insecticides

Advance researches also highlighted the wide application of nanosilver particles for insecticidal application to kill the mosquitoes and flies. Nanosilver particles possessed excellent antilice and mosquito larvicidal activity, having vital application in community health improvement. There are tremendous researches investigating the efficacies of plant-synthesized AgNPs as mosquito larvicidal agent against different species of mosquitoes, i.e., *Culex quinquefasciatus*, *Heteroscodra maculata*, *Rhipicephalus microplus*, and *Anopheles subpictus*, and suggesting that biologically produced silver nanoparticles showed the strong larvicidal activity against mosquito larvae and it can be used as an ideal eco-friendly approach for their control (Marimuthu et al. 2011; Suman et al. 2013; Mondal et al. 2014). Jayaseelan et al. (2011) studied on the pediculicidal and larvicidal activity of synthesized silver nanoparticles (from leaf extract of *Tinospora cordifolia*) against the head louse *Pediculus humanus* and larvae of *Anopheles subpictus* and *Culex quinquefasciatus* and showed maximum mortality. Rouhani et al. (2012) evaluated the insecticidal activity of silver nanoparticles against the *Aphis nerii*. Soni and Prakash (2015) have described the larvicidal and pupicidal properties of biologically produced silver nanoparticles (from fungal strain of *Aspergillus niger*) against the mosquito larvae of *Aedes aegypti*, *Culex quinquefasciatus*, and *Anopheles stephensi.*

12.2.4 Mechanistic Study of Anti-pathogenic Action of Silver Nanoparticles

Silver nanoparticle exhibits the inhibitory effects against microorganisms at different levels: (i) penetration of cell wall, (ii) turning of DNA into condensed form and losses their replication ability, and (iii) induced inactivation of bacterial protein by binding of its thiol group with Ag^+ ion (Clement and Jarrett 1994). It has been suggested that Ag^+ ion strongly interacts with thiol group of vital enzymes and inactivates them. By inactivating respiratory chain enzymes and proteins responsible for RNA and DNA replication, it causes bacterial cell death (Samuel and Guggenbichler 2004; Elchiguerra et al. 2005). Oxidative stress by excessive generation of reactive oxygen species (ROS) may be considered to be another mechanism by which the cells die. Silver nanoparticles act in catalysis and may catalyze reactions using oxygen (O_2) directing as oxidizing agent, which results into excessive ROS generation (Stohs and Bagchi 1995). Carlson et al. (2008) studied the mechanistic study of silver nanoparticles on eukaryotic cells and suggest that this could inhibit the antioxidant defense by minimizing the ratio of reduced glutathione (GSH) and oxidized glutathione (GSSG) and subsequently results to oxidative stress in the cell. Besides this, in prokaryotic cells silver ions induce generation of ROS by directly abrupting the superoxide dismutases and/or respiratory chain enzymes via intercalation with thiol groups. An overview of mechanism behind anti-pathogenic action of silver nanoparticle has been shown in Fig. 12.2.

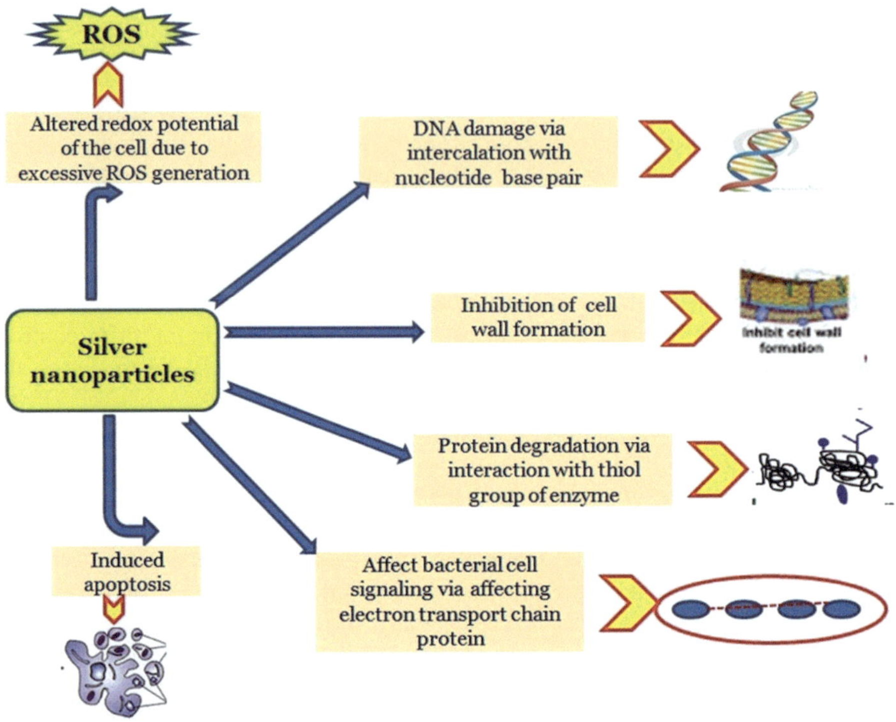

Fig. 12.2 Mechanism behind anti-pathogenic action of silver nanoparticles

12.2.5 Nanosilver Encapsulation for Controlled and Target Delivery of Agrochemicals: Nutrient Management

The wide use of anti-pathogenic agrochemicals in agriculture field has emphasized interests in scientists to clearly understand about their effects on nontarget pests upon interaction with these pesticides. Silver nanoparticles are very stable and biodegradable and it also displays slow release of agrochemicals. So it can be used for formation of nanocapsules for slow and optimized delivery of agrochemicals, pesticides, and fertilizers in agricultural practices (Chowdappa and Gowda 2013). Nanoencapsulated agrochemicals are designed to posses the desired properties including effective optimum concentration, time-controlled release, enhanced activity on target site, and least toxic effects (Tsuji 2001). It helps in slow release of agrochemical in controlled way to the particular host through dissolution, biodegradation, diffusion, and osmotic pressure with specific pH. Nanotagged agrochemicals reduce the damage to nontarget plant tissues and minimize risk of nonspecific chemical contamination in the surrounding environment (González-Melendi et al. 2008; Rai and Ingle 2012).

Combinations of inorganic fertilizer mainly supply three nutrients: nitrogen (N), potassium (K), and phosphorus (P) to various crops at different growing conditions.

To cope with increasing demands and limited supply of nutrient resources, such as phosphorus and potassium, more effective strategies are required to increase the yield and productivity of the agricultural crops. This brings out the idea of developing encapsulated fertilizers, in which NPK fertilizers are entrapped within nanosilver to optimize nutrient management. Fertilizer encapsulated by the nanoparticles may deliver nutrient more efficiently, allowing their slower and controlled release in the soil. So on, application of these nanobased slow-release fertilizers is an important aspect for the release of required amount of nutrients into the soil. It increases the efficiency of nutrient uptake in an optimized way, thus preventing nutrient loss as well as minimizing soil and water pollution. Therefore, nanoencapsulated fertilizers are relatively novel aspect, with potential commercial applications.

12.2.6 Role of Silver Nanoparticles in Wastewater Treatment and Water Purification

Nanoparticles can be effectively used for water purification because of their high surface area to volume ratio, its reactivity, strong functionalized approach, size-specific properties and higher affinity for various soil and waterborne contaminants, etc. Nanostructured membranes and filters have improved permeability, good flux rates, increased durability, reliability in purification, and reusability and thus are energy saving and cost-effective. In recent time, most of the products of water purification and disinfection systems use membranes impregnated with nanoscale silver particles. Silver nanoparticles have high efficiency of UV light photocatalysis. This characteristic helps in disinfecting/killing microbes during purification of drinking water (Dhakras 2011).

There are various studies in which experiments or proposed models concerning the behavior of silver nanoparticles in purification of drinking water and wastewater are reported (Kaegi et al. 2011; Fabrega et al. 2011; Quang et al. 2013; Gehrke et al. 2015). According to the report of Mpenyana-Monyatsi et al. (2012), combined cation-resin silver nanoparticle is an excellent and cheaper filter which can be used to remove most of the possible contaminants from drinking water and groundwater bodies.

12.2.7 Applicability of Nanosilver Particles in Pesticide Mineralization

Water pollution is a major and serious environmental problem in India. Due to heavy loads of pesticides in soil, its runoff from soil to groundwater is one of the major causes of drinking water contamination. Chemical-based pesticides are resistant to biodegradation and having carcinogenic effects on human health even at low levels. Traditional techniques of removal of pesticide (i.e., surface adsorption, photocatalysis, membrane separation, and biodegradation) are less effective, have long time consumption, and are expensive. As an alternative, nanosilver particles can be applied for the degradation of pesticides to overcome these problems. In modern nanotechnological research, applicability of nanosilver particles in pesticide mineralization is well reported. Nair and Pradeep (2003) demonstrated the halocarbon mineralization and catalytic destruction by means of silver and gold nanoparticles. The study reveals that halocarbon pollutants (methyl dichloride, methyl trichloride, carbon tetrachloride, and chlorofluorocarbons (CFCs) undergo catalytic destruction and mineralization with silver and gold nanoparticles in solution forming amorphous carbon and metal halides as reaction by-products. Results indicated that nanoparticle mediated complete destruction of halocarbon pollutants within 2 h and reaction occurred efficiently even at room temperature. Manimegalai et al. (2011) reported the applicability of silver nanoparticle for removal of pesticide "chlorpyrifos" and "malathion" from water. These nanoparticles have been shown to completely remove the pesticides as it actively anchored the pesticide to its inert surfaces. Silver nanoparticles are also efficiently used for removal of "endosulfan" from contaminated water bodies. Endosulfan adsorbs on the surface of nanoparticle and gets separated and precipitates out from the water.

12.2.8 Nanosilver as Biosolid for Agricultural Purpose

Industrial effluents containing nanosilver are drain out into wastewater bodies. Due to the small size of silver nanoparticle, these particles are unable to filter out during wastewater treatment, and these effluents are concentrated down as sewage sludge, which is subsequently dried and can be applied as a fertilizer "biosolids." Biosolids are the processed or refined sewage sludge effluents from wastewater treatment plants. Colman et al. (2013) used silver nanoparticles containing biowaste products as a fertilizer and/or additive nutrient supplement in soil of agricultural lands to improve its fertility. But in recent work its detrimental effect has also been shown. In such biosolids, Ag is predominantly present as Ag_2S form, and sulfidation of Ag potentially alters the properties of AgNPs including their surface charge and resulting to release Ag^+ ion which may exhibit toxicity to environment and biological life (Reinsch et al. 2012).

12.3 Toxicological Concern of Nanosilver Particles

Along with several beneficial aspects, nanosilver particles (AgNPs) are also an emerging environmental contaminant due to its wide application. With the tremendous uses of nanosilver (AgNPs)-based products, it is expected that it gets accumulated into the environment (Blaser et al. 2008). With the tremendous applicability of nanosilver particle (AgNP) as well as their possible release in the environment, there is an urgent need to critically discuss on potential health and environmental hazards of AgNPs, especially their impacts on biological systems that may help to facilitate the applicability of safer nanosilver products in the future. The potential risk of silver nanoparticles (AgNPs) toward human health is by means of its entry into human body via inhalation exposure route. Characterization of nanoparticle, i.e., size and surface, physicochemical properties, its crystalline nature, and the potential of release of silver ion are considered to be important factors in determining its toxicological effect on the environment. Strong oxidative potential of AgNP, surface area, and the release of ionic silver to surrounding environments trigger multiple negative effects on the structures and functions of cells, which finally induce cytotoxicity, genotoxicity, immunological responses, and even cell death. Water ecosystems are especially prone to exposure. Along with this, changes in microbial sensitivity to nanosilver particles are also the critical aspect of ecosystem risk.

12.3.1 Nanosilver Toxicity Response on Human Being

Tremendous use of nanoparticle-based product in world market leads to major global challenges about risk of its overexposure in human body (Asharani et al. 2010). On the basis of laboratory-based researches, silver nanoparticles are toxic to several cell lines, such as fibroblast, monocytes, liver cells, and germ line cells causing apoptosis, necrosis, and ROS generations that induced toxicity in the liver, kidney, and spleen, as main target organs. It also induces chronic respiratory problem in lungs, impairment and dysfunction of brain cells, and immunotoxicity via affecting olfactory mucosal cells and olfactory nerves. In recent studies, Kim and Shin (2014) reported on damaging effect of silver nanomaterials on human blood cells such as hemorheological, cell deformability and aggregation and coagulation of blood cells. Asharani et al. (2010)) reported that even small dose of silver nanoparticles have the potency to affect cytotoxicity and genotoxicity parameters of human being. Also, they suggested that the applications of nanosilver particles should be minimized because of its toxicity to normal cells.

12.3.2 Nanosilver Toxicity Effect on Soil Microorganism

There are number of researches exists in the sector of nanosilver toxicity, to show the detrimental impacts of nanosilver on eco-friendly soil microorganism (Silver and Phung 1996; Hänsch and Emmerling 2010). When nanoparticles are applied as a fertilizer on agricultural land, then it shows lethal effects to the soil microorganisms above the safe limit (Schlich et al. 2013). Nanosilver toxicity response on soil may alter ecosystem productivity and biogeochemistry by disturbing the soil microbes as they play the active role in management and recycling of nutrients in the ecosystem and optimizing fertility of soil, ecosystem sustainability. Plants also depend on soil bacteria and fungi for absorption of mineral and nutrients from the soil. Therefore, excess concentration of nanosilver affects the plants, whose growth is dependent on soil-dwelling microorganisms.

12.3.3 Nanosilver Toxicity Response on Plants

Nanosilver also induced the strong toxic effect on plant metabolism, i.e., growth traits, biochemical and physiological aspects, as well as genotoxic and cytotoxic effects due to silver-induced oxidative stress. Silver ion (Ag^+) from nanosilver particles is responsible for detrimental effects on plant metabolism (Stampoulis et al. 2009). The release of Ag^+ from nanosilver particles has been shown to restrict the activation of ethylene, plant hormones, and inhibit mitochondrial function (Knee 1992). In the past, various researchers evidenced that silver nanoparticle induced inhibitory effects in higher plants by inhibiting seed germination, root elongation, growth and metabolic processes (Yin et al. 2011; Hawthorne et al. 2012; Vannini et al. 2013). As reported, AgNPs strongly reduced the growth of the annual grass species (*Lolium multiflorum)* (Yin et al. 2011). El-Temsah and Joner (2012) showed the toxicity response of AgNP in plant species *Lolium perenne*, *Hordeum vulgare*, and *Linum usitatissimum* at 10 mg AgNP L^{-1} concentration. Kumari et al. (2009) reported the cytotoxic effect of nanosilver (AgNP) on *Allium cepa* cells demonstrating its lethal effect on chromosomal structure such as sticking and clumping of chromosomes, chromosomal breakage, and disturbance in metaphase stage of cell cycle as well as disintegration of cellular integrity. The observed effects of genotoxicity and cytotoxicity were significantly correlated with generation of superoxide radicals resulting in peroxidative damage of lipid membrane in these plant cells.

12.4 Conclusion

The benefits of nanosilver application in agro-practices as well as harmful aspects due to excessive utilization and accumulation in on soil, plant, and human life. Present review provides a short overview on applicability of silver nanoparticles in agricultural practices in various aspects along with related risk with overexposure of silver nanoparticles in environment and biological life. Nanosilver particles (AgNPs) have great attention in agricultural practices such as in crop protection and minimization of crop loss by acting as pesticides. Silver nanoparticles may also be used in genetic engineering of plant and need further advance researches. For example, targeted delivery of pest resistance genes are introduced in sensitive plants for developing resistant varieties, which will minimize expenses on agrochemicals required for disease control. In recent decades, advance researches are being carried out by using silver nanoparticle-based biosensors for detection and forecasting of pest and pathogens in agricultural crops. As silver (Ag) is the best conductor among metals and so Ag nanoparticles may facilitate more efficient electron transfer in biosensor. But apart from multiple positive aspects of silver nanoparticle for renovation and modern development of agricultural aspects, it's also noteworthy that detrimental effect of their overexposure due to silver (Ag^+) toxicity on environment and biological life cannot be overlooked.

Acknowledgments The author Rima Kumari acknowledges Science and Engineering Board (SERB), Department of Science and Technology (DST), New Delhi, for providing DST – Start-Up Young Scientist research grant to the project "Assessing the impact of silver nanoparticles on crop plants *V. radiata* and *Fagopyrum esculentum*: morphological, biochemical, genotoxic, and proteomics aspects" (Project F.No. SB/YS/LS-231/2013). We also thank the Head, Department of Environmental Science, BBAU, Lucknow.

References

Abbasi E, Milani M, Aval SF et al (2016) Silver nanoparticles: synthesis methods, bio-applications and properties. Crit Rev Microbiol 42:173–180

Asharani PV, Sethu S, Vadukumpully S et al (2010) Investigations on the structural damage in human erythrocytes exposed to silver, gold, and platinum nanoparticles. Adv Funct Mater 20:1233–1242

Blaser S, Scheringer M, Macleod M et al (2008) Estimation of cumulative aquatic exposure and risk due to silver: contribution of nano-functionalized plastics and textiles. Sci Total Environ 390:396–409

Carlson C, Hussan SM et al (2008) Unique cellular interaction of silver nanoparticles: size-dependent generation of reactive oxygen species. J Phys Chem B 112:13608–13619

Chen X, Schluesener HJ (2008) Nanosilver: a nanoproduct in medical application. Toxicol Lett 176:1–12

Chowdappa P, Gowda S (2013) Nanotechnology in crop protection: status and scope. Pest Manag Hortic Ecosyst 19:131–151

Chowdappa P, Shivakumar G (2013) Nanotechnology in crop protection: status and cope. Pest Manag Hortic Ecosys 19:131–151

Clement JL, Jarrett PS (1994) Antibacterial silver. Met Based Drugs 1:467–482

Colman BP, Arnaout CL, Anciaux S et al (2013) Low concentrations of silver nanoparticles in biosolids cause adverse ecosystem responses under realistic field scenario. PLoS One 8(2):e57189

Dhakras PA (2011) Nanotechnology applications in water purification and waste water treatment: a review Nanoscience, Engineering and Technology (ICONSET). 2011 International Conference on Date 28–30 Nov 2011, 285–291

Elchiguerra JL, Burt JL, Morones JR et al (2005) Interaction of silver nanoparticles with HIV-1. J Nanobiotechnol 3:6

Fabrega J, Luoma SN, Tyler CR et al (2011) Silver nanoparticles: behaviour and effects in the aquatic environment. Environ Int 37:517–531

Faunce T, Watal A (2010) Nanosilver and global public health: international regulatory issues. Nanomedicine 5(4):617–632

Franci G, Falanga A, Galdiero S et al (2015) Silver nanoparticles as potential antibacterial agents. Molecules 20:8856–8874

Gavanji S, Shams M, Shafagh N et al (2012) Destructive effect of silver nanoparticles on biocontrol agent fungi *Trichoderma viride* and *T. harzianum*. Casp J Appl Sci Res 1(12):83–90

Gehrke I, Geiser A, Somborn- Schulz A (2015) Innovations in nanotechnology for water treatment. Nanotechnol Sci Appl 8:1–17

González-Melendi P, Fernández-Pacheco R, Coronado MJ et al (2008) Nanoparticles as smart treatment-delivery systems in plants: assessment of different techniques of microscopy for their visualization in plant tissues. Ann Bot 101:187–195

Hänsch M, Emmerling C (2010) Effects of silver nanoparticles on the microbiota and enzyme activity in soil. J Plant Nutr Soil Sci 173:554–558

Hawthorne J, Musante C, Sinha SK et al (2012) Accumulation and phytotoxicity of engineered nanoparticles to *Cucurbita Pepo*. Int J Phytoremediation 14(4):429–442

Jayaseelan C, Rahaman AA, Rajkumar G et al (2011) Synthesis of pediculocidal and larvicidal silver nanoparticles by leaf extract from heart leaf moonseed plant, *Tinospora cordifolia* Miers. Parasitol Res 109:185–194

Jo YK, Kim BH, Jung G (2009) Antifungal activity of silver ions and nanoparticles on phytopathogenic fungi. Plant Dis 93:1037–1043

Kaegi R, Voegelin A, Sinnet B et al (2011) Behavior of metallic silver nanoparticles in a pilot wastewater treatment plant. Environ Sci Technol 45:3902–3908

Khan MR, Rizvi TF (2014) Nanotechnology: scope and application in plant disease management. Plant Pathol J 13:214–231

Kim MJ, Shin S (2014) Toxic effects of silver nanoparticles and nanowires on erythrocyte rheology. Food Chem Toxicol 67:80–86

Kim SW, Jung JH, Lamsal K et al (2012) antifungal effects of silver nanoparticles (AgNPs) against various plant pathogenic fungi. Mycobiology 40:53–58

Knee M (1992) Sensitivity of ATPases to silver ions suggests that silver acts outside the plasma membrane to block ethylene action. Phytochemistry 31:1093

Kumari M, Mukherjee A, Chadrasekaran N (2009) Genotoxicity of silver nanoparticle in *Allium cepa*. Sci Total Environ 407:5243–5246

Lamsal K, Kim SW, Jung JH et al (2011) Inhibition effects of silver nanoparticles against powdery mildew on cucumber and pumpkin. Mycobiol 39:26–32

Manimegalai G, Kumar SS, Sharma C (2011) Pesticide mineralization in water using silver nanoparticles. Int J Chem Sci 9:1463–1471

Marimuthu S, Rahuman AA, Rajkumar G et al (2011) Evaluation of green synthesized silver nanoparticles against parasites. Parasitol Res 108:1541–1549

Min JS, Kim KS, Kim SW et al (2009) Effects of colloidal silver nanoparticles on Sclerotium-forming phytopathogenic fungi. Plant Pathol J 25:376–380

Mondal NK, Chowdhury A, Dey U et al (2014) Green synthesis of silver nanoparticles and its application for mosquito control. Asian Pac J Trop Dis 4:S204–S210

Morones JR, Elechiguerra JL, Camacho A et al (2005) The bactericidal effect of silver nanoparticles. J Nanotechnol 16:2346–2353

Mpenyana-Monyatsi L, Mthombeni NH, Onyango MS et al (2012) Cost-effective filter materials coated with silver nanoparticles for the removal of pathogenic bacteria in groundwater. Int J Environ Res Public Health 9:244–271

Nair S, Pradeep T (2003) Halocarbon mineralization and catalytic destruction by metal nanoparticles. Curr Sci 84:12

Panyala NR, Pena Mendez EM, Havel J (2008) Silver or silver nanoparticle: a hazardous treat to the environment and human health? J Appl Med 6:117129

Park HJ, Kim SH, Kim HJ (2006) A new composition of nanosized silica-silver for control of various plant diseases. Plant Pathol 22:295–230

Pérez-de-Luque A, Rubiales D (2009) Nanotechnology for parasitic plant control. Pest Manag Sci 65:540–545

Quang Huy T, Quy N van, Anh-Tuan L (2013) Silver nanoparticles: synthesis, properties, toxicology, applications and perspectives. Adv Nat Sci Nanosci and Nanotechnol 4:033001

Rai M, Ingle A (2012) Role of nanotechnology in agriculture with special reference to management of insect pests. Appl Microbiol Biotechnol 94:287–293

Rai M, Yadav A, Gade A (2009) Silver nanoparticles as a new generation of antimicrobials. Biotechnol Adv 27:76–83

Rajesh KS,Malarkodi C (2014) In vitro antibacterial activity and mechanism of silver nanoparticles against foodborne pathogens. Bioinorg Chem Appl. Article ID 581890: 10 page

Raza MA, Kanwal Z, Rauf A et al (2017) Size- and shape-dependent antibacterial studies of silver nanoparticles synthesized by wet chemical routes. Nanomaterials 6(74):1–15

Reinsch BC, Levard C, Li Z et al (2012) Sulfidation of silver nanoparticles decreases escherichia coli growth inhibition. Environ Sci Technol 46:6992–7000

Rouhani M, Samih MA, Kalantri S (2012) Insecticidal effect of silica and silver nanoparticles on the cowpea seed beetle, *Callosobruchus maculatus* F (Col: Bruchidae). J Entomol Res 4:297–305

Salata OV (2004) Applications of nanoparticles in biology and medicine. J Nanobiotechnol 2:3

Samuel U, Guggenbichler JP (2004) Prevention of catheter-related infections: the potential of a new nano-silver impregnated catheter. Int J Antimicrob Agents 23S1:S75–S78

Schlich K, Klawonn T, Terytze K et al (2013) Hazard assessment of a silver nanoparticle in soil applied via sewage sludge. Environ Sci Eur 25:17

Silver S, Phung LT (1996) Annu Rev Microbiol 50:753–789

Singh S, Singh BK, Yadav SM et al (2015) Applications of nanotechnology in agricultural and their role in disease management. Res J Nanosci Nanotechnol 5:1–5

Soni N, Prakash S (2015) Antimicrobial and mosquitocidal activity of microbial synthesized silver nanoparticles. Parasitol Res 114:1023–1030

Stampoulis D, Sinha SK, White JC (2009) Assay-dependent phytotoxicity of nanoparticles to plants. Environ Sci Technol 43:9473–9479

Stephenson JB (2003) Pesticides on tobacco. Federal activities to assess risks and monitor residues. Report to the ranking minority member, committee on government reform, House of Representatives. GAO-03-485. 50 pp

Stohs SJ, Bagchi D (1995) Oxidative mechanisms in the toxicity of metal ions. Free Radic Biol Med 18:321–336

Suman TY, Elumali D, Kaleena PK (2013) GCMS analysis of bioactive components and synthesis of silver nanoparticle using *Ammannia baccifera* aerial extract and its larvicidal activity against malaria and filariasis vectors. Ind Crop Prod 47:239–245

Tsuji K (2001) Microencapsulation of pesticides and their improved handling safety. J Microencapsul 18:137–147

Vannini C, Domingo G, Onelli E et al (2013) Morphological and proteomic responses of Eruca sativa exposed to silver nanoparticles or silver nitrate. PLoS One 8:1–8

Yang Y, Quensen J, Mathieu J et al (2014) Pyrosequencing reveals higher impact of silver nanoparticles than Ag^{+} on the microbial community structure of activated sludge. Water Res 48:317–325

Yin L, Cheng Y, Espinasse B et al (2011) More than the ions: the effects of silver nanoparticles on *Lolium multiflorum*. Environ Sci Technol 45:2360–2367

Zhou Y, Kong Y, Kundu S et al (2012) Antibacterial activities of gold and silver nanoparticles against *Escherichia coli* and bacillus Calmette-Guérin. J Nanobiotechnol 10:19

The Significance of Plant-Associated Microbial Rhizosphere for the Degradation of Xenobiotic Compounds

13

Durgesh Kumar Jaiswal and Jay Prakash Verma

Abstract

Currently, remediation of xenobiotic compounds (heavy metals and hydrocarbons, pesticides, persistent organic pollutants (POPs) in the soil and water has become a major problem. Xenobiotic compounds in the soil exert alternations in the functionality of ecologically and agronomically important soil microflora. These chemicals get accumulated in lipid tissues of higher organisms and cause many problems to the human health (like immunosuppression, hormone disruption, reproductive abnormalities and cancer). Remediation of xenobiotic pollutants by the conventional approaches based on physicochemical methods is economically and technically challenging. But bioremediation techniques based on plant roots and their associated microbes are the most promising, efficient, cost-effective and sustainable technology. A variety of chemicals like organic acids, amino acids and phenolic compounds are secreted by such plants as root exudates. These compounds play a significant role in the interaction between plant root and microbes and also are helpful to stimulate the survival rate and the efficiency of microbes against xenobiotic pollutants. In this chapter, we describe how plant root-associated microbes help in the remediation of xenobiotic compounds and the impact of xenobiotic compounds on microbial community as well as their application feasibility on the basis of these attributes.

Keywords

Xenobiotic compounds • Root-microbe interaction • Remediation

D.K. Jaiswal • J.P. Verma (✉)
Institute of Environment and Sustainable Development, Banaras Hindu University, Varanasi 221005, Uttar Pradesh, India
e-mail: jpv.iesd@bhu.ac.in; verma_bhu@yahoo.co.in

A. Singh et al. (eds.), *Plant Responses to Xenobiotics*,
DOI 10.1007/978-981-10-2860-1_13

13.1 Introduction

Xenobiotics are artificially synthesized chemical compounds that are foreign to the biotic system. The major sources of xenobiotic compounds are (i) chemical, petrochemical and pharmaceutical industries that generate huge amounts of xenobiotic and synthetic polymers; (ii) pulp and paper bleaching (chlorinated organic compound); (iii) mining, which releases heavy metals into abiotic and biotic cycles (N_2 cycle, carbon cycle, phosphorus cycle, etc.); and (iv) chemical-based agricultural practices (like usage of pesticide and chemical fertilizer) (Thakur 2006). According to Esteve-Nunez et al. (2001), these organic compounds are not easily degenerated by microbial enzymes in soil and water, so they are accumulated in the environmental system. Having unnatural structural feature, they are resistant against biodegradation or undergo incomplete degradation or biotransformation. Remediation of xenobiotic pollutant by conventional approaches based on physicochemical methods (land filling, recycling, pyrolysis and incineration) is economically and technically challenging. But bioremediation techniques based on plant root and their associated microbes for the removal of pollutants are the most promising, efficient and cost-effective and sustainable technology (Verma et al. 2014). Plant-associated microbes, for example, endophytic bacteria (which occur naturally inside the plant) and rhizospheric bacteria (near the roots of plant), play an important role in the bioremediation of xenobiotic compounds (Verma et al. 2016) (Table 13.1). The rhizosphere region of plants play a significant role for making communication between plant roots and microbes to help in nutient translocation from soil to plants. Some organic compounds such as ions, free oxygen, enzymes, mucilage phenolic compound, sugar, amino acid and protein are secreted by the plants that play important role in the interaction of plant roots with microbe. It also boosts up the microbial activity against xenobiotic pollutants. On the basis of root-microbe interaction mechanism, the rhizosphere system plays a unique role for the remediation of contaminated soils. This type of remediation is called as rhizoremediation; in this process, rhizosphere microorganisms degrade the xenobiotic compounds in nontoxic organic compounds in the soil with the help of plant rhizosphere system; however, in this method, plants do not uptake degradable products resulting from remediation of xenobiotic compounds; they only provide a suitable habitat for rhizosphere microorganisms that proliferate and increase the survival rate of microbes in the contaminated soils.

13.2 The Fate of Xenobiotics in Soil

When a xenobiotic enters into the soil, it may be subjected to two basic processes (Cheng 1990).

13.2.1 Transfer Processes

When xenobiotic compounds enter into the environment, they are accumulated in the system without altering their structure through transfer process such as

Table 13.1 Xenobiotic compounds degrading various microbial strains

Microorganism	Target compound	Types of study	References
Endophytic microbes			
Herbaspirillum sp. K1	*PCBs, TCP*	Degradation of 2,3,4,6-tetrachlorophenol at low temperature and low dioxygen concentrations by phylogenetically different groundwater and bioreactor bacteria	Männistö et al. (2001)
Pseudomonas aeruginosa R75	*Chlorobenzoic acids*	Enzymatic activity in root exudates of Dahurian wild rye (*Elymusdauricus*) that degrades 2-chlorobenzoic acid	Siciliano et al. (1998)
*Pseudomonas savastanoi*CB35			
*Pseudomonas putida*VM1450	*Pesticide*	Bacterial endophyte-enhanced phytoremediation of the organochlorine herbicide 2,4-dichlorophenoxyacetic acid	Germaine et al. (2006)
	2,4-D		
Burkholderia cepacia G4	*VOCs*	Engineered endophytic bacteria improve phytoremediation of water-soluble, volatile	Barac et al. (2004)
*Pseudomonas putida*VM1441 (pNAH7)	*HCs*	Bacterial endophyte-mediated naphthalene phytoprotection and phytoremediation	Germaine et al. (2009)
	Naphthalene		
*Methylobacteriumpopuli*BJ001	*Explosives*	Biodegradation of nitro-substituted explosives 2,4,6-trinitrotoluene, hexahydro-1,3,5-trinitro-1,3,5-triazine inside poplar tissues (Populus deltoides x nigra DN-34)	Van Aken et al. (2004)
	TNT, RDX, HMX		
Rhizobacteria microbes			
Pseudomonas fluorescens	*PCBs*	Construction of a rhizosphere pseudomonad with potential to degrade polychlorinated biphenyls and detection of bph gene expression in the rhizosphere	Brazil et al. (1995)
Burkholderia cepacia	*Pesticides*	Plant protection and rhizosphere colonization of barley by seed-inoculated herbicide degrading *Burkholderia* (*Pseudomonas*) *cepacia* DB01 (pR0101) in 2,4-D contaminated soil	Jacobsen et al. (1997)
	2,4-D		
Pseudomonas fluorescens	*VOCs*	Rhizoremediation of trichloroethylene by a recombinant, root-colonizing *Pseudomonas fluorescens* stain expressing toluene ortho-monooxygenase constitutively	Yee et al. (1998)
	TCE		
Azospirillum brasilense Cd	*HCs*	A multiprocess phytoremediation system for removal of polyaromatic hydrocarbons from contaminated soils	Huang et al. (2004)
Enterobacter cloacae CAL 2	PAHs		
*Pseudomonas putida*UW3			

adsorption, retention by crops, runoff movements in dissolved or sorbed state, diffusion and vapour-phase diffusion, and sorption and desorption to soil colloid surfaces. During this process, sorption and desorption mechanisms are responsible for the interactions at interfaces between organic and inorganic soil colloids and xenobiotic. The abiotic soil components involved in the interaction with xenobiotic are pure and dirty clays, humic substances and humic-clay associations. The established interactions between organic and inorganic soil colloids and xenobiotic may affect the movement of xenobiotic, their availability for plant or microbial uptake, their transformation by abiotic or biotic agents and their influence on soil processes (Gianfreda and Rao 2008).

13.2.2 Degradation Processes

These processes include alteration of the chemical structure of the organic chemicals. It occurs through chemical, biological and photochemical transformations. The biotic components involved in the biological degradation of xenobiotic are microorganisms, plants and their enzymatic proteins as well as intra- or extracellular enzymes (Gianfreda and Rao 2008).

13.3 Rhizoremediation

It is a bioremediation technique based upon association of microbes with plant roots that intiate the removal of xenobiotic compounds from the contaminated sites. It is the most promising, efficient and cost-effective technology and sustainable technique. The growing plant secretes root exudates that play a role in root-microbe interaction and also help in the stimulation of the growth rate and tolerance or degrading activities of microbes against xenobiotic pollutants. A diverse range of compounds as root exudate (like organic, amino and fatty acids, carbohydrates, vitamins, nucleotides, phenolic compounds, polysaccharides and proteins), present in the rhizosphere system, provide a better habitat for the rhizosphere microbiomass. Subsequently, the presence of a diverse range of root exudates in rhizosphere system, responsible for the diversity and proliferation of microbiomes in rhizosphere system, is one of the significant factors behind the story of xenobiotic degradation.Zelenev et al. (2005) found that the percentage of cultivated microbes increases from <1 % in bulk soil to 2–7 % in the rhizosphere. Thus, root exudate is one of the most important factors to stimulate the growth of rhizospheric microbiomes (Rohrbacher and St-Arnaud 2016). Catabolism of aromatic compounds in bacteria has revealed that various enzymes (e.g. monooxygenase, dehydrogenase, hydrolase, oxygenase, isomerase, etc.) involve in the conversion of toxic chemicals of pesticide into various intermediate products such as protocatechuate and (substituted) catechols. These dihydroxylated intermediates (protocatechuate and catechols) directly enter into one of the two possible pathways, either a meta-cleavage-type or an ortho-cleavage-type pathway, thereafter supplied to

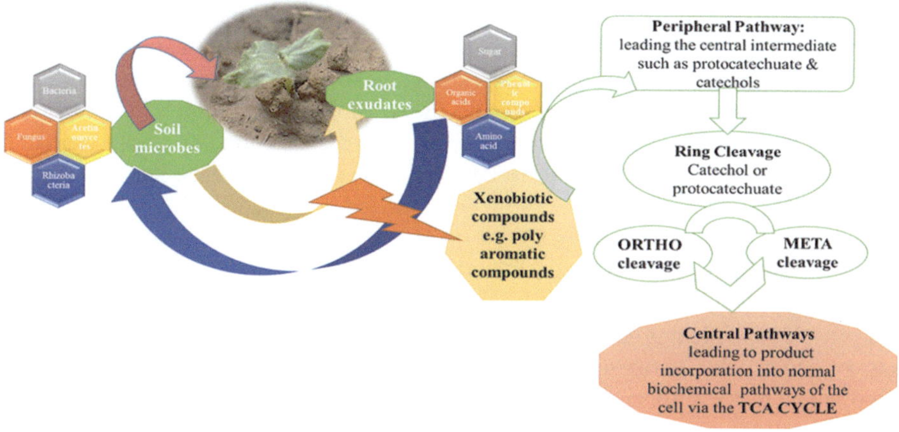

Fig. 13.1 Rhizoremediation of xenobiotic compounds

central metabolic routes such as the tricarboxylic acid cycle to form biochemical products (Verma et al. 2014) (Fig. 13.1)

13.3.1 Role of Root Exudates: Rhizoremediation of Xenobiotic Compounds

In rhizosphere system, root exudates like organic acids (citric, malic, succinic, oxalic and pyruvic), carbohydrates (glucose, xylose, fructose, maltose, sucrose, ribose), amino acids, fatty acids, proteins, enzymes, nucleotides and vitamins are the important nutrient resources for enhancing the survival rate of microorganisms under stress or polluted condition (Rohrbacher and St-Arnaud 2016). Typically, most microorganisms have sensory systems that guide them to interact with roots. The microbes obtain energy and nutrients for their survival through the root exudates. Corgié et al. (2003) reported that phenanthrene (PAH) biodegradation reached 86 % in the first 3 mm from the roots, 48 % between 3 and 6 mm and 36 % between 6 and 9 mm. They observed a parallel bacterial gradient, where high numbers of heterotrophs and PAH-degrading bacteria were close to the roots. Similarly, Corgie et al. (2004) found that, in the rhizosphere of perennial rye grass (*Lolium perenne* L.) growing in a petroleum hydrocarbon-contaminated soil, the highest rates of hydrocarbon degradation and the microbial degraders were mainly found within 3 mm of the root surface. In a phenanthrene-contaminated soil, the major phenanthrene had degraded by *Pseudoxanthomonas* sp. (Gammaproteobacteria) and *Microbacterium* spp.(Actinobacteria) (Cébron et al. 2011).

But when root exudates of rye grass were added, the population of phenanthrene degraders shifted mostly towards the actinobacterium, *Arthrobacter* sp., the Gammaproteobacteria *Pseudomonas stutzeri* and *Pseudoxanthomonas mexicana*. Consequently, Rentz et al. (2005) observed that the firmicutes *Bacillus* sp.,

Paenibacillus sp. and *Pseudomonas* sp. were able to use both root exudates and phenanthrene as carbon source. *Arthrobacter* sp. was shown to degrade hydrocarbons and more specifically phenanthrene (Radwan et al. 1998; Seo et al. 2006). Kozdrój and van Elsas (2000) also found *Pseudomonas* and *Arthrobacter* as dominant active phenanthrene degraders either in the presence of artificial root exudates or with phenanthrene alone (Rohrbacher and St-Arnaud 2016).

13.4 Factors Affecting Rhizoremediation

The physicochemical properties of rhizospheric soil compared to bulk soil affect the rate of degradation of organic compounds in soil, because its properties govern the sorption bioavailability and the persistence of xenobiotics in the soil (Pal et al. 2006; Zacharia 2011). Gold et al. (1996) reported that soil pH and clay content greatly affected the persistence of pesticide (chlorpyrifos, cypermethrin, permethrin, fenvalerate, etc.) under field conditions.

13.4.1 Soil Moisture

Water acts as a solvent for organic compound (xenobiotic) movement (Wardle and Parkinson 1992) and diffusion and is also essential for microbial function. The percentage of pesticide degradation increases with water content, because the rate of diffusion of atmospheric oxygen into soil is limited and anaerobic conversion of pesticide can succeed over aerobic degradation. However, poor oxygen transfer in high water content soil can retard or accelerate (Skopp et al. 1990) the pesticide degradation. Schroll et al. (2006) showed an optimum mineralization for isoproturon, benazolin-ethyl and glyphosate at the soil water potential of−0.015 MPa, whereas mineralization of pesticide was significantly reduced when soil moisture becomes near to water-holding capacity of soil. Soil moisture affects not only the activity but also the diversity of soil microbes (Bouseba et al. 2009).

13.4.2 Temperature

The effect of temperature on the rate of xenobiotic degradation depends on their molecular structure. Adsorption of pesticide in soil is affected by the temperature, due to the alteration in the solubility and hydrolysis of pesticide in soil (Burns 1975; Racke et al. 1997). Rani and Sud (2015) reported that increase in temperature is responsible for desorption process of absorbed pesticide, due to weak adsorption on tested soil. Thus, low level of temperature (below 20 °C) may be required for remediation process, because at this temperature pesticide is immobile in soil.

13.4.3 Soil pH

Soil pH plays an important role in abiotic/biotic degradation and also in the adsorption of ionic pesticide (Pal et al. 2006, Chaplain et al. 2011). Mostly pesticides are formulated in powder form and are soluble in water. Therefore, shifting of pH determines the degradation of pesticide. Commonly, insecticides are more liable to hydrolysis compared to fungicide and herbicide. According to Deer and Beard (2001), carbamate organophosphate insecticides are more susceptible than chlorinated hydrocarbon to hydrolysis. Thus, soil pH influences the sportive behaviour of pesticide molecules on clay and organic surface.

13.4.4 Soil Organic Matters

Soil organic matter plays a major role in the regulation of retention and mobility of pesticide in soil. According to some studies organic matter has dual activity such as it can enhance the microbial activities by co-metabolism (Pal et al. 2006) or it can decrease the microbial mediated pesticide degradation by stimulating adsorption process (Thom et al. 1997; Perucci et al. 2000; Pal et al. 2006). More than 1 % of organic matter is significant (Burns 1975) for the occurrence of an active autochthonous microbial population that can degrade the xenobiotics.

13.5 Challenges of Remediation Technology

According to Dixon (1996), in the United States alone, restoration of all contaminated sites will cost approximately $1.7 trillion. In addition, the conventional methods such as incineration, landfilling and excavation of soil are not sufficient to clean the contamination sites, and also these methods require much more cost and labour and also cause some environmental problems like water pollution and air pollution and produce toxic compounds. Therefore, demand of the time is to develop alternative methods to refurbish contaminated sites in a cost-effective, less labour-intensive, safe and environmental friendly way. One of such alternative methods is bioremediation, in which microbes work on environmental pollutants and degrade pollutant into nontoxic compounds (Caplan 1993; Dua et al. 2002). Bioremediation has main advantages like it can be applied in situ without the removal and transport of polluted soil and without the disturbance of the soil matrix and other bacterial degradation and usually results in complete mineralization of the organic pollutants.

13.5.1 Improvement in Rhizoremediation

During design of rhizoremediation process, researcher should focus on the following aspects:

- Screening of microorganism for remediation of xenobiotic compounds from polluted soils depends upon the presence of degradation ability with production of bio-surfactants properties (Płociniczak et al. 2011), because bio-surfactants facilitate the solubilization of xenobiotic compounds (Kuiper et al. 2004) and they also show a combined effect with chemotaxis processes, to proliferate and spread in polluted soil to increase the efficacy of rhizoremediation (Gerhardt et al. 2009).

- Formulation of the microbial consortium for degradation of xenobiotic compounds is found to be more efficient than the introduction of one single strain with the complete pathway, because each microbe of consortium plays a particular role in catabolic degradation pathway, involved in the degradation of a certain pollutant (Rahman et al. 2002).
- Selection of plant-microbe pairs, wherein a suitable rhizosphere strains is introduced together with a suitable plant (coating microbes on plant seed) to enhance the rhizoremediation, because introduced microbes might settle down on the root rhizosphere with indigenous population and enhance the efficiency of root colonization with addition of degrading microbes (Kuiper et al. 2001).

13.6 Conclusion

Currently, the remediation of xenobiotic compounds require attention to develop sustainable and green bioremediation technology at worldwide, because existing traditional technologies are very costly and labour-intensive and also act as a threat to the environment. Therefore, the application of rhizospheric microorganisms (like bacteria, fungi and actinomycetes) in contaminated soil help to make better association with plant roots for the degradation of xenobiotic compounds without causing any environmental problems and also provides efficient, economic and sustainable green remediation technology. So the need of the hour is to focus research initiatives on exploration of rhizospheric system in stress environment and find out the presence of specific gene, enzyme and soil's physical parameter (like soil pH, organic matter, temperature, water coefficient of soil) to enhance microbial quality that is responsible for biodegradation.

References

Barac T, Taghavi S, Borremans B, Provoost A, Oeyen L, Colpaert JV, van der Lelie D (2004) Engineered endophytic bacteria improve phytoremediation of water-soluble, volatile, organic pollutants. Nat Biotechnol 22(5):583–588

Bouseba B, Zertal A, Beguet J, Rouard N, Devers M, Martin C, Martin-Laurent F (2009) Evidence for 2, 4-D mineralisation in Mediterranean soils: impact of moisture content and temperature. Pest Manag Sci 65(9):1021–1029

Brazil GM, Kenefick L, Callanan M, Haro A, De Lorenzo V, Dowling DN, O'gara F (1995) Construction of a rhizosphere pseudomonad with potential to degrade polychlorinated biphenyls and detection of bph gene expression in the rhizosphere. Appl Environ Microbiol 61(5):1946–1952

Burns RG (1975) Factors affecting pesticide loss from soil. In: Paul EA, McLaren AD (eds) Soil biochemistry, vol 4. Marcel Dekker, Inc, New York, pp 103–141

Caplan JA (1993) The worldwide bioremediation industry: prospects for profit. Trends Biotechnol 11(8):320–323

Cébron A, Louvel B, Faure P, France-Lanor C, Chen Y, Murrell JC, Leyval C (2011) Root exudates modify bacterial diversity of phenanthrene degraders in PAH-polluted soil but not phenanthrene degradation rates. Environ Microbial 13(3):722–736

Chaplain V, Défossez P, Richard G, Tessier D, Roger-Estrade J (2011) Contrasted effects of no-till on bulk density of soil and mechanical resistance. Soil Tillage Res 111(2):105–114
Cheng HH (1990) Pesticides in the soil environment: processes, impacts, and modeling. Pesticides in the soil environment: processes, impacts, and modelling. Soil Science Society of America, Inc, Madison
Corgié SC, Joner EJ, Leyval C (2003) Rhizospheric degradation of phenanthrene is a function of proximity to roots. Plant Soil 257(1):143–150
Corgié SC, Beguiristain T, Leyval C (2004) Spatial distribution of bacterial communities and phenanthrene degradation in the rhizosphere of *Loliumperenne* L. Appl Environ Microbiol 70(6):3552–3557
Deer HM, Beard R (2001) Effect of water pH on the chemical stability of pesticides. AG/ Pesticides.14:1
Dixon B (1996) Bioremediation is here to stay. ASM News 62:527–528
Dua M, Singh A, Sethunathan N, Johri A (2002) Biotechnology and bioremediation: successes and limitations. Appl Microbiol Biotechnol 59(2–3):143–152
Esteve-Núñez A, Caballero A, Ramos JL (2001) Biological degradation of 2, 4, 6-trinitrotoluene. Microbiol Mol Biol Rev 65(3):335–352
Gerhardt KE, Huang XD, Glick BR, Greenberg BM (2009) Phytoremediation rhizoremediation of organic soil contaminants: potential and challenges. Plant Sci 176:20–30
Germaine KJ, Liu X, Cabellos GG, Hogan JP, Ryan D, Dowling DN (2006) Bacterial endophyte-enhanced phytoremediation of the organochlorine herbicide 2, 4-dichlorophenoxyacetic acid. FEMS Microbiol Ecol 57(2):302–310
Germaine KJ, Keogh E, Ryan D, Dowling DN (2009) Bacterial endophyte-mediated naphthalene phytoprotection and phytoremediation. FEMS Microbiol Lett 296(2):226–234
Gianfreda L, Rao MA (2008) Interactions between xenobiotics and microbial and enzymatic soil activity. Crit Rev Environ Sci Technol 38(4):269–310
Gold RE, Howell HN, Pawson BM, Wright MS, Lutz JL (1996) Persistence and bioavailability of termicides to subterranean termite from five soil types and location in Texas. Sociobiol 28:337–363
Huang XD, El-Alawi Y, Penrose DM, Glick BR, Greenberg BM (2004) A multi-process phytoremediation system for removal of polycyclic aromatic hydrocarbons from contaminated soils. Environ Pollut 130(3):465–476
Jacobsen CS (1997) Plant protection and rhizosphere colonization of barley by seed inoculated herbicide degrading *Burkholderia* (*Pseudomonas*) *cepacia* DBO1 (pRO101) in 2, 4-D contaminated soil. Plant Soil 189(1):139–144
Kozdrój J, van Elsas JD (2000) Response of the bacterial community to root exudates in soil polluted with heavy metals assessed by molecular and cultural approaches. Soil Biol Biochem 32:1405–1417
Kuiper I, Bloemberg GV, Lugtenberg BJJ (2001) Selection of a plant-bacterium pair as a novel tool for rhizostimulation of polycyclic aromatic hydrocarbon-degrading bacteria. Mol Plant-Microbe Interact 14:1197–1205
Kuiper I, Lagendijk EL, Bloemberg GV, Lugtenberg BJ (2004) Rhizoremediation: a beneficial plant microbe interaction. Mol Plant Microbe Interact 17:6–15
Männistö MK, Tiirola MA, Puhakka JA (2001) Degradation of 2, 3, 4, 6-tetrachlorophenol at low temperature and low dioxygen concentrations by phylogenetically different groundwater and bioreactor bacteria. Biodegradation 12(5):291–301
Pal R, Chakrabarti K, Chakraborty A, Chowdhury A (2006) Degradation and effects of pesticides on soil microbiological parameters-a review. Int J Agri Res 1(33):240–258
Perucci P, Dumontet S, Bufo SA, Mazzatura A, Casucci C (2000) Effects of organic amendment and herbicide treatment on soil microbial biomass. Biol Fertil Soils 32:17–23
Płociniczak MP, Płaza GA, Seget ZP, Cameotra SS (2011) Environmental applications of biosurfactants: recent advances. Int J Mol Sci 12:633–654

Racke KD, Skidmore MW, Hamilton DJ, Unsworth JB, Miyamoto J, Cohen SZ (1997) Pesticide fate in tropical soil. Pure and Appl Chem 69:1349–1371

Radwan SS, Al-Awadhi H, Sorkhoh NA, El-Nemr IM (1998) Rhizospheric hydrocarbon-utilizing microorganisms as potential contributors to phytoremediation for the oil Kuwaiti desert. Microbiol Res 153(3):247–251

Rahman KS, Rahman T, Lakshmanaperumalsamy P, Banat IM (2002) Occurrence of crude oil degrading bacteria in gasoline and diesel station soils. J Basic Microbiol 42:284–291

Rani S, Sud D (2015) Effect of temperature on adsorption-desorption behaviour of triazophos in Indian soils. Plant Soil Environ 61(1):36–42

Rentz JA, Alvarez PJ, Schnoor JL (2005) Benzo [a] pyrene co-metabolism in the presence of plant root extracts and exudates: Implications for phytoremediation. Environ Pollut 136(3):477–484

Rohrbacher F, St-Arnaud M (2016) Root exudation: the ecological driver of hydrocarbon rhizoremediation. Agronomy 6(1):19

Schroll R, Becher HH, Dorfler U, Gayler S, Grundmann S, Hartmann HP, Ruoss J (2006) Quantifying the effect of soil moisture on the aerobic microbial mineralization of selected pesticides in different soils. Environ Sci Technol 40(10):3305–3312

Seo JS, Keum YS, Hu Y, Lee SE, Li QX (2006) Phenanthrene degradation in Arthrobacter sp. P1-1: initial 1, 2-, 3, 4-and 9, 10-dioxygenation, and meta-and ortho-cleavages of naphthalene-1, 2-diol after its formation from naphthalene-1, 2-dicarboxylic acid and hydroxyl naphthoic acids. Chemosphere 65(11):2388–2394

Siciliano SD, Goldie H, Germida JJ (1998) Enzymatic activity in root exudates of Dahurian wild rye (*Elymusdauricus*) that degrades 2-chlorobenzoic acid. J Agri Food Chem 46(1):5–7

Skopp J, Jawson MD, Doran JW (1990) Steady-state aerobic microbial activity as a function of soil water content. Soil Sci Soc Am J 54(6):1619–1625

Thakur IS (2006) Xenobiotics: pollutants and their degradation-methane, benzene, pesticides, bioabsorption of metals. Environmental Microbiology. School of Environmental Sciences, Jawaharlal Nehru University. New Delhi-110 067

Thom E, Ottow JCG, Benckiser G (1997) Degradation of the fungicide difenoconazole in a silt loam soil as affected by pretreatment and organic amendment. Environ Pollut 96:409–414

Van Aken B, Yoon JM, Schnoor JL (2004) Biodegradation of Nitro-Substituted Explosives 2,4,6-Trinitrotoluene, Hexahydro-1,3,5-Trinitro-1,3,5-Triazine, and Octahydro-1,3,5,7-Tetranitro-1,3,5-Tetrazocine by a Phytosymbiotic Methylobacterium sp. Associated with Poplar Tissues (Populus deltoides×nigra DN34). Appl Environ Microbiol 70:1508–1517

Verma JP, Jaiswal DK, Sagar R (2014) Pesticide relevance and their microbial degradation: a-state-of-art. Rev Environ Sci Biotechnol 13(4):429–466

Verma JP, Jaiswal DK, Yadav J, Singh HB (2016) Cleaner production of agriculture due to beneficial interactions between plants and bacteria. Book Review of 'Beneficial plant-bacterial interactions', Glick, BR (2015), Springer, 243 ISBN: 978-3-319-13920-3

Wardle DA, Parkinson D (1992) Influence of the herbicides, 2,4-D and glyphosate on soil microbial biomass and activity a field experiment. Soil Biol Biochem 24:185–186

Yee DC, Maynard JA, Wood TK (1998) Rhizoremediation of trichloroethylene by a recombinant, root-colonizing Pseudomonas fluorescens strain expressing toluene ortho-monooxygenase constitutively. Appl Environ Microbiol 64(1):112–118

Zacharia JT (2011) Identity, physical and chemical properties of pesticides. In: Stoytcheva M (ed) Pesticides in the modern world – trends in pesticides analysis. In Tech, Rijeka, pp 1–18

Zelenev VV, Van Bruggen AHC, Semenov AM (2005) Modeling wave like dynamics of oligotrophic and copiotrophic bacteria along wheat roots in response to nutrient input from a growing root tip. Ecol Model 188(2):404–417

Biodegradable Polyhydroxyalkanoate Thermoplastics Substituting Xenobiotic Plastics: A Way Forward for Sustainable Environment

14

Laxuman Sharma, Janmejai K. Srivastava, and Akhilesh Kumar Singh

Abstract

Conventional plastics such as polyethylene, polypropylene, polystyrene, poly(vinyl chloride), and poly(ethylene terephthalate) are high-molecular-weight polymeric materials which vary from 50,000 to 1,000,000 Da. They have attained unique position in modern material technology. They are omnipresent in today's society with range from ordinary to high-tech, from vital to entirely lavish. These plastics have diverse feasible application in every field of industries/factories ranging from automobiles to medicine owing to their promising material properties, viz., lightweight, stability, long durability, economic viability, and feasibility to manipulate a range of strengths and shapes. The resistance to degradation, stability, and long durability are some miracle features associated with these plastic materials while in use. However, such properties become detrimental to the environment when out of usage, being synthetic polymers and exceptionally recalcitrant to microbial attack, i.e., nonbiodegradable (xenobiotic polymeric materials). To combat the menace posed by plastics to the environment, several efforts have been made for developing the products that are eco-friendly and degradable with comparable material properties as that of conventional plastics. This chapter presents a revolutionary insight with various technological strategies to overcome the detrimental effects of conventional plastics with special emphasis to completely biodegradable polyhydroxyalkanoate thermoplastics.

L. Sharma
Department of Horticulture, Sikkim University, Gangtok 737102, Sikkim, India
e-mail: lsharma@cus.ac.in

J.K. Srivastava • A.K. Singh (✉)
Amity Institute of Biotechnology, Amity University, Lucknow Campus 226028, Uttar Pradesh, India
e-mail: aksingh2@lko.amity.edu; akhiliit@yahoo.co.in

A. Singh et al. (eds.), *Plant Responses to Xenobiotics*,
DOI 10.1007/978-981-10-2860-1_14

Keywords
Conventional plastics • Xenobiotics • Polyhydroxyalkanoates • PHAs • SCL-PHAs • MCL-PHAs • LCL-PHAs

14.1 Introduction

14.1.1 Polymers

Polymers are the leading products of contemporary chemical industry that constitute the pillar of existing society. A polymer is a high-molecular-weight substance produced through the unification of vast number of small molecules that form the repeating entities known as monomers. The process by which the monomeric units are converted into polymeric substances is called polymerization (Verma et al. 2007). Day-to-day life have become comfortable and vibrant because of invention and wide-range uses of polymers, viz., plastic serving dishes, cups, nonstick pans; automobile tires and seat covers, raincoat, plastic pipes and fitting, plastic bags; TV, radio, and computer cabinets; synthetic glues, wide range of synthetic fibers for clothing; and flooring materials and materials for biomedical and surgical operations (Verma et al. 2007). Diverse applications of polymer in every walks of human being are now to such extent that no one imagines of the materialistic world without plastics. Among all the polymers, conventional plastics have distinctively categorized as miraculous polymer in contemporary material technology owing to most desirable material properties, i.e., physical (melting temperature, glass-transition temperature, thermal stability, and crystallinity) and mechanical properties (Young's modulus, tensile strength, and elongation to break). Table 14.1 depicted the properties of some common conventional plastics. These eye-catching

Table 14.1 Material properties of some common conventional plastics

Material property	Polypropylene	Low-density polyethylene	Polystyrene	Poly(ethylene terephthalate)
T_m (°C)	176	130	110	262
T_g (°C)	−10	−36	21	3400
Crystallinity (%)	50–70	20–50	–	–
Young's modulus (Gpa)	1.7	0.2	3.1	2.2
Tensile strength (Mpa)	38	10	50	56
Elongation to break (%)	400	620	–	7300
Thermal stability, $T_{d(5\ \%)}$	338	387	–	–

Singh et al. (2013) and Kumar et al. (2015)

Tm melting temperature, *Tg* glass-transition temperature, $Td_{(5\ \%)}$ temperature at 5 % weight loss properties facilitate conventional plastics for widespread applications in the medical field, constituents in automobiles, construction, sport and leisure equipment, packaging materials, routine home appliances, computer equipment, cell phones, printers, etc. (Kumar et al. 2015).

14.1.2 Conventional Plastics

The term "plastic" originates from the Greek word "plastikos" that means competency to precast into various forms/shapes (Joel 1995). The first true conventional plastic called Bakelite (a polymer of phenol and formaldehyde) was discovered in 1907 by Belgian chemist Leo Baekeland followed by successive production of numerous other conventional polymers (Thompson et al. 2009). Currently, the exploited conventional plastics are composed of organic and inorganic constituent, viz., carbon, silicon, hydrogen, nitrogen, oxygen, and chloride; the fossils fuel and natural gases are the precursors for synthesis of these plastics (Seymour 1989). The polymers that are manufactured in the laboratories and industries are termed as synthetic polymers or man-made polymers. Thus, high-molecular-weight conventional plastics, such as polyethylene, polypropylene, polystyrene, etc., are also synthetic polymers.

14.1.2.1 Classification of Polymers/Conventional Plastics

The mechanical properties of polymers are function of intermolecular forces, such as van der Waal's forces and hydrogen bonds persisting in the macromolecules and the simple molecules too. However, their impact in simple molecule is less effective than the macromolecules/polymers. Thus, lengthier the chain, stronger the impact of intermolecular forces (Verma et al. 2007).

Based on the nature of principal intermolecular force of attraction, polymers are broadly categorized into the following four types, i.e., elastomers, fibers, thermoplastics, and thermosetting polymers (Verma et al. 2007; Bernard 2014). Elastomers are the amorphous polymers, where polymer chains are held by feeblest force of attractions. In other words, they behave as rubber or rubberlike elastic materials. They are composed of arbitrarily coiled molecular chains with limited cross-links. Such polymer can be extended approximately ten times more than their normal length. Nevertheless, they regain their original state as soon as the stretching force is removed. In contrast, fibers are the polymers that involve reasonably strong interparticle forces such as hydrogen bonds with high tensile strength as well as high modulus. Nylon is one of the best examples of this type of polymers (Verma et al. 2007). Thermoplastics involve interparticle forces of attraction that are in between those of elastomers and fibers. These plastics, viz., polyethylene, polypropylene, polyvinyl chloride, polystyrene, etc. are common and could be easily precast into preferred shapes through heating followed by cooling at room temperature. These types of plastics soften on warming and transformed into fluids, which upon cooling become solid. Interestingly, such plastics involve no cross-linking between the polymer chains. They are proficient to undergo such reversible alterations upon warming and cooling repetitively (Verma et al. 2007). On the other hand, on heating the thermosetting plastics, Bakelite, for example, becomes hard and infusible as a result of excessive cross-linking among the polymer chains, thereby producing three-dimensional networks of bonds (Verma et al. 2007; Bernard 2014).

14.1.2.2 Methods for the Synthesis of Conventional Plastics

Addition and condensation reactions are the basis for synthesis of conventional plastics. Addition polymers involve the repetitively addition of monomers leading to the formation of long chains without the removal of any by-products (Verma et al. 2007). The monomers involved in the synthesis of such polymers are unsaturated in nature and are typically the derivatives of alkenes. Polyethylene, polypropylene, orlon, poly vinyl chloride, etc. are good examples of addition polymerization (Verma et al. 2007). In contrast, condensation polymer involves the formation of polymer as a result of condensation of two or more than two monomer units with the removal of simple molecules like water, alcohol or ammonia, etc. Dacron, Bakelite, and nylon-6,6 are examples of this type of polymer (Verma et al. 2007).

14.1.2.3 Application of Conventional Plastics

The exploitation of widespread uses of conventional plastics (as depicted from Table 14.2) is usually linked to their fundamental structures, desirable material properties (Table 14.1), possibility to regulate a range of strengths and shapes, hydrophobicity, lightweight, economic feasibility, and long durability (Bernard 2014; Kumar et al. 2015). These are made up of variable lengths of hydrocarbons as backbone and bound merely with hydrogen apart from other carbons. Remarkably, additional molecular constituents could be introduced into the backbone of conventional plastics that give rise to novel properties and thus opening viability towards producing wide range of useful commodities (Bernard 2014). These plastics are largely exploited as packaging materials for commodities/products, such as pharmaceuticals, food, cosmetics, and chemicals including detergents. Almost 30 % of the conventional plastics are exploited globally for packaging uses that further enhancing at a rate of 12 % per year (Sabir 2004; Shah et al. 2008). The extensively exploited conventional plastics for packaging involves polyethylene (low-density polyethylene and high-density polyethylene), polypropylene, polystyrene, polyvinyl chloride, polyurethane, poly(ethylene terephthalate), poly(butylene terephthalate), and nylons (Vona et al. 1965; Shah et al. 2008).

14.1.3 Conventional Plastics: Emerged as Alarming Xenobiotic Materials

The exploitation/utilization of conventional plastics for various application is found to be steady (Fig. 14.1) among the developed counties of the world including Europe (Plastics Europe 2015). However, the exploitation of these plastics in the developing counties has increasing trend (Fig. 14.1). China is the major manufacturer of conventional plastic materials followed by Europe (Fig. 14.2). It is anticipated that

Table 14.2 Industrial applications of conventional plastics

Conventional plastics	Repeating units (monomers)/precursors	Characteristics	Applications
Polyethylene	Ethylene molecules	Polyethylene classified into low-density polyethylene and high-density polyethylene. Low-density polyethylene composed of branched chain structure of polymer molecules with density and melting point of 0.92 g cm^{-3} and 383 K, respectively. It is transparent polymer and chemically inert, having moderate tensile strength and high toughness. On the other hand, high-density polyethylene made up of linear chain structure of polymer molecules with density and melting point of 0.97 g cm^{-3} and 403 K, respectively. It is a transluscent polymer and chemically inert with relatively greater toughness and high tensile strength compared to LDPE	Low-density polyethylene exploited as packing material and insulation for electrical wires and cables. However, high-density polyethylene utilized in the production of containers, pipes, bottles, toys, bags, etc.
Polypropylene	Propylene molecules	It is harder, lighter, and stronger compared to PE	It is used for making bottle caps, drinking straws, medicine bottles, car seats, car batteries, bumpers, disposable syringes, carpetbackings, gramophone records, ropes, etc.
Polystyrene	Styrene molecules	It is a white thermoplastic material that is transparent and floats on water	It is used for making disposable cups, packaging materials, laboratory ware, toys, combs, ceiling tiles, and certain electronic uses
Polyvinyl chloride	Vinyl chloride molecules	It is hard horny material with resistant to chemicals and heat	It is exploited as automobile seat covers, shower curtains, raincoats, bottles, visors, shoe soles, garden hoses, hand bags, toys, gramophone records, electrical insulations, floor covering, hosepipes, and electricity pipes

(continued)

Table 14.2 (continued)

Conventional plastics	Repeating units (monomers)/precursors	Characteristics	Applications
Polytetrafluoroethylene	Tetrafluoroethylene molecules	It is extremely tough and resistant to heat and chemical attack	It is used in various industrial applications such as specialized chemical plant, electronics, and bearings. It is met within the home as a coating on nonstick kitchen utensils, such as saucepans and frying pans
Nylon-66	Adipic acid and hexamethylenediamine	They are elastic material with high tensile strength and capable to resist abrasion	They are used in small bearings, speedometer gears, windshield wipers, water hose nozzles, football helmets, racehorse shoes, inks, clothing parachute fabrics, rainwear, carpets, bristles and brushes, fabrics in textile industry, cellophane, tire cords, and ropes
Nylon-6,10	Hexamethylenediamine and sebacoyl chloride		
Nylon-6 (Perlon)	Caprolactam		
Bakelite	Phenol and formaldehyde	It is a cross-linked thermosetting polymer	For making combs, electrical switches, handles of utensils, and computer disks
Polytetrafluoroethylene	Tetrafluoroethylene molecules	It is extremely tough and resistant to heat and chemical attack	It is used in various industrial applications such specialized chemical plant, electronics, and bearings. It is met within the home as a coating on nonstick kitchen utensils, such as saucepans and frying pans
Polyurethane	Diisocyanate and polyol	It is highly flexible with good electrical insulating properties. Also resistant to abrasion, heat, solvents, oil, acid, water, and grease with high tear resistance and tensile properties	It is exploited for making tires, gaskets, bumpers, in refrigerator insulation, sponges, furniture cushioning, and life jackets
Poly(ethylene terephthalate)	Terephthalic acid with ethylene glycol or dimethyl terephthalate with ethylene glycol	Very good barrier, crease resistance, solvent resistance, high melting point, resistance to fatigue, and high tenacity as either a film or a fiber	Used for carbonated soft drink bottles, processed meat packages, peanut butter jars, pillow and sleeping bag filling, textile fibers

Vona et al. (1965) and Verma et al. (2007)

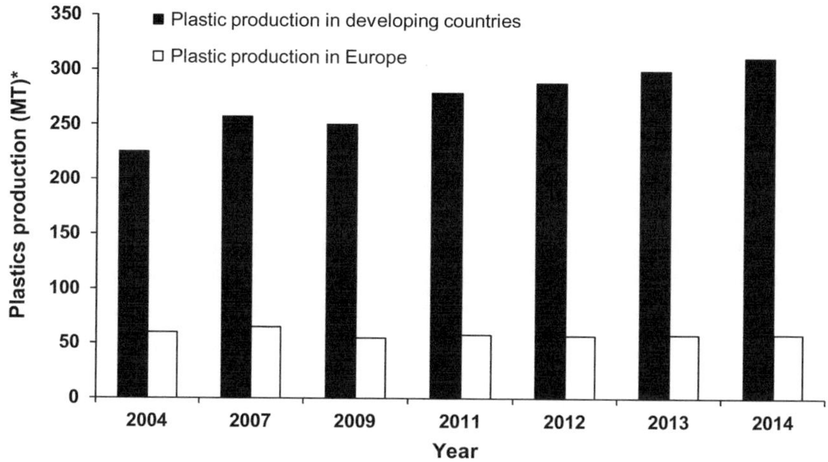

Fig. 14.1 Comparative account on production of conventional plastics in Europe and developing countries (Plastics Europe 2015) (*Involves plastics, viz., thermoplastics as well as polyurethanes including other plastics, such as thermosets, adhesives, coatings, and sealants but excluding the following fibers: polyethylene terephthalate, polyamide, polypropylene, and polyacryl fibers)

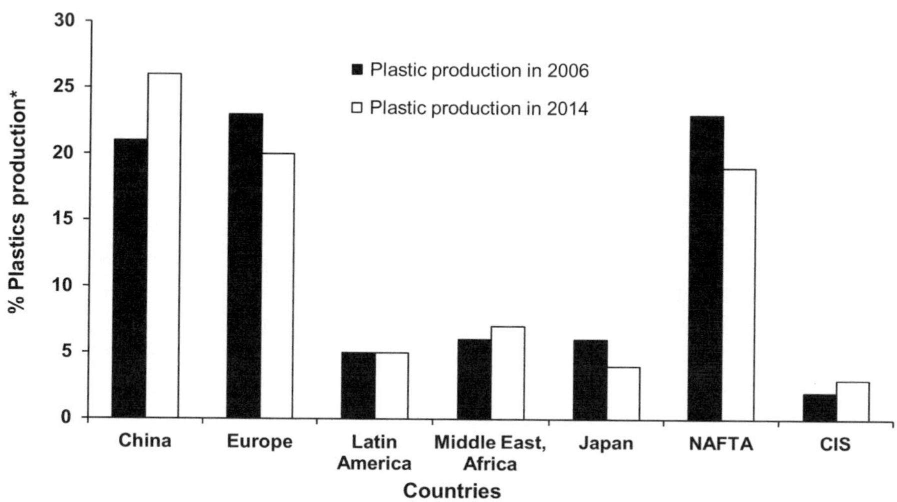

Fig. 14.2 Comparative account on worldwide production of conventional plastics (Plastics Europe 2015) (*Involves plastics, viz., thermoplastics as well as polyurethanes including other plastics, such as thermosets, adhesives, coatings, and sealants but excluding the following fibers: polyethylene terephthalate, polyamide, polypropylene, and polyacryl fibers; NAFTA: North American Free Trade Agreement; CIS: Commonwealth of Independent States)

there will be demand of 66.5 MT of plastics in Europe by 2020 and worldwide production of plastics could be increased up to threefold in 2050 (Plastics Europe 2010; European Commission 2013; Plastics Europe 2015).

Resistant to degradation, stability, and prolonged durability are the essential features when conventional plastics are in use. Nevertheless, such miracle possessions of plastic materials turn out to be detrimental to the environment when out of usage. They remain as such for several years when cast-off in environment as refuse/debris due to predominantly xenobiotic nature, i.e., biologically inactive or recalcitrant to microbial attack (Shah et al. 2008; Kumar et al. 2015). As a consequence, these plastics emerged as one of alarming xenobiotic materials worldwide. Xenobiotic compounds are anthropogenic materials present unnaturally in living or biological systems/environment uncommonly in high amounts. The feasible health hazard of a xenobiotic material is a function of its perseverance in the environment and the toxicity of the chemical class (Wilson et al. 1985). Therefore, the reason for the prolonged persistence of conventional plastics in the environment linked to their synthetic polymer-specific chemical structures that give rise to exceptional physicochemical features with a long-lasting existence still in adverse or harsh environments. The depolymerization is a main requisite for breakdown or degradation of polymeric material that result into elimination of physicochemical properties responsible for long-lasting life of the synthetic polymer/plastics (Kawai 2010). Moreover, the microorganisms, such as fungi and bacteria, are also incompetent to evolve proficient biological machinery systems for their degradation/metabolism (Bernard 2014). Hence, these polymeric materials not only emerged as wonder materials but also as indispensable evils.

Million tons of conventional plastics are produced worldwide, and substantial amounts of these polymeric materials access the environment especially as industrial refuses/wastes, where animals are killed due to clogging of cast-off plastics in the gut or entangled in waste plastics (Fiechter 1990; Shimao 2001). Considering these facts, the feasible ways for attaining plastic garbage-free environment involve dumping of plastic wastes at landfill sites, incineration, and recycling (Ray and Bousmina 2005). Table 14.3 summarizes methodologies for achieving plastic waste material-free environment. However, these viable approaches associated with some shortcomings (Table 14.3) owing to which the world is fueling towards the development of eco-friendly and sustainable plastics, i.e., biodegradable plastics/bioplastics.

14.1.4 Bioplastics: Eco-Friendly Alternative to Conventional Plastics

Bioplastics/biodegradable plastics defined as polymeric materials that are degradable as well as eco-friendly in nature leading to mineralization (Ray and Bousmina 2005). Bioplastics as innovative field of research is relatively young because the first proof of these polymeric substances was acknowledged less than a century ago (Lemoigne 1926). They are categorized into four types, i.e., photodegradable,

Table 14.3 Approaches for achieving plastic waste free environment

Approaches	Feasibility	References
Disposal of plastic wastes at landfill sites	One of the cost-effective approaches to dispose recalcitrant plastics is to exploit municipal landfills. Nevertheless, landfill sites are insufficient because of fast growth in human population and society along with accumulation of enormous quantity of recalcitrant plastic refuse	Ray and Bousmina (2005) and Suriyamongkol et al. (2007).
Incineration	Incineration can be exploited as other alternative approach to deal with these conventional plastic materials. Unfortunately, incineration of these plastics includes production of vast amount of carbon dioxide as well as seldom toxic gases, which is accountable towards global warming and global pollution, respectively. For example, incineration of polyvinyl chloride in municipal incinerators resulting into formation of extremely poisonous dioxins. Furthermore, polystyrene foam products are often found composed of chlorofluorocarbons and hydrochlorofluorocarbons, which are ozone-destroying and greenhouse chemicals	Chiras (1994) and Ray and Bousmina (2005)
Recycling	This approach is found to alter the material properties of the conventional plastic to a great extent due to which the quality of plastics is low-grade compared to plastic material manufactured from primary manufacturer. Thus, the usage of the recycled plastic material is restricted	Fiechter (1990) and Ray and Bousmina (2005)

semi-biodegradable, chemically synthesized, and polyhydroxyalkanoate thermoplastics. The backbone of the photodegradable plastics/polymers that are associated with light-sensitive groups is not used comprehensively because of their non-degradability in landfill. Such plastics fail to undergo degradation owing to scarcity of sunlight in landfill. Semi-degradable plastics are made up of starch and small fragments of conventional plastics such as polyethylene fragments. In such plastics, starch plays the role of filler as well as cross-linking agent, thereby, holding together the small fragments of conventional plastics (for instance, polyethylene). Biodegradation is linked with molecular size, i.e., the greater the molecular size, the lower the probability towards biodegradation. The properties of starch and polyethylene differ significantly that result into low compatibility of starch and polyethylene blend based semi-biodegradable plastics (Shujun et al. 2006). Microorganism found to proficiently degrade the starch but failed to attack and act on polyethylene fragments (Johnstone 1990). On the other hand, polyvinyl alcohol, polyethylene oxide, polyglycolic acid, poly(ε-caprolactone), and polylactic acid are the examples of chemically produced plastics, which are susceptible to enzymatic/microbial attack. Nevertheless, the material properties of these plastics are not akin to that of conventional plastics as a result of which not viable for wide range of uses (Khanna and Srivastava 2005). Interestingly, polyhydroxyalkanoate thermoplastics are merely fully biodegradable plastics composed of hydroxyl fatty acid units with material properties similar to conventional plastics. The attention-grabbing features associated with PHAs are thermoplastic, nontoxic, biodegradable, biocompatible, hydrophobic, elastomeric, piezoelectric, and optically active with wide range of industrial applications on account of desirable material properties. Considering these facts, polyhydroxyalkanoate thermoplastic is receiving increasing attention and commercial interest globally.

14.2 Role of PHA Thermoplastics in Developing a Sustainable Environment

Developments in science and technology have fueled the fast growth of modern society, which is certainly unsustainable due to the stress it places on present resources/environment. The energy and materials required to sustain the current society are mainly obtained from non-renewable fossil resources that will be exhaust in near future (Sudesh and Iwata 2008). Apart from this, the worldwide increase usage of fossil fuels is responsible for enhanced release of carbon dioxide into the atmosphere thereby, contributing towards global warming and climate change (Sudesh and Iwata 2008). Xenobiotic conventional plastics usually synthesized from finite and non-sustainable fossil fuels that not only persist in soil for a long period but also exhibit detrimental effects on environment (Fiechter 1990; Shimao 2001; Shah et al. 2008; Kumar et al. 2015). Nature's in-built mechanisms and self-regulation ability fail to tackle such pollutants as these are not familiar to it. Consequently, there has been growing public concern over the unfriendly conventional/petroleum-based plastics due to increased cost of solid waste disposal, decreasing capacity of municipal landfills, and the potential hazards from plastic waste incineration. All these issues lead to development of polyhydroxyalkanoate (PHA) thermoplastics as eco-friendly and sustainable plastics. These biodegradable plastics produced from renewable natural resources, i.e., microorganisms such as bacteria and cyanobacteria. They depict properties such as good strength, flexibility, nontoxicity, impermeability to oxygen, good moisture resistance, and stability including material properties comparable to petroleum-based plastics (Kumar et al. 2015). Moreover, biogenic synthesis of PHA polymeric materials exhibits benefits of saving fossil resources along with decrease of carbon dioxide release that makes them a significant innovation of sustainable development (Bugnicourt et al. 2014).

The main property that differentiates PHAs from conventional plastics is their biodegradability. PHAs are degraded upon exposure to soil, compost, or marine sediment. Biodegradability is defined as the ability of a material/substance to be broken down, particularly into harmless products, by the action of living things as microorganisms. In the natural world, bacteria and fungi are the main members in the biodegradation process. The breakdown of materials provides them with precursors for cell components and energy for energy-requiring processes. Biodegradation of PHAs under aerobic conditions produces carbon dioxide and water. However, in anaerobic conditions, the end/degradation products are carbon dioxide and methane. Being the product of renewable carbon sources, their production and uses

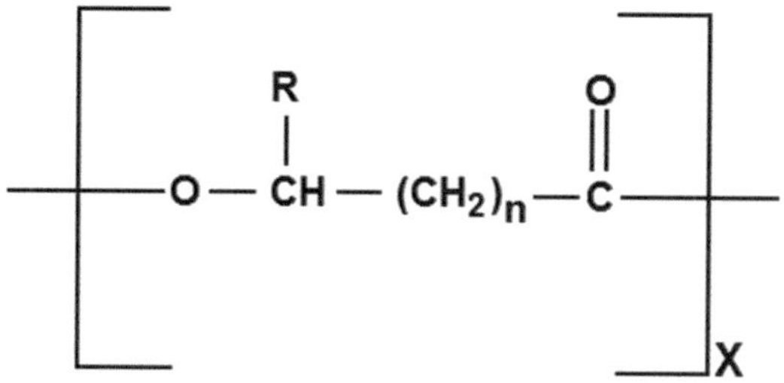

Fig. 14.3 General structure of polyhydroxyalkanoates

follow sustainable closed cycles (Braunegg et al. 1998). It is anticipated that in the near future, PHA thermoplastics, which is relatively young, will attain a unique position in modern material technology not only in terms of real commercial applications but also sustainability and environmental protection.

14.2.1 Polyhydroxyalkanoate (PHA) Thermoplastics

Naturally occurring polyhydroxyalkanoate (PHA) thermoplastics that manufactured in the cytoplasm as spherical and water-insoluble inclusions are polymeric materials of 0.2 ± 0.5 mm in diameter made up of a huge number (600–35,000) of hydroxyl fatty acid units (Dawes and Senior 1973; Madison and Huisiman 1999; Luengo et al. 2003; Khanna and Srivastava 2005). Each monomeric unit of PHA thermoplastics commonly includes saturated alkyl group as side chain. However, occasionally it can also hold unsaturated, branched, or substituted alkyl groups. So far 150 diverse forms of monomeric units of PHA thermoplastics have been recorded. The different PHA thermoplastics found to possess different monomeric units owing to variation in the supplementation of organic carbon substrates to the source microorganism as well as the metabolic pathways operating in the cell in which it is synthesized (Rehm 2003; Narayan 2006). PHA thermoplastics are represented by the common structural formula as presented in Fig. 14.3, where "X" can accept up to 35,000 and R-pendant group holds hydrogen atom or huge ranges of carbon chains (Lee 1995).

For example, when

n = 1	R = Hydrogen	Poly(3-hydroxypropionate)
	Methyl	Poly(3-hydroxybutyrate)
	Ethyl	Poly(3-hydroxyvalerate)
	Propyl	Poly(3-hydroxyhexanoate)
	Pentyl	Poly(3-hydroxyoctanoate)
	Nonyl	Poly(3-hydroxydodecanoate)
n = 2	R = Hydrogen	Poly(4-hydroxybutyrate)
n = 3	R = Hydrogen	Poly(5-hydroxyvalerate)

Uncommon PHA thermoplastics establish a specific group of polyoxo(thio) esters that associated with unusual monomer units (Lütke-Eversloh et al. 2001; Jain et al. 2010). Such unusual monomer units are introduced in the growing chain of PHA thermoplastic backbones merely by means of related carbon substrates supplemented to microorganisms. Interestingly, a novel type of sulfur incorporating PHA thermoplastic family has been reported that associated with thioester bonds (Lütke-Eversloh et al. 2001). Uncommon PHA thermoplastics are classified into two types: (a) microorganism originated PHA thermoplastics, where PHA polymers manufactured either from natural monomeric units or chemical derivatives of the natural ones, and (b) PHA thermoplastics resulting from chemical synthesis or through physical alterations of naturally existing polymers (Lütke-Eversloh et al. 2001; Jain et al. 2010). Uncommon PHA thermoplastics are categorized into four types based on their chemical structure (Lütke-Eversloh et al. 2001; Jain et al. 2010): (i) PHA thermoplastics with side chains bearing double or triple bonds or/and diverse functional groups such as methyl, methoxy, ethoxy, acetoxy, hydroxyl, epoxy, carbonyl, cyano, phenyl, nitrophenyl, phenoxy, cyanophenoxy, benzoyl, halogen atoms, etc.; (ii) PHA thermoplastics, where the monomer unit contributing towards oxoester bond undergoes alteration, i.e., the hydroxyl group to be esterified is not positioned at third carbon; (iii) PHA polymers, where some oxoester bonds substituted with thioester functions, i.e., thioester containing PHA thermoplastics; and (iv) PHA thermoplastics manipulated by chemical or physical means.

14.2.2 Classification of PHA Thermoplastics

Based on the size of monomeric units integrated in the polymeric chain (Anderson and Dawes 1990; Steinbuchel 1992; Steinbuchel et al. 1992; Madison and Huisiman 1999), PHA thermoplastic can be classified into five type, viz., (i) short-chain-length (SCL) PHA thermoplastics, which contain hydroxyacid (HA) monomeric units ranging from 3 to 5 carbon atoms (Nawrath et al. 1994; John 1997; Bohmert et al. 2000; Gouda et al. 2001; Nishioka et al. 2001; Borah et al. 2002; Thakor et al. 2003; Kahar et al. 2004; Toh et al. 2008; Sankhla et al. 2010; Bhati and Mallick 2012); (ii) medium-chain-length (MCL) PHA thermoplastics made up of HA monomer units with 6 to 14 carbon atoms (Mittendorf et al. 1998; Poirier 1999; Solaiman et al. 2006; Cerrone et al. 2014); (iii) long-chain-length (LCL) PHA thermoplastics composed of HA monomers with 15 or more than 15 carbon atoms (Singh and Mallick 2008 2009a, b; Singh et al. 2013 2015); (iv) SCL-MCL PHA thermoplastic copolymers comprised of SCL and MCL monomers, which are covalently linked in the same polyester molecules (Caballero et al. 1995; Lee et al. 1995a; Doi et al. 1995; Sheu and Lee 2004; Kahar et al. 2004; Phithakrotchanakoon et al. 2013); and (v) SCL-LCL-PHA thermoplastic copolymers, where SCL and LCL monomers are covalently associated in the same polymeric molecules (Singh and Mallick 2008, 2009a b; Sankhla et al. 2010; Singh et al. 2013, 2015). Bacterial species capable to synthesize either SCL- or MCL- or LCL-PHA thermoplastics but not normally copolymers of SCL-MCL- or SCL-LCL-PHA thermoplastics because of the substrate specificity of PHA synthases that can accept 3-hydroxyacids (3-HAs) of a specific range of carbon length (Anderson and Dawes 1990; Kato et al. 1996; Ashby et al. 2002; Singh et al. 2013, 2015; Kumar et al. 2015). On the other hand, the PHA synthase of *Alcaligenes eutrophus* can accept 3HAs with 3 to 5 carbon atoms, whereas that occur in *Bacillus cereus* ATCC14579, *Pseudomonas* sp. A33, *Pseudomonas fluorescens* GK13, *Pseudomonas marginalis* DSM50276, *Pseudomonas mendocina* DSM50017, *Aeromonas caviae*, recombinant *Pseudomonas putida* GPp104 PHA$^-$, recombinant *E. coli* ABC*Re*J1*Pp*, recombinant *E. coli* ABC2*Pp*J4*Pp*, recombinant *E. coli* ABC*Re*J4*Pp*, and *Pseudomonas oleovorans* can only recognize 3HAs with 6 to 14 carbon atoms (Caballero et al. 1995; Lee et al. 1995; Doi et al. 1995; Sheu and Lee 2004; Saharan et al. 2012; Phithakrotchanakoon et al. 2013). Interestingly, *Pseudomonas aeruginosa* MTCC 7925 can polymerize 3HAs holding 4 to18 carbon atoms (Singh and Mallick 2008, 2009a, b; Singh et al. 2013, 2015; Kumar et al. 2015).

Poly-3-hydroxybutyrate (PHB) is the general member of SCL-PHA thermoplastics. PHB is comparatively rigid as well as crystalline (Park and Lee 2004). However, MCL-/LCL-PHA thermoplastics are elastomeric polymeric materials with poor tensile strength. Thus, SCL-, MCL-, and LCL-PHA thermoplastics are not viable for wide range of commercial uses. Nevertheless, incorporation of MCL- or LCL- monomers in PHB backbone results into biosynthesis of SCL-MCL-PHA copolymer or SCL-LCL-PHA copolymer, respectively, that changes the material properties of the polymer, enabling it suitable for different application (Matsusaki et al. 2000; Singh and Mallick 2008, 2009a, b; Singh et al. 2013, 2015; Singh and Mallick 2015).

14.2.3 PHA Thermoplastics Production in Biological System

A critical review of literature perusal revealed that bacteria from almost 75 varied genera including both gram-positive and gram-negative involved in the biosynthesis of PHA thermoplastics (Reddy et al. 2003). The existence of PHA thermoplastics in cyanobacteria was reported since 1966, where PHA production pathway was found to depict resemblance with the heterotrophic bacteria (Wang et al. 2013; Drosg et al. 2015). However, in the case of plants, the first PHA production was reported in 1992 in the form of PHB thermoplastic in the cytoplasm of cells of *Arabidopsis thaliana* followed by production of diverse PHA thermoplastics in different species via introduction of novel metabolic pathways in the cytoplasm, plastid, or peroxisome (Poirier et al. 1992; Poirier 2001). PHA synthases are the main enzymes of PHA biosynthetic pathways, which utilize coenzyme A-thioester of (r)-hydroxy fatty acids as substrates (Jain et al. 2010). Depending on the primary structures, substrate specificity and the subunit composition, PHA synthases can be broadly categorized into following four types: (i) Class I/class II PHA synthases involve enzymes made up of merely one type of subunit (PhaC) having molecular weights ranging between 61 and 73 kDa (Qi and Rehm 2001). On the basis of substrate specificity, class I PHA synthases (*Cupriavidus necator*) specifically exploit CoA thioesters of different (*R*)-3-hydroxy fatty acids containing 3 to 5 carbon atoms. However, class II PHA synthases (*Pseudomonas aeruginosa*) specifically accept CoA thioester of different (*R*)-3-hydroxy fatty acids having 6 to 14 or 15 or more than 15 carbon atoms (Rehm 2003; Ren et al. 2000, Singh and Mallick 2008, 2009a, b; Singh et al. 2013, 2015). (ii) Class III PHA synthases (represented by *Allochromatium vinosum*) composed of enzymes with two different types of subunits, i.e., PhaC and PhaE subunits having molecular weight of approximately 40 and 40 kDa, respectively. PhaC subunit depicts amino acid sequence resemblance ranging from 21 to 28 % with respect to class I and II PHA synthases. Nevertheless, PhaE subunit exhibits no resemblance with PHA synthases. These PHA synthases accept CoA thioester of (*R*)-3-hydroxy fatty acids having 3 to 5 carbon atoms (Liebergesell et al. 1992; Liebergesell and Steinbüchel 1992). (iii) Class IV PHA synthases (represented by *Bacillus megaterium*) are found to depict resemblance with class III PHA synthases, where the subunit PhaE is substituted by PhaR having molecular weight of about 20 kDa (McCool and Cannon 1999, 2001).

The production of PHA thermoplastics in bacteria such as *Cupriavidus necator, Rhodopseudomonas palustris*, *Methylobacterium organophilum*, etc. and cyanobacteria such as *Synechococcus* sp. MA19, *Synechocystis* sp. PCC 6803, etc. was found to be stimulated as a result of inadequacy of an essential nutrient, viz., sulfate, ammonium, phosphate, potassium, iron, magnesium, or oxygen (Dawes and Senior 1973; Ward et al 1977; Steinbuchel and Pieper 1992; Byrom 1992; Nishioka et al. 2001; Reddy et al. 2003; Thakor et al. 2003; Panda and Mallick 2007; Bhati and Mallick 2012). The PHA thermoplastics pool could be improved through increasing the carbon to nitrogen ratio (Shi et al. 2007). Moreover, the accumulation of PHA thermoplastics observed under vigorous cell growth without involving restriction of any nutrients in bacteria, viz., *Alcaligenes latus* and recombinant *E. coli*, incorporated with PHA biosynthetic genes (Saharan et al. 2012). Table 14.4 summarized the production of PHA thermoplastics in various bacterial, cyanobacterial, and transgenic plant species.

Biosynthesis of PHA thermoplastics have been comprehensively investigated in the past several years, where acetyl-CoA acts as principal precursor to furnish 3-hydroxyalkanoyl-CoA of diverse lengths as substrates for PHA synthases of many specificities (Chen 2010). However, 3-hydroxyalkanoyl-CoA could also be generated as a result of β-oxidation of fatty acids of diverse chain lengths (Chen 2010). In addition, PHA thermoplastics also found to synthesize from acetyl-CoA via fatty acid biosynthetic pathway (Kumar et al. 2015; Singh and Mallick 2015). Various genes responsible for encoding different enzymes are straightforwardly or indirectly engaged in PHA production. The accumulation of PHA thermoplastics by different biosynthetic pathways is summarized in Table 14.5.

14.2.4 Material Properties of PHA Thermoplastics

The material properties such as melting temperature (T_m), glass-transition temperature (Tg), Young's modulus, elongation to break, and tensile strength are essential

Table 14.4 Summary of the PHA thermoplastics production in various bacterial species and transgenic plants including cyanobacteria

Organism	Type of organism	Substrate/condition	% PHA (dcw)	PHA composition	References
Ralstonia eutropha	Bacterial species	Glucose + propionate	70–80	P(3HB-*co*-3HV)	Byrom (1992)
**Pseudomonas* sp. 61–3		Glucose	26	P(3HB-*co*-3HO-*co*-3HD) + PHB	Kato et al. (1996)
		Fructose	17	P(3HB-*co*-3HHx-*co*-3HO-*co*-3HD-*co*-3HDD)	
Bacillus megaterium		Sugarcane molasses	46	PHB	Gouda et al. (2001)
Bacillus mycoides RLJ B-017		Sucrose	69	PHB	Borah et al. (2002)
Comamonas testosteroni		Naphthalene	85	PHB	Thakor et al. (2003)
Ralstonia eutropha H16		Soybean oil	76	PHB	Kahar et al. (2004)
*Recombinant *Ralstonia eutropha* H16 strain		Soybean oil	74	P(3HB-*co*-3HHx)	
*Recombinant *Aeromonas hydrophila* 4AK4		Lauric acid	53	P(3HB-*co*-3HHx)	Tian et al. (2005)
Methylobacterium sp. GW2		Methanol	30	P(3HB-*co*-3HV)	Yezza et al. (2006)
***Pseudomonas aeruginosa* MTCC 7925		Ethanol	69	P(3HB-*co*-3HV-*co*-3HHD-*co*-3HOD)	Singh and Mallick (2008)
***Pseudomonas aeruginosa* MTCC 7925		Palm oil + extract of palm oil cakes	75	P(3HB-*co*-3HV-*co*-3HHD-*co*-3HOD)	Singh and Mallick (2009a)
***Pseudomonas aeruginosa* MTCC 7925		Ethanol + glucose	78	P(3HB-*co*-3HV-*co*-3HHD-*co*-3HOD)	Singh and Mallick (2009b)
Brevibacillus invocatus MTCC 9039		Glucose + acetate + propionate	65	P(3HB-*co*-3HV)	Sankhla et al. (2010)
*Recombinant *E. coli* DH5α		Decanoate + glucose	7.3	P(3HB-*co*-3HHx-*co*-3HO-*co*-3HD)	Li et al. (2011)

(continued)

Table 14.4 (continued)

Organism	Type of organism	Substrate/condition	% PHA (dcw)	PHA composition	References
***Pseudomonas aeruginosa* MTCC 7925		Palm oil + extract of palm oil cakes	77	P(3HB-*co*-3HV-*co*-3HHD-*co*-3HOD)	Singh et al. (2013)
Recombinant *E. coli* XL1		Glucose	62	P(3HB-*co*-3HV)	Yang et al. (2014)
Serratia ureilytica		Volatile fatty acids	51	P(3HB-*co*-3HV)	Reddy and Mohan (2015)
Bacillus megaterium		Residual glycerol	52	PHB	Gómez Cardozo et al. (2016)
Arabidopsis thaliana	Transgenic plant species	Photoautotrophy	0.1	PHB	Poirier et al. (1992)
Arabidopsis thaliana			14	PHB	Nawrath et al. (1994)
Gossypium hirsutum			0.05	PHB	John (1997)
Arabidopsis thaliana			0.6	MCL-PHA	Mittendorf et al. (1998)
Arabidopsis thaliana			40	PHB	Bohmert et al. (2000)
Arabidopsis thaliana	Transgenic plant species	Photoautotrophy	8	PHB	Poirier and Gruys (2001)
Beta vulgaris			5.5	PHB	Menzel et al. (2003)
Nicotiana tabacum			0.14	PHB	Lossl et al. (2005)
Saccharum spp.			1.88	PHB	Petrasovits et al. (2007)
Arabidopsis thaliana			0.18	SCL-/MCL- PHA	Matsumoto et al. (2009)
Panicum virgatum			6.1	PHB	Somleva and Ali (2010)
Glycine max			0.36	PHB	Schnell et al. (2012)
Saccharum spp.			1.6	PHB	Petrasovits et al. (2013)

Nostoc muscorum	Cyanobacterial species	Dark + acetate	43	PHB	Sharma and Mallick (2005)
Synechocystis sp. PCC 6803		Phosphate deficiency + gas-exchange limitation + acetate	38	PHB	Panda and Mallick (2007)
Spirulina platensis UMACC 161		Acetate and CO_2	10	PHB	Toh et al. (2008)
Synechocystis sp. UNIWG		Acetate and CO_2	14	PHB	
Nostoc muscorum		Photoautotrophy	9	PHB	Bhati et al. (2010)
Nostoc muscorum Agardh		Phosphate deficiency + acetate + valerate	58	P(3HB-*co*-3HV)	Bhati and Mallick (2012)
		Nitrogen deficiency + acetate + valerate	60	P(3HB-*co*-3HV)	
Recombinant *Synechococcus* sp. PCC 6803		Photoautotrophy + CO_2	1.4	PHB	Osanai et al. (2013)
Aulosira fertilissima CCC 444		Fructose + valerate	77	P(3HB-*co*-3HV)	Samantaray and Mallick (2014)
Aulosira fertilissima CCC 444		Acetate + gas exchange limitation	49	PHB	Samantaray and Mallick (2015)

3HB: 3-hydroxybutyrate, 3HV: 3-hydroxyvalerate, 3HHx: 3-hydroxyhexanoate, 3HO: 3-hydroxyoctanoate, 3HD: 3-hydroxydecanoate, 3HDD: 3-hydroxydodecanoate, 3HHx: 3-hydroxyhexanoate, 3HHD: 3-hydroxyhexadecanoate, 3HOD: 3-hydroxyoctadecanoate; *SCL-MCL-PHA copolymer producers, **SCL-LCL-PHA copolymer producers

Table 14.5 Overview of various PHA thermoplastics biosynthetic pathways

Pathway	Abbreviation	Enzyme	Species	References
I	PhaA	β-Ketothiolase	*Ralstonia eutropha*	Sudesh et al. (2000)
	PhaB	NADPH-dependent acetoacetyl-CoA reductase		
	PhaC	PHA synthase		
Associated way	PhaZ	PHA depolymerase	*Aeromonas hydrophila* 4AK4	Sudesh et al. (2000)
		Dimer hydrolase	*Pseudomonas stutzeri* 1317	
		(*R*)-3-Hydroxybutyrate dehydrogenase	*Ralstonia eutropha*	
		Acetoacetyl-CoA synthetase	*Pseudomonas oleovorans*	
II	FabG	3-Ketoacyl-CoA reductase	*Pseudomonas putida* KT2442	Sudesh et al. (2000)
		Epimerase	*Aeromonas hydrophila* 4AK4,	Mittendorf et al. (1998)
	PhaJ	(*R*)-Enoyl-CoA hydratase/enoyl-CoA hydratase I	*Pseudomonas aeruginosa*	
		Acyl-CoA oxidase, putative		
		Enoyl-CoA hydratase I, putative		
III	PhaG	3-Hydroxyacyl-ACP-CoA transferase	*Pseudomonas mendocina*, recombinant *Escherichia coli*	Sudesh et al. (2000), Zheng et al. (2005) and Taguchi et al. (1999)
	FabD	Malonyl-CoA-ACP transacylase		
IV		NADH-dependent acetoacetyl-CoA reductase	*Rhizobium* (Cicer) sp. CC 1192	Chohan and Copeland (1998)
	SucD	Succinic semialdehyde dehydrogenase	*Clostridium kluyveri*	Valentin and Dennis (1997)
V	4hbD	4-Hydroxybutyrate dehydrogenase		
	OrfZ	4-Hydroxybutyrate-CoA:CoA transferase		
VI		Lactonase, putative	Mutants and recombinant of *Alcaligenes eutrophus*	Valentin and Steinbüchel (1995)
		Hydroxyacyl-CoA synthase, putative		
VII		Alcohol dehydrogenase, putative	*A. hydrophila 4AK4*	Xie and Chen (2008)

(continued)

Table 14.5 (continued)

Pathway	Abbreviation	Enzyme	Species	References
VIII	ChnA	Cyclohexanol dehydrogenase	*Acinetobacter* sp. SE19, *Brevibacterium epidermidis* HCU	Brzostowicz et al. (2002)
	ChnB	Cyclohexanone monooxygenase		
	ChnC	Caprolactone hydrolase		
	ChnD	6-Hydroxyhexanoate dehydrogenase		
	ChnE	6-Oxohexanoate dehydrogenase		
		Semialdehyde dehydrogenase, putative 6-hydroxyhexanoate		
		Dehydrogenase, putative		
		Hydroxyacyl-CoA synthase		

Chen (2010)

parameter to assess the effectiveness/usefulness of a PHA thermoplastic for a given commodity use. PHB depicts a number of features such as melting temperature (180 °C), tensile strength (40 Mpa), and Young's modulus (3.5 Gpa) analogous to conventional plastics, for example, polypropylene (PP) and low-density polyethylene (LDPP). Nevertheless, PHB thermoplastic is not only rigid, crystalline, and brittle but also exhibits low elongation to break that restricts its potential application (Table 14.6). Therefore, it is very important to improve the material properties of PHB thermoplastics so that desirable features could be attained for various applications. This can be achieved by co-monomer integration into the PHB thermoplastics backbone that results into biosynthesis of P(3HB-*co*-3HV), SCL-MCL-PHA, and SCL-LCL-PHA copolymers, i.e., PHA copolymers with improved material properties. Among PHA copolymers, copolymers of SCL-MCL-PHA or SCL-LCL-PHA thermoplastics revealed superior material properties over SCL-, MCL-, or LCL-PHAs or even P(3HB-*co*-3HV) copolymers marketed under the trade name of BIOPOL® (Byrom 1992; Lee 1995; Matsusaki et al. 2000; Singh and Mallick 2008, 2009a, b, Singh et al. 2013, 2015; Singh and Mallick 2015). These PHA copolymers depicted much improved material properties akin to that of conventional plastics such as polypropylene (PP) and polyethylene (PE) (Kumar et al. 2015; Singh and Mallick 2015). Table 14.6 showed comparison on the material properties of PHA polymers with common conventional plastics.

Table 14.6 Comparative account on the material properties of PHA thermoplastics with common conventional plastics

Property	PHB	P(3HB-*co*-3HV) (mol fraction 80:20)	P(3HB-*co*-3HA) (mol fraction 94: 06)	P(3HB-*co*-3HV-*co*-3HHD-*co*-3HOD) (Mol Fraction 84.8:7.2:3.1:4.9 – 95.7:1.0:1.8:1.5)	PP	LDPE
T_m (°C)	180	145	133	115 to131	176	130
T_g (°C)	4	−1	−8	−8 to −14	−10	−36
Crystallinity (%)	60	56	45	–	50–70	20–50
Young's modulus (Gpa)	3.5	0.8	0.2	0.2–0.3	1.7	0.2
Tensile strength (Mpa)	40	20	17	17–19	38	10
Elongation to break (%)	5	50	680	682–723	400	620

Singh et al. (2013), Kumar et al. (2015) and Singh and Mallick (2015)
Tm melting temperature, *Tg* glass-transition temperature, 3HA: [3-hydroxydecanoate (3 mol%), 3-hydroxydodecanoate (3 mol%), 3-hydroxyoctanoate (<1 mol%), and 3-hydroxy-*cis*-dodecenoate (<1 mol%)]; PP, polypropylene; LDPE, low-density polyethylene

14.2.5 Applications of PHA Thermoplastics

The interesting properties associated with PHA thermoplastics such nontoxicity, elastomeric, biocompatibility, piezoelectric, optically active, impermeable to gas, biodegradability, and hydrophobicity are accountable for various commercial uses (Chen 2010; Kumar et al. 2015; Singh and Mallick 2015). The PHA thermoplastics depict the following industrial application in the field of: (i) *PHA thermoplastics as packaging materials* (PHA thermoplastics were exploited for manufacture of bottles, packaging films for applications as containers and paper coatings, shopping bags, disposable substances as razors, diapers, feminine hygiene products, cosmetic containers and cup, upholstery, medical surgical garments, carpet, compostable bags and lids, tubs for thermoformed articles, tray for foods, utensils (Vincenzini and Philippis 1999; Chen 2010); (ii) *PHA thermoplastics as agricultural delivery carrier* (PHA thermoplastics could be exploited as sustainable platform for the controlled release of fertilizers, plant growth regulators pesticides as well as herbicides, seed encapsulation, and covering foils (Vincenzini and Philippis 1999); (iii) *PHA thermoplastics as Biomedical implant materials* (PHA thermoplastics have been utilized to make devices together with sutures, meniscus repair devices, suture fasteners, tacks, rivets, staples, surgical mesh, surgical pins, repair patches, screws, bone plates and bone plating systems, cardiovascular patches, slings, atrial septal defect repair devices, guided tissue repair and regeneration devices, bone marrow scaffolds, tendon repair devices, spinal fusion cages, dural substitutes, ocular cell implants, skin substitutes, articular cartilage repair devices, ligament and tendon grafts, meniscus regeneration devices, bone graft substitutes, vein valves, nerve guides, bulking and filling agents, orthopedic pins including bone filling augmentation material, pericardial patches, stents, adhesion barriers, bone dowels, hemostats, and wound dressings (Vincenzini and Philippis 1999; Chen 2010; Saharan et al. 2012); (iv) *PHA thermoplastics as drug delivery* (PHAs could be exploited as potential drug carrier and retarded drug release (Vincenzini and Philippis 1999). Homopolymers as well as copolymers of lactate and glycolate are extensively exploited towards commercially accessible sustained release products for drug delivery. Nevertheless, drug release cannot be completely regulated owing to degradation of lactate and glycolate copolymers through bulk hydrolysis (Pouton and Akhtar 1996). In this regard, PHA thermoplastics emerged as attractive alternative due to their unique properties such as biodegradability, biocompatibility, and degradation by surface erosion (Gould et al. 1987). Brophy and Deasy (1986) found that there was enhancement in liberation rate of sulfamethizole from irregularly shaped PHB microparticles as the molecular weight of polymer increased. To date, only PHB and P(3HB-*co*-3HV) copolymer were investigated for their potential towards controlled drug release. It is anticipated that other family members of PHA thermoplastic with different features will further improve the controlled release properties for the drug release (Chen 2010). We are at early stage in exploring such applications. Therefore, this field needs intensified research and innovation to explore the actual potential of PHA thermoplastics); (v) *PHA thermoplastics as biofuels* (it is anticipated that in the near feature the PHA thermoplastics act as potential precursor for the production of biofuels. For example, methyl esters of 3-hydroxybutyrate and MCL 3-hydroxyalkanoate produced through esterification of PHB and MCL-PHA could be exploited as biofuels (Zhang et al. 2009); and (vi) *PHA monomers as drug* (PHA monomers could be exploited as drug. This fact is evident from 3-hydroxybutyrate (3HB) and its derivatives that showed effect on cell apoptosis including calcium ion (Ca^{2+}) concentration in the cytoplasm of glial cells of mouse (Xiao et al. 2007). Under the influence of 3HB and its derivatives, the % cells experiencing apoptosis reduced considerably and was evident from flow cytometry. In vitro investigation, it was observed that 3HB derivatives intensely raised the Ca^{2+} concentration in the cytoplasm, where the extracellular and the intracellular Ca^{2+} together acted as pool for such Ca^{2+} concentration increase).

14.2.6 Current Commercial Status of PHA Thermoplastics

To date, the marketable production of PHA thermoplastic, i.e., 3-hydroxybutyrate and 3-hydroxyvalerate [P(3HB-*co*-3HV)] copolymer, is carried out by Metabolix, Inc. (USA) exploiting bacterium *Cupriavidus necator*, where 80 % of dry cell weight (dcw) P(3HB-*co*-3HV) copolymer synthesis have been reported (Byrom 1992). PHA thermoplastic market is at its initial phase of technology progression and still not achieved the cost-effective scale of polymer production (Kumar et al. 2015). The foremost hurdles stopping the marketable use of bacterial PHA is its high production price contributed through supplementation of expensive carbon substrates during bacterial fermentation. Specifically, almost 30–50 % of the overall

PHA production cost is due to the organic carbon substrate (Lynd et al. 1999; Choi and Lee 1999; Braunegg et al. 2004). For example, the recombinant *E. coli* with a PHB thermoplastic quantity of 157 g l^{-1} and a PHB thermoplastic of 77 % (dcw), the cost of carbon substrate is raised by 38 % of the entire cost when the production scale was assumed to be 1,00,000 tons $year^{-1}$ (Choi and Lee 1999). Therefore, to diminish this shortcoming, photoautotrophic hosts (plants and cyanobacteria) and/or alternate economical carbon substrates are getting the present consideration. However, plant-based PHA production undergoes some shortcomings, viz., low expression level (Table 14.4), long growth period, etc. which make the system less preferred for large-scale exploitation (Singh and Mallick 2015). Thus, cyanobacteria are observed as eye-catching hosts for the production of PHAs due to their negligible nutrient need, competence to convert "greenhouse gas" into biodegradable plastics through photosynthetically, and non-requirements of fertile lands for their cultivation as most of them are aquatic in nature. Regardless of these benefits, the notable obstacle in photosynthetic synthesis of PHB thermoplastic in cyanobacteria is the shortage of a cost-effective viable mass cultivation system. It was reported that ongoing mass cultivation of cyanobacteria have productivity up to 10–15 g dry wt/m^2 day^{-1} in sunlight covering area. Furthermore, the cost-effective harvesting of these tiny cyanobacterial cells/filaments is additional drawback related to cyanobacterial production of PHA (Belay 2004).

Exploration of economical substrates is found to be another possibility for PHA thermoplastic production. The utilization of organic wastes, viz., swine waste liquor, palm oil mill effluents, and vegetable including fruit wastes, is being observed as alternative substrates for decreasing the manufacture cost of PHA thermoplastic (Hassan et al. 1996, 1997a, b; Meesters 1998; Reis et al. 2003; Salehizadeh and van Loosdrecht 2004). Wastewaters, spent washes, pressed muds, and sludges generated from food and brewery industries, municipal sewage, etc. hold significantly large quantity of organic materials, for example, volatile fatty acids (VFAs). The VFAs such as acetic, propionic, and butyric acids are precursors of PHA thermoplastics (Ruan et al. 2003). Combined benefits of decreasing disposal price as well as manufacture of valuable products could be anticipated if waste components/products could be exploited as carbon source for PHA thermoplastics production. Thus, various investigators conducted study on PHA thermoplastics production from activated sludge supplemented with malt waste, acetate, volatile fatty acids, paper mill wastewater or rice grain-based distillery spent wash, municipal wastewaters alone as well as supplemented with carbon sources, fermented food waste, synthetic wastewater, and also anaerobic wastewaters (Singh and Mallick 2015). Furthermore, prices for process development exploiting mixed cultures could be considerably decreased since simple, non-sterilizable reactors with inexpensive substrates require less process control compared to a pure culture (Satoh et al. 1998; Reis et al. 2003; Salehizadeh and van Loosdrecht 2004). Exploitation of open mixed microbial cultures enables the use of diverse substrates as the population of microorganisms is capable to acclimatize constantly with fluctuating substrate (Reis et al. 2003). Consequently, by means of supplementation of activated sludges and wastes under suitable conditions might be an encouraging alternative for PHA thermoplastic production that warrants more research as no exhaustive investigation on the optimization of nutrient removal and PHA production by mixed cultures are available (Singh and Mallick 2015; Kumar et al. 2015). However, such mixed culture system too showed few limitations such as little PHA thermoplastics content rate and cell density, insufficient substrate exploitation ratio, and long acclimatizing time for activated sludge. Considering these, economical substrates such as whey, nonedible plant oils, hydrolysates of starch (corn and tapioca), molasses, as well as cellulose could be exceptional carbon sources for bacteria for manufacturing PHA thermoplastics that could contribute towards significant cost-effective advantages (Singh and Mallick 2015; Kumar et al. 2015). Many investigators found to conduct study on PHA thermoplastic production using diverse bacterial strains such as *Azotobacter chroococcum*, recombinant *E. coli*, *Ralstonia eutropha*, *Pseudomonas cepacia*, *Brevundimonas vesicularis*, *Bacillus megaterium*, *Pseudomonas aeruginosa*, *Burkholderia cepacia*, *Pseudomonas putrefaciens*, *Haloferax* mediterranei, *Hydrogenophaga pseudoflava*, *Serratia ureilytica*, etc. to convert cost-effective substrates into PHA thermoplastics (Singh and Mallick 2015; Kumar et al. 2015). It is anticipated that increasing and advanced research work on PHA thermoplastic production using bacterial strains with cost-effective substrates could be utmost cost-effective approach pertaining to economical PHA polymers production and effective commercialization of PHA thermoplastic in future (Singh and Mallick 2015; Kumar et al. 2015).

14.3 Conclusions

Polymers are the leading products of modern chemical industry that constitutes the backbone of existing society. Among these polymers, conventional plastics emerged as unique and extraordinary polymer in modern material technology because of their anticipated material properties including resistant to degradation, stability, and prolonged durability that enable them for various industrial uses. Nevertheless, such wonder properties associated with these plastics turn out to be harmful to environment/ecosystem when out of usage especially because of primarily xenobiotic nature. Disposal of plastic wastes at landfill sites, incineration, and recycling are viable approaches for acquiring plastic garbage-free environment depicted limitations. As a result, global research is powered towards the development of eco-friendly and sustainable bioplastics such as polyhydroxyalkanoate (PHA). Owing to material properties comparable to conventional plastics, PHA thermoplastics open up various commercial applications. However, the notable hurdles preventing the marketable exploitation of PHA are its high production price especially contributed by exploitation of expensive carbon substrates. Consequently, photoautotrophic hosts and alternate cost-effective substrates are receiving the current attention. Unfortunately, photoautotrophic-based PHA production suffers some limitations. Thus, exploitation of various and novel inexpensive substrates for PHA thermoplastics accumulating microorganism should be encouraged as a way towards cost-effective PHA thermoplastic production. Overall, as these thermoplastics are in their initial phase of commercial development, therefore, the fruitful, cost-effective, and sustainable biogenic production of PHA thermoplastics requires further rigorous research on recombinant microbial species, mixed cultures, proficient fermentation strategies, recovery, and separation processes with the exploitation of novel economical substrates for commercial-level production and marketing.

References

Anderson AJ, Dawes EA (1990) Occurrence, metabolism, metabolic role, and industrial uses of bacterial polyhydroxyalkanoates. Microbiol Rev 54:450–472

Ashby RD, Solaiman DKY, Foglia TA (2002) The synthesis of short and medium chain-length poly(hydroxyalkanoate) mixtures from glucose- or alkanoic acid-grown *pseudomonas oleovorans*. J Ind Microbiol Biotechnol 28:147–153

Belay A (2004) Mass culture of *Spirulina* outdoors – the Earthrise farms experience. In: Vonshak A (ed) *Spirulina platensis (Arthrospira):* physiology, cell-biology and biotechnology. Taylor and Francis, London, pp 131–158

Bernard M (2014) Industrial potential of polyhydroxyalkanoate bioplastic: a brief review. Univ Sask Undergr Res J 1:1–14

Bhati R, Mallick N (2012) Production and characterization of poly(3-hydroxybutyrate-*co*-3-hydroxyvalerate) co-polymer by a N_2-fixing cyanobacterium, *Nostoc muscorum* Agardh. J Chem Technol Biotechnol 87:505–512

Bhati R, Samantaray S, Sharma L, Mallick N (2010) Poly-β-hydroxybutyrate accumulation in cyanobacteria under photoautotrophy. Biotechnol J 5:1181–1185

Bohmert K, Balbo I, Kopka J, Mittendorf V, Nawrath C, Poirier Y, Tischendorf G, Tretchewey RN, Willmitzer L (2000) Transgenic *Arabidopsis* plants can accumulate polyhydroxybutyrate up to 4 % of their fresh weight. Planta 211:841–845

Borah B, Thakur PS, Nigam JN (2002) The influence of nutritional and environmental conditions on the accumulation of poly-β-hydroxybutyrate in *Bacillus mycoides* RLJ B-017. J Appl Microbiol 92:776–783

Braunegg G, Gilles L, Klaus F (1998) Polyhydroxyalkanoates biopolyesters from renewable resources: physiological and engineering aspects. J Biotechnol 65:127–161

Braunegg G, Bona R, Koller M (2004) Sustainable polymer production. Polym-Plast Technol Eng 43:1779–1793

Brophy MR, Deasy PB (1986) In vitro and in vivo studies on biodegradable polyester microparticles containing sulfamethizole. Int J Pharm 29:223–231

Brzostowicz PC, Blasko MS, Rouvière PE (2002) Identification of two gene clusters involved in cyclohexanone oxidation in *Brevibacterium epidermidis* strain HCU. Appl Microbiol Biotechnol 58:781–789

Bugnicourt E, Cinelli P, Lazzeri A, Alvarez V (2014) Polyhydroxyalkanoate (PHA): review of synthesis, characteristics, processing and potential applications in packaging. Express Polym Lett 8:791–808

Byrom D (1992) Production of poly-*β*-hydroxybutyrate and poly-*β*-hydroxyvalerate copolymers. FEMS Microbiol Rev 103:247–250

Caballero KP, Karel SF, Register RA (1995) Biosynthesis and characterization of hydroxybutyrate-hydroxycaproate copolymers. Int J Biol Macromol 17:86–92

Cerrone F, Choudhari SK, Davis R et al (2014) Medium chain length polyhydroxyalkanoate (mcl-PHA) production from volatile fatty acids derived from the anaerobic digestion of grass. Appl Microbiol Biotechnol 98:611–620

Chen G-Q (2010) Plastics completely synthesized by bacteria: polyhydroxyalkanoates. In: Chen G-Q (ed) Plastics from bacteria: natural functions and applications, Microbiology monographs. Springer, Berlin/Heidelberg, pp 17–38

Chiras DD (1994) Environmental science. The Benjamin/Cumming Publishing Company, Inc., Redwood, p 611

Chohan SN, Copeland L (1998) Acetoacetyl coenzyme A reductase and polyhydroxybutyrate synthesis in *Rhizobium* (Cicer) sp. strain CC 1192. Appl Environ Microbiol 64:2859–2863

Choi J, Lee SY (1999) Efficient and economical recovery of poly(3-hydroxybutyrate) from recombinant *Escherichia coli* by simple digestion with chemicals. Biotechnol Bioeng 62:546–553

Dawes EA, Senior PJ (1973) The role and regulation of energy reserve polymers in microorganisms. Adv Microbiol Physiol 10:135–266

Doi Y, Kitamura S, Abe H (1995) Microbial synthesis and characterization of poly(3-hydroxybutyrate-*co*-3-hydroxyhexanoate). Macromolecules 28:4822–4828

Drosg B, Fritz I, Gattermayr F, Silvestrini L (2015) Photo-autotrophic production of poly(hydroxyalkanoates) in cyanobacteria. Chem Biochem Eng Q 29:145–156

European Commission (2013) Green paper-on a European strategy on plastic waste in the environment. ec.europa.eu/environment/waste/studies/pdf/green_paper_plastic.pdf

Fiechter A (1990) Plastics from bacteria and for bacteria: poly (β-hydroxyalkanoates) as natural, biocompatible, and biodegradable polyesters. Springer, New York, pp 77–93

Gómez Cardozo JR, Mora Martínez AL, Yepes Pérez M, Correa Londoño GA (2016) Production and characterization of Polyhydroxyalkanoates and native microorganisms synthesized from fatty waste. Int J Polym Sci. Article ID 6541718. http://dx.doi.org/10.1155/2016/6541718

Gouda MK, Swellam AE, Omar SH (2001) Production of PHB by a *Bacillus megaterium* strain using sugarcane molasses and corn steep liquor as sole carbon and nitrogen sources. Microbiol Res 156:201–207

Gould PL, Holland SJ, Tighe BJ (1987) Polymers for biodegradable medical devices. 4-Hydroxybutyrate valerate copolymers as non-disintegrating matrices for controlled-release oral dosage forms. Int J Pharm 38:231–237

Hassan MA, Shirai N, Kusubayashi N, Abdul Karim MI, Nakanishi K, Hashimoto K (1996) Effect of organic acid profiles during anaerobic treatment of palm oil mill effluent on the production of polyhydroxyalkanoates by *Rhodobacter sphaeroides*. J Ferment Bioeng 82:151–156

Hassan MA, Shirai N, Kusubayashi N, Abdul Karim MI, Nakanishi K, Hashimoto K (1997a) The production of polyhydroxyalkanoates from anaerobically treated palm oil mill effluent by *Rhodobacter sphaeroides*. J Ferment Bioeng 83:485–488

Hassan MA, Shirai N, Kusubayashi N, Abdul Karim MI, Nakanishi K, Hashimoto K (1997b) Acetic acid separation from anaerobically treated palm oil mill effluent for the production of polyhydroxyalkanoate by *Alcaligenes eutrophus*. Biosci Biotechnol Biochem 61:1465–1468

Jain R, Kosta S, Tiwari A (2010) Polyhydroxyalkanoates: a way to sustainable development of bioplastics. Chron Young Sci 1:10–15

Joel FR (1995) Polymer science & technology: introduction to polymer science, 3rd edn. Prentice Hall PTR Inc., Upper Saddle River 07458, pp 4–9

John ME (1997) Cotton crop improvement through genetic engineering. Crit Rev Biotechnol 17:185–208

Johnstone B (1990) A throw away answer. Far East Econ Rev 147:62–63

Kahar P, Tsuge T, Taguchi K, Doi Y (2004) High yield production of polyhydroxyalkanoates from soybean oil by *Ralstonia eutropha* and its recombinant strain. Polym Degrad Stabil 83:79–86

Kato M, Bao HJ, Kang C-K, Fukui T, Doi Y (1996) Production of novel copolyester of 3-hydroxybutyric acid and medium-chain-length 3-hydroxyalkanoic acids by *Pseudomonas* sp. 61–3 from sugars. Appl Microbiol Biotechnol 45:363–370

Kawai F (2010) The biochemistry and molecular biology of xenobiotic polymer degradation by microorganisms. Biosci Biotechnol Biochem 74:1743–1759

Khanna S, Srivastava AK (2005) Recent advances in microbial polyhydroxyalkanoates. Process Biochem 40:607–619

Kumar A, Srivastava JK, Mallick N, Singh AK (2015) Commercialization of bacterial cell factories for the sustainable production of polyhydroxyalkanoate thermoplastics: progress and prospects. Recent Pat Biotechnol 9:4–21

Lee SY (1995) Bacterial polyhydroxyalkanoates. Biotechnol Bioeng 49:1–4

Lee EY, Jendrossek D, Schirmer A, Choi CY, Steinbuchel A (1995) Biosynthesis of copolyesters consisting of 3-hydroxybutyric acid and medium-chain-length 3-hydroxyalkanoic acids from 1,3-butanediol or from 3-hydroxybutyrate by *Pseudomonas* sp. A33. Appl Microbiol Biotechnol 42:901–909

Lemoigne M (1926) Products of dehydration and of polymerization of β-hydroxybutyric acid. Bull Soc Chem Biol 8:770–782

Li QA, Chen QA, Li MJ, Wang FS, Qi QS (2011) Pathway engineering results the altered polyhydroxyalkanoates composition in recombinant *Escherichia coli*. New Biotechnol 28:92–95

Liebergesell M, Steinbüchel A (1992) Cloning and nucleotide sequences of genes relevant for biosynthesis of poly(3-hydroxybutyric acid) in *Chromatium vinosum* strain D. Eur J Biochem 209:135–150

Liebergesell M, Schmidt B, Steinbüchel A (1992) Isolation and identiication of granule-associated proteins relevant for poly(3-Hydroxyalkanoic Acid) biosynthesis in *Chromatium vinosum* D. FEMS Microbiol Lett 99:227–232

Lossl A, Bohmert K, Harloff HJ, Eibl C, Muhlbauer S, Koop HU (2005) Inducible trans-activation of plastid transgenes: expression of the *R. eutropha* phb operon in transplastomic tobacco. Plant Cell Physiol 46:1462–1471

Luengo JM, Garcia B, Sandoval A, Naharro G, Olivera ER (2003) Bioplastics from microorganisms. Curr Opin Microbiol 6:251–260

Lütke-Eversloh T, Bergander K, Luftmann H, Steinbüchel A (2001) Identification of a new class of biopolymer: bacterial synthesis of a sulfur-containing polymer with thioester linkages. Microbiology 147:11–19

Lynd LR, Wyman CE, Gerngross TU (1999) Biocommodity engineering. Biotechnol Prog 15:777–793

Madison LL, Huisiman GW (1999) Metabolic engineering of poly(3-hydroxyalkanoates): from DNA to plastic. Microbiol Mol Biol Rev 63:21–53

Matsusaki H, Abe H, Doi Y (2000) Biosynthesis and properties of poly(3-hydroxybutyrate-co-3-hydroxyalkanoates) by recombinant strains of *Pseudomonas* sp. 61–3. Biomacromolecules 1:17–22

Matsumoto K, Murata T, Nagao R, Nomura CT, Arai S, Arai Y et al (2009) Production of short-chain-length/medium-chain-length polyhydroxyalkanoate (PHA) copolymer in the plastid of *Arabidopsis thaliana* using an engineered 3-ketoacyl-acyl carrier protein synthase III. Biomacromolecules 10:686–690

McCool GJ, Cannon MC (1999) Polyhydroxyalkanoate inclusion body-associated proteins and coding region in *Bacillus megaterium*. J Bacteriol 181:585–592

McCool GJ, Cannon MC (2001) PhaC and PhaR are required for polyhydroxyalkanoic acid synthase activity in *Bacillus megaterium*. J Bacteriol 183:4235–4243

Meesters KHP (1998) Production of poly (3 hydroxyalkanoates) from waste streams. Report of Technical University of Delft, Delft

Menzel G, Harloff HJ, Jung C (2003) Expression of bacterial poly(3-hydroxybutyrate) synthesis genes in hairy roots of sugar beet (*Beta vulgaris* L.). Appl Microbiol Biotechnol 60:571–576

Mittendorf V, Robertson EJ, Leech RM, Krüger N, Steinbuchel A, Poirier Y (1998) Synthesis of medium-chain-length polyhydroxyalkanoates in *Arabidopsis thaliana* using intermediates of peroxisomal fatty acid β-oxidation. Proc Natl Acad Sci 95:13397–13402

Narayan R (2006) Biobased and biodegradable polymer materials: rationale, drivers, and technology exemplars. In: Khemani K, Scholz C (eds) Degradable polymers and materials: principles and practice. American Chemical Society, Washington, DC, pp 282–306

Nawrath C, Poirier Y, Somerville C (1994) Targeting of the polyhydroxybutyrate biosynthetic pathway to the plastids of *Arabidopsis thaliana* results in high levels of polymer accumulation. Proc Natl Acad Sci 91:12760–12764

Nishioka M, Nakai K, Miyake M, Asada Y, Taya M (2001) Production of poly-β-hydroyxybutyrate by thermophilic cyanobacterium, *Synechococcus* sp. MA19, under phosphate limitation. Biotechnol Lett 23:1095–1099

Osanai T, Numata K, Oikawa A, Kuwahara A, Iijima H, Doi Y et al (2013) Increased bioplastic production with an RNA polymerase sigma factor SigE during nitrogen starvation in *Synechocystis* sp. PCC 6803. DNA Res 20:525–535

Panda B, Mallick N (2007) Enhanced poly-*β*-hydroxybutyrate accumulation in a unicellular cyanobactrium, *Synechocystis* sp. PCC 6803. Lett Appl Microbiol 44:194–198

Park SJ, Lee SY (2004) Biosynthesis of poly (3-hydroxybutyrate-co-3-hydroxyalkanoates) by metabolically engineered *Escherichia coli* strains. Appl Biochem Biotechnol 113⁄116:335–346

Petrasovits LA, McQualter RB, Gebbie LK, Blackman DM, Nielsen LK, Brumbley SM (2013) Chemical inhibition of acetyl coenzyme A carboxylase as a strategy to increase polyhydroxybutyrate yields in transgenic sugarcane. Plant Biotechnol J 11:1146–1151

Phithakrotchanakoon C, Champreda V, Aiba S, Pootanakit K, Tanapongpipat S (2013) Engineered *Escherichia coli* for short-chain-length medium-chain-length polyhydroxyalkanoate copolymer biosynthesis from glycerol and dodecanoate. Biosci Biotechnol Biochem 77:1262–1268

Plastics Europe (2010) Plastics – the facts 2010, an analysis of European plastics production, demand and recovery for 2009. Association of Plastics Manufactures. http//plasticseurope.com

Plastics Europe (2015) Plastics-the facts 2014/2015: an analysis of European plastics production, demand and waste data. http//plasticseurope.com

Poirier Y (1999) Production of new polymeric compounds in plants. Curr Opin Biotechnol 10:181–185

Poirier Y (2001) Production of poylesters in transgenic plants. In: Babel W, Steinbuchel A (eds) Biopolyesters. Springer, Berlin, pp 209–240

Poirier Y, Gruys KJ (2001) Production of polyhydroxyalkanoates in transgenic plants. In: Doi Y, Steinbuchel A (eds) Biopolyester. Wiley-VCH, Weinheim, pp 401–435

Poirier Y, Dennis DE, Klomparens K, Somerville C (1992) Polyhydroxybutyrate, a biodegradable thermoplastic, produce in transgenic plants. Science 256:520–523

Pouton CW, Akhtar S (1996) Biosynthetic polyhydroxyalkanoates and their potential in drug delivery. Adv Drug Deliv Rev 18:133–162

Qi Q, Rehm BHA (2001) Polyhydroxybutyrate biosynthesis in Caulobacter crescentus: molecular characterization of the polyhydroxybutyrate synthase. Microbiology 147:3353–3358

Ray SS, Bousmina M (2005) Biodegradable polymers and their layered silicate nanocomposites: in greening the 21st century materials world. Prog Mater Sci 50:962–1079

Reddy MV, Mohan SV (2015) Polyhydroxyalkanoates production by newly isolated bacteria *Serratia ureilytica* using volatile fatty acids as substrate: bio-electro kinetic analysis. J Microb Biochem Technol 7:26–32

Reddy CSK, Ghai R, Rashmi KVC (2003) Polyhydroxyalkanoates: an overview. Biores Technol 87:137–146

Rehm BH (2003) Polyester synthases: natural catalysts for plastics. Biochem J 376:15–33

Reis MAM, Serafim LS, Lemos PC, Ramos AM, Aguiar FR, Van Loosdrecht MCM (2003) Production of polyhydroxyalkanoates by mixed microbial cultures. Biopro Biosyst Eng 25:377–385

Ren Q, De Roo G, Kessler B, Witholt B (2000) Recovery of active medium-chain-length-poly-3hydroxyalkanoate polymerase from inactive inclusion bodies using ion-exchange resin. Biochem J 349:599–604

Ruan W, Chen J, Lun S (2003) Production of biodegradable polymer by *A. eutrophus* using volatile fatty acids from acidified wastewater. Process Biochem 39:295–299

Sabir I (2004) Plastic industry in Pakistan. http://www.jang.com.pk/thenews/investors/nov2004/index.html

Saharan BS, Ankita, Sharma D (2012) Bioplastics for sustainable development: a review. Int J Microbial Res Technol 1:11–23

Salehizadeh H, Van Loosdrecht MCM (2004) Production of polyhydroxyalkanoates by mixed culture: recent trends and biotechnological importance. Biotechnol Adv 22:261–279

Samantaray S, Mallick N (2014) Production of poly(3-hydroxybutyrate-*co*-3-hydroxyvalerate) co-polymer by the diazotrophic cyanobacterium *Aulosira fertilissima* CCC 444. J Appl Phycol 26:237–245

Samantaray S, Mallick N (2015) Impact of various stress conditions on Poly-β-Hydroxybutyrate (PHB) accumulation in *Aulosira fertilissima* CCC 444. Curr Biotechnol 4:366–372

Sankhla SS, Bhati R, Singh AK, Mallick N (2010) Poly(3-hydroxybutyrate-*co*-3- hydroxyvalerate) co-polymer production from a local isolate, *Brevibacillus invocatus* MTCC 9039. Biores Technol 101:1947–1953

Satoh H, Iwamoto Y, Mino T, Matsuo T (1998) Activated sludge as a possible source of biodegradable plastic. Water Sci Technol 38:103–109

Schnell J, Treyvaud-Amiguet V, Arnason J, Johnson D (2012) Expression of polyhydroxybutyric acid as a model for metabolic engineering of soybean seed coats. Transgenic Res 21:895–899

Seymour RB (1989) Polymer science before & after 1899: notable developments during the lifetime of Maurtis Dekker. J Macromol Sci Chem 26:1023–1032

Shah AA, Hasan F, Hameed A, Ahmed S (2008) Biological degradation of plastics: a comprehensive review. Biotechnol Adv 26:246–265

Sharma L, Mallick N (2005) Accumulation of poly-*β*-hydroxybutyrate in *Nostoc muscorum*: regulation by pH, light–dark cycles, N and P status and carbon sources. Biores Technol 96:1304–1310

Sheu D-S, Lee C-Y (2004) Altering the substrate specificity of polyhydroxyalkanoates synthase 1 derived from *Pseudomonas putida* GPo1 by localized semirandom mutagenesis. J Bacteriol 186:4177–4184

Shi HP, Lee CM, Ma WH (2007) Influence of electron acceptor, carbon, nitrogen, and phosphorus on polyhydroxyalkanoate (PHA) production by *Brachymonas* sp. P12. World J Microbiol Biotechnol 23:625–632

Shimao M (2001) Biodegradation of plastics. Curr Opin Biotechnol 12:242–247

Shujun W, Jiugao Y, Jinglin Y (2006) Preparation and characterization of compatible and degradable thermoplastic starch/polyethylene film. J Polym Environ 14:1

Singh AK, Mallick N (2008) Enhanced production of SCL-LCL-PHA co-polymer by sludge-isolated *Pseudomonas aeruginosa* MTCC 7925. Lett Appl Microbiol 46:350–357

Singh AK, Mallick N (2009a) Exploitation of inexpensive substrates for production of a novel SCL–LCL-PHA co-polymer by *Pseudomonas aeruginosa* MTCC 7925. J Ind Microbiol Biotechnol 36:347–354

Singh AK, Mallick N (2009b) SCL-LCL-PHA copolymer production by a local isolate, *Pseudomonas aeruginosa* MTCC 7925. Biotechnol J 4:703–711

Singh AK, Mallick N (2015) Biological system as a reactor for production of biodegradable thermoplastics, Polyhydroxyalkanoates. In: Thangadurai D, Sangeetha J (eds) Industrial biotechnology: sustainable production and bioresource utilization. CRC Press/Taylor and Francis, USA, pp 281–323

Singh AK, Bhati R, Samantaray S, Mallick N (2013) *Pseudomonas aeruginosa* MTCC 7925: producer of a novel SCL-LCL-PHA co-polymer. Curr Biotechnol 2:81–88

Singh AK, Ranjana B, Mallick N (2015) *Pseudomonas aeruginosa* MTCC 7925 as a biofactory for production of the novel SCL-LCL- PHA thermoplastic from non-edible oils. Curr Botechnol 4:65–74

Solaiman DKY, Ashby RD, Hotchkiss AT Jr, Foglia TA (2006) Biosynthesis of medium-chain-length poly(hydroxyalkanoates) from soy molasses. Biotechnol Lett 28:157–162

Somleva M, Ali A (2010) Propagation of transgenic plants. International patent application WO/2010/102220

Steinbuchel A (1992) Biodegradable plastics. Curr Opin Biotechnol 3:291–297

Steinbuchel A, Pieper U (1992) Production of copolyesters of 3-hydroxybutyric acid and 3-hydroxyvaleric acid by a mutant of *Alcaligenes eutrophus* from single unrelated carbon sources. Appl Microbiol Biotechnol 37:1–6

Steinbuchel A, Hustede E, Liebergesell M, Pieper U, Timm A, Valentin H (1992) Molecular basis for biosynthesis and accumulation of polyhydroxyalkanoic acids in bacteria. FEMS Microbiol Rev 103:217–230

Sudesh K, Iwata T (2008) Sustainability of biobased and biodegradable plastics. CLEAN – Soil Air Water 36:433–442

Sudesh K, Abe H, Doi Y (2000) Synthesis, structure and properties of polyhydroxyalkanoates: biological polyesters. Prog Polym Sci 25:1503–1555

Suriyamongkol P, Weselake R, Narine S, Moloney M, Shah S (2007) Biotechnological approaches for the production of polyhydroxyalkanoates in microorganisms and plants – a review. Biotechnol Adv 25:148–175

Taguchi K, Aoyagi Y, Matsusaki H, Fukui T, Doi Y (1999) Over-expression of 3-ketoacyl-ACP synthase III or malonyl-CoA-ACP transacylase gene induces monomer supply for polyhydroxybutyrate production in *Escherichia coli* HB101. Biotechnol Lett 21:579–584

Thakor NS, Patel MA, Trivedi UB, Patel KC (2003) Production of poly(β-hydroxybutyrate) by *Comamonas testosteroni* during growth on naphthalene. World J Microbiol Biotechnol 19:185–189

Thompson RC, Swan SH, Moore CJ, vomSaal FS (2009) Our plastic age. Phil Trans R Soc B 364:1973–1976

Tian SJ, Lai WJ, Zheng Z, Wang HX, Chen GQ (2005) Effect of over-expression of phasin gene from *Aeromonas hydrophila* on biosynthesis of copolyesters of 3-hydroxybutyrate and 3-hydroxyhexanoate. FEMS Microbiol Lett 244:19–25

Toh PSY, Jau MH, Yew SP, Abed RMM, Sudesh K (2008) Comparison of polyhydroxyalkanoates biosynthesis, mobilization and the effects on cellular morphology in *Spirulina platensis* and *Synechocystis* sp. UNIWG. J Biosci 19:21–38

Valentin HE, Dennis D (1997) Production of poly(3-hydroxybutyrate-co-4-hydroxybutyrate) in recombinant *Escherichia coli* grown on glucose. J Biotechnol 58:33–38

Valentin HE, Steinbüchel A (1995) Accumulation of poly(3-hydroxybutyric acid-co-3-hydroxyvaleric acid-co-4-hydroxyvaleric acid) by mutants and recombinant strains of *Alcaligenes eutrophus*. J Environ Polym Degrad 3:169–175

Verma NK, Khanna SK, Kapila B (2007) Comprehensive chemistry XII, vol 1. Laxi Publications Pvt. Ltd., 113, Golden house, Daryaganj, New Delhi, India, pp 1581–1608

Vincenzini M, De Philippis R (1999) Polyhydroxyalkanoates. In: Cohen Z (ed) Chemicals from microalgae. Taylor and Francis Inc., USA, pp 292–312

Vona IA, Costanza JR, Cantor HA, Roberts WJ (1965) Manufacture of plastics, vol 1. Wiley, New York, pp 141–142

Wang B, Pugh S, Nielsen DR, Zhang W, Meldrum DR (2013) Engineering cyanobacteria for photosynthetic production of 3-hydroxybutyrate directly from CO_2. Metab Eng 16:68–77

Ward AC, Rowley BI, Dawes EA (1977) Effect of oxygen and nitrogen limitation on poly-*β*-hydroxybutyrate biosynthesis in Ammonium-grown *Azotobacter beijerinckii*. J Gen Microbiol 102:61–68

Wilson JT, McNabb JF, Cochran JW et al (1985) Influence of microbial adaptation on the fate of organic pollutants in ground water. Environ Toxicol Chem 4:721–726

Xiao XQ, Zhao Y, Chen GQ (2007) The effect of 3-hydroxybutyrate and its derivatives on the growth of glial cells. Biomaterials 28:3608–3616

Xie WP, Chen GQ (2008) Production and characterization of terpolyester poly(3-hydroxybutyrate-co-4-hydroxybutyrate-co-3-hydroxyhexanoate) by recombinant *Aeromonas hydrophila* 4AK4 harboring genes *phaPCJ*. Biochem Eng J 38:384–389

Yang JE, Choi YJ, Lee SJ et al (2014) Metabolic engineering of *Escherichia coli* for biosynthesis of poly(3-hydroxybutyrate-*co*-3-hydroxyvalerate) from glucose. Appl Microbiol Biotechnol 98:95–104

Yezza A, Fournier D, Halasz A, Hawari J (2006) Production of polyhydroxyalkanoates from methanol by a new methylotrophic bacterium *Methylobacterium* sp. GW2. Appl Microbiol Biotechnol 73:211–218

Zhang XJ, Luo RC, Wang Z, Deng Y, Chen GQ (2009) Applications of (*R*)-3-hydroxyalkanoate methyl esters derived from microbial polyhydroxyalkanoates as novel biofuel. Biomacromolecules. doi:10.1021/bm801424e

Zheng LZ, Li Z, Tian HL, Li M, Chen GQ (2005) Molecular cloning and functional analysis of (*R*)3-hydroxyacyl-acyl carrier protein:coenzyme A transacylase from *Pseudomonas mendocina* LZ. FEMS Microbiol Lett 252:299–307

MIX
Papier aus verantwortungsvollen Quellen
Paper from responsible sources
FSC® C105338

If you have any concerns about our products,
you can contact us on
ProductSafety@springernature.com

In case Publisher is established outside the EU,
the EU authorized representative is:
Springer Nature Customer Service Center GmbH
Europaplatz 3, 69115 Heidelberg, Germany

Printed by Libri Plureos GmbH
in Hamburg, Germany